Physics for Diagnostic Radiology
Third Edition

Series in Medical Physics and Biomedical Engineering

Series Editors: John G Webster, Slavik Tabakov, Kwan-Hoong Ng

Other recent books in the series:

Series in Medical Physics and Biomedical Engineering

Physics for Diagnostic Radiology
Third Edition

P P Dendy, B Heaton

With contributions by
O W E Morrish, S J Yates, F I McKiddie, P H Jarritt, K E Goldstone,
A C Fairhead, T A Whittingham, E A Moore, and G Cusick

CRC Press
Taylor & Francis Group
Boca Raton London New York

CRC Press is an imprint of the
Taylor & Francis Group, an **informa** business

A TAYLOR & FRANCIS BOOK

CRC Press
Taylor & Francis Group
6000 Broken Sound Parkway NW, Suite 300
Boca Raton, FL 33487-2742

© 2012 by Taylor & Francis Group, LLC
CRC Press is an imprint of Taylor & Francis Group, an Informa business

No claim to original U.S. Government works

Printed in the United States of America on acid-free paper
Version Date: 20110623

International Standard Book Number: 978-1-4200-8315-6 (Hardback)

Visit the Taylor & Francis Web site at
http://www.taylorandfrancis.com

and the CRC Press Web site at
http://www.crcpress.com

Contents

About the Series

The Series in Medical Physics and Biomedical Engineering describes the applications of physical sciences, engineering, and mathematics in medicine and clinical research.

The series seeks (but is not restricted to) publications in the following topics:

- Artificial organs
- Assistive technology
- Bioinformatics
- Bioinstrumentation
- Biomaterials
- Biomechanics
- Biomedical engineering
- Clinical engineering
- Imaging
- Implants
- Medical computing and mathematics
- Medical/surgical devices

- Patient monitoring
- Physiological measurement
- Prosthetics
- Radiation protection, health physics, and dosimetry
- Regulatory issues
- Rehabilitation engineering
- Sports medicine
- Systems physiology
- Telemedicine
- Tissue engineering
- Treatment

The *Series in Medical Physics and Biomedical Engineering* is an international series that meets the need for up-to-date texts in this rapidly developing field. Books in the series range in level from introductory graduate textbooks and practical handbooks to more advanced expositions of current research.

The *Series in Medical Physics and Biomedical Engineering* is the official book series of the International Organization for Medical Physics.

The International Organization for Medical Physics

The International Organization for Medical Physics (IOMP), founded in 1963, is a scientific, educational, and professional organization of 76 national adhering organizations, more than 16,500 individual members, several corporate members, and four international regional organizations.

IOMP is administered by a council, which includes delegates from each of the adhering national organizations. Regular meetings of the council are held electronically as well as every three years at the World Congress on Medical Physics and Biomedical Engineering. The president and other officers form the executive committee, and there are also committees covering the main areas of activity, including education and training, scientific, professional relations, and publications.

Objectives

- To contribute to the advancement of medical physics in all its aspects
- To organize international cooperation in medical physics, especially in developing countries
- To encourage and advise on the formation of national organizations of medical physics in those countries which lack such organizations

Activities

Official journals of the IOMP are *Physics in Medicine and Biology* and *Medical Physics and Physiological Measurement*. The IOMP publishes a bulletin, *Medical Physics World*, twice a year, which is distributed to all members.

A World Congress on Medical Physics and Biomedical Engineering is held every three years in cooperation with IFMBE through the International Union for Physics and Engineering Sciences in Medicine (IUPESM). A regionally based international conference on medical physics is held between world congresses. IOMP also sponsors international conferences, workshops, and courses. IOMP representatives contribute to various international committees and working groups.

The IOMP has several programs to assist medical physicists in developing countries. The joint IOMP Library Programme supports 69 active libraries in 42 developing countries, and the Used Equipment Programme coordinates equipment donations. The Travel Assistance Programme provides a limited number of grants to enable physicists to attend the world congresses. The IOMP website is being developed to include a scientific database of international standards in medical physics and a virtual education and resource center. Information on the activities of the IOMP can be found on its website at www.iomp.org.

Acknowledgements

We are grateful to many persons for constructive comments on and assistance with the production of this book. In particular, we wish to thank Dr J Freudenberger, G Walker, Dr G Buchheim, Dr K Bradshaw, Mr M Bartley, Professor G Barnes, Dr A Noel, D Goodman, Dr A Parkin, Dr I S Hamilton, D A Johnson, I Wright, S Yates, E Hutcheon, M Streedharen and K Anderson.

Figures 2.9, 2.15 and 2.28 are reproduced by permission of Siemens AG. Dura and STRATON are registered trademarks of Siemens.

Figures 2.24 and 2.25 are reproduced by permission of Philips Electronics UK Ltd.

Figures 5.18 and 5.20 are reproduced by permission of Medical Physics Publishing.

Figure 5.26 is reproduced by permission of the British Institute of Radiology.

Figure 7.2a is reproduced by permission of Artinis Medical Systems.

Figures 7.3, 7.19, 7.20, 7.23, 7.24, 7.25 and 7.26 are reproduced by permission of the British Institute of Radiology.

Figure 7.4 is reproduced by permission of The Royal Society, London UK (Campbell F W, *Phil Trans Soc B*, 290, 5–9, 1980).

Figure 7.11 is reproduced by permission of Elsevier Publishing and Professor P N T Wells (*Scientific Basis of Medical Imaging* edited by P N T Wells, 1982, figure 1.19, p. 18).

Figure 7.21 is reproduced by permission of the International Commission on Radiation Units and Measurements.

Figure 7.22 is reproduced by permission of the Radiological Society of North America Inc (Macmohon H, Vyborny C J, Metz C E et al. *Radiology*, 158, 21–26, figure 5, 1986).

Figure 9.13 is reproduced by permission of Dr J N P Higgins and H Szutowicz.

Figure 9.16 is reproduced by permission of Professor G T Barnes.

Figure 9.19 is reproduced by permission of Oxford University Press.

Figure 11.2 is reproduced by permission of Elsevier Publishing (Cherry S R, Sorenson J A & Phelps M E. *Physics in Nuclear Medicine*, 3rd edition, 2003).

Figure 12.12 is reproduced by permission of the Radiological Society of North America (Boyce J D, Land C E, & Shore R E. Risk of breast cancer following low-dose radiation exposure, *Radiology*, 131, 589–597, 1979).

Figures 12.14 and 12.17 and Table 12.2 are reproduced by permission of the British Institute of Radiology.

Tables 13.1, 13.3, 13.4, 13.5 and 13.16 are reproduced by permission of the Health Protection Agency, UK.

Tables 13.2, 13.8, 13.9, 13.10, 13.11, 13.12, 13.14 and 13.15 are reproduced by permission of the British Institute of Radiology.

Introduction to the Third Edition

> Learn from yesterday, live for today, hope for tomorrow. The important thing is not to stop questioning.

Albert Einstein

The first edition of this book, published in 1987, was written in response to a rapid development in the range of imaging techniques available to the diagnostic radiologist over the previous 20 years and a marked increase in the sophistication of imaging equipment. There was a clear need for a textbook that would explain the underlying physical principles of all the relevant imaging techniques at the appropriate level.

Since that time, there have been major developments in imaging techniques and the physical principles behind them. Some of these were addressed in the second edition, published in 1999, notably the much greater importance attached to patient doses, the increasingly widespread use of digital radiography, the importance of both patient dose and image quality in mammography, the increasing awareness of the need to protect staff and related legislation. The chapters on ultrasound and magnetic resonance imaging (MRI) were completely rewritten.

The past decade has seen yet more advances, and parts of the second edition are no longer 'state of the art'. In this third edition all the chapters have been revised and brought up-to-date, with major additions in the following areas:

- The image receptor—new material on digital receptors
- The radiological image—emphasising the differences between analogue and digital images
- Computed tomography—multi-slice CT and three-dimensional resolution, dual energy applications, cone beam CT
- Special radiographic techniques—especially subtraction techniques and interventional radiology
- Positron emission tomography—a new chapter including aspects of multi-modality imaging (PET/CT)
- Radiation doses and risks to patients
- Data handling in radiology—a new chapter covering picture archiving and communication systems (PACS), teleradiology, networks, archiving and related factors

The second evolutionary change since 1987 has been in the scope of the anticipated readership. Radiologists in training are still a primary target, and there are many reasons to emphasise the importance of physics education as a critical component of radiology training. As an imaging technique becomes more sophisticated it is essential for radiologists to know 'how it works', thus providing them with a unique combination of anatomical, physiological and physical information. This helps to differentiate the expertise of radiologists from that of other physicians who read images and helps to position radiology as a science-based practice. There is a need for substantial additional educational resources in physics and better integration of physics into clinical training (Hendee 2006; Bresolin et al. 2008).

However, experience with the first and second editions of the book has shown that it is a useful text for other groups, including radiographers/technicians engaged in academic training and undergraduates in new courses in imaging sciences. It is a good introductory text for master's degree courses in medical physics and for physicists following the training programme in diagnostic radiology recommended by the European Federation of Organisations in Medical Physics (EFOMP—in preparation). It will also be of value to teachers of physics to radiologists and radiographers.

Many features of the first and second editions have been retained:

- The material is presented in a logical order. After an introductory chapter of basic physics, Chapters 2 to 7 follow through the X-ray imaging process—production of X-rays, interaction with the patient, radiation measurement, the image receptor, the radiological image and the assessment of image quality. Chapters 8 and 9 cover more advanced techniques with X-rays and Chapters 10 and 11 cover imaging with radioactive materials. Chapters 12 through 14 deal with radiobiology and risk and radiation protection. Chapters 15 and 16 cover imaging with non-ionising radiation (ultrasound and MRI) and finally Chapter 17 discusses data handling in a modern, electronic radiology department.

- Extensive cross-referencing is used to acknowledge the fact that much of the subject matter is very interactive, without the need for undue repetition.

- Lateral thinking has been encouraged wherever possible, for example, pointing out the similarities in the use of the exponential in radioactive decay, attenuation of X-rays and MRI.

- There are exercises at the end of each chapter and, at the end of the book, there are multiple choice questions (MCQs), at an appropriate level and sometimes drawing on material from more than one chapter, to assist readers in assessing their understanding of the basic principles. The MCQs are not designed to provide comprehensive coverage of any particular syllabus because other books are available for this purpose.

- Text references and recommendations for further reading are given at the end of each chapter.

There are two major changes in the layout:

1. Each chapter begins with a summary of the main teaching points.
2. To accommodate some variation in the background knowledge of readers, some insights have been included. These are not essential to a first reading but cover more subtle points that may involve ideas presented later in the book or require a somewhat greater knowledge of physics or mathematics.

And finally—why the quotation from Einstein? Digital imaging, molecular imaging and functional imaging have great potential in medicine, but as they develop they will inevitably require a better knowledge of physics and become more quantitative. We have tried to show the way forward to both radiologists and scientists who are prepared to ask the question, Why?

References

Bresolin L, Bissett III GS, Hendee WR and Kwakwa FA, Methods and resources for physics education in radiology residence programmes, *Radiology,* 249, 640–643, 2008.

EFOMP, Guidelines for education and training of medical physicists in radiology—in preparation, European Federation of Organisations for Medical Physics.

Hendee WR, An opportunity for radiology, *Radiology,* 238, 389–394, 2006.

Contributors

G Cusick
Medical Physics and Bioengineering
UCL Hospitals NHS Foundation Trust
London, United Kingdom

P P Dendy (Retired)
East Anglian Regional Radiation
    Protection Service
Cambridge University Hospitals NHS
    Foundation Trust
Cambridge, United Kingdom

A C Fairhead
Department of Biomedical Physics and
    Bioengineering
NHS Grampian
Aberdeen, United Kingdom

K E Goldstone
East Anglian Regional Radiation
    Protection Service
Cambridge University Hospitals NHS
    Foundation Trust
Cambridge, United Kingdom

B Heaton
Aberdeen Radiation Protection Services
Aberdeen, United Kingdom

P H Jarritt
Clinical Director of Medical Physics and
    Clinical Engineering
Department of Medical Physics and
    Clinical Engineering
Cambridge University Hospitals NHS
    Foundation Trust
Cambridge, United Kingdom

F I McKiddie
Department of Nuclear Medicine
NHS Grampian
Aberdeen, United Kingdom

Elizabeth A Moore
Philips Healthcare
Best, the Netherlands

O W E Morrish
East Anglian Regional Radiation
    Protection Service
Cambridge University Hospitals NHS
    Foundation Trust
Cambridge, United Kingdom

T A Whittingham
Regional Medical Physics Department
Newcastle General Hospital
Newcastle upon Tyne, United Kingdom

S J Yates
East Anglian Regional Radiation
    Protection Service
Cambridge University Hospitals NHS
    Foundation Trust
Cambridge, United Kingdom

1

Fundamentals of Radiation Physics and Radioactivity

P P Dendy and B Heaton

SUMMARY

- Why some atoms are unstable is explained.
- The processes involved in radioactive decay are presented.
- The concepts of physical and biological half-life and the mathematical explanation of secular equilibrium are addressed.
- The basic physical properties of X and gamma photons and the importance of the K shell electrons in diagnostic radiology are explained.
- The basic concepts of the quantum nature of electromagnetic (EM) radiation and energy, the inverse square law and the interaction of radiation with matter are introduced.

CONTENTS

1.1 Structure of the Atom

All matter is made up of atoms, each of which has an essentially similar structure. All atoms are formed of small, dense, positively charged nuclei, typically some 10^{-14} m in diameter, orbited at much larger distances (about 10^{-10} m) by negatively charged, very light particles. The atom as a whole is electrically neutral. Note that because matter consists mainly of empty space, radiation may penetrate many atoms before a collision results.

The positive charge in the nucleus consists of a number of *protons* each of which has a charge of 1.6×10^{-19} coulombs (C) and a mass of 1.7×10^{-27} kilograms (kg). The negative charges are *electrons*. An electron carries the same numerical charge as the proton, but of opposite sign. However, an electron has only about 1/2000th the mass of the proton (9×10^{-31} kg). Each element is characterised by a specific number of protons, and an equal number of orbital electrons. This is called the *atomic number* and is normally denoted by the symbol Z. For example, Z for aluminium is 13, whereas for lead Z = 82.

Electrons are most likely to be at fairly well-defined distances from the nucleus and are described as being in 'shells' around the nucleus (Figure 1.1). More important than the distance of the electron from the nucleus is the electrostatic force that binds the electron to the nucleus, or the amount of energy the electron would have to be given to escape from the field of the nucleus. This is equal to the amount of energy a free electron will lose when it is captured by the electrostatic field of a nucleus. It is possible to think in terms

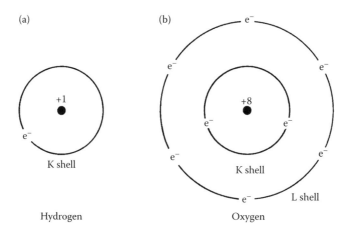

FIGURE 1.1
Examples of atomic structure. (a) Hydrogen with one K shell electron. (b) Oxygen with two K shell electrons and six L shell electrons.

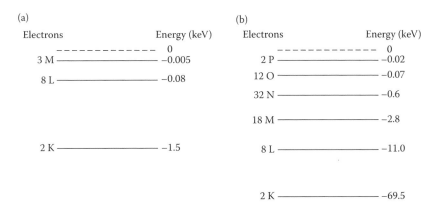

FIGURE 1.2
Typical electron energy levels. (a) Aluminium (Z = 13). (b) Tungsten (Z = 74).

of an energy 'well' that gets deeper as the electron is trapped in shells closer and closer to the nucleus.

The unit in which electron energies are measured is the *electron volt* (eV)—this is the energy one electron would gain if it were accelerated through 1 volt of potential difference. One thousand electron volts is a kilo electron volt (keV) and one million electron volts is a mega electron volt (MeV). Some typical electron shell energies are shown in Figure 1.2. Note that

1. If a free electron is assumed to have zero energy, all electrons within atoms have negative energy—that is, they are bound to the nucleus and must be given energy to escape.

2. The energy levels are not equally spaced and the difference between the K shell and the L shell is much bigger than any of the other differences between shells further away from the nucleus. Shells are distinguished by being given a letter. The innermost shell is the K shell and subsequent shells follow in alphabetical order. When a shell is full (e.g. the M shell can only hold 18 electrons) the next outer shell starts to fill up.

The K shell energies of many elements are important in several aspects of the physics of radiology and a table of their various values and where they are used for different aspects of radiology is given in Table 1.1. This table will be useful for reference when reading the subsequent chapters.

The X-ray energies of interest in diagnostic radiology are between 10 and 120 keV. Below 10 keV too many X-rays are absorbed in the body, above 120 keV too few X-rays are stopped by the image receptor. However, higher energy gamma photons are used when imaging with radioactive materials where the imaging process is quite different.

Insight

K Shell Energies

The most important energy level in imaging is the K shell energy. The L shell energies are small (lead 15.2 keV, tungsten 12.1 keV, caesium 5.7 keV, for example) and are mostly outside the energy range of interest in radiology (we have set the lower limit at 10 keV, some L shell energies are

TABLE 1.1

K Shell Energies for Various Elements and the Aspect of Radiology Where They
Are Important

Element	Area of Application	Z Number	K Shell Energy (keV)
Carbon	(a)	6	0.28
Oxygen	(a)	8	0.53
Aluminium	(b)	13	1.6
Silicon	(c)	14	1.8
Phosphorus	(a)	15	2.1
Sulphur	(a)	16	2.5
Calcium	(a)	20	4.0
Copper	(f)	29	9.0
Germanium	(c)	32	11.1
Selenium	(c)	34	12.7
Molybdenum	(b and e)	42	20.0
Rhodium	(e)	45	23.2
Palladium	(e)	46	24.4
Caesium	(c)	55	36.0
Barium	(d and f)	56	37.4
Iodine	(c and d)	53	33.2
Gadolinium	(c)	54	50.2
Erbium	(b)	68	57.5
Ytterbium	(c)	70	61.3
Tungsten	(e)	74	69.5
Lead	(f)	82	88.0

(a) Body tissue components—but the X-rays associated with these K shells have too low an
    energy to have any external effect and are absorbed in the body.
(b) Used to filter the beam emerging from the X-ray tube.
(c) Used as a detector (in a monitor) or an image receptor of X-ray photons.
(d) Used as a contrast agent to highlight a part of the body.
(e) Used to influence the spectral output of an X-ray tube.
(f) Used as shielding from X-ray photons.

slightly above this). Since the (negative) K shell energy is a measure of how tightly bound these two
electrons are held by the positive charge on the nucleus, the binding energy of the K shell increases
as the atomic number increases as can be seen in Table 1.1. As noted in Table 1.1 K shell energies
have important applications in the shape of the X-ray spectra (Section 2.2), filters (Section 3.8),
intensifying screens, scintillation detectors and digital receptors (Chapter 5) and contrast agents
(Section 6.3.4).

1.2 Nuclear Stability and Instability

If a large number of protons were forced together in a nucleus they would immediately
explode owing to electrostatic repulsion. Very short-range attractive forces are therefore
required within the nucleus for stability, and these are provided by *neutrons*, uncharged
particles with a mass almost identical to that of the proton.

The total number of protons and neutrons, collectively referred to as *nucleons*, within the nucleus is called the *mass number*, usually given the symbol A. Each particular combination of Z and A defines a *nuclide*. One notation used to describe a nuclide is ^A_ZN.

The number of protons Z defines the element N, so for hydrogen Z = 1, for oxygen Z = 8 and so on, but the number of neutrons is variable. Therefore an alternative and generally simpler notation that carries all necessary information is N-A. The notation ^A_ZN will only be used for equations where it is important to check that the number of protons and the number of nucleons balance.

Nuclides that have the same number of protons but different numbers of neutrons are known as *isotopes*. Thus O-16, the most abundant isotope of oxygen, has 8 protons (by definition) and 8 neutrons. O-17 is the isotope of oxygen which has 8 protons and 9 neutrons. Since isotopes have the same number of protons and hence when neutral the same number of orbital electrons, they have the same chemical properties.

The number of neutrons required to stabilise a given number of protons lies within fairly narrow limits and Figure 1.3a shows a plot of these numbers. Note that for many elements of biological importance the number of neutrons is equal to the number of protons, but the most abundant form of hydrogen, which has one proton but no neutrons, is an important exception. At higher atomic numbers the number of neutrons begins to increase faster than the number of protons—lead, for example, has 126 neutrons but only 82 protons.

An alternative way to display the data is to plot the sum of neutrons and protons against the number of protons (Figure 1.3b). This is essentially a plot of nuclear mass against

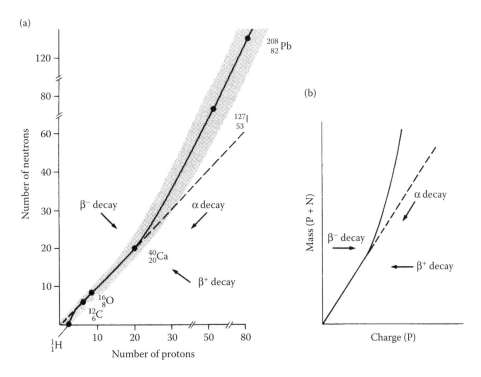

FIGURE 1.3
Graphs showing the relationship between number of neutrons and number of protons for the most abundant stable elements. (a) Number of neutrons plotted against number of protons. The *dashed line* is at 45°. The *cross-hatched area* shows the range of values for which the nucleus is likely to be stable. (b) Total number of nucleons (neutrons and protons) plotted against number of protons. On each graph the changes associated with β⁺, β⁻ and α decay are shown.

nuclear charge (or the total charge on the orbiting electrons). This concept will be useful when considering the interaction of ionising radiation with matter, and in Section 3.4.3 the near constancy of mass/charge (A/Z is close to 2) for most of the biological range of elements will be considered in more detail.

If the ratio of neutrons to protons is outside narrow limits, the nuclide is radioactive or a *radionuclide*. For example, H-1 (normal hydrogen) is stable, H-2 (deuterium) is also stable, but H-3 (tritium) is radioactive. A nuclide may be radioactive because it has too many or too few neutrons.

A simple way to make radioactive nuclei is to bombard a stable element with a flux of neutrons in a reactor. For example, radioactive phosphorus may be made by the reaction shown below:

$$^{31}_{15}P + {}^{1}_{0}n = {}^{32}_{15}P + {}^{0}_{0}\gamma$$

(the emission of a gamma ray as part of this reaction will be discussed later). However, this method of production results in a radionuclide that is mixed with the stable isotope since the number of protons in the nucleus has not changed and not all the P-31 is converted to P-32. Radionuclides that are 'carrier free' can be produced by bombarding with charged particles such as protons or deuterons, in a cyclotron; for example, if sulphur is bombarded with protons,

$$^{34}_{16}S + {}^{1}_{1}p = {}^{34}_{17}Cl + {}^{1}_{0}n$$

The radioactive product is now a different element and thus may be separated by chemical methods.

The *activity* of a source is a measure of its rate of decay or the number of disintegrations per second. In the International System of Units it is measured in becquerels (Bq) where 1 Bq is equal to one disintegration per second. The becquerel has replaced the older unit of the curie (Ci), but since the latter is still encountered in textbooks and older published papers and is still actively used in some countries, it is important to know the conversion factor.

$$1 \text{ Ci} = 3.7 \times 10^{10} \text{ Bq}$$

Hence,

   1 mCi (millicurie) = 3.7×10^7 Bq (37 megabecqerels or MBq)
   1 μCi (microcurie) = 3.7×10^4 Bq (37 kilobecquerels or kBq)

1.3 Radioactive Concentration and Specific Activity

These two concepts are frequently confused.

1.3.1 Radioactive Concentration

This relates to the amount of radioactivity per unit volume. Hence it will be expressed in Bq ml⁻¹. It is important to consider the radioactive concentration when giving a bolus injection. If one wishes to inject a large activity of technetium-99m (Tc-99m) in a small volume, perhaps for a dynamic nuclear medicine investigation, it is preferable to elute a 'new' molybdenum-technetium generator when the yield might be 8 GBq (200 mCi) in a 10 ml eluate [0.8 GBq ml^{-1} (20 mCi ml^{-1})] rather than an old generator when the yield might be only about 2 GBq (50 mCi) [0.2 GBq ml^{-1} (5 mCi ml^{-1})]. For a fuller discussion of the production of Tc-99m and its use in nuclear medicine see Section 1.7 and Chapter 10.

1.3.2 Specific Activity

This relates to the proportion of nuclei of the element of interest that are actually labelled. Non-radioactive material, for example iodine-127 (I-127) in a sample of I-125 may be present as a result of the preparation procedure or may have been added as carrier. The unit for the total number of atoms or nuclei present is the mole so the proportion that are radioactive or the *specific activity* can be expressed in Bq mol^{-1} or Bq kg^{-1}. The specific activity of a preparation should always be checked since it determines the total amount of the element being administered. Modern radiopharmaceuticals generally have a very high specific activity so the total amount of the element administered is very small, and problems such as iodine sensitivity do not normally arise in diagnostic nuclear medicine.

Insight

Pure Radionuclides

The number of molecules in one gram-molecular weight is 6.02×10^{23} (Avogadro's number). Very few radionuclide solutions or solids are pure radionuclide. Most consist of radioactivity mixed with some form of non-radioactive carrier.

1.4 Radioactive Decay Processes

Three types of radioactive decay that result in the emission of charged particles will be considered at this stage.

1.4.1 β⁻ Decay

A negative β particle is an electron. Its emission is actually a very complex process but it will suffice here to think of a change *in the nucleus* in which a neutron is converted into a proton. The particles are emitted with a range of energies. Note that although the process results in emission of electrons, it is a *nuclear* process and has nothing to do with the orbiting electrons.

The mass of the nucleus remains unchanged but its charge increases by one, thus this change is favoured by nuclides which have too many neutrons.

1.4.2 β⁺ Decay

A positive β particle, or positron, is the anti-particle to an electron, having the same mass and an equal but opposite charge. Again, its precise mode of production is complex but it can be thought of as being released when a proton in the nucleus is converted to a neutron. Note that a positron can only exist while it has kinetic energy. When it comes to rest it spontaneously combines with an electron.

The mass of the nucleus again remains unchanged but its charge decreases by one, thus this change is favoured by nuclides which have too many protons.

1.4.3 α Decay

An α particle is a helium nucleus, thus it comprises two protons and two neutrons. After α emission, the charge is reduced by two units and the mass by four units.

The effects of β^-, β^+ and α decay are shown in Figure 1.3. Note that emission of α particles only occurs for the higher atomic number nuclides.

1.5 Exponential Decay

Although it is possible to predict from the number of protons and neutrons in the nucleus which type of decay might occur, it is not possible to predict how fast decay will occur. One might imagine that nuclides that were furthest from the stability line would decay fastest. This is not so and the factors which determine the rate of decay are beyond the scope of this book.

However, all radioactive decay processes do obey a very important rule. This states that the only variables affecting the number of nuclei ΔN decaying in a short interval of time Δt are the number of unstable nuclei N present and the time interval Δt. Hence

$$\Delta N \propto N \Delta t$$

If the time interval is very short, the equation becomes

$$dN = -kN \, dt$$

where the constant of proportionality k is characteristic of the radionuclide, known as its *decay constant* or *transformation constant*, and the negative sign has been introduced to show that, mathematically, the number of radioactive nuclei actually decreases with elapsed time.

The equation may be integrated (see Insight) to give the well-known exponential relationship

$$N = N_0 \exp(-kt)$$

where N_0 is the number of unstable nuclei present at $t = 0$.

Insight

Mathematics of the Exponential Equation

$$dN = -kN \, dt$$

Rearranging

$$\frac{dN}{N} = -k\,dt$$

Integrating between limits $N = N_0$ at $t = 0$ and $N = N_t$ at $t = t$

$$\int_{N_0}^{N_t} \frac{dN}{N} = -k \int_0^t dt$$

$$[\ln N]_{N_0}^{N_t} = -kt$$

$$\ln \frac{N_t}{N_0} = -kt$$

$$\frac{N_t}{N_0} = e^{-kt} \quad \text{(from the definition of Naperian logarithms)}$$

$$N = N_0 e^{-kt} \quad \text{(the sub t is usually dropped from N_t)}$$

Since the activity of a source A is equal to the number of disintegrations per second,

$$A = \frac{dN}{dt} = -kN = kN_0 \exp(-kt)$$

when $t = 0$,

$$\left(\frac{dN}{dt}\right) = kN_0 = A_0,$$

so

$$A = A_0 \exp(-kt) \tag{1.1}$$

Thus the activity also decreases exponentially.

1.6 Half-Life

An important concept is the *half-life* or the time $(T_{1/2})$ after which the activity has decayed to half its original value.

If A is set equal to $A_0/2$ in Equation 1.1,

$$\tfrac{1}{2} = \exp(-kT_{1/2}) \quad \text{or} \quad kT_{1/2} = \ln 2$$

Hence

$$A = A_0 \exp\left(\frac{-\ln 2\,t}{T_{1/2}}\right) = \exp\left(\frac{-0.693}{T_{1/2}}\right) \tag{1.2}$$

since ln 2 = 0.693. Equally,

$$N = N_0 \exp\left(\frac{-0.693}{T_{\frac{1}{2}}}\right)$$

Two extremely important properties of exponential decay must be remembered.

1. The idea of half-life may be applied from any starting point in time. Whatever the activity at a given time, after one half-life, the activity will have been halved.
2. The activity never becomes zero, since there are many millions of radioactive nuclei present, so their number can always be halved to give a residue of radioactivity.

Clearly, if the value of $T_{\frac{1}{2}}$ is known, and the rate of decay is known at one time, the rate of decay may be found at any later time by solving equation 1.2 given above. However, the activity may also be found, with sufficient accuracy, by a simple graphical method.

Proceed as follows:

1. Use the y-axis to represent activity and the x-axis to represent time.
2. Mark the x-axis in equal units of half-lives.
3. Assume the activity at $t = 0$ is 1. Hence the first point on the graph is (0,1).
4. Now, apply the half-life rule. After one half-life, the activity is ½, so the next point on the graph is (1,½).
5. Apply the half-life rule again to obtain the point (2,¼) and successively $(3,\frac{1}{8})(4,\frac{1}{16})$ $(5,\frac{1}{32})$.

See Figure 1.4a.

Note that, so far, the graph is quite general without consideration of any particular nuclide, half-life or activity. To answer a specific problem, it is now only necessary to re-label the axes with the given data, for example, 'The activity of an oral dose of I-131 is 90 MBq at 12 noon on Tuesday, 4 October. If the half-life of I-131 is 8 days, when will the activity be 36 MBq?' Figure 1.4b shows the same axes as Figure 1.4a re-labelled to answer this specific problem. This quickly yields the answer of 10½ days, that is, at 12 midnight on 14 October.

This graphical approach may be applied to any problem that can be described in terms of simple exponential decay.

Insight

Solving this problem using equation 1.2:

$$36 = 90 \exp\left(\frac{-\ln 2 \cdot t}{8}\right) \quad \text{where } t \text{ is the required time in days.}$$

$$\ln\left(\frac{90}{36}\right) = \left(\frac{\ln 2 \cdot t}{8}\right) \quad \text{from which } t = 10.6 \text{ days.}$$

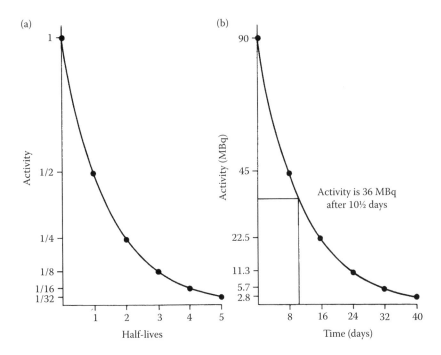

FIGURE 1.4

Simple graphical method for solving any problem where the behaviour is exponential. (a) A basic curve that may be used to describe any exponential process. (b) The same curve used to solve the specific problem on radioactive decay set in the text.

1.7 Secular and Transient Equilibrium

As already explained, radioactive decay is a process by which the nucleus attempts to achieve stability. It is not always successful at the first attempt and further decay processes may be necessary. For example, two major decay schemes occur in nature each of which involves a long sequence of decay processes, terminating finally in one of the stable lead isotopes.

In such a sequence the nuclide which decays is frequently called the *parent* and its decay product the *daughter*. If both the parent and daughter nuclides are radioactive, and the parent has a longer half-life than the daughter, the rate of decay of the daughter is determined not only by its own half-life but also by the rate at which it is produced. As a first approximation, assume that the activity of the parent remains constant, or is constantly replenished so that the rate of production of the daughter remains constant. If none of the daughter is present initially, its rate of production will at first exceed its rate of decay and equilibrium will be reached when the rate of production is just equal to the rate of decay (Figure 1.5a).

The curve is of the exponential type so the activity never actually reaches equilibrium. The rate of approach to equilibrium depends on the half-life of the daughter and after 10 half-lives the activity will be within 0.1% of equilibrium (see Insight). The equilibrium activity is governed by the activity of the parent.

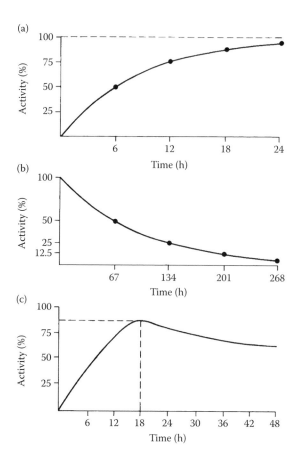

FIGURE 1.5
(a) Increase in activity of a daughter when the activity of the parent is assumed to be constant. Generation of Tc-99m from Mo-99 has been taken as a specific example, but with the assumption that the supply of Mo-99 is constantly replenished. (b) The decay curve for Mo-99 which has a half-life of 67 h. (c) Increase in activity of a daughter when the activity of the parent is decreasing. Generation of Tc-99m from Mo-99 has been taken as a specific example. Curves (a) and (b) are multiplied to give the resultant activity of Tc-99m.

Insight

Secular Equilibrium

Mathematically, the shape of Figure 1.5a is given by

$$N = N_{max}\left[1 - \exp\left(\frac{-\ln 2 \cdot t}{T_{1/2}}\right)\right]$$

where $T_{1/2}$ is now the half-life of the daughter radionuclide.
    Thus after n half-lives

$$N = N_{max}\left[1 - \exp(-n \ln 2)\right] = N_{max}[1 - (1/2)^n]$$

Substituting $n = 10$ then $(1/2)^{10} = 1/1024$ so N differs from its maximum value by less than 1 part in 1000.

Two practical situations should be distinguished.

1. The half-life of the parent is much longer than the half-life of the daughter; for example, radium-226, which has a half-life of 1620 years, decays to radon gas which has a half-life of 3.82 days. For most practical purposes the activity of the radon gas reaches a constant value, only changing very slowly as the radium decays. This is known as *secular equilibrium*.

2. The half-life of the parent is not much longer than that of the daughter. The most important example for radiology arises in diagnostic nuclear medicine and is molybdenum-99 (Mo-99) which has a half-life of 67 h before decaying to technetium-99m (Tc-99m) which has a half-life of 6 h. Now the growth curve for Tc-99m when the Mo-99 activity is assumed constant (Figure 1.5a) must be multiplied by the decay curve for Mo-99 (Figure 1.5b). The resultant (Figure 1.5c) shows that an actual maximum of Tc-99m activity is reached after about 18 h. By the time the 10 half-lives (60 h) required for Tc-99m to come to equilibrium with Mo-99 have elapsed, the activity of Mo-99 has fallen to half its original value.

This is known as *transient equilibrium* because although the Tc-99m is in equilibrium with the Mo-99, the activity of the Tc-99m is not constant. It explains why the amount of activity that can be eluted from a Mo-Tc generator (see Section 1.3 and Chapter 10) is much higher when the generator is first delivered than it is a week later.

1.8 Biological and Effective Half-Life

When a radionuclide is administered, either orally or by injection, in addition to the reduction of activity with time due to the physical process of decay, activity is also lost from the body as a result of biological processes. Generally speaking, these processes also show exponential behaviour so the concentration of substance remaining at time t after injection is given by

$$C = C_0 \exp\left(\frac{-\ln 2 \cdot t}{T_{\frac{1}{2}\text{biol}}}\right)$$

(cf. Equation 1.2), where $T_{\frac{1}{2}\text{biol}}$ is the biological half-life.

When physical and biological processes are combined, the overall loss is the product of two exponential terms and the activity at any time after injection is given by

$$A = A_0 \exp\left(\frac{-\ln 2 \cdot t}{T_{\frac{1}{2}\text{phys}}}\right) \cdot \exp\left(\frac{-\ln 2 \cdot t}{T_{\frac{1}{2}\text{biol}}}\right)$$

$$= A_0 \exp\left[-\ln 2 \cdot t \left(\frac{1}{T_{\frac{1}{2}\text{phys}}} + \frac{1}{T_{\frac{1}{2}\text{biol}}}\right)\right]$$

To find the effective half-life $T_{\frac{1}{2}\text{eff}}$, set

$$A = A_0 \exp\left(\frac{-0.693t}{T_{\frac{1}{2}\text{eff}}}\right)$$

Hence, by inspection,

$$\frac{1}{T_{\frac{1}{2}\text{eff}}} = \frac{1}{T_{\frac{1}{2}\text{phys}}} + \frac{1}{T_{\frac{1}{2}\text{biol}}}$$

Note that if $T_{\frac{1}{2}\text{phys}}$ is much shorter than $T_{\frac{1}{2}\text{biol}}$, the latter may be neglected, and vice versa. For example, if $T_{\frac{1}{2}\text{phys}} = 1$ h and $T_{\frac{1}{2}\text{biol}} = 20$ h,

$$\frac{1}{T_{\frac{1}{2}\text{eff}}} = 1 + \frac{1}{20} = 1.05$$

and $T_{\frac{1}{2}\text{eff}} = 0.95$ h or almost the same as $T_{\frac{1}{2}\text{phys}}$.

1.9 Gamma Radiation

Some radionuclides emit radioactive particles to gain stability. Normally, in addition to the particle, the nucleus also has to emit some energy, which it does in the form of gamma radiation. Note that emission of gamma rays as a mechanism for losing energy is very general and, as shown in Section 1.2, may also occur when radionuclides are produced. Although the emission of the particle and the gamma ray are, strictly speaking, separate processes, they normally occur very close together in time. However, some nuclides enter a metastable state after emitting the particle and emit their gamma ray some time later. When the two processes are separated in time in this way, the second stage is known as an *isomeric transition*. An important example in nuclear medicine is technetium-99m (the 'm' stands for metastable) which has a half-life of 6 h. This is long enough for it to be separated from the parent molybdenum-99 and the decay is then by gamma ray emission only which is particularly suitable for *in vivo* investigations (see Chapter 10).

Just as electrons in shells around the nucleus occupy well-defined energy levels, there are also well-defined energy levels in the nucleus. Since gamma rays represent transitions between these levels, they are monoenergetic. However, gamma rays with more than one well-defined energy may be emitted by the same nuclide, for example, indium-111 emits gamma rays at 163 keV and 247 keV.

Insight

Decay Schemes

It should be noted that in some radionuclides all disintegrations do not produce all the possible gamma photon energies or β particles with the same maximum energy that the radionuclide can produce. However, for a large number of disintegrations the ratio of gamma photons at one energy to those of another energy is always constant. This is illustrated in Table 1.2.

1.10 X-rays and Gamma Rays as Forms of Electromagnetic Radiation

The propagation of energy by simultaneous vibration of electric and magnetic fields is known as EM radiation. Unlike sound, which is produced by the vibration of molecules

TABLE 1.2

Decay Data for Molybdenum-99

Type of Particle	Maximum Energy of Decay (MeV)	Percentage of Disintegrations (%)	Gamma Photon Energies (MeV)	Percentage of Disintegrations (%)
β⁻	0.45	14	0.04	1.0
β⁻	0.87	~1	0.14	4.6
β⁻	1.23	85	0.18	4.5
			0.37	1.0
			0.74	10.0
			0.78	4.0

TABLE 1.3

The Different Parts of the Electromagnetic Spectrum Classified in Terms of Wavelength, Frequency and Quantum Energy

	Radio Waves	Infra-red	Visible Light	Ultra Violet	X-rays and Gamma Rays
Wavelength (m)	$10^3 - 10^{-2}$	$10^{-4} - 10^{-6}$	5×10^{-7}	5×10^{-8}	$10^{-9} - 10^{-13}$
Frequency (Hz)	$3 \times 10^5 - 3 \times 10^{10}$	$3 \times 10^{12} - 3 \times 10^{14}$	6×10^{14}	6×10^{15}	$3 \times 10^{17} - 3 \times 10^{21}$
Quantum energy (eV)	$10^{-9} - 10^{-4}$	$10^{-2} - 1$	2	20	$10^3 - 10^7$

and therefore requires a medium for propagation (see Chapter 15), EM radiation can travel through a vacuum. However, like sound, EM radiation exhibits many wave-like properties such as reflection, refraction, diffraction and interference and is frequently characterised by its wavelength. EM waves can vary in wavelength from 10^{-13} m to 10^3 m and different parts of the EM spectrum are recognised by different names (see Table 1.3).

X-rays and gamma rays are both part of the EM spectrum and an 80 keV X-ray is identical to, and hence indistinguishable from, an 80 keV gamma ray. To appreciate the reason for the apparent confusion, it is necessary to consider briefly the origin of the discoveries of X-rays and gamma rays. As already noted, gamma rays were discovered as a type of radiation emitted by radioactive materials. They were clearly different from α rays and β rays, so they were given the name gamma rays. X-rays were discovered in quite a different way as 'emission from high energy machines of radiations that caused certain materials, such as barium platino-cyanide to fluoresce'. It was some time before the similar identity of X-rays produced by machines and gamma rays produced by radioactive materials was confirmed.

For a number of years, X-rays produced by machines were of lower energy than gamma rays, but with the development of linear accelerators and other high-energy machines, this distinction is no longer useful.

No distinction between X-rays and gamma rays is totally self-consistent, but it is reasonable to describe gamma rays as the radiation emitted as a result of nuclear interactions, and X-rays as the radiation emitted as a result of events out with the nucleus. For example, one method by which nuclides with too few neutrons may approach stability is by *K-electron capture*. This mode of radioactive decay has not yet been discussed. The nucleus 'steals' an electron from the K shell to neutralise one of its protons. The K shell vacancy is filled by electrons from outer shells and the energy that has to be lost in this process is emitted

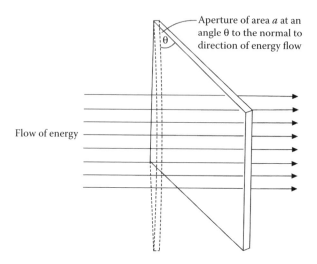

Aperture of area *a* at an
angle θ to the normal to
direction of energy flow

Flow of energy

FIGURE 1.6
A simple representation of the meaning of intensity.

as EM radiation (characteristic radiation) which is referred to as X-rays, even though they result from radioactive decay.

An important concept is the *intensity* of a beam or X or gamma rays. This is defined as the amount of energy crossing unit area placed normal to the beam in unit time. In Figure 1.6, if the total amount of radiant energy passing though the aperture of area *a* in time *t* is *E*, the intensity $I = E/(a \cos \theta \cdot t)$ where $a \cos \theta$ is the cross-sectional area of the aperture normal to the beam. If *a* is in m², *t* in s and *E* in joules, the units of *I* will be J m⁻²s⁻¹.

1.11 Quantum Properties of Radiation

As well as showing the properties of waves, short-wavelength EM radiation, such as X and gamma rays, can sometimes show particle-like properties. Each particle is in fact a small packet of energy and the size of the energy packet (ε) is related to frequency (*f*) and wavelength (λ) by the fundamental equations

$$\varepsilon = hf = \frac{hc}{\lambda}$$

where *h* is the *Planck constant* and *c* is the speed of EM waves. Taking $c = 3 \times 10^8$ m s⁻¹ and $h = 6.6 \times 10^{-34}$ J s

$$\varepsilon \text{ (in joules)} = \frac{2 \times 10^{-25}}{\lambda \text{ (in metres)}}$$

Thus, the smaller the value of λ, the larger the value of the energy packet. For a typical diagnostic X-ray wavelength of 2×10^{-11} m, the value of ε in joules for a single photon is inconveniently small, so the electron volt, a unit of energy that has already been introduced (see Section 1.1), is used where 1 eV = 1.6×10^{-19} J. A wavelength of 2×10^{-11} m corresponds to a photon energy of 62 keV.

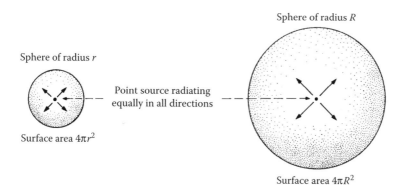

FIGURE 1.7
A diagram showing the principle of the inverse square law.

1.12 Inverse Square Law

Before considering the interaction of radiation with matter, one important law that all radiations obey under carefully defined conditions will be introduced. This is the *inverse square law* which states that for a point source, and in the absence of attenuation, the intensity of a beam of radiation will decrease as the inverse of the square of the distance from that source.

The law is essentially just a statement of conservation of energy, since if the rate at which energy is emitted as radiation is E, the energy will spread out equally (isotropically) in all directions and the amount crossing unit area per second at radius r, $I_r = E/4\pi r^2$ (Figure 1.7). Similarly the intensity crossing unit area at radius R, $I_R = E/4\pi R^2$. Thus the intensity is decreasing as $1/(\text{radius})^2$.

1.13 Interaction of Radiation with Matter

As a simple model of the interaction of radiation with matter, consider the radiation as a stream of fast moving particles (alphas, betas or photons) and the medium as an array of nuclei each with a shell of electrons around it (Figure 1.8). As the particle tries to penetrate the medium, it will collide with atoms. Sometimes it will transfer energy of excitation during a collision. This type of interaction will be considered in more detail in Chapter 3. The energy is quickly dissipated as heat. Occasionally, the interaction will be so violent that one of the electrons will be torn away from the nucleus to which it was bound and become free. *Ionisation* has occurred because an ion pair has been created. Sometimes, as in interaction C, the electron thus released has enough energy to cause further ionisations and a cluster of ions is produced.

The amount of energy required to create an ion pair is about 34 eV. Charged particles of interest in medicine invariably possess this amount of energy. For EM radiation, a quantum of X or gamma rays always has more than 34 eV but a quantum of, say, ultraviolet or visible light does not. Hence the EM spectrum may be divided into ionising and non-ionising radiations.

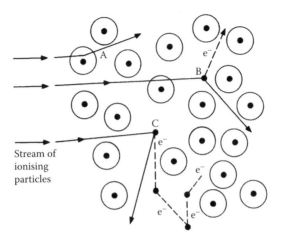

FIGURE 1.8

Simple model of the interaction of radiation with matter. Interaction A causes excitation, interaction B causes ionisation, and interaction C causes multiple ionisations. At each ionisation an electron is released from the nucleus to which it was bound. Recall the comment in Section 1.1 that matter consists mostly of empty space. Hence the chance of a collision is much smaller in practice than this diagram suggests.

TABLE 1.4

Approximate Ranges of Electrons in Soft Tissue

Electron Energy (keV)	Approximate Range (mm)
20	0.01
40	0.03
100	0.14
400	1.3

The aforementioned, very simple model may also be used to predict how easily different types of radiation will be attenuated by different types of material. Clearly, as far as the stopping material is concerned, a high density of large nuclei (i.e. high atomic number) will be most effective for causing many collisions. Thus gases are poor stopping materials, but lead (Z = 82) is excellent and, if there is a special reason for compact shielding, even depleted uranium (Z = 92) is sometimes used.

With regard to the bombarding particles, size (or mass) is again important and since the particle is moving through a highly charged region, interaction is much more probable if the particle itself is charged and, therefore, likely to come under the influence of the strong electric fields associated with the electron and nucleus. Since X and gamma ray quanta are uncharged and have zero rest mass, they are difficult to stop and higher energy photons require dense material such as lead to cause appreciable attenuation.

The β⁻ particle is more massive and is charged so it is stopped more easily—a few mm of low atomic number materials such as perspex will usually suffice. Since it will be shown in Chapter 3 that the mechanism of energy dissipation by X and gamma rays is via secondary electron formation, a table of electron ranges in soft tissue will be helpful (Table 1.4).

Protons and α particles are more massive than β^- particles and are charged, so they are stopped easily. α particles, for example, are so easily stopped, even by a sheet of paper, that great care must be taken when attempting to detect them to ensure that the detector has a thin enough window to allow them to enter the counting chamber. Neutrons are more penetrating because, although of comparable mass to the proton, they are uncharged.

One final remark should be made regarding the ranges of radiations. Charged particles eventually become trapped in the high electric fields around nuclei and have a finite range. Beams of X or gamma rays are stopped by random processes, and as shown in Chapter 3, are attenuated exponentially. This process has many features in common with radioactive decay. For example, the rate of attenuation by a particular material is predictable but the radiation does not have a finite range.

1.14 Linear Energy Transfer

Beams of ionising radiation are frequently characterised in terms of their linear energy transfer (LET). This is a measure of the rate at which energy is transferred to the medium and hence of the density of ionisation along the track of the radiation. Although a difficult concept to apply rigorously, it will suffice here to use a simple definition, namely that LET is the energy transferred to the medium per unit track length. It follows from this definition that radiations which are easily stopped will have a high LET, those which are penetrating will have a low LET. Some examples are given in Table 1.5.

1.15 Energy Changes in Radiological Physics

Energy cannot be created or destroyed but can only be converted from one form to another. Therefore, it is important to summarise the different forms in which energy may appear. Remember that *work* is really just another word for energy—stating that body A does work on body B means that energy is transferred from body A to body B.

TABLE 1.5

Approximate Values of Linear Energy Transfer for Different Types of Radiation

Radiation	LET (keV μm^{-1})
1 MeV γ rays	0.5
100 kVp X-rays	6
20 keV β^- particles	10
5 MeV neutrons	20
5 MeV α particles	50

Examples of different forms of energy are given in the insight.

Insight

Different Forms of Energy

Mechanical Energy
This can take two well-known forms.

1. *Kinetic energy,* $\frac{1}{2}mv^2$, where *m* is the mass of the body and *v* its velocity.
2. *Potential energy, mgh,* where *g* is the gravitational acceleration and *h* is the height of the body above the ground.

Kinetic energy is more relevant than potential energy in the physics of X-ray production and the behaviour of X-rays.

Electrical Energy
When an electron, charge *e*, is accelerated through a potential difference *V*, it acquires energy *eV*. Thus if there are *n* electrons they acquire total energy *neV*. Note:

1. Current (*i*) is rate of flow of charge. Thus $i = ne/t$ where *t* is the time. Hence, rearranging an alternative expression for the energy in a beam of electrons is *Vit*.
2. Just as *Vit* is the amount of energy gained by electrons as they accelerate through a potential difference *V*, it is also the amount of energy lost by electrons (usually as heat) when they fall through a potential difference of *V*, for example, when travelling through a wire that has resistance *R*.
3. If the resistor is 'ohmic', that is to say it obeys *Ohms law*, then $V = iR$ and alternative expressions for the heat dissipated are V^2/R or i^2R. Note, however, that many of the resistors encountered in the technology of X-ray production are non-ohmic.

Heat Energy
When working with X-rays, most forms of energy are eventually degraded to heat and when a body of mass *m* and specific heat capacity *s* receives energy *E* and converts it into heat, the rise in temperature ΔT will be given by

$$E = ms\Delta T$$

Excitation and Ionisation Energy
Electrons are bound in energy levels around the nucleus of the atom. If they acquire energy of excitation they may jump into a higher energy level. Sometimes the energy may be enough for the electrons to escape from the energy well referred to in Section 1.1 (ionisation). Note that if this occurs the electron may also acquire some kinetic energy in addition to the energy required to cause ionisation.

Radiation Energy
Radiation represents a flow of energy. This is usually expressed in terms of beam intensity *I* such that $I = E/(a \cdot t)$ where *E* is the total energy passing through an area *a* placed normal to the beam, in time *t*.

Quantum Energy
X and gamma radiation frequently behave as exceedingly small energy packets. The energy of one quantum is *hf* where *h* is the Planck constant and *f* is the frequency of the radiation. The energy of one quantum is so small that the joule is an inconveniently large unit so the electron volt is introduced where $1 \text{ eV} = 1.6 \times 10^{-19} \text{ J}$.

Mass Energy

As a result of Einstein's work on relativity, it has become apparent that mass is just an alternative form of energy. If a small amount of matter, mass *m,* is converted into energy, the energy released $E = mc^2$ where *c* is the speed of EM waves. This change is encountered most frequently in radioactive decay processes. Careful calculation, to about one part in a million, shows that the total mass of the products is slightly less than the total mass of the starting materials, the residual mass having been converted to energy according to the above equation. Annihilation of positrons (see Section 3.4.4) is another good example of the equivalence of mass and energy.

As an example of the importance of conservation of energy in diagnostic radiology, consider the energy changes in the production and attenuation of X-rays and registration on photographic film. First, electrical energy is converted into the kinetic energy of the electrons in the X-ray tube. When the electrons hit the anode, their kinetic energy is destroyed. The majority is converted into heat, a little into X-rays. As the X-rays penetrate the body, some of their energy is absorbed, more in bone than in soft tissue, and causes ionisation before eventually being converted into heat. Finally the X-rays which strike the intensifying screen cause excitation and the emission of visible light quanta and these lower energy quanta stimulate the physico-chemical processes in photographic film leading eventually to blackening.

1.16 Conclusion

Ionising and non-ionising radiation may be used for imaging without a detailed mathematical understanding of the underlying physics. However, to obtain the best images or quantitative information from them in the safest possible manner, a full understanding of their physical properties is imperative. The subsequent chapters in this book build on the basic background information contained in this chapter to allow the maximum benefits to be achieved.

Further Reading

Allisy-Roberts P and Williams J Farr 2008. *Physics for Medical Imaging,* 2nd edition. Saunders, Elsevier, pp 1–21, Chapter 1.

Bushberg J T, Seibert A J A, Leidholdt E M and Boone J M 2002. *The essential physics of medical imaging,* 2nd edition. Lippincott, Williams and Wilkins, Philadelphia, pp 17–29.

Johns H E and Cunningham J R 1983. *The Physics of Radiology,* 4th edition. Thomas Springfield, Chapter 2.

Meredith W J and Massey J B 1977. *Fundamental Physics of Radiology,* 3rd edition. Wright Bristol.

Exercises

1. Describe in simple terms the structure of the atom and explain what is meant by atomic number, atomic weight and radionuclide.

2. What is meant by the binding energy of an atomic nucleus? Define the unit in which it is normally expressed and indicate the order of magnitude involved.

3. Describe the different ways in which radioactive disintegration can occur.

4. What is meant by the decay scheme of a radionuclide and radioactive equilibrium?

5. What is a radionuclide generator?

6. A radiopharmaceutical has a physical half-life of 6 h and a biological half-life of 20 h. How long will it take for the activity in the patient to fall to 25% of that injected?

7. The decay constant of iodine-123 is 1.34×10^{-5} s^{-1}. What is its half-life and how long will it take for the radionuclide to decay to one-tenth of its original activity?

8. Investigate whether the values of radiation intensity given below decrease exponentially with time :

Intensity $(Jm^{-2}s^{-1})$	100	70	50	33	25	20	10	6.7	5.0	4.0
Time (s)	0	1.0	2.0	3.0	4.0	5.0	10.0	15.0	20.0	25.0

9. Radionuclide A decays into a nuclide B which has an atomic number one less than that of A. What types of radiation might be emitted either directly or indirectly in the disintegration process? Indicate briefly how they are produced.

10. Give typical values for the ranges of α particles and β^- particles in soft tissue. Why is the concept of range not applicable to gamma rays?

11. For an unknown sample of radioactive material explain how it would be possible to determine by simple experiment

    a. The types of radiation emitted

    b. The half-life

12. State the inverse square law for a beam of radiation and give the conditions under which it will apply exactly.

13. A surface is irradiated uniformly with a monochromatic beam of X-rays of wavelength 2×10^{-11} m. If 20 quanta fall on each square cm of the surface per second, what is the intensity of the radiation at the surface? (Use data given in Section 1.11).

14. Place the following components in order of the power of dissipation:

    a. A fluorescent light

    b. An X-ray tube

    c. An electric fire

    d. A pocket calculator

    e. An electric iron

2

Production of X-Rays

P P Dendy and B Heaton

SUMMARY

- The photons emitted by an X-ray tube are not all of the same energy (wavelength). There are two components: a continuous spectrum and one or more line spectra—the origins of each are explained.
- The important difference between *quantity* of X-rays and *quality* of X-rays produced is emphasised and the various factors affecting each of them are discussed.
- Design features of the X-ray tube that are essential for high quality performance in radiology are analysed.
- A good radiograph requires the field of view to be uniformly exposed to X-rays. This is not a trivial requirement and both the limitations and methods for optimisation are discussed.
- An X-ray tube is subject to both electrical and thermal rating limits. These will be discussed with particular reference to design features of the tube.
- The elements of good quality assurance required to ensure that excessive radiation is not used for each exposure and to keep repeat examinations to a minimum, will be outlined.

CONTENTS

2.1 Introduction

When electrons are accelerated to energies in excess of 5 keV and are then directed onto a target surface, X-rays may be emitted. The X-rays originate principally from rapid deceleration of the electrons when they interact with the nucleus of the target atoms. These X-rays are known as 'Bremsstrahlung' or braking radiation.

The essential features of a simple, low output, X-ray tube are shown in Figure 2.1 and comprise:

1. A heated metal filament to provide a copious supply of electrons by thermionic emission and to act as a cathode
2. An evacuated chamber across which a potential difference can be applied

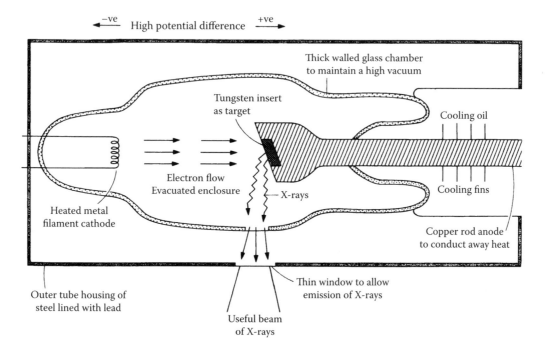

FIGURE 2.1
Essential features of a simple, stationary anode X-ray tube. Note that for simplicity only X-ray photons passing through the window are shown. In practice they are emitted in all directions.

3. A metal anode (the target) with a high efficiency for conversion of electron energy into X-ray photons

4. A thinner window in the chamber wall that will be transparent to most of the X-rays

Insight

Electrical Supply to the Tube

This tube will be energised by an electrical supply generator with the following features:

- An input transformer to adjust primary and secondary voltages
- A high voltage transformer to provide up to 150 kV
- A rectifier system to convert alternating current (AC) to direct current (DC)
- A low voltage transformer providing 8–12 V for the filament heating
- A timing mechanism for terminating exposures

In this chapter the mechanisms of X-ray production will be considered in detail and the main components of a modern X-ray tube will be described. X-ray tubes may be used clinically for either radiography—the creation of still images with a short pulse of X-rays, or for fluoroscopy—the production of images in real time using continuous X-ray exposure. Physical factors affecting the design and performance of X-ray sets and the implications for obtaining high quality images will be discussed for both applications.

2.2 The X-ray Spectrum

If the accelerating voltage across the X-ray tube shown in Figure 2.1 were about 100 kV, the spectrum of radiation that would be used for radiology might be something like that shown in Figure 2.2. The various features of this spectrum will now be discussed.

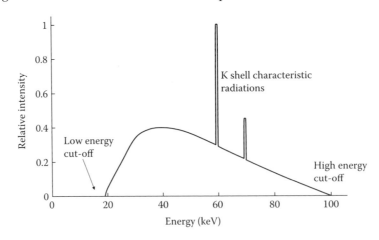

FIGURE 2.2
Spectrum of radiation incident on a patient from an X-ray tube operating at 100 kVp using a tungsten target and 2.5 mm aluminium filtration.

2.2.1 The Continuous Spectrum

When a fast-moving electron strikes the anode, several things may happen. The most common is that the electron will suffer a minor interaction with an orbital electron as depicted at (a) in Figure 2.3. This will result in the transfer of a small amount of energy to the target which will appear eventually as heat. At diagnostic energies, at least 99% of the electron energy is converted into heat and the dissipation of this heat is a major technical problem that will be considered in Section 2.5.

Occasionally, an electron will come close to the nucleus of a target atom, where it will suffer a much more violent change of direction because the charge and mass of the nucleus are much greater than those of an electron (example b). The electron does not penetrate the nucleus because the energy barrier presented by these positive charges in the nucleus is far in excess of the electron energy. This results in the electron being deviated around it. The interaction results in a change in kinetic energy of the electron and the emission of electromagnetic radiation, that is in the X-ray range of the spectrum. The amount of energy lost by the electron in such an interaction is very variable and hence the energy given to the X-ray photon can take a wide range of values. Note that X-ray emission may occur after two or three earlier slight deviations (example c). Therefore not all emissions occur from the surface of the anode. This factor is important when the spatial distribution of X-ray emission is considered.

2.2.2 The Low and High Energy Cut-Off

These parts of the spectrum are simply explained. Low energy electromagnetic radiation is easily attenuated and below a certain energy is so heavily attenuated—by the materials

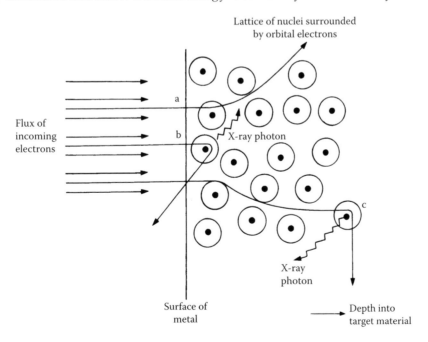

FIGURE 2.3
Schematic representation of the interaction of electrons with matter. (a) Interaction resulting in the generation of low energy electromagnetic radiations (infra red, visible, ultraviolet and very soft X-rays). All these are rapidly converted into heat. (b) Interaction resulting in the production of an X-ray. (c) Production of an X-ray after previous interactions that resulted only in heat generation.

of the anode, by the window of the X-ray tube and by any added filtration—that the intensity emerging is negligible. X-ray attenuation is discussed in detail in Chapter 3.

The high energy cut-off occurs because all the energy of an electron may, very occasionally, be used to produce a single X-ray photon. Hence for any given electron energy, determined by the accelerating voltage across the X-ray tube, there is a well-defined maximum X-ray energy equal to the energy of a single electron. This corresponds to a minimum X-ray wavelength. Note that it is not possible, by quantum theory, for the energy of several electrons to be stored up in the anode to produce a jumbo-sized X-ray quantum of energy.

It is useful to calculate the electron velocity, the maximum X-ray photon energy and the minimum X-ray photon wavelength associated with a given tube kilovoltage. To avoid complications associated with relativistic effects, a tube operating at only 30 kV is considered.

The energy of each electron is given by the product of its charge (e coulombs) and the accelerating voltage (V volts)

$$eV = 1.6 \times 10^{-19} \text{ (C)} \times 3 \times 10^4 \text{ (V)}$$

$$= 4.8 \times 10^{-15} \text{ J (or } 3 \times 10^4 \text{ electron volts, i.e. 30 keV)}$$

Note the distinction between an accelerating voltage, measured in kV, and the electron energy after passing from the anode to the cathode measured in keV.

The electron velocity can be obtained from the fact that its kinetic energy is $\frac{1}{2}m_e v_e^2$ where m_e is the mass of the electron and v_e its velocity. Hence

$$\tfrac{1}{2}m_e v_e^2 = 4.8 \times 10^{-15} \text{ J}.$$

Since $m_e = 9 \times 10^{-31}$ kg, v_e is approximately 10^8 m s^{-1} which is one-third the speed of light. It can be seen from this example that relativistic effects are important even at quite low tube kilovoltages.

From above, the maximum X-ray photon energy ε_{Imax} is 4.8×10^{-15} J and the minimum wavelength is obtained by substitution in

$$\varepsilon = hf = \frac{hc}{\lambda}$$

(h is the Planck constant, c the speed of light and λ the wavelength of the resulting X-ray.) Hence

$$\lambda = \frac{hc}{\varepsilon} = \frac{6.6 \times 10^{-34} \times 3 \times 10^8}{4.8 \times 10^{-15}} = 4.1 \times 10^{-11} \text{ m (or 0.041 nm)}$$

Note that calculations giving the maximum X-ray photon energy and minimum wavelength are valid even when the electrons travel at relativistic speeds.

2.2.3 Shape of the Continuous Spectrum

A detailed treatment of the continuous spectrum is beyond the scope of this book, but the following approach is helpful since it involves some other important features of the X-ray production process. First, imagine a very thin anode, and consider the production of X-rays, not the X-rays that finally emerge. It may be shown by theoretical arguments that the intensity of X-rays produced will be constant up to a maximum X-ray energy determined by the energy of the electrons (see Figure 2.4a).

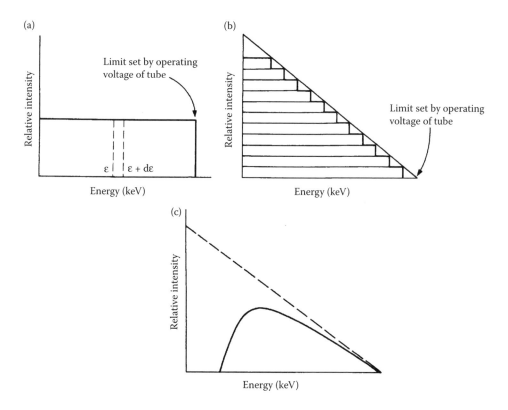

FIGURE 2.4

A simplified explanation of the shape of the continuous X-ray spectrum. (a) Production of X-rays from a very thin anode. Note that the intensity of the beam in the small range ε to $\varepsilon + d\varepsilon$ will be equal to the number of photons per square metre per second multiplied by the photon energy. Fewer high energy photons are produced but their energy is higher and the product is constant. (b) Production of X-rays from a thicker anode treated as a series of thin anodes. (c) X-ray emission (solid line) compared with X-ray production (dotted line).

A thick anode may now be thought of as composed of a large number of thin layers. Each will produce a similar distribution to that shown in Figure 2.4a, but the maximum photon energy will gradually be reduced because the incident electrons lose energy as they penetrate the anode material. Thus, the composite picture for X-ray production might be as shown in Figure 2.4b.

However, before the X-rays emerge, the intensity distribution will be modified in two ways. First, X-rays produced deep in the anode will be attenuated in reaching the surface of the anode and second, all the X-rays will be attenuated in penetrating the window of the X-ray tube. Both processes reduce the intensity of the low energy radiation more than that of the higher energies so the resultant is the solid curve in Figure 2.4c.

Insight

Effective Energy

Two properties of the spectrum that are sometimes mentioned are the photon energy at which the intensity is maximum (ε_{Imax}) and the mean energy (ε_{mean}). Since the spectrum is not symmetrical, they are not the same and neither has much practical significance.

A more useful quantity is the *effective energy* (ε_{eff}). This is defined as the energy of a narrow beam of monochromatic radiation that would have the same penetrating power (measured in terms of half value layer (HVL) or linear attenuation coefficient) as the mixed energy spectrum. For a well-filtered beam (see Section 3.9) ε_{eff} will be close to, but not identical with ε_{lmax} and approximately one-third ε_{max}.

2.2.4 Line or Characteristic Spectra

Superimposed on the continuous spectrum there may be a set of line spectra which result from an incoming electron interacting with a bound orbital electron in the target. If the incoming electron has sufficient energy to overcome the binding energy, it can remove the bound electron creating a vacancy in the shell. The probability of this happening is greatest for the innermost shells. This vacancy is then filled by an electron from a higher energy level falling into it and the excess energy is emitted as an X-ray. Thus, if, for example, the vacancy is created in the K shell, it may be filled by an electron falling from either the L shell, the M shell or the outer shells. Even a free electron may fill the vacancy but the most likely transition is from the L shell. The process is summarised in Figure 2.5.

As discussed in Section 1.1, orbital electrons must occupy well-defined energy levels and these energy levels are different for different elements. Thus the X-ray photon emitted when an electron moves from one energy level to another has an energy equal to the difference between the two energy levels in that atom and hence is characteristic of that element.

Reference to Figure 1.2b shows that the K series of lines for tungsten (Z = 74) will range from 58.5 keV (for a transition from the L shell to the K shell) to 69.5 keV (if a free electron fills the K shell vacancy). Transitions to the L shell are of little practical importance in diagnostic radiology since the maximum energy change for tungsten is 11 keV and photons of this energy are normally absorbed before they leave the tube.

Lower atomic number elements produce characteristic X-rays at lower energies. The K shell radiations from molybdenum (Z = 42) at circa 19 keV are important in mammography (see Section 9.2). Note that characteristic radiation cannot be produced unless the operating kV of the X-ray tube is high enough to give the accelerating electrons sufficient energy to remove the relevant bound electrons from the anode target atoms.

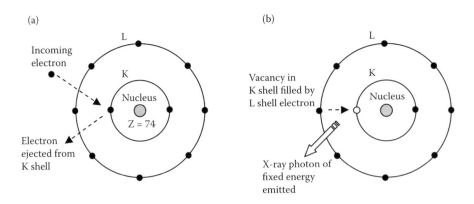

FIGURE 2.5
Production of a line spectrum. (a) The incoming electron removes an electron from the K shell of a tungsten atom. (b) An electron from the L shell falls into the K shell potential well and an X-ray photon with a well-defined energy $\varepsilon_L - \varepsilon_K$ is emitted. Only the K and L shell electrons are shown.

2.2.5 Factors Affecting the X-ray Spectrum

If the spectrum changes in such a manner that its shape remains unaltered, that is, the intensity or number of photons at every photon energy changes by the same factor, there has been a change in radiation *quantity*. If on the other hand, the intensities change such that the shape of the spectrum also changes, there has been a change in radiation *quality* (the penetrating power of the X-ray beam). A number of factors that affect the X-ray spectrum may be considered.

2.2.5.1 Tube Current, I_T

This determines the number of electrons striking the anode. Thus the emitted X-ray energy or exposure E is proportional to tube current, but only the quantity of X-rays is affected ($E \propto I_T$).

2.2.5.2 Time of Exposure

This again determines the number of electrons striking the anode so exposure is proportional to time but only the quantity of X-rays is affected ($E \propto t$).

2.2.5.3 Applied Voltage

If other tube operating conditions are kept constant, the output increases approximately as the square of tube kilovoltage ($E \propto kV^2$). Two factors contribute to this increase. First the electrons have more energy to lose when they hit the target. Second, as shown in Table 2.1, the efficiency of conversion of electrons into X-rays rather than into heat also increases with tube kilovoltage by a small amount over diagnostic energy changes (the change associated with a large increase in kV is shown to emphasise the effect). Thus both the flux of X-ray photons and their mean energy increases.

Increasing the tube kilovoltage also alters the radiation *quality* since the high energy cut-off has now increased. The position of any characteristic lines will not change.

Insight

Effect of kV on Exposure Factors

Note that the effect of increasing kV on the amount of radiation (mAs) required for an exposure is generally greater than that implied by $E \propto kV^2$. At higher kV the radiation penetrates the patient more easily and detector sensitivity varies with kV. For a film-screen receptor a very approximate relationship between kV_1 and kV_2

$$(kV_1)^4 \times mAs_1 = (kV_2)^4 \times mAs_2$$

TABLE 2.1

Efficiency of Conversion of Electron Energy into X-rays as a Function of Tube Kilovoltage

Tube Kilovoltage (kV)	Heat (%)	X-rays (%)
60	99.5	0.5
200	99	1.0
4000	60	40

2.2.5.4 Waveform of Applied Voltage

So far it has been assumed that the X-ray tube is operating from DC with a fixed voltage. In fact the voltage is generated by rotating a coil of wire in a magnetic field and changes in magnitude as the position of the coil changes relative to the direction of the field. Thus the magnitude of the voltage changes sinusoidally with time (Figure 2.6a) and produces AC. Since in Figure 2.1 one end of the X-ray tube must act as a 'cathode' and the other end as 'anode', no current flows when an alternating potential is applied during the half cycle when the cathode is positive with respect to the anode. Half wave rectification (Figure 2.6b) may be achieved by inserting a rectifier in the anode circuit (see Section 2.3.4) but since X-rays are only emitted for half the cycle, output is poor. Improved output can be achieved by full wave rectification obtained by using a simple bridge circuit. However, the tube is still not emitting X-rays all the time (Figure 2.6c). Furthermore, the majority of X-rays are emitted at a kilovoltage below the peak value (kVp).

A more constant voltage will improve the quality of the radiation and this can be achieved by using a three phase supply. This uses three rotating coils arranged such that, at any instant they are in different positions relative to the magnetic field. The X-ray tube is now driven by three separate voltage supplies, each of which has been fully rectified. The three supplies are 60° out of phase and switching circuits ensure that each supply only drives the X-ray tube when the voltage is near to peak value. The resultant voltage profile (Figure 2.6d) shows only about 15% variation. If the cathode supply is also three phase and is arranged to be 30° out of phase with the anode supply, a 12 peak generator is possible and fluctuations can be reduced to about 6%. With high frequency generators (see Section 2.3.4)

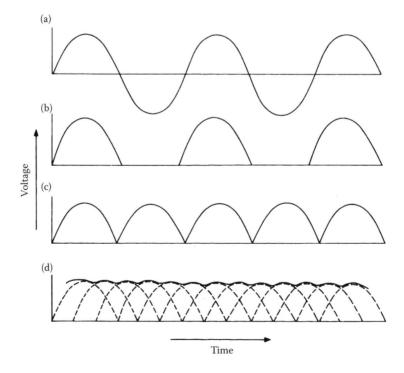

FIGURE 2.6
Examples of different voltage waveforms. (a) Mains supply; (b) Half wave rectification; (c) Full wave rectification using a bridge rectifier; (d) Three phase supply (with rectification).

used routinely nowadays, an almost constant voltage output can be achieved after rectification and smoothing.

Variation in the voltage supply is known as the *voltage ripple* and is defined as $V_{max} - V_{min}/V_{max}$. Variations in V will affect the instantaneous output. For general radiography a ripple of 5% may be acceptable, but when highly uniform X-ray output is essential, for example, in computed tomography (CT), ripple must be reduced to less than 1%. This subject is considered in more detail in Section 2.3.4. In future, in accordance with standard practice, operating voltages will be expressed in kVp to emphasise that the peak voltage with respect to time is being given.

2.2.5.5 Filtration

This also has a marked effect on both the quantity and quality of the X-ray beam, not only reducing the overall output but also reducing the proportion of low energy photons. Special filters (K-edge filters) can be used to create a window of transmitted X-ray energies and thus reduce the number of both high and low energy photons. The effect of beam filtration is considered in detail in Section 3.9.

2.2.5.6 Anode Material

Choice of anode material affects the efficiency of bremsstrahlung production (the continuous spectrum) with unfiltered output increasing approximately proportional to the atomic number of the target. The position of the characteristic lines also moves to higher energies as atomic number increases, for further detail see Section 2.3.2.

2.3 Components of the X-ray Tube

2.3.1 The Cathode

The cathode assembly normally consists of two parts—(a) an electron source (emitter) and (b) an auxiliary electrode surrounding it. The electron emitter is usually a coiled wire filament 0.2–0.3 mm in diameter of reasonably high resistance R. Therefore, for a given filament heating current I_F (typically in the region of 5A), effective ohmic heating ($I_F^2 R$) and minimum heat losses will occur. A metal is chosen for the cathode that will give a copious supply of electrons by thermionic emission at temperatures where there is very little evaporation of metal atoms into the vacuum (e.g. tungsten, melting point of 3370°C). These electrons form the tube current I_T. Note that the emission current response time is slow so controlling electron emission is not a practical means of controlling the time of X-ray exposure.

The tube current increases exponentially with increasing filament current and might, typically, rise by four orders of magnitude if the filament temperature increased from 2000 K to 3000 K (but note the adverse effect on tube lifetime—see Section 2.5.2). Also if I_F and I_T are too high it will be difficult to focus the electron beam. Thus the filament is kept well below its melting point.

Between exposures, the filament is kept warm on stand-by because although its resistance may be typically 5 Ω at 2000°K, at room temperature it falls to about 0.1 Ω. Thus a large current would be required to heat the filament rapidly from room temperature to its working temperature. Surrounding the filament is a cloud of negatively charged electrons, commonly called the *space charge*. The number of electrons in the space charge tends to a self-limiting constant value dependent on the filament temperature.

For reasons related primarily to geometrical unsharpness in the image, a small target for electron bombardment on the anode is essential. However, unless special steps are taken, the random thermally induced velocities and mutual repulsion of the electrons leaving the cathode will cause a broad beam to strike the anode. Therefore the filament is surrounded by an auxiliary electrode, or focussing cup (the Wehnelt electrode), typically made of nickel. This electrode provides an electric field which exercises a focusing action on the electrons by changing the equipotential lines and pressing the electrons together to produce a small focal spot on the anode. Originally, the Wehnelt electrode was maintained at the same potential as the emitter but smaller spots can be obtained by making its potential slightly more negative.

If the Wehnelt electrode is made about 2 kV more negative than the filament, or an additional electrode or 'grid' is used to provide this voltage, the electron beam will be stopped completely. This technique, known as *grid control*, can be used to improve the output profile of the X-rays since pulsed control of the current switches the beam on and off with very little inertia. A more recent alternative method of output control is *primary pulsing*. This is one of the benefits of recent developments in high frequency generation (see Section 2.3.4) and permits the direct modulation of tube voltage. Output profiles with steep rise and fall times are important when rapid pulses or very short exposures (a few milliseconds) are required, for example in fluorography, digital subtraction angiography, CT and paediatrics.

Most diagnostic X-ray tubes have a choice of focal spot size. A smaller focal spot produces sharper images (see Section 6.9.1) but places greater demands on heat dissipation in the anode (see Section 2.5.3). Some tubes have a dual filament assembly, each filament having its own focusing cup producing two spots of different sizes. Alternatively, the negative voltage bias on the Wehnelt electrode may be varied to refocus the electron output from a single filament electrostatically.

Note that spot size does vary somewhat with tube current and tube kilovoltage since the focusing action cannot be readily adjusted to compensate for variations in the mutual electrostatic repulsion between electrons when either their density or their energy changes. The effect may not be apparent if tube current is increased from 100 mA to 300 mA at 140 kVp but at 80 kVp the focal spot size may increase by a factor of two or more.

The effective or apparent size of the focal spot on the anode is smaller than the actual focal spot because of the anode angle. The smaller the anode angle the smaller the apparent focal spot size (see Sections 2.4 and 6.9.1).

2.3.2 The Anode Material

The material chosen for the anode should satisfy a number of requirements. It should have

1. A high conversion efficiency for electrons into X-rays. High atomic numbers are favoured since the X-ray intensity is proportional to Z. At 100 keV, lead (Z = 82) converts 1% of the energy into X-rays but aluminium (Z = 13) converts only about 0.1%.

2. A high melting point so that the large amount of heat released causes minimal damage to the anode.

3. A high conductivity so that the heat is removed rapidly.

4. A low vapour pressure, even at very high temperatures, so that atoms are not boiled off from the anode.

5. Suitable mechanical properties for anode construction.

In stationary anodes the target area is pure tungsten (W) ($Z = 74$, melting point 3370°C) set in a metal of higher conductivity such as copper. Originally, rotating anodes were also made of pure tungsten. However, at the high temperatures generated in the rotating anode (see Section 2.3.3), deep cracks developed at the point of impact of the electrons. The deleterious effects of damaging the target in this way are discussed in Sections 2.3.5 and 2.4. The addition of 5%–10% rhenium (Rh) ($Z = 75$, melting point 3170°C) greatly reduced the cracking by increasing the ductility of tungsten at high temperatures. The wear resistant rhenium alloy in the focal spot path ensures minimal ageing, thus high and constant exposure values for a long life. However, pure W/Rh anodes would be extremely expensive so molybdenum is now chosen as the base metal. Molybdenum ($Z = 42$, melting point 2620°C) stores twice as much heat, weight for weight, as tungsten, but the anode volume is now greater because molybdenum has a smaller density than tungsten. As shown in Figure 2.7a only a thin layer of W/Rh is used to prevent distortion that might arise from the differences in thermal expansion of the different metals.

2.3.3 Anode Design

The two principal requirements of anode design are first to make adequate arrangements for dissipation of the large quantity of heat generated and second to ensure a good spatial

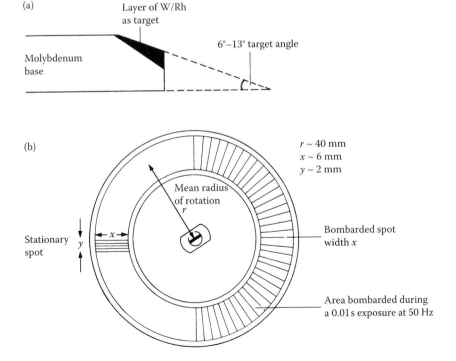

FIGURE 2.7
(a) Detail of the target area on a modern rotating anode. (b) Principle of the rotating anode showing the area bombarded in a 0.01 s exposure at 50 Hz.

distribution of X-rays. Design features related primarily to heat dissipation are discussed below, the spatial distribution of X-rays is considered later.

2.3.3.1 Stationary Anode

Figure 2.1 showed the design of a relatively simple X-ray tube with a stationary anode. When low outputs suffice, for example in dental and small mobile units, stationary anodes may still be used. Tungsten in the form of a small disc about 1 mm thick and 1 cm in diameter is embedded in a large block of copper which protrudes through the tube envelope into the surrounding oil. Heat is transferred from the tungsten to copper by conduction and thence to the oil by convection. The cooling fins assist the convection process. The oil transfers this heat to the X-ray tube shield by conduction and it is eventually removed by air in the X-ray room by convection. A stationary anode tube is simple, robust, very reliable and has a long lifetime, but its low power makes it unsuitable for many radiographic applications.

2.3.3.2 Rotating Anode

For reasons related almost entirely to heating effects in the anode, the majority of diagnostic X-ray tubes currently use rotating anodes. Such a tube is shown schematically in Figure 2.8 and the discussion will focus primarily on this design. The principle of the rotating anode is very simple (Figure 2.7b) but its design has two important features to assist heat dissipation. First, as shown in Figure 2.7a, the anode surface is steeply angled to the electron beam. If the required focal spot size on the target is, say 2 mm × 2 mm, for an anode angled to the beam at about 16° the dimensions of the area actually bombarded by electrons are about 6 mm × 2 mm (see Section 2.4 for further detail). Second the area over which heat is dissipated can be increased considerably by arranging for the tungsten target to be an annulus of material which rotates rapidly (see Figure 2.7b).

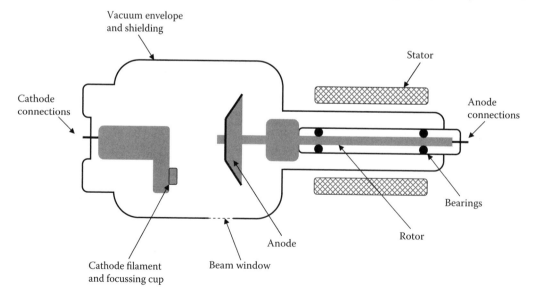

FIGURE 2.8
Design features of a rotating anode X-ray tube.

It may be shown that, if the exposure time is long enough for the anode to rotate at least once:

$$\frac{\text{Effective area for heat absorption with rotating anode}}{\text{Effective area for heat absorption with stationary anode}} = \frac{2\pi r \cdot x}{y \cdot x}$$

For the 80 mm diameter anode suitable for general radiographic work shown in Figure 2.7(b), this is an improvement of about $6 \times 40/2 = 120$ times and the heat input can be increased considerably (although not by a factor of 120). When high *loading* is required, that is high heating rates and high heat storage capacity, anodes up to 200 mm in diameter may now be used.

In addition to anode size, surface area, disc mass and rate of rotation all affect the loading. If a graphite block is brazed onto the back of an anode, its low mass and high melting point increase the heat storage capacity, and the heat radiating efficiency is increased because of the bigger anode surface and the better emission coefficient of black graphite.

Rotation rates range from 3000–3600 rpm with 50–60 Hz mains supply, up to 9–10,000 rpm with a 3-phase supply, ensuring that the anode rotates several times during even the shortest exposure, thus maximising the area over which the heat is distributed. However, this does create some problems with respect to the type of mounting and cooling mechanism. Adequate electrical contact is maintained via bearings on which the anode rotates, but the area of contact is quite insufficient for adequate heat conduction. Either ball bearings or sleeve bearings may be used (see Insight).

Insight

Bearing Systems

Such systems must overcome very challenging requirements:

- They form a connection between a very hot anode plate and a cold environment.
- They must operate in a vacuum.
- They must withstand high turning speeds and weight loads.

Lubrication of ball bearings cannot be by oil or grease because of the vacuum required in the X-ray tube. However, the lubricants must be soft, deformable materials that are stable at high temperatures and have low vapour pressure under vacuum. Dry metallic lubricants such as silver paste are used.

A recent development is the introduction of liquid, sliding bearings which utilise the aquaplaning effect of liquid metals. A good analogy is a locked car wheel on a wet road surface. Water accumulates between the tyre and road and forms a bow wave. As pressure builds a wedge is created between them and eventually a film of water is forced in, separating the tyre and road along the whole contact area. For sliding bearings a liquid metal (e.g. an eutectic of gallium, indium and tin which melts at –10°C) is used in a 20 μm gap. An important advantage of the new design is the extra anode cooling (1–2 kW) by fast heat flux from the anode through the liquid metal into the cooling system.

Since the anode is an evacuated tube, there are no heat losses by convection. The initial mode of heat transfer from the anode to the cooling oil must therefore be primarily by radiation at a rate proportional to (anode temperature)^4 – (oil temperature)^4. With a rotating anode, heat loss by conduction along the anode support is actually minimised since it might result in overheating of the bearings. Thus the rotating anode is mounted on a thin rod of low conductivity material such as molybdenum. Care must be taken that the length and method of support of this rod ensure that the anode remains stable when rotating.

Although radiation remains the primary mechanism of heat loss, other developments have improved heat dissipation and heat storage.

1. In modern tubes which use sleeve bearings, it is possible to include a means of oil circulation along the axle, providing an additional mechanism of heat loss by conduction and forced convection, further improving the rating of the tube.

2. Sliding bearings provide an additional cooling pathway through the liquid metal—see insight above.

3. X-ray tubes with a metal/ceramic casing rather than glass are now being produced—see Section 2.3.5. This has several advantages, one of which is that the anode can be supported on a shaft with bearings at both ends allowing much larger anodes to be installed. These have much better thermal properties allowing higher tube currents to be used.

2.3.3.3 Rotating Envelope

In a recent development which permits even more efficient heat dissipation, the anode is an integral part of the tube envelope and both of them rotate. The whole tube rotates around a symmetry axis with a single electron emitter at the centre of rotation. The electron beam is continuously deflected by magnetic fields to a fixed spot in space on the anode plate. The anode still rotates so heat is distributed around the focal ring and performance for brief high-power X-ray generation is similar to that of a cold rotating conventional anode. However, since the rear of the anode is now directly exposed to cooling agent (c.f. the stationary anode) it cools quickly and no extra heat storage is required. Enhanced cooling compared to conventional tubes is shown in Figure 2.9. Note that this extra heat loss is by conduction not radiation.

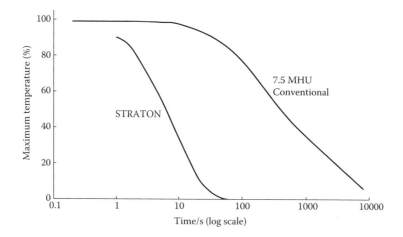

FIGURE 2.9
Cooling curves for a rotating envelope tube (Siemens STRATON) compared with cooling curves for conventional tubes. (MHU = mega heat unit (see Section 2.5.5)) (With permission from Oppelt A (ed.) *Imaging systems for medical diagnostics, chapter 12, X-ray components and systems*, Siemens, Erlangen, 2005, 264–412. Dura and STRATON are registered trademarks of Siemens.)

2.3.4 Electrical Circuits

Only brief details will be given of the essential electrical components of an X-ray generator. For further details see Dowsett et al. (2006), but the reader should be aware that some aspects of the workings of a modern generator are much more complicated than as described there.

2.3.4.1 The Transformer

This provides a method of converting high ACs at low potential difference to low ACs at high potential difference. It consists of two coils of wire which are electrically insulated from one another but are wound on the same soft iron former (Figure 2.10). If no energy is lost in the transformer

$$\frac{V_{out}}{V_{in}} = \frac{n_2}{n_1}$$

where n_1 and n_2 are the number of turns on the primary and secondary coils, respectively. Note

1. By making a connection at different points, different output voltages may be obtained. For example, the alternating potential difference across AB is given by $V_{AB} = V_{out}/3$

2. Since, in the ideal case, all power is transferred from the input circuit to the output circuit $V_{in}I_{in} = V_{out}I_{out}$

   Hence if V_{out} and I_{out} to the X-ray set are 100 kVp and 50 mA, respectively, since V_{in} will be the 240 V AC supply,

$$I_{in} = \frac{100 \times 10^3 \times 50 \times 10^{-3}}{240} = 20A$$

   so input currents are very high. (Hence the requirement for battery-powered or capacitor discharge mobile X-ray units (see Section 2.6))

3. Power loss occurs in all transformers, and the amount depends on working conditions, especially I_{in}. Hence V_{out} and I_{out} also vary and auxiliary electrical circuits are required to stabilise outputs from an X-ray set.

4. The efficiency of a transformer increases as the frequency of operation rises and consequently its size decreases.

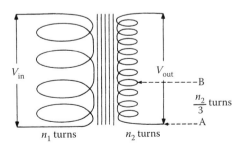

FIGURE 2.10
Essential features of a simple transformer.

Figure 2.14 (see later) also shows, on the extreme left, an autotransformer. An autotransformer comprises one winding only and works on the principle of self-induction. Since the primary and secondary circuits are in contact, it cannot transform high voltages or step up from low to high voltages. However, it does give a variable secondary output on the low voltage side of the transformer and hence controls kV directly.

2.3.4.2 Generating Different Voltage Wave Forms

As explained in Section 2.2.5, the alternating potential must be rectified before it is applied to an X-ray set. The X-ray tube can act as its own rectifier (self-rectification) since it will only pass current when the anode is positive and the cathode is negative. However, this is a very inefficient method of X-ray production because if the anode gets hot, it will start to release electrons by thermionic emission. These electrons will be accelerated towards the cathode filament during the half cycle when the cathode is positive and will damage the tube. Thus the voltage supply is rectified independently.

If a gas-filled diode valve or a solid state p-n junction diode rectifier is placed in the anode circuit, half wave rectification (Figure 2.6b) is obtained. Historically, a gas-filled diode, a simplified X-ray tube, comprising a heated cathode filament and an anode in an evacuated enclosure, was used. Electrons may only flow from cathode to anode but the diode differs from the X-ray set in that it is designed so that only a small proportion of the electrons boiled off the cathode travel to the anode. In terms of Figure 2.22 (see later), which shows the effect of tube kilovoltage on tube current, the diode operates on the rapidly rising portion of the curve, whereas the X-ray tube operates on the near-saturation portion.

The design and mode of operation of a p-n junction diode will be considered in Section 4.7.2 when its use as a radiation detector is discussed. It has many advantages over the gas-filled diode as a rectifier, including its small size, long working lifetime, and robustness. It is also easy to manufacture in bulk, is inexpensive, requires no filament heating circuit, has a low heat dissipation and a fast response time. For rectification, silicon rectifiers have a number of advantages over selenium, including a negligible forward voltage drop and a very high reverse resistance resulting in negligible reverse current flow. They can also withstand high reverse bias voltages so only a few hundred silicon rectifiers are required rather than a few thousand if made of selenium, and they can work up to 200°C if required.

The essential features of a full wave rectified supply are shown in Figure 2.11.

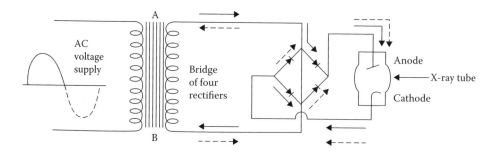

FIGURE 2.11

Essential features of a full wave rectified supply. (Solid) and (dotted) arrows show that irrespective of whether A or B is at a positive potential, the current always flows through the X-ray tube in the same direction.

2.3.4.3 Medium and High Frequency Generators

In these generators a frequency converter—a combination of a rectifier and inverting rectifier sometimes called an 'inverter' or 'thyrister'—is used to convert an AC of one frequency into an AC at a much higher frequency. The first generation of thyrister-based generators typically operated at 5–15 kHz and have become known as medium frequency generators. High frequency generators, with switching frequencies above the audible range, mostly in the range 50–100 kHz, are now the norm. Very high frequency generators up to 500 kHz are being developed. Figure 2.12 summarises the stages in the process.

1. Full wave rectification of the line AC voltage u_1 at frequency f_1.
2. This provides a DC voltage u_0 (after smoothing).
3. Rapid chopping of this DC voltage by the inverter to provide an alternating voltage u_2 comparable to u_1 at a much higher frequency f_2.
4. A high frequency transformer now transforms this to a higher voltage u_3 and, after again being rectified and smoothed this voltage is fed to the X-ray tube.

Advantages of the medium and high frequency generators are as follows:

1. *High output*—The output is very high and comparable with a three phase, 12-peak generator. At 100 kV (virtually DC), 0.1 A gives 10 kW so for a 0.1 s exposure there are 1000 J of energy to dissipate.
2. *Reduced voltage ripple*—At this high frequency, because pulses are very short the kV never falls very far below its peak value. Thus the average energy of the X-ray photons is higher than for a three phase supply and the output is very constant.
3. *Compact size*—The transformer equation u/fnA = constant where n is the number of turns on the transformer and A its cross-sectional area, shows that if f is increased by a factor of 100, say from 50 Hz to 5 kHz, nA may be reduced by a similar amount. Since the efficiency has been improved the transformer is much smaller, perhaps one-tenth the size of a three phase 12 peak generator.
4. *Rapid response*—The high voltage is switched on and off, and its level may be regulated even during exposure, under feed-back control of the inverter. The rise time of the tube voltage can be less than 200 μs.
5. *Long-term stability*—The tube current is more stable at the higher frequency f_2 and is independent of the voltage.
6. *Timer precision*—The precision of the exposure timer can be improved.
7. *Voltage range*—The generator may be used across the full kV range from mammography to CT.

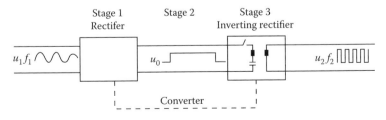

FIGURE 2.12
Schematic representation of frequency converter.

2.3.4.4 Action of Smoothing Capacitors

A capacitor in parallel with the X-ray unit will help to smooth out any variations in applied potential (Figure 2.13).

Consider, for example, the full wave rectified supply shown in Figure 2.6c. When electrons are flowing from the bridge circuit, some of them flow onto the capacitor plates and are stored there. When the potential across the bridge circuit falls to zero, electrons flow from the capacitor to maintain the current through the X-ray tube.

Voltage ripple (Section 2.2.5) has decreased considerably as generator design has improved—see Table 2.2.

2.3.4.5 Tube Kilovoltage and Tube Current Meters

These are essential components of the circuit and are shown in relation to other components in Figure 2.14. Note that the voltmeter is placed in the primary circuit so that a reading may be obtained before the exposure key is closed. There are two ammeters.

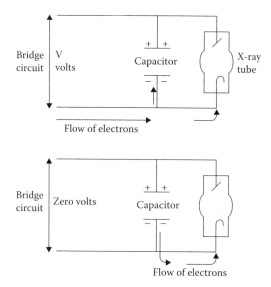

FIGURE 2.13
Illustration of the use of a capacitor for voltage smoothing. For explanation see text.

TABLE 2.2

Nominal Values of Voltage Ripple for Different Generators

Voltage Wave Form	Ripple (%)
Half wave/full wave	100
Full wave + capacitor smoothing	20
Three phase – 6 pulse unsmoothed	15
Three phase – 12 pulse unsmoothed	6
High frequency generator[a] (typical)	1–4
High frequency generator[a] (attainable but expensive)	0.1

[a] Note that a high frequency output has to be smoothed. Otherwise it would be just a half wave rectified profile at high frequency.

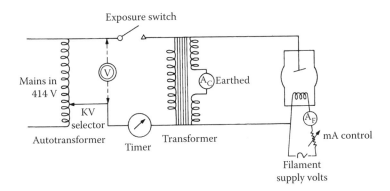

FIGURE 2.14
Simplified representation of the position of kV and current meters in the electrical circuit.

A_F measures the filament supply current (I_F) which may be adjusted to give the required thermionic emission before exposure starts. The actual tube current flowing during exposure (I_c) is measured by ammeter A_c.

2.3.5 The Tube Envelope and Housing

2.3.5.1 The Envelope

The envelope must be strong enough to withstand atmospheric pressure, resistant to the considerable amount of heat produced by the anode and able to transfer this heat away efficiently. Historically, it was constructed from thick-walled borosilicate glass under very clean conditions to a high precision so as to provide adequate insulation between the cathode and anode. The X-rays were emitted through a thinner glass window.

In modern tubes the envelope is frequently metal-walled. These tubes are more compact, have greater mechanical stability, have ceramic material to provide better electrical insulation between the anode/cathode connections and have good absorption of X-rays not passing through the window—now beryllium or titanium. They are also more efficient heat exchangers because of a high absorption coefficient for radiation on the vacuum side and good conductivity.

Insight

Borosilicate Glass versus Metal/Ceramic Envelope

Glass tubes are more likely to break during manufacture and it is difficult to adjust mechanical tolerances inside the glass housing. However, they are cheaper and so are still used for standard applications, that is, radiography with smaller power demands.

Metal/ceramic envelopes allow better mechanical precision, less manufacturing wastage and can resist larger mechanical stress. The stress tolerance is particularly important in CT and fast-moving 3D imaging in angiography.

The envelope also provides a vacuum seal to the metallic components that protrude through it. Great care must be taken at the manufacturing stage to achieve a very high level of vacuum before the tube is finally sealed. The electrons have a mean free path of several metres. If residual gas molecules are bombarded by electrons, the electrons may be scattered and strike the walls of the envelope, thereby causing reactions that result in release of gas molecules and further reduction of the vacuum.

The presence of atoms or molecules of gas or vapour in the vacuum, whatever their origin, is likely to have a deleterious effect on the performance of the tube. For example, metal evaporation from the anode can cause a conducting film across a glass envelope, thereby distorting the pattern of charge across the tube. This can change the output characteristics since it is assumed that the flow of electrons from cathode to anode will be influenced by the repulsive effect of a static layer of charge on the tube envelope. If this charge is not static, the electrons in the beam are not repelled by the tube envelope and deviate to it. This diversion of current may significantly reduce tube output. Metal enclosures repel ion deposits so they are less susceptible to build-up of tungsten ions from tungsten vapour. They also collect electrons scattered from the anode, thereby reducing extrafocal radiation (see Section 2.4).

Both residual gas and anode evaporation cause a form of tube instability which may occasionally be detected during screening as a kick on the milliammeter as discharges take place. In the extreme case, the tube goes 'soft' and arcs over during an exposure.

2.3.5.2 *The Tube Housing*

This has various functions which may be summarised as follows:

- Shields against stray X-rays because it is lined with lead—leakage must not exceed 1 mGy in 1 h at 1 m (see Section 14.5.2)
- Provides an X-ray window—which filters out some low energy X-rays
- Contains the anode rotation power source
- Provides high voltage terminals
- Insulates the high voltage
- Allows precise mounting of the X-ray tube envelope
- Provides a means for mounting the X-ray tube
- Provides a reference and attaching surface for X-ray beam collimation devices
- Contains the cooling oil

The advantages of filling this housing with oil are as follows:

(a) High voltage insulation.
(b) Effective conduction of heat from the X-ray inset tube.
(c) Since the oil expands, an expansion diaphragm can be arranged to operate a switch when the oil reaches its maximum safe temperature.

2.3.6 Switching and Timing Mechanisms

Most switching takes place in the primary circuit of the high voltage transformer where, although the currents are high, the voltage is low. Switching in the high voltage secondary circuit is only undertaken if very short exposures are required or exposures must be taken in quick succession.

2.3.6.1 *Primary Switches*

These are now almost all based on a thyrister which is a solid state controlled rectifier turned on and off by a logic pulse. A small positive logic pulse allows the thyrister to avalanche and a large current to flow which operates a relay closing the primary circuit. The switching is very rapid and is suitable for most radiographic exposures. The circuit used to drive the device is, however, quite complicated and beyond the scope of this book.

Switching the high voltage side of the transformer can be undertaken in two ways. The large electrical power that has to be accommodated (up to 100 kW) precludes the use of solid state devices and triode valves have to be used. Alternatively the exposure can be regulated with the Wehnelt electrode (grid control)—see Section 2.3.1. This grid is used to switch the tube on and off very rapidly by changing its voltage from negative to just positive relative to the cathode. When negative, even though the full kVp is applied across the tube, electrons cannot move from the cathode to the anode.

2.3.6.2 Timing Mechanisms

Accurate timing of an exposure is more complex than simply recording the 'on' time and 'off' time. As shown in Figure 2.15, there is a lag time before the kV builds to its full value, a brief delay before the kV responds to the 'exposure off' command and then a further lag as the kV decays.

2.3.6.3 The Electronic Timer

For simple adult radiography the 'on' time is long compared to the uncertainties and can be measured adequately with an electronic timer. If a capacitor C is charged to a fixed potential V_o, either positive or negative, and then placed in series with a resistor R, the rate

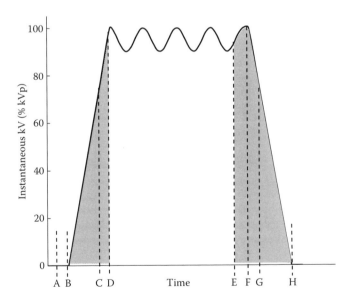

FIGURE 2.15
Timing uncertainties due to lag in switching mechanism. (A) Exposure 'on' command; (B) X-ray output starts; (C) 75% maximum kVp; (D) 100% of maximum kVp; (E) Exposure 'off' command; (F) Response to 'off' command; (G) 75% of maximum kVp; (H) X-ray output terminates. C to G is the International Electrotechnical Commission (IEC) definition of irradiation time. There is a corresponding uncertainty in output (patient dose), shown shaded. The shorter the exposure, the greater the percentage uncertainty. (With permission from Oppelt A (ed.) *Imaging systems for medical diagnostics, chapter 12, X-ray components and systems*, Siemens, Erlangen, 2005, 264–412.)

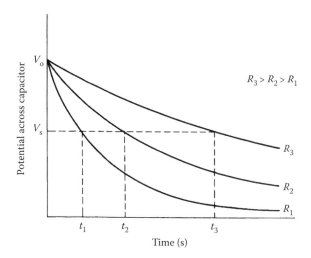

FIGURE 2.16
Curves showing the rate of discharge of a capacitor through resistors of different resistance. The time taken to reach V_S when the switching mechanism would operate, depends on the value of R.

of discharge of the capacitor depends on the values of C and R. A family of curves for fixed C and variable R is shown in Figure 2.16. Note that a large resistance reduces the rate of flow of charge so the rate of fall of V is slower.

These curves may be used as the basis for a timer if a switching device is arranged to operate when the potential across C reaches say V_s.

2.3.6.4 Frequency or Pulse Counting Timers

When very short exposures or a series of short exposures are required (e.g. in paediatrics and digital subtraction angiography, respectively) errors in the exposure profile may be important and the electronic timer responds too slowly. New high frequency inverter systems based on transistor technology (see Section 2.3.4) have greatly reduced exposure uncertainties associated with switching. Furthermore with a medium frequency generator, pulses are being generated at a rate of 5–15 kHz. Thus an alternative way to think of mAs is charge/pulse × number of pulses. Hence for a preset mAs the exposure can be controlled simply by counting the required number of pulses.

This pulse counting timer would be accurate to ±1 pulse, or at a frequency f = 5 kHz 1/10,000 s (there are two pulses/cycle).

Since the error in charge (ΔQ) would be the current flowing I multiplied by the time of one pulse $\Delta Q = I/2f$.

For a current of 1000 mA

$$\Delta Q = 1000 \times \frac{1}{1000} \quad \text{or} \quad \pm 0.1 \text{ mAs}$$

2.3.6.5 The Photo Timer

The weakness of any timer that predetermines the exposure is that a change in any factor which affects the amount of radiation actually reaching the receptor, notably patient attenuation, will alter the response. For a digital detector it may be possible to adjust for this by altering the window level (see Section 6.3.3) but for film the blackening will change. Thus the skill of the radiographer in estimating the thickness of the patient and choosing the correct exposure is of great importance. In the photo timer the exposure is linked more directly to the amount of radiation reaching the receptor. This is known as *automatic exposure control* or AEC.

One design, used especially when film is the receptor, places small ionisation chamber monitors in the cassette tray system between the patient and the film-screen combination. The amount of radiation required to produce a given degree of film blackening with a given film-screen combination under standard development conditions is known, so when the ion chamber indicates that this amount of radiation has been received, the exposure is terminated. This type of exposure control does not need to be 'set' before each exposure, but some freedom of adjustment is provided to allow for minor variations in film blackening if required. Adjustment will also be required if screens of different sensitivity are used.

As an alternative to ion chambers, photomultiplier tubes (see Section 4.8) may be used after the X-rays have been converted to light by a phosphor. Some have the disadvantage, however, of being X-ray opaque so they must be placed behind the cassette, where the X-ray intensity is low, and special radiolucent cassettes must be used. Others use phosphor-coated lucite, which can be placed in front of the cassette as it does not attenuate the X-ray beam. The light produced is internally reflected to the side where the photomultiplier can view it. The energy response of the phosphor probably differs from that of the receptor and some form of compensation must be built into the circuitry by using software control.

A weakness with some types of phototimer is that the ion chamber or photomultiplier tube only monitors the radiation reaching a small part of the receptor and this may not be representative of the radiation reaching the rest of the receptor. This problem can be partially overcome by using several small ion chambers, usually three, and controlling the exposure with the one, that is, closest to the region of greatest interest on the resulting image.

There must also be a backup exposure timer so that should the AEC fail, the patient exposure, although longer than necessary, will be terminated without intervention by the operator. It should be noted that this backup time should *not* be set at the thermal limit of the tube as this could result in a large radiation dose to the patient.

2.3.7 Electrical Safety Features

A number of features of the design of X-ray sets are primarily for safety and should be summarised briefly.

2.3.7.1 *The Tube Housing*

As already indicated, this provides a totally enclosing metallic shield that can be firmly earthed thereby contributing to electrical safety.

2.3.7.2 *High Tension Cables*

High tension cables are constructed so that they can operate up to potentials of at least 150 kV. Since the outermost casing metal braid of the cable must be at earth potential for safety, a construction of multiple coaxial layers of rubber and other insulators must be used to provide adequate resistance between the innermost conducting core and the outside to prevent current flow across the cable.

It is essential that high tension cables are not twisted or distorted in any other manner that might result in breakdown of the insulation. They must not be load bearing.

2.3.7.3 Electrical Circuits

These are designed in such a way that the control panel and all meters on the control panel are at earth potential. Nevertheless, it is important to appreciate that many parts of the equipment are at very high potential and the following simple precautions should be observed.

- Ensure that equipment is installed and maintained regularly by competent technicians.
- Record and report to the service engineer any evidence of excessive mechanical wear, especially to electrical cables, plugs and sockets.
- Similarly, report any equipment malfunction.
- Adopt all other safety procedures that are standard when working with electrical equipment.

2.4 Spatial Distribution of X-rays

When 40 keV electrons strike a thin metal target, the directions in which X-rays are emitted are as shown in Figure 2.17a. Most X-rays are emitted at angles between 45° and 90° to the direction of electron travel. The more energetic X-rays travel in a more forward direction (smaller value of θ). It follows that if the mean X-ray energy is increased by increasing the energy of the electrons, the lobes are tilted in the direction of the electron flow.

When electrons strike a thick metal target, the situation is more complicated because X-ray production may occur from the surface or it may occur at depth in the target. Also, the spatial distribution of X-rays will now depend on the angle presented by the anode to the incoming electron beam. Consider the anode shown in Figure 2.17b with an *anode angle* of 30°. (Note carefully which angle is defined as the *anode angle*—the angle between the anode surface and the electron beam is 90-α or 60⁰ in this instance.) X-rays produced in the direction B are much more heavily attenuated than those produced in the direction A because they travel further through anode material. This is clearly a disadvantage since a primary objective of good X-ray tube design is to ensure that the field of view is uniformly exposed to radiation. Only if this is achieved can variations in receptor response, for example, film blackening, be attributed to variations in scatter and absorption within the patient. Variation in intensity across the field is minimised by carefully selecting the angle at which the anode surface is inclined to the vertical (Figure 2.18).

Note the following additional points:

1. The radiation intensity reaching the detector is still not quite uniform, being maximum near the centre of the field of view. This is due to
   (a) An inverse square law effect—radiation reaching the edges of the field has to travel further
   (b) A small obliquity effect—beams travelling through the patient at a slight angle must traverse a greater thickness of the patient and are thus more attenuated

   Neither of these factors is normally of great practical importance.

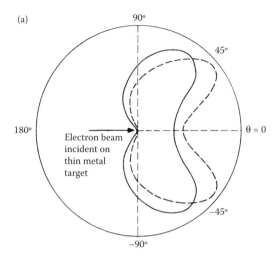

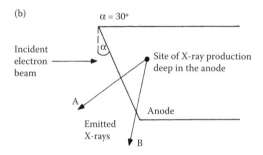

FIGURE 2.17

(a) Approximate spatial distribution of X-rays generated from a thin metal target bombarded with 40 keV electrons. This figure is known as a polar diagram: the distance of the curve from the origin represents the relative intensity of X-rays emitted in that direction. The polar diagram that might be obtained with 100 keV electrons is shown dotted. (b) The effect of self-absorption within the target on X-ray production from a thick anode.

2. The anode angle selected does not remove the asymmetry completely and this is known as the *heel effect*. The effect of X-ray absorption in the target, which results in a bigger exposure at A than at B, is more important than asymmetry in X-ray production, which would favour a bigger exposure at B.

3. Some compensation for the heel effect can be achieved by tilting the filter. The left hand edge of the beam will pass through a smaller thickness of filter than the right hand edge. This modification is being used in some mammography tubes.

4. No such asymmetry exists in a direction normal to that of the incident electron beam so if careful comparison of the blackening on the two sides of the film is essential the patient should be positioned accordingly although care must be taken balancing a tall patient at right angles to the table.

5. The shape of the exposure profile is critically dependent on the quality of the anode surface. If the latter is pitted owing to overheating by bombarding electrons, much greater differences in exposure may ensue.

6. An angle of about 13°–16° is frequently chosen for general radiography and this has one further benefit. One linear dimension of the effective spot for the production of

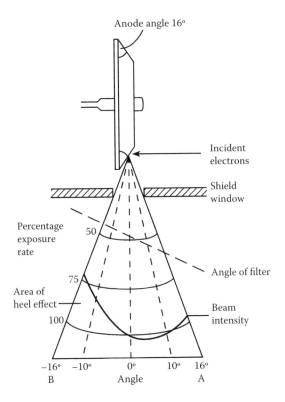

FIGURE 2.18
Variation of X-ray intensity across the field of view for a typical anode target angle.

X-rays is less than the dimension of the irradiated area by a factor equal to sin α. Sin 13° is about 0.2, so angling the anode in this way allows the focusing requirement on the electron beam to be relaxed whilst ensuring a good focal spot for X-ray production (Figure 2.19). This is known as the *line focus principle*. If a very small focal spot (~0.3 mm) is required, a smaller angle, perhaps only 6° may be used. Note that with a small anode angle, the heel effect greatly restricts the field size. This may not be a problem if the field of view is inherently small but, in general, the only compensation is to increase the focus-receptor distance. For example, if the minimum acceptable variation in optical density across the field of view is 0.2, for a film size of 43 cm × 35 cm the minimum focus-film distance increases from about 110 cm for a 16° angle to 150 cm for a 12° angle. There is a consequent loss of intensity at the film due to the inverse square law. Some X-ray tubes have used anodes with two angles so that the best angle for the focal spot size chosen can be used, but this is rare now.

Even with a well-designed anode, a certain amount of *extrafocal radiation* arises from regions of the anode out with the focal spot. These X-rays may be the result of poorly collimated electrons but are more usually the consequence of secondary electrons bouncing off the target and then being attracted back to the anode remote from the focal spot. Note that extrafocal radiation is not scattered radiation. Extrafocal X-rays may contribute as much as 10%–15% of the total output exposure of the tube but are of lower average energy. Many of them will fall outside the area defined by the light beam diaphragm and under extreme conditions may cast a shadow of the patient (Figure 2.20).

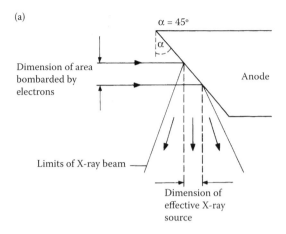

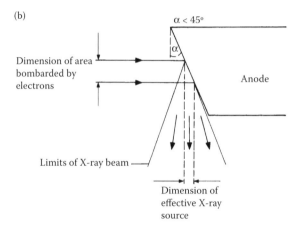

FIGURE 2.19
The effect of the anode angle on the effective focal spot size for X-ray production. (a) If the anode is angled at 45°, the effective spot for X-ray production is equal to the target bombarded by electrons. (b) If the anode is angled at less than 45° to the vertical the effective spot for X-ray production is less than the bombarded area. Note that in all cases these areas are measured normal to the X-ray beam. The actual area of the impact of electrons on the anode will be greater for reasons explained in the text.

Over the region of interest, the principal effect of extrafocal radiation is that it creates a uniform low level X-ray intensity. This contributes to the reduction in contrast produced by scattered radiation (see Section 6.5). Since this reduces image quality and hence, indirectly, increases the dose for imaging, it remains an important consideration. An additional effect is an increase in geometric blurring (enlarged effective spot size).

Insight

Secondary Collimation

Note that some of the X-ray photons from off-focus sources can be stopped by secondary collimation—a second set of collimators placed below the first set (Figure 2.21). This double collimation acts somewhat like a parallel hole collimator in a gamma camera (see Section 10.2.1).

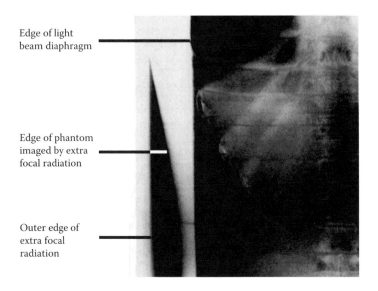

FIGURE 2.20
Radiograph showing the effect of extra focal radiation. The field of view defined by the light beam diaphragm is shown on the right but the outer edge of the 'patient' (a phantom in this instance) is also radiographed on the left by the extra focal radiation.

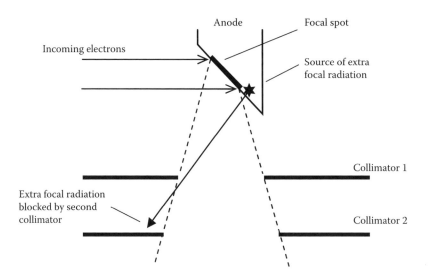

FIGURE 2.21
Reduction of extra focal radiation by double collimation.

With rotating anode tubes extrafocal radiation is generated on the anode plate—that is, in a strip perpendicular to the tube axis, so the effect is most clearly visible on edges of the radiograph that are parallel to the axis of the X-ray tube. In the metal cased tubes discussed in Section 2.3.5 the case is at earth potential and attracts off-focus electrons. The amount of off-focus radiation produced is reduced but not eliminated.

Extrafocal radiation is a potentially serious problem in image intensifier fluoroscopy because if unattenuated radiation reaches the input screen of the image intensifier the very bright fluorescent areas reduce contrast by light scattering and light conduction effects. The effect of extrafocal radiation on receptors in digital radiology probably merits investigation.

2.5 Rating of an X-ray Tube

2.5.1 Introduction

The production of a good radiograph depends on the correct choice of tube kVp, current, exposure time and focal spot size. In many situations a theoretical optimum would be to use a point source of X-rays to minimise geometrical blurring (see Section 6.9.1), and a very short exposure time, say 1 ms, to eliminate movement blurring (see Section 6.9.2). However, these conditions would place impossible demands on the power requirement of the set. For example, an exposure of 50 mAs would require a current of 50 A. Even if this current could be achieved, the amount of heat generated in such a small target area in such a short time would cause the anode to melt. This condition must be avoided by increasing the focal spot size or the exposure time, generally in practice the latter. Furthermore, during prolonged exposures, for example, in current applications of CT or interventional procedures, a secondary limitation may be placed on the total amount of heat generated in the tube and shield.

Thus the design of an X-ray tube places both electrical and thermal constraints on its performance and these are frequently expressed in the form of *rating charts*, which recommend *maximum* operating conditions to ensure a reasonably long tube life when used in equipment that is properly designed, installed, calibrated and operated. Note that lower ratings should be used whenever possible to maximise tube life.

2.5.2 Electrical Rating

Electrical limits are not normally a problem for a modern X-ray set but are summarised here for reference.

2.5.2.1 Maximum Voltage

This will be determined by the design, especially the insulation, of the set and the cables. It is normally assumed that the high voltage transformer is centre grounded (see Figure 2.14), so that the voltages between each high voltage tube terminal and ground are equal. A realistic upper limit is 150 kVp.

2.5.2.2 Maximum Tube Current

This is determined primarily by the filament current. Very approximately, the tube current (I_C) will be about one-tenth of the filament current (I_F). In other words only about one-tenth of the electrons passing through the filament coil are 'boiled off' from it. A modern X-ray tube may be designed to operate with a tube current of up to 1000 mA but under normal conditions it will be less than half this value. The lifetime of the tube can be significantly extended by a small reduction in current. The lifetime of a filament operating at 4.3 A is about 10 times that of one operating at 4.8 A.

If the voltage is increased at fixed filament current, the tube current will change as shown in Figure 2.22a. At low voltages, the tube current increases as the kV is increased because

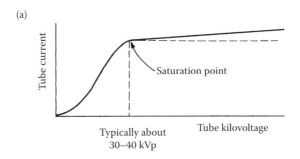

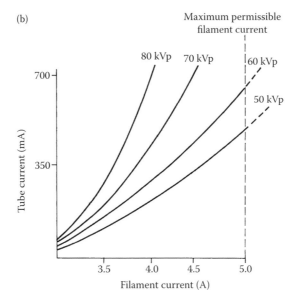

FIGURE 2.22
(a) The effect of increasing tube kilovoltage on the tube current for a fixed filament current. (b) A family of curves relating tube current to filament current for different applied voltages.

more and more electrons from the space charge around the cathode are being attracted to the anode. In theory, the tube current should plateau when the voltage is large enough to attract all electrons to the anode. In practice there is always a cloud of electrons (the space charge) around the cathode and as the potential difference is increased, a few more electrons are attracted to the anode. The result is that, as the tube kV is increased, the maximum tube current attainable also increases. Hence a typical family of curves relating tube current to filament current might be as in Figure 2.22b. Modern X-ray tubes contain several compensating circuits one of which stabilises the tube current against the effect of changes in voltage.

2.5.2.3 Generator Power

The relationship between the total output of X-rays required for a successful radiograph (*E*) and the exposure settings of kilovoltage (*kV*), current (*I*) and time (*t*) can be expressed approximately in the form

$$E = (kV)^4 \times I \times t$$

The required value of E will depend on the body parts being radiographed—for example, low for dental work, high for lateral lumbar spine. However, the required power in the generator (kV × mA) will depend also on t since short exposures for a given E will require higher kV and/or mA values.

The nominal operating power is specified as the kW that can be delivered at 100 kV for 0.1 s. Thus a 30 kW tube allows 300 mA, a 100 kW tube allows 1000 mA. Typical maximum powers for different applications are shown in Table 2.3.

2.5.3 Thermal Rating—Considerations at Short Exposures

When electrons strike the anode of a diagnostic tube, 99% of their energy is converted into heat. If this heat cannot be adequately dissipated, the anode temperature may quickly rise to a value at which damage occurs due to excessive evaporation, or the anode may melt which is even worse. The amount of heat the anode can absorb before this happens is governed by its thermal rating.

Exposure times between 0.02 s and 10 s are, somewhat arbitrarily, regarded as short exposures. The primary thermal consideration is that the area over which the electrons strike the anode should not overheat, sometimes referred to as short-term loadability. This is achieved by dissipating the heat over the anode surface as much as possible. The factors that determine heat dissipation will now be considered.

2.5.3.1 Effect of Cooling

It is important to appreciate that when the maximum heat capacity of a system is reached, any attempt to achieve acceptable exposure factors by increasing the exposure time is dependent on the fact that during a protracted exposure some cooling of the anode occurs. Consider the extreme case of a tube operating at its anode thermal rating limit for a given exposure. If the exposure time is doubled in an attempt to increase film blackening then, in the absence of cooling, the tube current must be halved. This is because at a given kVp the energy deposited in the anode is directly proportional to the product of the current and the period of exposure. However, in the presence of cooling, longer exposure times do permit greater power dissipation as shown in Figure 2.23.

2.5.3.2 Target Spot Size

For fixed kVp and exposure time, the maximum permitted current increases with target spot size because the heat is absorbed over a larger area. For very small spots (~0.3 mm) the maximum current is approximately proportional to the area of the spot since this determines the volume in which heat is generated. For larger spots (~2 mm) the maximum

TABLE 2.3

Typical Maximum Generator Powers for Different Applications

Application	Typical Maximum Power (kW)
Dental and mammography	5
Mobiles	30
General purpose fixed units	60–80
Angiography and CT	100

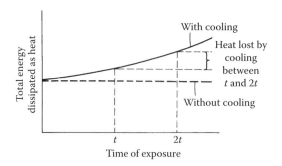

FIGURE 2.23
Total energy dissipated as heat for different exposure times with and without cooling.

current is more nearly proportional to the perimeter of the spot since the rate at which heat is conducted away becomes the most important consideration.

The larger focal spot, although allowing short exposure times, increases geometrical unsharpness (Section 6.9.1).

2.5.3.3 Anode Design

The main features that determine the instantaneous rating of a rotating anode are

- Its radius, which will determine the circumference of the circle on which the electrons fall
- Its rate of rotation
- The anode angle
- The focal spot size

The last two are closely related since the critical factor for heating is the area of the electron bombardment spot. For the same electron bombardment area a large angle anode will give a large focal spot, a small anode angle will give a small spot.

Rather old rating curves showing the maximum permissible tube current for different exposure times for anodes of different design are shown in Figure 2.24. A small anode angle and rapid rotation give the highest rating but note that the differences between the curves become progressively less as the exposure time is extended.

Note that

1. For the first complete rotation of the anode surface electrons are falling on unheated metal. For an anode rotation frequency of 50 Hz (the mains supply) one rotation requires 0.2 s so for even shorter exposures electrons fall on only part of the target length and the maximum tube current is independent of exposure time. The maximum permissible tube current for a stationary anode operating under similar conditions would be much lower.

2. For these old X-ray tubes the permitted tube current was very low (a rating curve for a high performance modern tube is shown later in Figure 2.28).

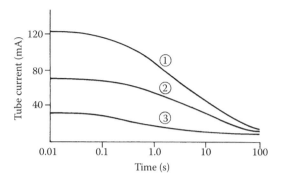

FIGURE 2.24

Historical rating curves showing the maximum permissible tube current at different exposure times for anodes of different design. Each tube is operating at 100 kVp three phase with a 0.3 mm focal spot. (1) Type PX 410 4 inch diameter anode with a 10° target angle and 150 Hz stator. (2) Type PX 410 4 inch diameter anode with a 10° target angle and 50 Hz stator. (3) Type PX 410 4 inch diameter anode with a 15° target angle and 50 Hz stator. Curves (1) and (2) show the effect of increasing the speed of rotation of the anode. Curves (2) and (3) show the effect of changing the target angle. (From Waters G, *J Soc X-ray Tech.* Winter 1968/69, 5, with permission of Philips Healthcare.)

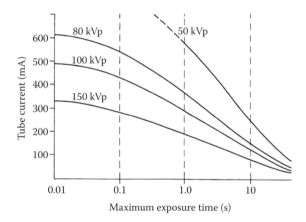

FIGURE 2.25

Maximum permissible tube current as a function of exposure time for various tube kilovoltages for a rotating anode. Type PX 306 tube with a 3 inch diameter anode, 15° target angle operating on a single phase with a 60 Hz stator and a 2 mm focal spot—circa 1982. Note: the dotted line indicates that the maximum permissible filament current would probably be exceeded under these conditions. (Commercial Rating Curves for PX 306 X-Ray Tube, with permission of Philips Healthcare.)

2.5.3.4 Tube Kilovoltage

As the kVp increases, the maximum permissible tube current for a fixed exposure time decreases (Figure 2.25). This is self-evident if a given power dissipation is not to be exceeded.

Such a rating chart may be used to determine if a given set of exposure conditions is admissible with a particular piece of equipment. For example, is an exposure of 400 mA at 70 kVp for 0.2 s allowed? Reference to Figure 2.25 shows that the maximum permissible exposure time for 400 mA at 70 kVp is about 1.0 s so the required conditions can be met. Note that for very long exposures the product kVp × mA × time is converging to the same value for all curves and the heat storage capacity of the anode then becomes the limiting factor (see Section 2.5.5).

Insight

Full Wave Rectified and Three Phase Supplies

When full wave rectified and three phase supply rating charts are compared at the same kVp, all other features of anode design being kept constant, the curves actually cross (Figure 2.26). For very short exposures higher currents can be used with a three phase than with a single phase supply, but the converse holds at longer exposures.

To understand why this is so, consider the voltage and current waveforms for two tubes with the same kVp and mA settings (Figure 2.27). Note:

1. The current does not follow the voltage in the full wave rectified tube. As soon as the potential difference is sufficient to attract all the thermionically emitted electrons to the anode, the current remains approximately constant.
2. The three phase current remains essentially constant throughout.
3. The peak value of the current must be higher for the full wave rectified tube than for the three phase tube, if the average values as shown on the meter are to be equal.

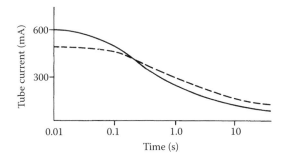

FIGURE 2.26
Maximum permissible current as a function of exposure time for 80 kVp single phase full wave rectified (dotted line) and 80 kVp three phase supplies (solid line).

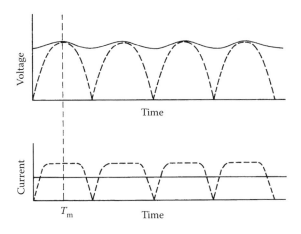

FIGURE 2.27
Voltage and current profiles for two tubes with the same kVp and mA settings but with three phase (solid line) or full wave rectified (dotted line) supplies.

For very short exposures, instantaneous power is important. This is maximum at T_m and since the voltages are then equal, power is proportional to instantaneous current and is higher for the full wave rectified system. Inverting the argument, if power dissipation cannot exceed a predetermined maximum value, the average current limit must be lower for the full wave rectified tube.

For longer exposures, average values of kV and mA are important. Average values of current have been made equal, so power is proportional to average voltage and this is seen to be higher for the three phase supply. Hence, again inverting the analysis, the current limit must be lower for the three phase tube as predicted by the rating curve graphs.

It is left as an exercise to the reader to explain, by similar reasoning, why the rating curve for a full wave rectified tube will always be above the curve for a half wave rectified tube.

2.5.4 Overcoming Short Exposure Rating Limits

If a desired combination of kVp, mA, time and focal spot size is unattainable owing to rating limits, several things can be done, although all may degrade the image in some way. Increasing the focal spot size or the time of exposure has already been mentioned. The other possibility is to increase the kVp. At first inspection, this appears to give no benefit. Suppose the rating limit has been reached and the kVp is increased by 10%. The current will have to be reduced by 10% otherwise the total power dissipated as heat will increase. The gain from increasing the kVp appears to be negated by a loss due to reduced mA. However, although X-ray output will fall by 10% as a result of reducing mA, it will increase by about 20% as a result of the 10% increase in kVp (see Section 2.2.5). Furthermore, X-ray transmission through the patient is better at the higher kVp and, in the diagnostic range, if film is the receptor, film sensitivity increases with kVp. Thus film blackening, which is ultimately the relevant criterion, is increased about 40% by the 10% increase in kVp and reduced by only about 10% due to reduction in mA, yielding a net positive gain. Some image degradation may occur as a result of loss of contrast at the higher kVp (see Section 6.3) but patient dose would be reduced because there is less attenuation of X-rays in the body.

The most effective way to overcome short exposure rating limits, in the longer term, is by improved anode design. Especially in X-ray CT, the combined requirements for quick scanning sequences, dynamic serial studies and spiral CT have necessitated a high tube current being maintained for several seconds. Some interventional procedures, for example, sequential images of blood vessels where the blood flow is very rapid, also put severe demands on the anode.

An important design improvement was the introduction of anodes mounted on spiral groove bearings with a liquid metal alloy as lubricant (see Section 2.3.3). In addition to allowing the anode to rotate all day once power is applied to the unit, this design permits good thermal contact so a significant amount of heat may be lost by conduction and the load bearing is greater allowing an anode of greater diameter to be used.

The design improvements noted here and elsewhere in this chapter, coupled with more effective heat transfer from the anode and a greatly increased overall heat storage capacity (see Section 2.5.5) have resulted in rating curves such as those shown in Figure 2.28a. It is instructive to compare these curves for modern tubes with those given in Figure 2.24 (circa 1966) and Figure 2.25 (circa 1982). The maximum tube current is now much higher (for comparable focal spot size) and is being maintained for much longer. The pair of curves in Figure 2.28b shows the effect of focal spot size, not previously illustrated.

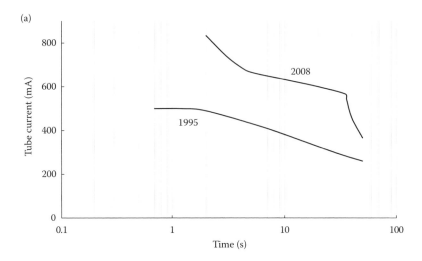

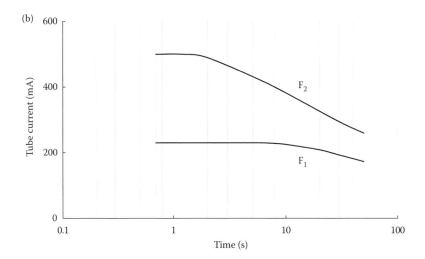

FIGURE 2.28
More modern rating curves: (a) Curves at 120 kVp for a Siemens 502 MC tube for computed tomography with a 0.8 mm × 1.1 mm focal spot (circa 1995) and a Siemens STRATON tube with a similar focal spot size (2008). (b) Curves for the Siemens 502 MC tube with a small, 0.6 mm × 0.6 mm, spot (F$_1$) and a larger 0.8 mm × 1.1 mm spot (F$_2$), both at 120 kVp. (Characteristics Rating Curves for a Siemens 502 MC Tube Designed for Computerised Tomography 1995. Reproduced by permission of Siemens Healthcare. Dura and Straton are registered trademarks of Siemens.)

2.5.5 Multiple or Prolonged Exposures

If too many exposures are taken in a limited period of time, the tube may overheat for three different reasons:

1. The surface of the target can be overheated by repeated exposures before the surface heat has time to dissipate into the body of the anode.

2. The entire anode can be overheated by repeating exposures before the heat in the anode has had time to radiate into the surrounding oil and tube housing.

3. The tube housing can be overheated by making too many exposures before the tube shield has had time to lose its heat to the surrounding air.

The second and third problems can also arise during continuous fluoroscopy. Although tube currents are now low, 1–5 mA, compared to 500 mA or more in radiography, exposure times can be very long. The surface of the target will not overheat but heat dissipation from the entire anode or tube housing may still require consideration.

The heat capacity of the total system, or of parts of the system, is sometimes expressed in heat units (HU). By definition, 1.4 HU are generated when 1 J of energy is dissipated.

The basis of this definition can be understood for a full wave rectified supply

$$HU = 1.4 \times \text{energy}$$
$$= 1.4 \times \text{root mean square (rms) kV} \times \text{average mA} \times \text{s}$$

But

$$\text{rms kV} = 0.71 \times \text{kVp}$$

Hence

$$HU = \text{kVp} \times \text{mA} \times \text{s}$$

Thus the HUs generated in an exposure are just the product of (voltage) × (current) × (time) shown on the X-ray control panel, so the introduction of the HU was very convenient for single phase generators.

Unfortunately, this simple logic does not hold for modern generators. The mean kV is now much higher, perhaps 0.95 kVp or better, so

$$HU = 1.4 \times 0.95 \text{ kVp} \times \text{mA} \times \text{s} = 1.35 \times \text{kVp} \times \text{mA} \times \text{s}$$

Hence for three phase supply the product of kVp and mAs as shown on the meters must be multiplied by 1.35 to obtain the heat units generated. With the near universal use of three phase and high frequency generators, joules are becoming the preferred unit.

The rating charts already discussed may be used to check that the surface of the target will not overheat during repeat exposures. This cannot occur provided that the total heat units of a series of exposures made in rapid sequence does not exceed the heat units permissible, as deduced from the radiographic rating chart, for a single exposure of equivalent total exposure duration.

When the time interval between individual exposures exceeds 20 s there is no danger of focal track overheating. The number and frequency of exposures is now limited either by the anode or by the tube heat storage capacity. A typical set of anode thermal characteristic curves is shown in Figure 2.29. Two types of curve are illustrated:

1. Input curves showing the heat stored in the anode after a specified, long period of exposure. Also shown, dotted, is the line for 700 watt input power in the absence of cooling. This line is a tangent to the curve at zero time since the anode is initially cold and loses no heat. At constant kVp the initial slope is proportional to the current. As the anode temperature increases, the anode starts to lose heat and the curve is no longer linear.

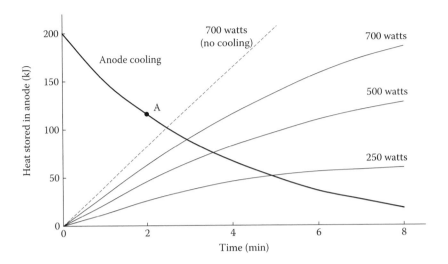

FIGURE 2.29
Typical anode thermal characteristic curves, showing the heat stored in the anode as a function of time for different input powers.

2. A cooling curve showing the heat stored in the anode after a specified period of cooling. Note that if the heat stored in the anode after exposure is only 120 kJ, the same cooling curve may be used but the point A must be taken as $t = 0$.

Two other characteristics of the anode are important. First, the maximum anode heat storage capacity, which is 200 kJ here, must be known. For low screening currents, the heat stored in the anode is always well below its heat limit, but for higher input power the maximum heat capacity is reached and screening must stop.

The second characteristic is the maximum anode cooling rate. This is the rate at which the anode will dissipate heat when near its maximum temperature (800 watts) and gives a measure of the maximum current, for given a kVp, at which the tube can operate continuously. Note that under typical modern screening conditions, say 2 mA at 80 kVp, the rate of heat production is only $2 \times 10^{-3} \times 80 \times 10^3 = 160$ W.

During screening, or a combination of short exposures and screening, the maximum anode heat storage capacity must not be exceeded. Exercises in the use of this rating chart are given at the end of this chapter. Note that with the increased use of microprocessors to control X-ray output, the system will not allow the operator to make an exposure that might exceed a rating limit.

When the total time for a series of exposures exceeds the time covered by the anode thermal characteristic chart, a tube shield cooling chart must be consulted. This is similar to the anode chart except that the cooling time will extend (typically) to 100 min and the maximum tube shield storage capacity in some modern units may be as high as 4×10^6 J. Note that cooling of the housing can be enhanced with a heat exchanger. For example, by pumping oil or water through a set of tubes in the housing the time taken for the housing to lose 90% of its heat might be reduced by 50%.

As a final comment on thermal rating, it is worth noting that a significant amount of power is required to set the anode rotating and this is also dissipated eventually as heat. In a busy accident department taking many short exposures in quick succession, three times as much heat may arise from this source as from the X-ray exposures themselves.

2.5.6 Falling Load Generators

One way to keep exposure times as short as possible for AEC exposures, without exceeding rating limits, is to use the falling load principle. This method of operation uses the fact that the rate of heat loss from the anode is greatest when the anode is at its maximum working temperature, so the current through the tube is kept as high as possible without this maximum temperature being exceeded. If the tube current is high for the initial part of the exposure while the anode is cold, but reduced during the exposure, the same mAs can be achieved in a shorter time than with constant current. This is illustrated in Figure 2.30.

The anode temperature is monitored and if it reaches the maximum allowed, a motor driven rheostat introduces an extra resistance into the filament circuit thereby reducing the tube current in a step-wise manner. Because the transformer is not ideal, this lowering of tube current causes an increase in the kVp, and this has to be compensated for by increasing the resistance in the primary circuit in the transformer.

The exposure must be controlled using a meter calibrated in milliampere seconds or terminated using a phototimer for AEC (see Section 2.3.6). Note, however, that the falling load generator will be of little value for short exposure times (say 0.4 s) because there will be insufficient time for the current to fall through many steps. Also there is wear on the tube at high current so lifetime is shortened by falling load operation. Thus a falling load generator might be a possibility for a busy orthopaedic clinic examining spines with heavy milliampsecond loadings and long exposure times. For chest work it would be useless.

2.5.7 Safety Interlocks

These are provided to ensure that rating limits are not exceeded on short exposures. If a combination of kVp, mA, exposure time and spot size is selected that would cause anode overheating, a 'tube overload' warning light will appear and the tube cannot be energised.

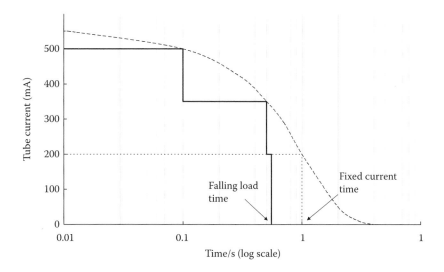

FIGURE 2.30
Illustration of the falling load principle. To achieve 200 mAs at a fixed mA requires a one second exposure. Using the falling load principle, the tube can operate at 500 mA for 0.1 s = 50 mAs, then drop to 350 mA for a further 0.4 s = 140 mAs, and finally to 200 mA for 0.05 s = 10 mAs, a total of 200 mAs in only 0.55 s.

During multiple exposures a photoelectric cell may be used to sample radiant heat from the anode and thereby determine when the temperature of the anode disc has reached a maximum safe value. A visual or audible warning is then triggered. In modern systems the tube loading is under computer control. Anode temperature is continuously calculated from a knowledge of heat input and cooling characteristics. When the rating limit is reached, generator output is automatically reduced.

2.5.8 X-ray Tube Lifetime

The life of an X-ray tube can be extended by taking steps to avoid thermal stress and other problems associated with heating. For example, the anode is very brittle when cold and if a high current is used in this condition, deep cracks may develop. Thus at the start of operations several exposures at approximately 75 kVp and 400 mAs (200 mA for 2 s) should be made at 1 min intervals. Ideally, if the generator is idle for periods exceeding 30 min, the process should be repeated.

Keeping the 'prepare' time to a minimum will reduce filament evaporation onto the surface of the tube and also bearing wear in the rotating anode. The generator should be switched off when not in use.

The tube should be operated well below its rating limits whenever possible.

2.6 Mobile X-ray Generators

There are a number of situations in which it is not practicable to take the patient to the X-ray department and the X-ray set must be taken to the patient. These include routine bedside procedures in intensive care, emergency, paediatric departments and operating theatres. Although mobile units generally take planar images, either film or, increasingly, digital images, there is also a role for mobile image intensifiers both in theatres, for example, in orthopaedic practice, and on the wards, for example, to monitor the progress of an endoscope as it is inserted into the patient. The manoeuvrability of a C-arm image intensifier linked via a fibre optic directly to a charge coupled device (see Section 5.13.4) to produce digital images in real time allows the radiologist or radiographer to position the arm so that only the region of interest is irradiated, thereby reducing the dose to adjacent organs. Note, however, that a screening procedure should not be adopted if the same information can be obtained from a few short conventional exposures because the latter will almost certainly result in a lower dose to the patient.

Historically, mobile units operated with single phase full wave rectified generators. These had all the limitations on output and rating limits imposed by such a voltage profile. Nowadays there are just two types of unit in use, battery powered and capacitor discharge.

2.6.1 Battery Powered Generators

Battery powered generators typically use a nickel-cadmium battery which can store a charge equivalent to about 10,000 mAs at normal operating voltages. The DC voltage of approximately 130 V from this battery must be converted to an alternating voltage before it can be used to supply the generator transformer. This conversion is carried out by an inverter (see Section 2.3.4). As with the medium/high frequency generator described there, the AC voltage produced differs significantly from the mains AC voltage in that it is at 5 kHz (or higher), some 100 times greater than the normal mains frequency. At this high frequency the transformer is more efficient and as a consequence can be much smaller.

Although the transformer output is basically single phase full wave rectified, at 5 kHz the capacitance inherent in the secondary circuit smoothes the output to a waveform that is essentially constant. The tube current is also constant for the whole of the exposure which, as well as simplifying design, allows exposures to be calculated with ease. The battery is depleted and the voltage falls slightly from one exposure to the next so some form of compensation must be applied until the unit is recharged. This compensation can be applied either automatically or manually. Recharging takes place, when necessary, at a low current from any hospital 13 A mains supply.

Typical values, or range of values, for the specification of a battery powered mobile unit are shown in Table 2.4. The output is very stable and low power machines are very suitable for chest radiography or premature baby units.

2.6.2 Capacitor Discharge Units

In a capacitor discharge unit, the capacitor is connected directly to a grid-controlled X-ray tube (see Section 2.3.1). Recall that grid control (a) can be turned on and off independently providing instantaneous control of the X-ray tube current and very precisely timed exposures; (b) is useful in any X-ray unit where extremely short exposures (a few milliseconds) are required; (c) is also useful where rapid repeat exposures are required because the switching mechanism is without inertia.

The capacitor can be charged, at low current, from any hospital 13 A mains supply. Charging will continue until the capacitor reaches a preset kilovoltage. The capacitor must be charged immediately before use, any significant delay will allow charge to leak away from the capacitor. Once an exposure has started the mAs delivered must be monitored and the exposure is terminated as required. Unlike the battery powered generator where the kV falls only very slightly during exposure, the operating kV of a capacitor discharge unit falls a lot, being high at the start of the exposure and relatively low at the end (Figure 2.31). This is because the kilovoltage across the tube is reduced as charge is taken off the capacitor. If the capacitor C has a value of 1 μF, the reduction in the kilovoltage is 1 kV per 1 mAs of charge, Q, lost as exposure (from $V = Q/C$). The output falls accordingly during the exposure and failure to realise this, and the cause, has often led to mistakes being made in the setting of exposure factors with these machines. Consequently they have acquired an unjustified reputation of lacking in output.

TABLE 2.4

Typical Specifications for a Range of
Battery Operated Mobile Units

I_{max}	100–400 mA
kVp	50–100 kV
Shortest exposure	1 ms
Focal spot	0.7 mm
Power	10–30 kW
Tube heat storage	100–200 kJ

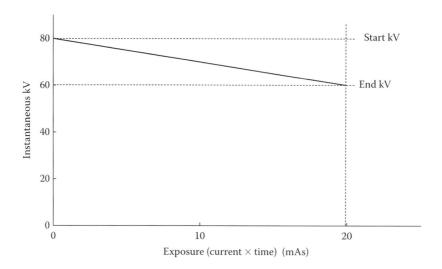

FIGURE 2.31
Variation of kV with exposure for a capacitor discharge mobile unit.

Evans et al. (1985) were perhaps the first to propose an empirical formula for the equivalent kilovoltage of a capacitor discharge unit, that is the setting on a battery powered constant potential kilovoltage machine which would produce the equivalent radiographic effect. The equivalent kV is approximately equal to the starting voltage minus one-third of the fall in tube voltage which occurs during the exposure. For a 1 µF capacitor the tube voltage drop is numerically equal to the mAs selected, hence

$$\text{equivalent kV} = \text{starting kV} - \text{mAs}/3$$

If, therefore, one has a radiographic exposure setting of 85 kV and 30 mAs, the equivalent voltage is 75 kV. If this exposure is insufficient and an under-exposed radiograph is produced, simply increasing the mAs may not increase the blackening on the film sufficiently. For example, suppose the exposure is increased to 50 mAs at the same 85 kV. The equivalent kV will now be only 68 kV. Since the equivalent kV is less, some of the increase in mAs will be used to provide soft radiation which does not contribute to the radiograph. The appropriate action is to change the exposure factors to 92 kV and 50 mAs, thereby maintaining the equivalent kV at 75 kV but increasing the exposure to 50 mAs.

Insight

Under-exposure and Digital Receptors

Note that with a digital receptor the software will compensate for this under-exposure and produce an image within the desired grey scale range. However, this is bad radiographic practice since the quantum noise in the image will be increased and may reach an unacceptable level.

If it is desired to reduce the starting kV after the capacitor has been charged or when radiography is finished, the capacitor much be discharged. When the 'discharge' button is depressed, an exposure takes place at a low mA for several seconds until the required charge

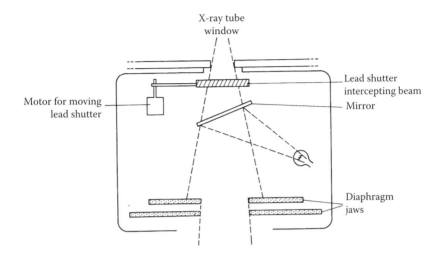

FIGURE 2.32
Arrangement of the lead shutter for preventing exposure during capacitor discharge and the light beam diaphragm assembly on a capacitor discharge mobile unit.

has been lost. During this exposure the tube produces unwanted X-rays. These are absorbed by a lead shutter across the light beam diaphragm. An automatic interlock ensures that the tube cannot discharge without the lead shutter in place to intercept the beam (Figure 2.32). It should be noted, however, that this shutter does not absorb all the X-rays produced, especially when the discharge is taking place from a high kV. Neither patient nor image receptor should be underneath the light beam diaphragm during the discharge operation.

New special capacitors, sometimes called *ultra caps* or *boost caps*, have been developed with smaller-volume storage units than normal electrolytic capacitors. Capacitor discharge units require more operator training to ensure optimum performance than other mobile units, but when used under optimum conditions they have a sufficiently high output to permit acceptably short exposures for most investigations.

2.7 Quality Assurance of Performance for Standard X-ray Sets

It is sometimes difficult to identify the boundary between quality assurance and radiological protection in diagnostic radiology. This is because the primary purpose of good quality assurance is to obtain the best diagnostic image required at the first exposure. If this is achieved successfully, not only is the maximum information obtained for the radiation delivered, but also the radiation dose to both the patient and to staff is minimised.

A number of performance checks should be carried out at regular intervals. These will, for example, confirm that

- The tube kVp is correctly produced
- The mAs reading is accurate
- The exposure timer is accurate
- The radiation output is reproducible during repeat exposures

All these factors affect the degree of film blackening and the level of contrast. With a digital detector the system software will correct for variations and deterioration may become quite serious before it is detected in the image.

Tube kV may be measured directly by the service engineer. However, for more frequent checks a non-invasive method is to be preferred. The usual method is based on the fact that different materials show different attenuation properties at different beam energies (see Chapter 3). When this approach was first suggested by Ardran and Crooks (1968), a copper step wedge was used to compensate for the difference in sensitivity between a fast and a slow film screen. With suitable calibration the wedge thickness that gave an exact match in film blackening could be converted into tube kilovoltage. This involves exposure to the special cassette at each kVp with an appropriate mAs.

An early method for checking exposure times was to interpose a rotating metal disc with a row of holes or a radial slit in it between the X-ray tube and the film. Inspection of the pattern of blackening allowed the exposure time to be deduced. Both of the above methods are described more fully in older text books—see, for example, Dendy and Heaton (1987).

Later development simplified both these checks, using a digital timer meter and a digital kV meter attached to a fast responding storage cathode ray oscilloscope (CRO). Several balanced photo detectors can be used under filters of different materials and different thicknesses. By using internally programmed calibration curves, a range of kVps may be checked with the same filters. Accuracy to better than 5% should be achievable. This is particularly important for mammography because the absorption coefficient of soft tissue falls rapidly with increasing kV at low energies. Note that the CRO also displays the voltage profile so a fairly detailed analysis of the performance of the X-ray tube generator is possible. For example, any delay in reaching maximum kV could affect output at short exposure times. Note that these devices are only accurate at a set tube filtration and readings must be corrected if the tube filtration is different. In modern equipment semiconductor technology has further simplified these measurements.

Consistency of output may be checked by placing an ion chamber in the direct beam and making several measurements. Both the effect of changes in generator settings (kV, mAs) on the reproducibility of tube output and the repeatability between consecutive exposures at the same setting can be checked. Measurements should be made over a range of clinical settings to confirm a linear relationship between tube output (mGy in air for a fixed exposure time) and the preset mA. It is not normal to make an absolute calibration of tube current.

For AEC systems a film and suitable attenuating material, for example, perspex or a water equivalent slab, should be used to determine the film density achieved when an object of uniform density is exposed under automatic control. A check should be made to ensure that the three ion chambers are matched, so that irrespective of which one is selected to control the exposure, the optical density is similar and repeatable. For a digital system a semi-conductor detector may be used to check that the dose at the image receptor is consistent and within specified limits.

Low energy X-rays in the spectrum would be absorbed in the patient, increasing the dose but contributing nothing to the image. They must be selectively removed by filtration, see Section 3.9. Recommended values for the total beam filtration are 1.5 mm Al for units operating up to 70 kVp and 2.5 mm Al for those operating above 70 kVp and these should be checked. To do so an ion chamber is placed in the direct beam and readings are obtained with different known thickness of aluminium in the beam. The HVL, that is the thickness of aluminium, which reduces beam intensity to half, may be obtained by trial and error or graphically. Note that the HVL obtained in this way is *not* the beam filtration (although when expressed in mm of Al the values are sometimes very similar) and the filtration must be obtained from a look-up table (Table 2.5).

It is left as an exercise to the reader, after a careful study of Chapter 3, to explain why the look-up table will be different at other tube voltages and for other voltage profiles.

Since all operators are urged to use the smallest possible field sizes, it is important to ensure that the optical beam, as defined by the light beam diaphragm, is in register with the X-ray beam. This may be done by placing an unexposed X-ray film on the table and using lead strips or a wire rectangle to define the optical beam. An exposure is made, at a very low mAs because there is no patient attenuation, and the film developed. The exposed area should correspond to the radiograph of the lead strips to better than 1 cm at a focus-film distance of 1 m (Figure 2.33). At the same time a check can be made that the axis of the X-ray beam is vertical by arranging two small (2 mm) X-ray opaque spheres vertically one above the other about 20 cm apart in the centre of the field of view. If their images are not superimposed on the developed radiograph, the X-ray beam axis is incorrectly aligned.

Few centres check focal spot size on general radiographic equipment regularly, perhaps because there is evidence from a range of routine X-ray examinations that quite large changes in spot size are not detectable in the quality of the final image. However, focal spot size is one of the factors affecting tube rating and significant errors in its value could affect

TABLE 2.5

Typical Relationship between the Beam Filtration and Half Value Layer for a Full Wave Rectified X-ray Tube Operating at 70 kVp

Half value layer (mm A1)	1.0	1.5	2.0	2.5	3.0
Total filtration (mm A1)	0.6	1.0	1.5	2.2	3.0

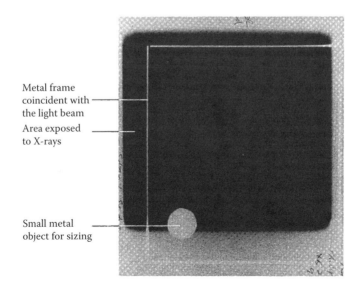

FIGURE 2.33
Radiograph showing poor alignment of the X-ray beam and the light beam diaphragm as defined by the metal frame. The small coin is used for orientation and sizing.

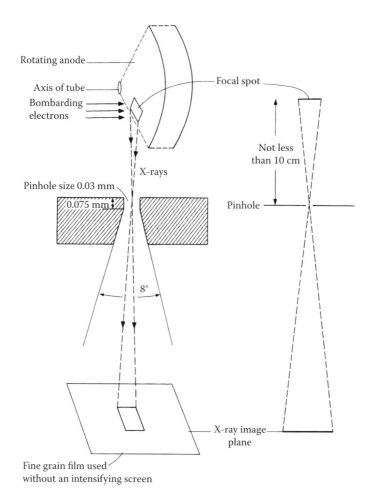

FIGURE 2.34
Use of a pin-hole technique to check focal spot size.

the performance of the tube generator. The pin-hole principle illustrated in Figure 2.34 may be used to measure the size of the focal spot. The drawing is not to scale but typical dimensions for a 1 mm spot are shown.

By similar triangles

$$\frac{\text{Size of image}}{\text{Size of focal spot}} = \frac{\text{Pin-hole film distance}}{\text{Focus to pin-hole distance}}$$

and for a pin-hole of this size this ratio is usually about 3. Note that the pin-hole must be small—its size affects the size of the image and hence the apparent focal spot size. The 'tunnel' in the pin-hole must be long enough for X-rays passing through the surrounding metal to be appreciably attenuated. This principle is used in the slit camera.

Focal spot size can also be measured and information may be obtained on uniformity of output within the spot, if required, by using a star test pattern. For fuller details see, for example, Curry et al. 1990. Such information would be important, for example, if one were attempting to image, say, a 0.4 mm blood vessel at two times magnification because image quality would then be very dependent on both spot size and shape. Similarly, the high resolution required in mammography (see Section 9.2) requires regular checks on the focal spot size.

For further information on quality control, the reader is referred to 'further reading' given at the end of the chapter.

2.8 Conclusions

In this chapter the basic principles of X-ray production have been discussed. Since the output from the X-ray set is the starting point for high quality radiographic images, there are important principles to establish at this early stage in the book.

1. An X-ray tube output consists of a continuous spectrum determined by the operating kVp and, if the kVp is high enough, characteristic line spectra determined by the atomic number of the target material.

2. Distinguish carefully between radiation quantity, which is related to the overall intensity of X-rays produced, and radiation quality which requires a more detailed consideration of the distribution of X-ray intensities with photon energy. The former depends on a number of factors such as tube kilovoltage, beam filtration, time of exposure and atomic number of the target anode. Tube current, kilovoltage and beam filtration, which will be considered in detail in Chapter 3, affect radiation quality. Use of a three phase supply or high frequency generator maintains the tube kilovoltage close to maximum throughout the exposure and both the quantity and quality of X-rays are thereby enhanced.

3. There has been steady progress over the past 40 years in X-ray tube technology (see Table 2.6) and the high performance of modern X-ray equipment relies on careful design and construction of many components both in the X-ray tube itself and in the associated circuitry. Two features of anode design are particularly important. The first is a consequence of the fact that only about 0.5% of the electron energy is converted into X-rays, whilst the remainder appears as heat which must be removed. Considerable progress has been made in devising ways to remove this heat, especially from the electron bombardment target on the anode. The second is the requirement for the X-ray exposure to be as uniform as possible over the irradiated field. This is achieved by careful attention to the anode shape and particularly the angle at which it is presented to the electron flux.

4. Notwithstanding substantial improvements in anode design, generation of heat imposes constraints on X-ray tube performance especially when very short exposures with small focal spot sizes are attempted. The limiting conditions are usually expressed in the form of rating curves. If a rating limit is exceeded, either the duration of exposure or the focal spot size must be increased. Occasionally the desired result may be achieved by increasing the tube kV.

5. To minimise the need for repeat X-rays, thereby increasing the overall radiation body burden to the population, careful quality control of the performance of X-ray sets at regular intervals is essential.

TABLE 2.6

Milestones in the Development of X-ray Tube
Technology

1950s	Rotating X-ray anode
	Control of exposure timer
1960s	High duty X-ray tubes (200 kHU)
	Large anode target disc
	High speed rotation
	Molybdenum/tungsten target
1970s	Rhenium/tungsten target
	Three phase 12-peak generators
	Short exposure times (1 ms)
1980s	High frequency generators
	Even higher duty X-ray tubes (500 kHU)
	Microprocessor control
1990s	Anodes mounted on spiral tube bearings
	Variable focal spots
	Liquid metal bearings
	Metal/ceramic housing
	Enhanced filtration (see Section 3.9)
2000s	Constant potential generators
	Magnetic beam deflection
	Rotating tube envelope

References

Ardran G M and Crooks H E. Checking diagnostic X-ray beam quality. *Br J Radiol 41*, 193–198, 1968.

Curry T S, Dowdey J E and Murray R C Jr. *Christensen's Introduction to the Physics of Diagnostic Radiology*, 4th edn. Lea & Febiger, Philadelphia, 1990.

Dendy P and Heaton B. *Physics for Radiologists*, 1st edn. Blackwell Scientific Publications, Oxford, 1987, p 60.

Evans S A, Harris L, Lawinski C P and Hendra I R F. Mobile X-ray generators: a review. *Radiography 51*, 89–107, 1985.

Further Reading

AAPM. *Quality control in diagnostic radiology.* Report no. 74, Task Group 12, Diagnostic Imaging Committee, Medical Physics Publishing, Madison, July, 2002.

Bushberg J T, Seibert J A, Leidholdt E M Jr, and Boone J M. *The Essential Physics of Medical Imaging*, 2nd edn. Lippincott, Williams and Wilkins, Philadelphia, 2002.

Dowsett D J, Kenny P A and Johnson R E. *The Physics of Diagnostic Imaging*, 2nd edn. Hodder Arnold, London, 2006, pp 71–112.

IPEM. *Measurement of the performance characteristics of diagnostic X-ray systems used in medicine. Part I X-ray tubes and generators* (2nd ed). Report no 32—Institute of Physics and Engineering in Medicine, York, YO24 1ES, 1996.

IPEM. *Recommended standards for the routine performance testing of diagnostic X-ray imaging systems.* Report no 91—Institute of Physics and Engineering in Medicine, York, YO24 1ES, 2005.

Oppelt A (ed.) *Imaging systems for medical diagnostics, chapter 12, X-ray Components and Systems,* Siemens, Erlangen, 2005, pp 264–412.

Exercises

1. Explain why the X-ray beam from a diagnostic set consists of photons with a range of energies rather than a monoenergetic beam.

2. What is meant by 'characteristic radiation'? Describe very briefly three processes in which characteristic radiation is produced.

3. Describe, with the aid of a diagram, the two physical processes that give rise to the production of X-rays from energetic electrons. How would the spectrum change if the target were made thin?

4. Explain why there is both an upper and a lower limit to the energy of the photons emitted by an X-ray tube.

5. What is the source of electrons in an X-ray tube and how is the number of electrons controlled?

6. The cathode of an X-ray tube is generally a small coil of tungsten wire.

    (a) Why is it a small coil?

    (b) Why is the material tungsten?

7. Figure 2.8 shows a rotating anode tube. Explain the functions of the various parts and the advantages of the materials used.

8. How would the output of an X-ray tube operating at 80 kVp change if the tungsten anode (Z = 74) were replaced by a tin anode (Z = 50)?

9. What is the effect on the output of an X-ray set of

    (a) Tube kilovoltage?

    (b) The voltage profile (ripple)?

10. It is required to take a radiograph with a very short exposure. Explain carefully why it may be advantageous to increase the tube kilovoltage.

11. What advantages does a rotating anode offer over a stationary anode in an X-ray tube?

12. Discuss the effect of the following on the rating of an X-ray tube:

    (a) Length of exposure

    (b) Profile of the voltage supply as a function of time

    (c) Previous use of the tube

13. Discuss the factors that determine the upper limit of current at which a fixed anode X-ray tube can be used.

14. What do you understand by the thermal rating of an X-ray tube? Explain how suitable anode design may be used to increase the maximum permissible average beam current for

    (a) Short exposures

    (b) Longer exposures

15. A technique calls for 550 mA, 0.05 s with the kV adjusted in accordance with patient thickness. If the rating chart of Figure 2.25 applies, what is the maximum kVp that may be used safely?

16. A technique calls for 400 mAs at 90 kVp. If the possible milliampere values are 500, 400, 300, 200, 100 and 50 and the rating chart in Figure 2.25 applies, what is the shortest possible exposure time?

17. An exposure of 400 mA, 100 kVp, 0.1 s is to be repeated at the rate of six exposures per second for a total of 3 s. Is this technique safe if the rating chart of Figure 2.25 applies?

18. A radiographic series consisting of six exposures of 200 mA, 75 kVp and 1.5 s has to be repeated. What is the minimum cooling time that must elapse before repeating the series if the rating chart of Figure 2.29 applies?

19. If the series of exposures in exercise 18 is preceded by fluoroscopy at 100 kVp and 3 mA, for how long can fluoroscopy be performed before radiography?

20. Discuss the developments in X-ray tube technology since 1950 with specific reference to the production of high quality X-ray images of patients.

21. Suggest reasons why radionuclides do not provide suitable sources of X-rays for medical radiography.

3

Interaction of X-Rays and Gamma Rays with Matter

B Heaton and P P Dendy

SUMMARY

- In this chapter the fundamental physics of the interaction of X-rays and gamma rays with matter is explained.
- The meaning of 'bound' and 'free' electrons is discussed and a careful distinction is made between attenuation, absorption and scatter.
- The four interaction processes that are important in the diagnostic energy range—elastic scattering, the photoelectric effect, the Compton effect and pair production—are discussed.
- The concept of linear attenuation coefficient is introduced and the distinction between narrow and broad beam attenuation is explained.
- The implications of beam filtration (the selective attenuation of different parts of the X-ray spectrum) by different materials is considered.
- Finally, the chapter concludes with a review of the optimum operating kVps for some standard radiographic procedures.

CONTENTS

3.1 Introduction

The radiographic process depends on the fact that when a beam of X-rays passes through matter its intensity is reduced by an amount that is determined by the physical properties, notably thickness, density and atomic number, of the material through which the beam passes. Hence it is variations in these properties from one part of the patient to another that create detail in the final radiographic image. These variations are often quite small, so a full understanding of the way in which they affect X-ray transmission under different circumstances, especially at different photon energies, is essential if image detail is to be optimised. Note that one of the causes of inappropriate X-ray requests is because the clinical symptoms do not suggest there will be any informative changes in thickness, tissue density or atomic number in the affected region (RCR 2007).

In this chapter an experimental approach to the problem of X-ray beam attenuation in matter will first be presented and then the results will be explained in terms of fundamental processes. Finally some implications of particular importance to radiology will be discussed.

3.2 Experimental Approach to Beam Attenuation

X-rays and gamma rays are indirectly ionising radiation. When they pass through matter they are absorbed by processes which set electrons in motion and these electrons produce ionisation of other atoms or molecules in the medium. The electrons have short, finite ranges (see Table 1.4) and their kinetic energy is rapidly dissipated first as ionisation and excitation, eventually as heat. With diagnostic X-rays these electrons do not have enough energy to produce 'Bremsstrahlung' (cf X-ray production) especially in low atomic number materials where the process is inefficient.

Conversely, the X-rays and gamma rays themselves do not have finite ranges, whatever their energy. If a fairly well collimated beam (a narrow beam) passes through different thicknesses of absorbing material, it will be found that equal thicknesses of stopping material reduce the beam intensity to the same fraction of its initial value, but the beam intensity is never reduced to zero. For simplicity a monoenergetic beam of gamma rays will be considered at this stage. The fact that an X-ray beam comprises photons with a range of energies introduces complications that will be considered towards the end of the chapter.

Referring to Figure 3.1a, if a thickness x of material reduces the beam intensity by a fraction α, the beam intensity after crossing a further thickness x will be

$$\alpha \times (\alpha I_0) = \alpha^2 I_0.$$

If, as shown in Figure 3.1b, $\alpha = \frac{1}{2}$ it may readily be observed that the variation of intensity with thickness is closely similar to the variation of radioactivity with time discussed in Section 1.5. In other words the intensity decreases exponentially with distance, as expressed by the equation $I = I_0 e^{-\mu x}$ where I_0 is the initial intensity and I the intensity after passing through thickness x. μ is a property of the material and is known as the *linear attenuation coefficient*. If x is measured in mm, μ has units mm^{-1}.

A quantity analogous to the half-life of a radioactive material is frequently quoted. This is the *half value layer* (HVL) or *half value thickness* (HVT) $H_{\frac{1}{2}}$ and is the thickness of material that will reduce the beam intensity to a half.

The analogy with radioactive decay is shown in Table 3.1.

Attenuation behaviour may be described in terms of either μ or $H_{\frac{1}{2}}$, since there is a simple relationship between them. If the value of $H_{\frac{1}{2}}$ is known, or is calculated from μ, then

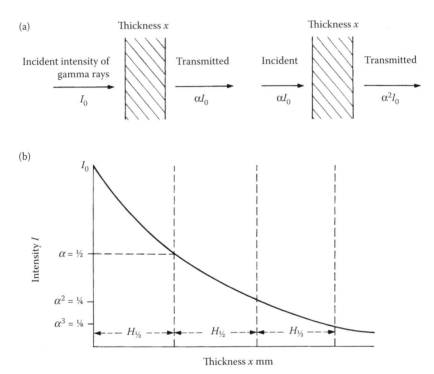

FIGURE 3.1
(a) Transmission of a monoenergetic beam of gamma rays through layers of attenuating medium of different thickness. (b) Variation of intensity with thickness of attenuator.

TABLE 3.1

Analogy between Beam Attenuation and Radioactive Decay

Radioactive Decay	Attenuation of a Monoenergetic Gamma Ray Beam
$A = A_0 e^{-\alpha t}$	$I = I_0 e^{-\mu x}$
$\alpha = \dfrac{\ln 2}{T_{\frac{1}{2}}} = \dfrac{0.693}{T_{\frac{1}{2}}}$	$\mu = \dfrac{\ln 2}{H_{\frac{1}{2}}} = \dfrac{0.693}{H_{\frac{1}{2}}}$
$A = A_0 e^{-0.693 t / T_{\frac{1}{2}}}$	$I = I_0 e^{-0.693 x / H_{\frac{1}{2}}}$

TABLE 3.2

Typical Values of μ and $H_{\frac{1}{2}}$

Energy (keV)	Material	Atomic Number	Density (kg m^{-3})	μ (mm^{-1}) (Narrow Beam)	$H_{\frac{1}{2}}$ (mm)
30	Water	7.5[a]	10^3	0.036	19
60				0.02	35
200				0.014	50
30	Bone	12.3[a]	1.65×10^3	0.16	4.3
60				0.05	13.9
200				0.02	35
30	Lead	82	11.4×10^3	33	2×10^{-2}
60				5.5	0.13
200				1.1	0.6

[a] These values are effective atomic numbers. For a discussion of the calculation of effective atomic numbers for mixtures and compounds see Section 3.5.

Source: Adapted from Johns H E and Cunningham J R. *The Physics of Radiology*, 4th ed. Thomas, Springfield, 1983.

the graphical method described in Section 1.6 may be used to determine the reduction in beam intensity caused by any thickness of material. Conversely, the method may be used to find the thickness of material required to provide a given reduction in beam intensity. This is important when designing adequate shielding. The smaller the value of μ, the larger the value of $H_{\frac{1}{2}}$ and the more penetrating the radiation. Table 3.2 gives some typical values of μ and $H_{\frac{1}{2}}$ for monoenergetic radiations.

The following are the main points to note:

(1) In the diagnostic range μ decreases ($H_{\frac{1}{2}}$ increases) with increasing energy, that is the radiation becomes more penetrating.

(2) μ increases ($H_{\frac{1}{2}}$ decreases) with increasing density. The radiation is less penetrating because there are more molecules per unit volume available for collisions in the stopping material.

(3) Variation of μ with atomic number is complex although it clearly increases quite sharply with atomic number at very low energies. In Table 3.2 some trends are obscured by variations in density.

(4) For water, which for the present purpose has properties very similar to those of soft tissue, $H_{\frac{1}{2}}$ in the diagnostic range is about 30 mm. Thus in passing through a water phantom the intensity of a narrow X-ray beam will be reduced by a factor of 2 for every 30 mm travelled. If a phantom is 18 cm across this represents six HVTs so the intensity is reduced by 2^6 or 64 times.

(5) At similar energies, $H_{\frac{1}{2}}$ for lead is 0.1 mm or less so quite a thin layer of lead provides perfectly effective shielding for, say, the door of an X-ray room.

It is sometimes convenient to separate the effect of density ρ from other factors. This is achieved by using a *mass attenuation coefficient*, μ/ρ, and then the equation for beam intensity is rewritten

$$I = I_0 e^{-(\mu/\rho)\rho x}$$

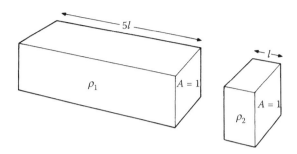

FIGURE 3.2
Demonstration that the mass attenuation coefficient of a gas is independent of its density.

When the equation is written in this form, it may be used to show that the stopping power of a fixed mass of material per unit area is constant, as one would expect since the gamma rays encounter a fixed number of atoms. Consider, for example, two containers, each filled with the same gas and each with the same area A, but of length $5l$ and l (Figure 3.2). Let the densities of the two gases be ρ_1 and ρ_2. Furthermore let the mass of gas be the same in each container. Then ρ_2 will be equal to $5\rho_1$, since the gas in container 2 occupies only one-fifth the volume of the gas in container 1.

If the simple expression $I = I_0 e^{-\mu x}$ is used, both μ and x will be different for the two volumes. If $I = I_0 e^{-(\mu/\rho)\rho x}$ is used, then since the product ρx is constant, μ/ρ is also the same for both containers, thus showing that it is determined by the types of molecule and not their number density. Of course both equations will show that beam attenuation is the same in both volumes. The dimensions of mass attenuation coefficient are $m^2 kg^{-1}$.

It is important to emphasise that in radiological imaging the linear attenuation coefficient is the more relevant quantity.

3.3 Introduction to the Interaction Processes

To understand why μ and $H_{\frac{1}{2}}$ vary with photon energy and atomic number in the manner shown in Table 3.2, it is necessary to consider in greater detail the nature of the interaction processes between X- and gamma rays and matter. A large number of different processes have been postulated. However, only four have any relevance to diagnostic radiology and need be considered here. As shown in Chapter 1, these interactions are essentially collisions between electromagnetic photons and the orbital electrons surrounding the nuclei of matter through which the radiation is passing.

Before considering any interactions in detail, it is useful to discuss some general ideas.

3.3.1 Bound and Free Electrons

All electrons are 'bound' in the sense that they are held by positive attractive forces to their respective nuclei, but the binding energy is very variable, being much higher for the K shell electrons than for electrons in other shells. When an interaction with a passing photon occurs, the forces of interaction between the electron and the photon may be smaller than the forces holding the electron to the nucleus, in which case the electron will remain 'bound' to its nucleus and will behave accordingly. Conversely, the forces of interaction may be much greater than the binding forces, in which case the latter may be discounted and the electron behaves as if it were 'free'. Since for the low atomic number elements found in the body, the energy of one photon of X-rays is much higher than the binding energy of even K shell electrons, most electrons can behave as if they are free when the interaction is strong enough. However, the interaction is frequently much weaker, a sort of glancing blow by the photon which involves only a fraction of its energy, and thus in many interactions electrons behave as if they were bound. Hence interactions that involve both bound and free electrons will occur under all circumstances and it is frequently the relative contribution of each type of interaction that is important.

This simple picture allows two general statements to be made. First, the higher the energy of the bombarding photons, the greater the probability that the interaction energy will exceed the binding energy. Thus the proportion of interactions involving free electrons can be expected to increase as the quantum energy of the radiation increases. Second, the higher the atomic number of the bombarded atom, the more firmly its electrons are held by electrostatic forces. Hence interactions involving bound electrons are more likely when the mean atomic number of the stopping material is high.

3.3.2 Attenuation, Scatter and Absorption

It is important to distinguish between these three processes and this can also be done on the basis of the simple model of interaction that has already been described (Figure 1.8).

When a beam of collimated X-ray photons interacts with matter, some of the X-rays may be *scattered*. This simply means they no longer travel in the same direction as the collimated beam. Some photons have the same energy after the interaction as they had before and lose no energy to the medium. Other photons lose some energy when they are scattered. This energy is transferred to electrons and, as already noted, is dissipated locally. Such a process results in energy *absorption* as well as scattering in the medium. Finally, some photons may undergo interactions in which they are completely destroyed and all their energy is transferred to electrons in the medium. Under the conditions normally obtaining in diagnostic radiology, all this energy is usually dissipated locally and thus the process is one of *total absorption*. Both scatter and absorption result in beam *attenuation*, that is a reduction in the intensity of the collimated beam.

If a new term, the *mass absorption coefficient* μ_a/ρ is introduced, it follows that μ_a/ρ is always less than the mass attenuation coefficient μ/ρ although the difference is small at low photon energies. From the viewpoint of good radiology, only absorption is desirable since scatter results in uniform irradiation of the image receptor. As shown in Section 6.5, scatter reduces contrast on film. It also has an adverse effect in digital imaging systems. Note also that only the absorbed energy contributes to the radiation dose to the patient and μ_a/ρ is the most relevant quantity for dosimetry. This is an undesirable but unavoidable side-effect if good radiographic images are to be obtained.

3.4 The Interaction Processes

Four processes will be considered. Two of these, the photoelectric effect and the Compton effect are the most important in diagnostic radiology. However, it may be more helpful to discuss the processes in a more logical order, starting with one that is only important at very low photon energies and ending with the one that dominates at high photon energies. Low photon energies are sometimes referred to as 'soft' X-rays, higher photon energies as 'hard' X-rays.

3.4.1 Elastic Scattering

When X-rays pass close to an atom, they may cause electrons to take up energy of vibration. The process is one of resonance such that the electron vibrates at a frequency corresponding to that of the X-ray photon. This is an unstable state and the electron quickly re-radiates this energy in all directions and at exactly the same frequency as the incoming photons. The process is one of scatter and attenuation without absorption.

The electrons that vibrate in this way must remain bound to their nuclei, thus the process is favoured when the majority of the electrons behave as bound electrons. This occurs when the binding energy of the electrons is high, that is the atomic number of the scattering material is high, and when the quantum energy of the bombarding photons is relatively low. The probability of elastic scattering can be expressed by identifying a mass attenuation coefficient with this particular process, say ε/ρ. Numerically, ε/ρ is expressed as a cross-section area. If the effective cross-section area for elastic scattering is high, the process is more likely to occur than if the effective cross-section area is low. ε/ρ increases with increasing atomic number of the scattering material ($\varepsilon/\rho \propto Z^2$) and decreases as the quantum energy of the radiation increases ($\varepsilon/\rho \propto 1/hf$).

Although a certain amount of elastic scattering occurs at all X-ray energies, it never accounts for more than 10% of the total interaction processes in diagnostic radiology.

Also the very low energy scattered radiation is heavily absorbed in the patient so the contribution to image formation is very low (less than 1%).

3.4.2 Photoelectric Effect

At the lower end of the diagnostic range of photon energies, the photoelectric effect is the dominant process. From an imaging view point this is the most important interaction that can occur between X-rays and bound electrons. In this process the photon is completely absorbed, dislodging an electron from its orbit around a nucleus. Part of the photon energy

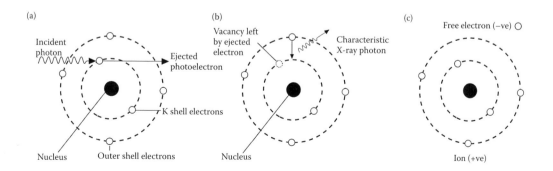

FIGURE 3.3
Schematic representation of the photoelectric effect. (a) Incident photon ejects electron leaving vacancy in shell. (b) Higher energy electron fills vacancy and characteristic X-ray photon emitted. (c) Positively charged ion and free electron (ion pair) produced.

is used to overcome the binding energy of the electron, the remainder is given to the electron as kinetic energy and is dissipated locally (Figure 3.3a) (see Section 3.2). Although the photoelectric interaction can happen with electrons in any shell, it is most likely to occur with the most tightly bound electron the photon is able to dislodge. The following equation describes the energy changes

$$hf \quad = \quad W \quad + \quad \tfrac{1}{2}m_e v^2$$

photon   binding energy   kinetic energy

energy   of electron to nucleus   of electron

However, this now leaves the residual atom in a highly excited state since there is a vacancy in one of its orbital electron shells often the K shell. One possibility is that the vacancy will be filled by an electron of higher energy and characteristic X-ray radiation will be produced (Figure 3.3b) in exactly the same way as characteristic radiation is produced as part of the X-ray spectrum from, say, a tungsten anode. Remaining behind is an atom minus one electron which is referred to as a positively charged ion with one extra proton. The liberated electron and this positively charged ion are sometime referred to as an ion pair (Figure 3.3c). The number of X-rays emitted, expressed as a fraction of the number of primary vacancies created in the atomic electron shells is known as the *fluorescence yield*.

In high atomic number materials the fluorescence yield is high and quite appreciable re-radiation of characteristic radiation in a manner similar to X-ray production may occur. This factor is important in the choice of suitable materials for X-ray beam filtration (see Section 3.9).

In lower atomic number materials any X-rays that are produced are of low energy (corresponding to low K shell energy) and are absorbed locally. Also the fluorescence yield is low. Production of Auger electrons released from the outer shell of the atom is now more probable. These electrons have energies ranging from a few to several hundred electron volts, so their ranges in tissue are short and the photoelectric interaction process now results in total absorption of the energy of the initial photon. Note that the dense shower of Auger electrons that is emitted deposits its energy in the immediate vicinity of the decay site. The resulting high local energy density can equal or even exceed that along the track of an α particle with corresponding radiobiological damage (see Section 12.4).

Since the process is again concerned with bound electrons, it is favoured in materials of high mean atomic number and the photoelectric mass attenuation coefficient τ/ρ is proportional to Z^3. The process is also favoured by low photon energies with τ/ρ proportional to $1/(hf)^3$. Notice that, as a result of the Z^3 factor, at the same photon energy lead ($Z = 82$) has a 300 times greater photoelectric coefficient than bone ($Z = 12.3$). This explains the big difference in μ values for these two materials at low photon energies as shown in Table 3.2.

The cross-section for a photoelectric interaction falls steeply with increasing photon energy although the decrease is not entirely regular because of absorption edges (see Section 3.8).

Thus the photoelectric effect is the major interaction process at the low end of the diagnostic X-ray energy range.

3.4.3 The Compton Effect

The most important effect in radiology involving unbound electrons is inelastic scattering or the Compton effect. The Compton effect mass attenuation coefficient is represented by σ/ρ. This process may be thought of most easily in terms of classical mechanics in which

Compton Effect

① $\Delta\lambda \neq m_e$ change in wavelength not dependence on incident energy

② $\Delta E \sim m_e$ Loss of energy dependent on incident photon energy

$\phi = 60°$

↳ ΔE 2%   @   20 keV

ΔE 9%   @   200 keV

ΔE 50%   @   1 MkeV

- Photon has energy ($h\gamma$) and momentum $h\gamma/C$
- Photon collides with free electron; both energy and momentum conserved
- Proportion of energy & momentum transferred to scattered photon & electron determined by θ and ϕ
- Energy of electron dissipated by ionisation and excitation (& heat)
- Photon has lowered energy

* Scatter & partial absorption of energy

$h\gamma$ ～～～→ ○ $h\gamma'$ (ϕ)

(θ)

compton electron

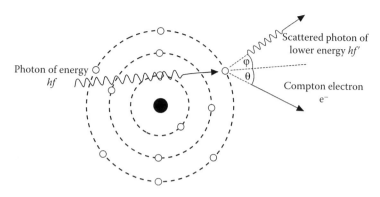

FIGURE 3.4
Schematic representation of the Compton effect.

the photon has energy hf and momentum hf/c and makes a billiard ball type collision with a stationary free electron, with both energy and momentum conserved (Figure 3.4).

The proportions of energy and momentum transferred to the scattered photon and to the electron are determined by θ and ϕ. The kinetic energy of the electron is rapidly dissipated by ionisation and excitation and eventually as heat in the medium and a scattered photon of lower energy than the incident photon emerges from the medium—assuming no further interaction occurs. Thus the process is one of scatter and partial absorption of energy.

The equation used most frequently to describe the Compton process is

$$\lambda' - \lambda = \frac{h}{m_e c^2} (1 - \cos \phi) \tag{3.1}$$

where λ' is the wavelength of the scattered photon and λ is the wavelength of the incident photon. This equation shows that the change in wavelength $\Delta\lambda$ when the photon is scattered through an angle ϕ is independent of photon energy. However, it may be shown that the change in energy of the photon ΔE is given by

$$\Delta E = \frac{E^2}{m_e c^2} (1 - \cos \phi) \tag{3.2}$$

Thus the loss of energy by the scattered photon does depend on the incident photon energy. For example, when the photon is scattered through 60°, the proportion of energy taken by the electron varies from about 2% at 20 keV to 9% at 200 keV and 50% at 1 MeV.

Remaining behind is a positively charged ion as in the photoelectric effect.

Insight

Energy Sharing in the Compton Process

The mathematical derivation of Equation 3.2 from Equation 3.1 is as follows:

If $\lambda = hc/E$ (see Section 1.11)

Then $\lambda + \Delta\lambda = hc/(E - \Delta E)$

where $\Delta\lambda$ is a small change in wavelength of the photon and ΔE is the corresponding energy change.

So

$$\Delta\lambda = \frac{hc}{E - \Delta E} - \frac{hc}{E} \simeq \frac{hc}{E^2} \cdot \Delta E$$

Rearranging

$$\Delta E = \frac{E^2}{hc} \cdot \Delta\lambda = \frac{E^2}{m_e c^2}(1 - \cos\phi) \quad \text{(from Equation 3.1)}$$

Note that

$$m_e c^2 = 9 \times 10^{-31} \times 9 \times 10^{16} = 81 \times 10^{-15}\,\text{J}$$

$$1\,\text{J} = 6.3 \times 10^{18}\,\text{eV}$$

So

$$m_e c^2 = 81 \times 10^{-15} \times 6.3 \times 10^{18}\,\text{eV}$$

$$= 510 \times 10^3\,\text{eV or 510 keV}$$

(the more exact value is 511 keV, a number that will re-appear in Chapter 11 as the photon energy of importance in positron emission tomographic imaging, PET).

Example 1 A 20 keV photon is scattered through 30°

$$\Delta E = \frac{20 \times 20}{510}(1 - \cos 30)$$

$$= \frac{400 \times 0.14}{510}$$

$$= 0.11\,\text{keV}$$

The photon loses only one tenth of a keV of energy which is taken away by the electron.

Example 2 A 100 keV photon is back scattered through 180°

$$\Delta E = \frac{100 \times 100}{510}(1 - \cos 180)$$

$$= \frac{1000 \times 2}{510}$$

$$= 39\,\text{keV}$$

Thus the photon is now losing 39% of its energy to the electron.

Since the process is one of attenuation with partial absorption, the variation in the amount of energy absorbed in the medium, averaged over all scattering angles, with initial photon energy depends on the following:

(1) The probability of an interaction (Figure 3.5a)
(2) The fraction of the energy going to the electron (Figure 3.5b)
(3) The fraction of the energy retained by the photon (Figure 3.5b)

To find out how much energy is absorbed in the medium, the Compton cross-section must be multiplied by the percentage of energy transferred to the electron (Figure 3.5c). This shows that there is an optimum X-ray energy for energy absorption by the Compton effect. However, it is well above the diagnostic range.

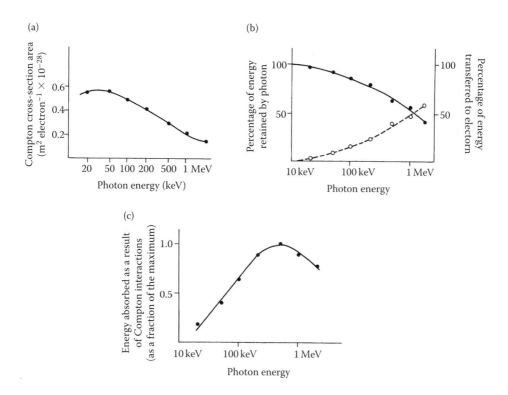

FIGURE 3.5

(a) Variation of Compton cross-section with photon energy. (b) Percentage of energy transferred to the electron (dotted line) and percentage retained by the photon (solid line) per Compton interaction as a function of photon energy. (c) Product of (a) and (b) to give variation in total Compton energy absorption as a function of photon energy.

As shown by Equation 3.2, the amount of energy transferred to the electron depends on the scattering angle ϕ. At diagnostic energies, the proportion of energy taken by the electron, that is absorbed, is always quite small. For example, even a head-on collision ($\phi = 180°$) only transfers 8% of the photon energy to the electron at 20 keV. Thus at low energies Compton interactions cause primarily scattering and this will have implications when the effects of scattered radiation on image contrast are considered. Although the energy of scattered X-rays is always lower than the energy of the primary beam, in very low energy work, for example mammography, the difference can be quite small.

One further consequence of Equation 3.2 is that when E is small, quite large values of ϕ are required to produce appreciable changes ΔE. This is important in nuclear medicine where pulse height analysis is used to detect changes in E and hence to discriminate against scattered radiation.

3.4.3.1 Direction of Scatter

After Compton interactions, photons are scattered in all directions and this effect may be displayed by using polar diagrams similar to those used in Section 2.4 to demonstrate the directions in which X-rays are emitted from the anode (Figure 3.6). As the photon energy increases, the scattered photons travel increasingly in the forward direction but this change is quite small in the diagnostic energy range where a significant proportion of X-rays may be back scattered. Note that, as discussed above, the mean energy of back-scattered photons is lower than the mean energy of forward scattered photons.

For thicker objects, for example a patient, the situation is further complicated by the fact that both the primary beam and the scattered radiation will be attenuated. Thus, although in Figure 3.7 the polar diagrams of Figure 3.6 could be applied to each slice in turn, because of body attenuation the X-ray intensity on slice Z may be only 1% of that on slice A. In the example shown in Figure 3.7 the intensity of radiation scattered back at 150° is 4 times higher than that scattered forward at 30° and is comparable with the intensity

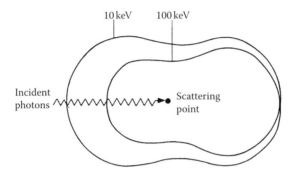

FIGURE 3.6
Polar diagrams showing the spatial distribution of scattered X-rays around a free electron at two different energies.

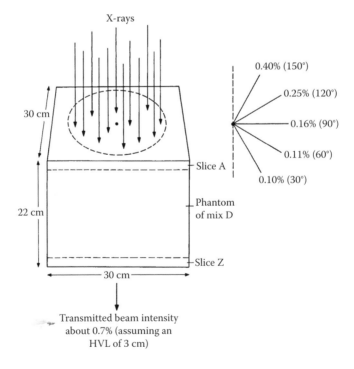

FIGURE 3.7
Distribution of scattered radiation around a patient-sized phantom of tissue equivalent material. The phantom measured 30 × 30 cm by 22 cm deep and a 400 cm² field was exposed to 100 kVp X-rays. Intensities of scattered radiation at 1 m are expressed as a percentage of the incident surface dose. (Adapted from McVey, G. The effect of phantom type, beam quality, field size and field position on X-ray scattering simulated using Monte Carlo techniques. *Br. J. Radiol.* 79, 130–141, 2006.)

in the primary transmitted beam. This work by McVey (2006) confirms the slightly earlier work of Sutton and Williams (2000) on a RANDO phantom which produced essentially the same factor of 4 difference.

Hence a high proportion of the scattered radiation emerging from the patient travels in a backwards direction and this has implications for radiation protection, for example when using an over-couch tube for fluoroscopy.

Insight

Factors Controlling the Effect of Compton-Scattered Photons on Image Quality

In very unfavourable conditions, scattered photons reaching the image receptor could contribute as much as ten times radiation to the image as the primary beam. Scatter has a seriously adverse effect on contrast (see Section 6.5).

It is not possible to eliminate scatter completely but some measures can be taken. Working through the imaging process, factors affecting scatter are as follows:

- *Tube kV*—The kVp has only a small effect on scatter *production*. As can be seen from Figure 3.8 (see page 90) the production of scattered radiation by the Compton effect is relatively constant throughout the diagnostic range. However, the scattered radiation reaching the imaging receptor will rise appreciably as the kVp rises because as more scattered photons move in a forward direction and they have a higher energy so they are less attenuated by the patient and are therefore not self absorbed. Overall, as the kVp rises so does the scatter but the dose to the patient to produce the image falls.
- *Beam collimation*—The effect of field size is an important factor determining scatter but in practical terms is often more relevant to patient dose (see Chapter 13) than actually affecting the scattered radiation reaching the imaging plane. As the size of the X-ray field on the body increases from a very small value, the quantity of scattered radiation reaching the imaging plane rises rapidly at first but then slowly reaches a maximum as the field continues to increase in size. Increasing the field size beyond this point does not change the amount of scatter reaching the image receptor. Although the total amount of scattered photons continues to rise as the beam size is increased, self absorption within the body stops scattered photons travelling through to the image receptor. In practice this saturation point is at a field size of about 30 × 30 cm.
- *Body thickness*—The total number of scattered photons will increase as the thickness of the body in the beam increases but the amount reaching the image receptor reaches a constant value for a given kVp as the scatter produced in the upper part of the body is absorbed by the body. The thickness of the body part being imaged can only be influenced on a very few occasions by the use of compression (see Section 6.7).
- *Grids*—This is the most effective form of scatter reduction at the receptor—see Section 6.8.
- *Air gap technique*—This may also be useful in some circumstances, especially in high kV radiography—see Section 9.3.3.

The effect of scattered radiation on image quality is considered in detail in Chapter 6.

3.4.3.2 *Variation of the Compton Coefficient with Photon Energy and Atomic Number*

For free electrons, the probability of a Compton interaction decreases steadily as the quantum energy of the photon increases. However, when the electrons are subject to the forces

TABLE 3.3

Values of Charge/Mass Ratio for the Atoms of Various Elements in the Periodic Table

	H	C	N	O	P	Ca	Cu	I	Pb
Z	1	6	7	8	15	20	29	53	82
A	1	12	14	16	31	40	63	127	208
Z/A	1	0.5	0.5	0.5	0.5	0.5	0.5	0.4	0.4

of other atoms, low energy interactions frequently do not give the electron sufficient energy to break away from these other forces. Thus in practice the Compton mass attenuation coefficient is approximately constant in the diagnostic range and only begins to decrease $(\sigma/\rho \propto 1/hf)$ for photon energies above about 100 keV.

σ/ρ is almost independent of atomic number. To understand why this should be so, recall the information from Chapter 1 on atomic structure. The Compton effect is proportional to the number of electrons in the stopping material. Thus if the Compton coefficient is normalised by dividing by density, σ/ρ should depend on electron density. Now for any material the number of electrons is proportional to the atomic number Z and the density is proportional to the atomic mass A. Hence

$$\frac{\sigma}{\rho} \propto \frac{Z}{A}$$

Examination of Table 3.3 shows that Z/A is almost constant for a wide range of elements of biological importance, decreasing slowly for higher atomic number elements.

Hydrogen is an important exception to the 'rule' for biological elements. Materials that are rich in hydrogen exhibit elevated Compton interaction cross-sections and this explains, for example, the small but measurable difference in mass attenuation coefficients in the Compton range of energies between water and air, even though their mean atomic numbers of approximately 7.5 and 7.8 (see Section 3.5), respectively are almost identical.

The Compton effect is a major interaction process at diagnostic X-ray energies, particularly at the upper end of the energy range.

3.4.4 Pair Production

When a photon with energy in excess of 1.02 MeV passes close to a heavy nucleus, it may be converted into an electron and a positron. This is one of the most convincing demonstrations of the equivalence of mass and energy. The well-defined threshold of 1.02 MeV is simply the energy equivalence of the electron and positron masses m_{e^-} and m_{e^+}, respectively according to the equation:

$$E = m_{e^-} c^2 + m_{e^+} c^2$$

Any additional energy possessed by the photon is shared between the two particles as kinetic energy. Each particle can receive any fraction between all and nothing.

The electron dissipates its energy locally and has a range given by Table 1.2. The positron dissipates its kinetic energy but when it comes to rest it undergoes the reverse of the formation reaction, annihilating with an electron to produce two 0.51 MeV gamma rays which fly away simultaneously in opposite directions.

$$e^+ + e^- \rightarrow 2\gamma \ (0.51 \text{ MeV})$$

These gamma rays (or *annihilation radiation*) are penetrating radiations which escape from the absorbing material and 1.02 MeV of energy is re-irradiated. Hence the pair production process is one of attenuation with partial absorption.

Pair production is the only one of the four processes considered that shows a steady increase in the chance of an interaction with increasing photon energy above 1.02 MeV (but see Section 3.8). Since a large, heavy nucleus is required to remove some of the photon momentum, the process is also favoured by high atomic number materials (the pair production mass attenuation coefficient $\pi/\rho \propto Z$).

Although pair production has no direct relevance in diagnostic radiology, the subsequent annihilation process is important when positron emitters are used for *in vivo* imaging in positron emission tomography (see Section 11.3.1). The characteristic 0.51 MeV gamma rays are detected and since two are emitted simultaneously, coincidence circuits may be used to discriminate against stray background radiation.

3.5 Combining Interaction Effects and Their Relative Importance

Table 3.4 summarises the processes that have been considered.

Each of the processes occurs independently of the others. Thus for the photoelectric effect

$$I = I_0 e^{-(\tau/\rho)\rho x}$$

and for the Compton effect

$$I = I_0 e^{-(\sigma/\rho)\rho x}$$

and so on.

TABLE 3.4

A Summary of the Four Main Processes by Which X-rays and Gamma Rays Interact with Matter

Process	Normal Symbol for Process	Type of Interaction	Variation with Photon Energy (*hf*)	Variation with Atomic Number (Z)
Elastic	ε/ρ	Bound electrons	$\propto 1/hf$	$\propto Z^2$
Photoelectric	τ/ρ	Bound electrons	$\propto 1/(hf)^3$	$\propto Z^3$
Compton	σ/ρ	Free electrons	Almost constant 10–100 keV; $\propto 1/hf$ above 100 keV	Almost independent of Z
Pair production	π/ρ	Promoted by heavy nuclei	Rapid increase above 1.02 MeV	$\propto Z$

Hence effects can be combined simply by multiplying the exponentials to give

$$I = I_0 e^{-(\tau/\rho) \cdot \rho x} e^{-(\sigma/\rho) \cdot \rho x \cdots}$$

or

$$I = I_0 e^{-(\tau/\rho + \sigma/\rho + \ldots) \cdot \rho x}$$

leading to the simple relationship

$$\frac{\mu}{\rho} = \frac{\tau}{\rho} + \frac{\sigma}{\rho} + \cdots$$

where additional interaction coefficients can be added if they contribute significantly to the value of μ/ρ. Hence the total mass attenuation coefficient is equal to the sum of all the component mass attenuation coefficients obtained by considering each process independently.

In the range of energies of importance in diagnostic radiology, the photoelectric effect and the Compton effect are the only two interactions that need be considered (as shown above). Since the latter process generates unwanted scattered photons of lower energy but the former does not, and the former process is very dependent on atomic number but the latter is not, it is clearly important to know the relative contributions of each in a given situation. Figure 3.8a, b shows photoelectric and Compton cross-sections for nitrogen, which has

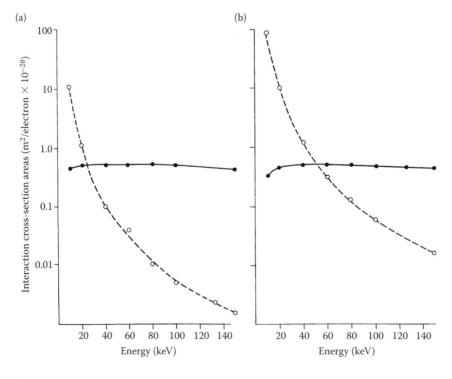

FIGURE 3.8
(a) Compton (•) and photoelectric (○) interaction coefficients for nitrogen (Z = 7). (b) Compton (•) and photoelectric (○) interaction coefficients for aluminium (Z = 13).

approximately the same atomic number as soft tissue, and for aluminium, with approximately the same atomic number as bone, respectively. Because the photoelectric coefficient is decreasing rapidly with photon energy (note the logarithmic scale) there is a sharp transitional point. By 30 keV the Compton effect is already the more important process in soft tissue.

Understanding the difference in relative importance of these two effects at different energies is quite fundamental to appreciating the origins of radiological image contrast that is caused by differences in atomic number in body materials. The photoelectric effect is very dependent on Z, the Compton effect is not. Thus at energies where the photoelectric effect dominates, for example, mammography, small changes in both mean atomic number and in kV can have a big effect on contrast. At much higher energies, for example, 120 keV where the Compton effect dominates, image contrast will be relatively insensitive to differences in atomic number and small changes in kV. Note that contrast which is caused by differences in density between different structures, is not susceptible to this effect.

For the higher atomic number material the photoelectric curve is shifted to the right but the Compton curve remains almost unchanged and the cross-over point is above 50 keV. This trend continues throughout the periodic table. For iodine which is the major interaction site in a sodium iodide scintillation detector, the cross-over point is about 300 keV and for lead it is about 500 keV. Note that even for lead, pair production does not become comparable with the Compton effect until 2 MeV and for soft tissues not until about 20 MeV.

Note that although calculating the combined effect of different interaction processes to estimate attenuation is fairly straightforward, it is not so easy to work out an *effective atomic number*. This is a useful concept in both radiological imaging and in radiation dosimetry when dealing with a mixture or compound. One wishes to quote a single atomic number for a hypothetical material that would interact in exactly the same way as the mixture or compound. Unfortunately, the effective atomic number of a mix of elements will change as the photon energy increases from the region where the photoelectric effect dominates to the region where the Compton effect dominates. This is because in the photoelectric region the effective atomic number must be heavily weighted in favour of the high Z components. No such weighting is necessary in the Compton region. For further discussion on this point see Johns and Cunningham (1983).

Finally, it is now possible to interpret more fully the data in Table 3.2. For any material, the linear attenuation coefficient will decrease with increasing energy, initially because the photoelectric coefficient is decreasing and subsequently because the Compton coefficient is decreasing. At low energy there is a very big difference between the μ values for water and lead, partly because of the density effect but also because the photoelectric effect dominates and this depends on Z^3.

3.6 Broad Beam and Narrow Beam Attenuation

As already explained, the total attenuation coefficient is simply the sum of all the interaction processes and this should lead to unique values for μ, the linear attenuation coefficient, and $H_{1/2}$, the HVL or HVT.

However, if a group of students were each asked to measure $H_{1/2}$ for a beam of radiation in a given attenuating material, they would probably obtain rather different results. This is because the answer would be very dependent on the exact geometrical arrangement used and whether any scattered radiation reached the detector.

Two extremes, narrow beam and broad beam conditions, are illustrated in Figure 3.9a, b. For narrow beam geometry it is assumed that the primary beam has been collimated so that the scattered radiation misses the detector. For broad beam geometry, radiation scattered out of primary beam that would otherwise have reached the detector, labelled A and A', is not recorded. However, radiation such as B and B' which would normally have missed the detector is scattered into it and multiple scattering may cause a further increase in detector reading. Hence a broad beam does not appear to be attenuated as much as a narrow beam and, as shown in Figure 3.10, the value of $H_{1/2}$ will be different. Absorbed radiation is of course stopped equally by the two geometries. When stating $H_{1/2}$ values, the conditions (especially broad or narrow beam) should always be specified. It should be assumed that modern data is for a narrow beam.

In radiology broad beam geometry is used when the image receptor is either film or a two dimensional digital detector. Therefore an important consequence of the inferior broad beam attenuation is the recommendation to reduce the field size to the smallest value consistent with the required image. If a broader beam than necessary is used extra scattered radiation reaches the receptor thereby reducing contrast (see Section 6.5). Reducing the field size of course also reduces the total radiation energy absorbed in the patient and, in interventional procedures, radiation scattered to staff.

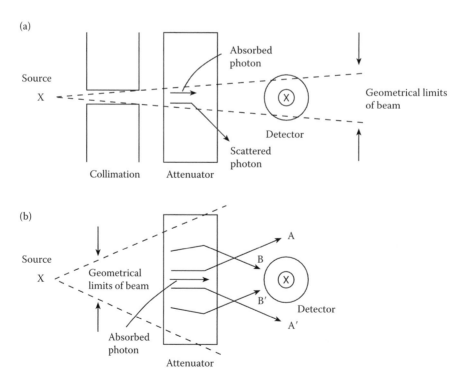

FIGURE 3.9
Geometrical arrangement for study of (a) narrow beam and (b) broad beam attenuation.

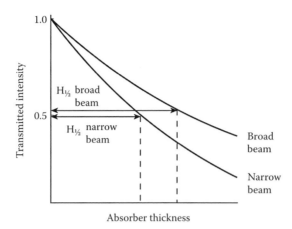

FIGURE 3.10
Attenuation curves and values of $H_{\frac{1}{2}}$ corresponding to the geometries shown in Figure 3.9.

TABLE 3.5

Effect of Broad and Narrow Beams on Transmission of 100 kVp
X-ray Photons (with 1 mm Cu Filtration) through 8 cm of Perspex

	Narrow Beam	Broad Beam
HVL cm	2.2	3.3 (5)
Transmitted intensity (%)	11	16

Insight

More on Half Value Layers

As explained in the main text, because of the variable contribution from scattered radiation, the measured value of HVL, with the same radiation beam and absorber will vary depending on the geometry of the measuring system. It is important to be aware of this when using HVL measurements to predict some other parameter such as the filtration in an X-ray beam. Graphical methods of finding the HVL will normally produce sufficiently accurate values to use for these predictions if carried out carefully. Narrow beam geometry (as in Figure 3.9a) is the easiest to reproduce but is not representative of the very broad beams used in radiology with most imaging systems.

Because of the exponential nature of the attenuation process this difference can have a big effect on transmission. To illustrate this the transmission of an X-ray beam operating at 100 kVp with 1mm Cu filtration and perspex sheets acting as an attenuator , under broad and narrow beam conditions, is shown in Table 3.5.

Figures for transmitted intensity show the big contribution of scattered radiation under broad beam conditions to the beam intensity after attenuation. This can be very important when calculating the shielding in the walls of a radiation enclosure and the fundamental formula $I = I_0 e^{-\mu x}$ is changed to $I = B I_0 e^{-\mu x}$ where B is a 'Build up factor', the magnitude of which is generally obtained from tables of shielding data for different materials. The value of B is a function of the photon energy, the thickness of the attenuator, the distance from the attenuator to the measuring device and the area of the beam. It can never be less than 1. As well as being important in shielding calculations it is also important in quantitative calculations in digital imaging, for example, CT.

Note: This variation in HVL is quite different from the variation resulting from filtration and beam hardening (see Section 3.9).

3.7 Consequences of Interaction Processes when Imaging Patients

It will be useful at this point to summarise the implications of some of the points made in this chapter for practical radiological imaging. Variations in X-ray intensity, or flux of X-ray photons, leaving different parts of the patient carry the information required to form the image. In part these are desirable variations which, as discussed above, depend on the relative contributions of the Compton and photoelectric effects (see Figure 3.8), the effective atomic number of body constituents and the densities of body parts. Relevant data is given in Table 3.6.

At low photon energies the photoelectric effect predominates, consequently the much higher effective atomic number of bone will be important. Compton effect interactions are largely independent of atomic number but differences in density affect contrast at all energies. The contrast in high kV images, such as in CT, is almost entirely due to density differences (for a more rigorous treatment of contrast see Section 5.5.2).

In addition there are undesirable variations in the beam leaving different parts of the patient due to non-uniformity of the X-ray beam (see Section 2.4) and in patient thickness. The situation is further complicated by the scattered radiation produced by Compton interactions.

Overall, a high percentage of the incident beam is either absorbed in the patient or scattered. As shown in Figure 3.7 quite a lot is back scattered. For an anterior-posterior (AP) or lateral view of the head, only about 1% of the incident photons are transmitted. Transmission for lung, breast and extremities will be higher but may be as low as 0.1% for a lumber spine radiograph. In most views, the scattered radiation will comprise at least 50% of the transmitted beam and can be as high as 75%.

3.8 Absorption Edges

Whenever the photon energy is just slightly greater than the energy required to remove an electron from a particular shell around the nucleus, there is a sharp increase in the photoelectric absorption coefficient. This is known as an absorption edge, and absorption edges associated with K shell electrons have a number of important applications in radiology (see Table 1.1). The edges associated with the L shell and subsequent outer shells are at energies that are too low to be of any practical significance.

TABLE 3.6

Effective Atomic Numbers and Densities for Body Constituents

Constituents	Effective Atomic Number	Density (kg m^{-3})
Air	7.8	1.2
Fat	6.6	0.9×10^3
Water	7.5	1.0×10^3
Muscle/soft tissue	7.6	1.0×10^3
Bone	13.8	1.8×10^3
Lung	7.4	2.4×10^2

As shown in Figure 3.11 there will be a substantial difference in the attenuating properties of a material on either side of the absorption edge. There are two reasons for the sudden increase in absorption with photon energy. First the number of electrons available for release from the atom increases. However, in the case of lead the number available only increases from 80 to 82 since the K shell only contains two electrons, and the increase in absorption is proportionately much bigger than this. Thus a more important reason is that a resonance phenomenon occurs whenever the photon energy just exceeds the binding energy of a given shell. Since at 88–90 keV the photon energy is almost exactly equal to that required to remove K shell electrons from lead, a disproportionately large number of K shell interactions will occur and absorption by this process will be high.

Because of absorption edges, there will be limited ranges of photon energies for which a material of low atomic number actually has a higher absorption coefficient than a material of higher atomic number and this has a number of practical applications. For example the presence of K absorption edges has an important influence on the selection of suitable materials for intensifying screens, where a high absorption efficiency by the photoelectric process is required. Although tungsten in a calcium tungstate screen has a higher atomic number than the rare earth elements and therefore has an inherently higher mass absorption coefficient, careful comparison of the appropriate absorption curves (Figure 3.12) shows that in the important energy range from 40 to 70 keV where for many investigations there will be a high proportion of photons, absorption by rare earth elements is actually higher than for tungsten.

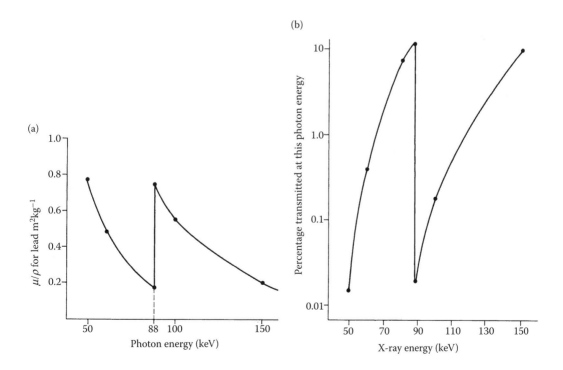

FIGURE 3.11
(a) Variation in the mass attenuation coefficient for lead across the K-edge boundary. (b) Corresponding variation in transmission through 1 mm of lead. Note the use of logarithmic scale; at the absorption edge the transmitted intensity falls by a factor of about 500.

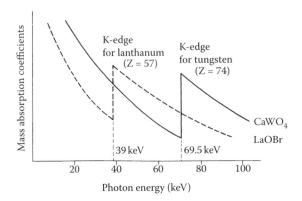

FIGURE 3.12
Curves showing the relative absorptions of lanthanum oxybromide (LaOBr) and calcium tungstate (CaWO₄) as
a function of X-ray energy in the vicinity of their absorption edges (not to scale).

Other examples of the application of absorption edges are as follows:

1. In the use of iodine (Z = 53, K-edge = 33 keV) and barium (Z = 56, K-edge = 37 keV)
   as contrast agents. Typical diagnostic X-ray beams contain a high proportion of photons at or just above these energies, thus ensuring high absorption coefficients.
2. The use of an erbium filter (Z = 68, K-edge = 57.5 keV) for paediatric radiography.
3. The presence of absorption edges also has a significant effect on the variation of
   sensitivity of photographic film with photon energy (see Section 14.7.1).

An important consequence of the absorption edge effect is that a material is relatively
transparent to radiation that has a slightly lower energy than the absorption edge, including the material's own characteristic radiation. This factor is important when choosing
materials for X-ray beam filtration (see Section 3.9). It should also be considered when
examining the properties of materials for shielding. For example, although lead is normally used for shielding, if a particular X-ray beam contains a high proportion of photons
of energy approaching but just less than the K-edge for lead (Z = 82, K-edge = 88 keV), some
other material, for example, tin (Z = 50, K-edge = 29 keV), may be a more effective attenuator on a weight for weight basis.

3.9 Filtration and Beam Hardening

Thus far in this chapter, attenuation has, for ease of explanation, mainly been discussed
in terms of a beam of photons of a single energy. As a reminder the description 'gamma
rays' has frequently been used. Radiation from an X-ray set has a range of energies and
these radiations will not all be attenuated equally. Since, in the diagnostic range, the lower
energy radiations are the least penetrating, they are removed from the beam more quickly.
In other words the attenuating material acts like a filter (Figure 3.13).

The choice of a suitable filter is important to the performance of an X-ray set because
this provides a mechanism for reducing the intensity of low energy X-ray photons. These
photons would be absorbed in the patient, thereby contributing nothing to the image but

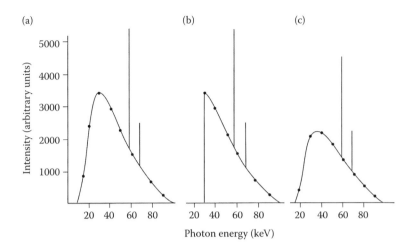

FIGURE 3.13
Curves showing the effect of filters on the quality (spectral distribution) of an X-ray beam. (a) Spectrum of the emergent beam generated at 100 kVp with inherent filtration equivalent to 0.5 mm Al. (b) Effect of an ideal filter on this spectrum. (c) Effect of total filtration of 2.5 mm aluminium on the spectrum.

increasing the dose to the patient. The effect of an ideal low energy filter is shown in Figure 3.13a, b, but, in practice, no filter completely removes the low energy radiation or leaves the useful component unaffected.

The position of the K absorption edge must also be considered when choosing a filter material. For example, tin, with a K-edge of 29 keV, will transmit 25–29 keV photons rather efficiently and these would be undesirable in, for example, a radiographic exposure of the abdomen. Aluminium (Z = 13, K-edge = 1.6 keV) is the material normally chosen for filters in the diagnostic range. Aluminium is easy to handle, and 'sensible' thicknesses of a few mm are required. Photons of energy less than 1.6 keV, including the characteristic radiation from aluminium, will either be absorbed in the X-ray tube window or in the air gap between filter and patient. The effect of 2.5 mm of aluminium on a 100 kV beam from a tungsten target is shown in Figure 3.13c.

Finally, the thickness of added filtration will depend on the operating kVp and the inherent filtration. The inherent filtration is the filtration caused by the glass envelope, insulating oil and window of the X-ray tube itself and is usually equivalent to about 0.5–1 mm of aluminium. Note that the thickness of aluminium that is equivalent to this inherent filtration will vary with kV. The total filtration should be at least 1.5 mm Al for tubes operating up to 70 kVp (e.g. dental units) and at least 2.5 mm Al for tubes capable of operating at higher kVp. Note that if a heavily filtered beam is required at high kV 0.5 mm copper (Z = 29) may be preferred. However, characteristic X-rays at 9 keV will now be produced as a result of the photoelectric effect (Section 3.4.2) so aluminium will be required too as these could contribute to the patient skin dose.

When a beam passes through a filter, it becomes more penetrating or 'harder' and its $H_{1/2}$ increases. If log (intensity) is plotted against the thickness of absorber, the curve will not be a straight line as predicted by $I = I_0e^{-\mu x}$. It will fall rapidly at first as the 'soft' radiation is removed and then more slowly when only the harder, high energy component remains (Figure 3.14a). Note that the beam never becomes truly monochromatic but after about four HVLs, that is when the intensity has been reduced to about one-sixteenth of its original value, the spread of photon energies in the beam is quite small. The value of $H_{1/2}$ then

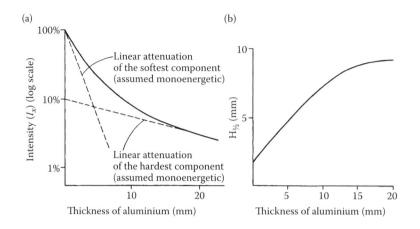

FIGURE 3.14

(a) Variation of intensity (I_x) with absorber thickness (x) as an heterogeneous beam of X-rays becomes progressively harder on passing through attenuating material (plotted on a log scale). (b) Corresponding change in the value of $H_{1/2}$.

TABLE 3.7

Skin Dose and Exposure Time for Comparable Density Radiographs of a Pelvic Phantom (18 cm thick) Using a 60 kVp Beam with Different Filtration

Aluminium Filtration (mm)	Skin Dose in Air (mGy)	Exposure Time at 100 mAs
None	20.7	1.4 (1)
0.5	16.1	1.6 (1)
1.0	11.0	1.6 (4)
3.0	4.1	2.1 (4)

Source: Adapted from Trout E D, Kelley J P and Cathey G A. The use of filters to control radiation exposure to the patient in diagnostic radiology. *Am. J. Roentgenol.* 67, 946–963, 1952.

becomes constant within the accuracy of measurement and *for practical purposes* the beam is monochromatic (Figure 3.14b).

The effect of filtration on skin dose is shown in Table 3.7. There is clearly a substantial benefit to be gained in terms of patient dose from the use of filters. Note, however, the final column of the table which shows that there is a price to be paid. Although heavy filtration (3 mm Al) reduces the skin dose even more than 1 mm filtration, the X-ray tube output begins to be affected and this is reflected in the increased exposure time. A prolonged exposure may not be acceptable; but if an attempt is made to restore the exposure time to its unfiltered value by increasing the tube current, there may be problems with the tube rating.

Insight

More on Filtration and Tube Rating

It should be noted that the results in Table 3.7 were obtained using an X-ray tube that would now be considered of an old design, which would have been a single phase unit with, at the best, full rectification. For a modern unit there would be two important differences:

(1) The technical advances in X-ray tube design discussed in Chapter 2 have resulted in powerful X-ray tubes with much greater intensity. The loss of intensity because of filtration then

becomes less of a problem—typically, with a two and a half times increase in tube power an additional 0.5 mm of Cu filtration may be used without loss of image quality during screening but with a consequent saving in patient dose.

(2) The effect of filtration on the spectrum for a modern three-phase unit or high frequency unit would also be less. This is because an older unit, with high voltage ripple, would produce many of the photons whilst the tube was operating below its kVp. With almost zero ripple in the kVp these X-ray photons would be absent. Note that the inherent low energy photons arising from the mechanism of X-ray production itself would still be present.

At the higher kVps currently used for pelvic radiographs, increasing the filtration from 1 mm to 3 mm aluminium would have a very much smaller effect on exposure time.

Thus far the discussion has been concerned with the use of filtration to remove the low energy part of the spectrum. Occasionally, for example, when scatter is likely to be a problem, it may b e desirable to remove the high energy component of the spectrum. This is difficult because the general trend for all materials is for the linear attenuation coefficient to decrease with increasing keV. However, it is sometimes possible to exploit the absorption edge discussed in Section 3.8. The K shell energy and hence the absorption edge increases with increasing atomic number and for elements in the middle of the periodic table, for example, gadolinium (Z = 64, K-edge = 50 keV) and erbium (Z = 68, K-edge = 57.5 keV), this absorption may remove a substantial proportion of the higher energies in a conventional spectrum. Figure 3.15 shows the effect of a 0.25 mm thick gadolinium filter on the spectrum shown in Figure 3.13c.

As with low energy filtration, the output of the useful beam is reduced so there is an adverse effect on tube loading. Thus the technique is perhaps best suited to thin body parts where scatter is a problem and a grid is undesirable because of the increased dose to the patient. Paediatric radiology is a good example.

A K-edge filter may also sometimes enhance the effect of a contrast agent, for example, iodine. Refer again to Figure 3.12 and imagine the lanthanum curve to be replaced by the absorption curve for iodine (Z = 53, K-edge = 33 keV) and the tungsten curve replaced by

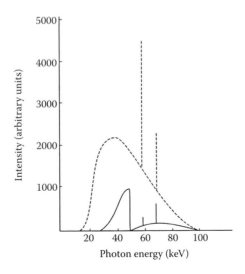

FIGURE 3.15
The effect of a 0.25 mm gadolinium filter on a conventionally filtered X-ray spectrum (solid curve). The dotted curve is taken from Figure 3.13c.

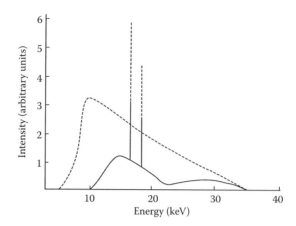

FIGURE 3.16
The effect of a 0.05 mm molybdenum filter on the spectrum from a tube operating at 35 kV constant potential using a molybdenum target (dotted line = no filter; solid line = with filter).

the curve for erbium. The curves would be very similar with both edges displaced slightly to the left. Thus in the energy range from 33 to 57.5 keV, the erbium filter would be transmitting X-rays freely, whilst absorbing higher energies, whereas the iodine contrast in the body would absorb heavily in the 33–57.5 keV range.

Because materials are relatively transparent to their own characteristic radiation, the effect of filtration can be rather dramatic when the filter is of the same material as the target anode producing the X-rays. This effect is exploited in mammography and Figure 3.16 shows how a 0.05 mm molybdenum filter changes the spectrum from a tube operating at 35 kV constant potential using a molybdenum target. Note that the output is not only near monochromatic, it contains a high proportion of characteristic radiation. This component of the spectrum does not drift with kVp, for example, with generator performance, and this helps to keep the soft tissue contrast constant. For further discussion see Section 9.2.2.

3.10 Conclusions

In this chapter both experimental and theoretical aspects of the interaction of X-rays and gamma rays with matter have been discussed. In attenuating materials the intensity of such beams decreases exponentially, provided they are monochromatic or near monochromatic, at a rate determined by the density and mean atomic number of the attenuator and the photon energy. In the diagnostic X-ray range the situation is more complex as the linear attenuation coefficient of the material falls as the beam is hardened when the lower energy (soft) X-ray photons are absorbed.

The two most important interaction processes are the photoelectric effect and the Compton effect. The former is primarily responsible for differences in attenuation (contrast) at low photon energies and its effect is very dependent on atomic number. However, the photoelectric effect decreases rapidly with increasing photon energy and when the Compton effect dominates, only differences in density cause any appreciable difference in attenuation.

TABLE 3.8

Optimum kVp for Different Procedures

Procedure	Optimum kVp (± 5)
Mammography	30
Extremities	55
Dental	65
Thoracic spine	70
Lateral lumber spine	80
Chest PA	85

For imaging, the photoelectric effect is the more desirable because it discriminates on the basis of tissue density and atomic number. The Compton effect is less desirable (though unavoidable) because it discriminates only on the basis of density. It also generates scattered lower energy radiation which is undesirable because it reduces contrast in the radiograph and constitutes a radiation hazard to staff.

The ideal kVp is one which allows sufficient intensity of X-rays to penetrate the patient to form the image with adequate variation between tissues to give the required contrast. Since tissue penetration increases with tube kV but contrast decreases (and scatter increases) choice of kVp for a particular imaging procedure is a compromise. A further complication is introduced when patient dose also has to be taken into account. Filtering the beam to remove the low energy photons hardens the beam.

Suggested optimum kVps for some standard radiographic procedures are shown in Table 3.8.

Note

1. For contrast studies a kVp should be chosen ideally so that the peak intensity in the X-ray photons is as close to the energy of the K shell absorption edge of the contrast material as possible. The requirement to penetrate the patient sometimes means this cannot be adhered to.

2. The choice of optimum kVp may have to be modified because of the energy response of the image receptor (see Section 5.4.1).

References

Johns H E and Cunningham J R. 1983 *The Physics of Radiology,* 4th ed. Thomas, Springfield.

McVey G. 2006 The effect of phantom type, beam quality, field size and field position on X-ray scattering simulated using Monte carlo techniques. *Br. J. Radiol.* 79: 130–141.

RCR Working Party. 2007 *Making the best use of Clinical Radiology Services* The Royal College of Radiologists, London.

Sutton D G and Williams J R. 2000 *Radiation Shielding for Diagnostic X-Rays* British Institute of Radiology.

Trout E D, Kelley J P and Cathey G A. 1952 The use of filters to control radiation exposure to the patient in diagnostic radiology. *Am. J. Roentgenol.* 67: 946–963.

Further Reading

Bushberg J T, Seibert A J A, Leidholt E M and Boone J M. 2002 *The essential physics of medical imaging 2nd ed.* Lippincott, Williams and Wilkins, Philadelphia, pp 31–60, Chapter 3.
Dowsett D J, Kenny P A and Johnson R E. 2006 *The physics of diagnostic radiology 2ⁿᵈ edition* Hodder Arnold, London, pp 113–142, Chapter 5.

Exercises

1. Explain the terms
   (a) Inverse square law
   (b) Linear attenuation coefficient
   (c) Half value thickness
   (d) Mass absorption coefficient
   Indicate the relationships between them, if any.

2. Describe the process of Compton scattering, explaining carefully how both attenuation and absorption of X-rays occur.

3. Describe the variation of the Compton attenuation coefficient and Compton absorption coefficient with scattering angle in the energy range 10–200 keV.

4. How does the process of Compton scattering of X-rays depend on the nature of the scattering material and upon X-ray energy? What is the significance of the process in radiographic imaging?

5. An X-ray beam loses energy by the processes of absorption and/or scattering. Discuss the principles involved at diagnostic X-ray energies and explain how the relative magnitude of the processes is modified by different types of tissue.

6. Explain why radiographic exposures are usually made with an X-ray tube voltage in the range 50–110 kVp.

7. If the mass attenuation coefficient of aluminium at 60 keV is 0.028 m^2kg^{-1} and its density is 2.7×10^3 kg m^{-3}, estimate the fraction of a monoenergetic incident beam transmitted by 2 cm of aluminium.

8. A parallel beam of monoenergetic X-rays impinges on a sheet of lead. What is the origin of any lower energy X-rays which emerge from the other side of the sheet travelling in the same direction as the incident beam?

9. What is meant by characteristic radiation? Describe briefly three situations in which characteristic radiation is produced.

10. How would a narrow beam of 100 kV X-rays be changed as it passed through a thin layer of material? What differences would there be if the layer were
    (a) 1 mm lead (Z = 82, $\rho = 1.1 \times 10^4$ kg m^{-3})
    (b) 1 mm aluminium (Z = 13, $\rho = 2.7 \times 10^3$ kg m^{-3})

11. Before the X-ray beam generated by electrons striking a tungsten target is used for radiodiagnosis it has to be modified. How is this done and why?

12. What factors determine whether a particular material is suitable as a filter for diagnostic radiology?

13. A narrow beam of X-rays from a diagnostic set is found experimentally to have a half value thickness of 2 mm of aluminium. What would happen to

    (a) The half value of thickness of the beam

    (b) The exposure rate

    if an additional filter of 1 mm aluminium were placed close to the X-ray source?

14. Discuss the advantages and disadvantages of using aluminium as the filter material in X-ray sets at 20, 80 and 110 kVp generating potentials.

15. Compare the output spectra produced by a tungsten target and a copper target operating at 60 kVp. What would be the effect on these spectra of using

    (a) An aluminium filter

    (b) A lead filter

	Atomic number	K shell (keV)	L shell (keV)
Al	13	1.6	-
Cu	29	9.0	1.1
W	74	69.5	11.0
Pb	82	88.0	15.0

16. The dose rate in air at a point in a narrow beam of X-rays is 0.3 Gy min^{-1}. Estimate, to the nearest whole number, how many half value thicknesses of lead are required to reduce the dose rate to 10^{-6} Gy min^{-1}. If $H_{\frac{1}{2}}$ at this energy is 0.2 mm, what is the required thickness of lead?

17. What is the 'lead equivalence' of a material?

18. Explain what you understand by the homogeneity of an X-ray beam and describe briefly how you would measure it.

19. What is meant by an inhomogeneous beam of X-rays and why does it not obey the law of exponential attenuation with increasing filtration?

4

Radiation Measurement

B Heaton and P P Dendy

SUMMARY

- Measurement of the intensity of an X-ray beam must be made in terms of measurable physical, chemical or biological changes the X-rays may cause.
- The importance of ionisation in air as the primary radiation standard is explained.
- Measuring instruments that depend on the principle of ionisation in air are described.
- The relationship between exposure in air and absorbed dose is explained.
- Radiation monitors which depend on other physical principles (semi-conductors and scintillation crystals) are described.
- Measurement of radiation spectra is described and the variation in sensitivity of some detectors with photon energy is explained.

CONTENTS

4.1 Introduction

Lord Kelvin (1824–1907), who is probably best remembered for the absolute thermo-dynamic scale of temperature, is reported to have stated on one occasion: 'Anything that cannot be expressed in numbers is valueless'. In view of the potentially harmful effect of X-rays it is particularly important that methods should be available to 'express in numbers' the 'strength' or intensity of X-ray beams.

With respect to measurement, three separate features of an X-ray beam must be identified. The first consideration is the flux of photons travelling through air from the anode towards the patient. The ionisation produced by this flux is a measure of the *radiation exposure*. If expressed per unit area per second it is the *intensity*. Of more fundamental importance as far as the biological risk is concerned is the *absorbed dose of radiation*. This is a measure of the amount of energy deposited as a result of ionisation processes. Finally, it may be important to know about the energy of the individual photons. Because of the mechanism of production, an X-ray beam will contain photons with a wide range of energies. A complete specification of the beam would require determination of the full spectral distribution as shown in Figure 2.2. This represents information about the *quality* of the X-ray beam.

Clearly the intensity of an X-ray beam must be measured in terms of observable physical, chemical or biological changes that the beam may cause, so it will be useful to review briefly relevant properties of X-rays. Two of them are sufficiently fundamental to be classified as *primary properties*—that is to say measurements can be made without reference to a standard beam.

1. *Heating effect.* X-rays are a form of energy which can be measured by direct conversion into heat. Unfortunately, the energy associated with X-ray beams used in diagnostic radiology is so low that the temperature rise can scarcely be measured (see Section 12.2.1).

2. *Ionisation.* In the diagnostic energy range X-rays cause ionisation by photoelectric and Compton interactions in any material through which they pass. Pair production is only important at higher energies. The number of ions produced in a fixed volume under standard conditions of temperature and pressure will be fixed.

   A number of other properties of X-rays can, and often are, used for dosimetry. In all these situations, however, it is necessary for the system to be calibrated by first measuring its response to beams of X-rays of known intensity so these are usually called *secondary properties*.

3. *Physical effects.* When X-rays interact with certain materials, visible light is emitted. The light may either be emitted immediately following the interaction

(fluorescence); after a time interval (phosphorescence); or, for some materials, only upon heating (thermoluminescence).

4. *Physico-chemical effects.* The action of X-rays on photographic film is well-known and widely used.

5. *Chemical changes.* X-rays have oxidising properties, so if a chemical such as ferrous sulphate is irradiated, the free ions that are produced oxidise some Fe^{++} to Fe^{+++}. This change can readily be detected by shining ultraviolet light through the solution. This light is absorbed by Fe^{+++} but not by Fe^{++}.

6. *Biochemical changes.* Enzymes rely for their action on the very precise shape associated with their secondary and tertiary structure. This is critically dependent on the exact distribution of electrons, so enzymes are readily inactivated if excess free electrons are introduced by ionising radiation.

7. *Biological changes.* X-rays can kill cells and bacteria, so, in theory at least, irradiation of a suspension of bacteria followed by an assay of survival could provide a form of biological dosimeter.

Unless specifically stated otherwise, in the remainder of this chapter references to X-rays apply equally to gamma rays of the same energy.

4.2 Ionisation in Air as the Primary Radiation Standard

There are a number of important prerequisites for the property chosen as the basis for radiation measurement.

1. It must be accurate and unequivocal, that is personal, subjective judgement must play no part.

2. It must be very sensitive to producing a large response for a small amount of radiation energy.

3. It must be reproducible.

4. The measurement should be independent of intensity, that is an intensity I for time t must give the same answer as an intensity $2I$ for time $t/2$. This is the *Law of Reciprocity.*

5. The method must apply equally well to very large and very small doses.

6. It must be reliable at all radiation energies.

7. The answer must convert readily into a value for the absorbed energy in biological tissues or 'absorbed dose' since this is the single most important reason for wishing to make radiation measurements.

None of the properties of ionising radiation satisfies all these requirements perfectly but ionisation in air comes closest and has been internationally accepted as the basis for radiation dosimetry. There are two good reasons for choosing the property of ionisation.

1. Ionisation is an extremely sensitive process in terms of energy deposition. Only about 34 eV is required to form an ion pair, so if a 100 keV photon is completely

absorbed, almost 3000 ion pairs will have been formed when all the secondary ionisation has taken place.

2. As shown in Chapter 12, the extreme sensitivity of biological tissues to radiation is directly related to the process of ionisation so it has the merit of relevance as well as sensitivity.

There are also good reasons for choosing to make measurements in air:

1. It is readily available.
2. Its composition is close to being universally constant.
3. More important, for medical applications, the mean atomic number of air ($Z = 7.6$) is very close to that of muscle/soft tissue ($Z = 7.4$). Thus, provided ionisation and the associated process of energy absorption is expressed per unit mass by using mass absorption coefficients rather than linear absorption coefficients, results in air will be closely similar to those in tissue.

The unit of *radiation exposure* (X) is defined as that amount of radiation which produces in air ions of either sign equal to 1 C (coulomb) kg^{-1}. Expressed in simple mathematical terms:

$$X = \frac{\Delta Q}{\Delta m}$$

where ΔQ is the sum of all the electrical charges on all the ions of one sign produced in air when all the electrons liberated in a volume of air whose mass is Δm are completely stopped in air. The last few words ('are completely stopped in air') are extremely important. They mean that if the electron generated by say a primary photoelectric interaction is sufficiently energetic to form further ionisations (normally it will be), all the associated ionisations must occur within the collection volume and all the electrons contribute to ΔQ.

The older, obsolescent unit of radiation exposure is the roentgen (R). One roentgen is that exposure to X-rays which will release one electrostatic unit of charge in one cubic centimetre of air at standard temperature and pressure (STP). Hence 1 R = 2.58×10^{-4} C kg^{-1}.

4.3 The Ionisation Chamber

Figure 4.1 shows a direct experimental interpretation of the definition of radiation exposure. The diaphragm, constructed of a heavy metal such as tungsten, defines an X-ray beam of accurately known cross-section *A*. This beam passes between a pair of parallel plates in an air-filled enclosure. The upper plate is maintained at a high potential relative to the lower and, in the electric field arising from the potential difference between the plates, all the ions of one sign produced in the region between the dashed lines move to the collecting electrode. This is generally referred to as a free air ionisation chamber.

Either the current flow (exposure rate) or the total charge (exposure) may be measured using the simplified electrical circuits shown in Figure 4.2a and b.

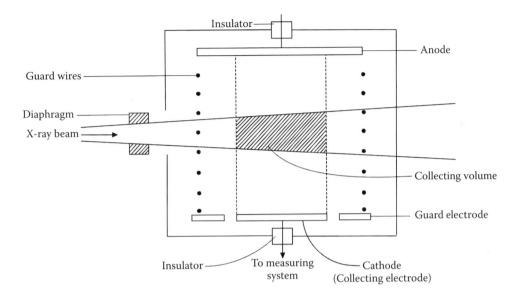

FIGURE 4.1
The free air ionisation chamber.

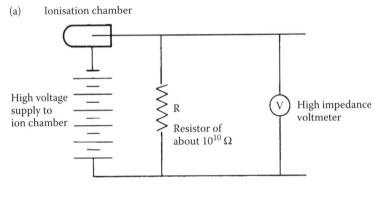

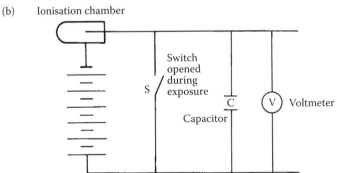

FIGURE 4.2
Simplified electrical circuits for measuring (a) current flow (exposure rate), (b) total charge (exposure).

Since 1 ml of air weighs 1.3×10^{-6} kg at STP, a chamber of capacity 100 ml contains 1.3×10^{-4} kg of air. A typical exposure rate might be 2.5 µC kg^{-1} h^{-1} (a dose rate of approximately 0.1 mGy h^{-1}, see Section 4.5) which corresponds to a current flow of $(2.5 \times 10^{-6}/3600) \times 1.3 \times 10^{-4}$ Cs^{-1} or about 10^{-13} A.

If $R = 10^{10} \Omega$, since $V = IR$ then $V = 1$ mV which is not too difficult to measure. However, the voltmeter must have an internal resistance of at least $10^{13} \Omega$ so that no current flows through it and this is quite difficult to achieve.

Since the free air ionisation chamber is a primary standard for radiation measurement, accuracy better than 1% (i.e. more precision than the figures quoted here) is required.

Although it is a simple instrument in principle, great care is required to achieve such precision and a number of corrections have to be applied to the raw data.

Insight

Corrections to the Ionisation Chamber Reading

Corrections must be made if the air in the chamber is not at STP. For air at pressure P and temperature T, the true reading R_T is related to the observed reading by

$$R_T = R_0 \left(\frac{P_0}{P}\right) \cdot \left(\frac{T}{T_0}\right)$$

where P_0, T_0 are STP values.

Rather than trying to memorise this equation, the reader is advised to refer to first principles. Ion pairs are created because X-ray photons interact with air molecules. If the air pressure increases above normal atmospheric pressure, the number of air molecules will increase, the number of interactions will increase, and the reading will be artificially high. Changes in temperature may be considered similarly.

The requirement for precision also creates design difficulties. For example, great care must be taken, using guard rings and guard wires (see Figure 4.1), to ensure that the electric field is always precisely normal to the plates. Otherwise, electrons from within the defined volume may miss the collecting plate or, conversely, may reach the collecting plates after being produced outside the defined volume.

Major difficulties arise as the X-ray photon energy increases, especially above about 300 keV, because of the ranges of the secondary electrons (see Table 1.4). Recall that all the secondary ionisation must occur within the air volume. If the collecting volume is increased, eventually it becomes impossible to maintain field uniformity.

Thus the free air ionisation chamber is very sensitive in the sense that one ion pair is created for the deposition of a very small amount of energy. However, it is insensitive when compared to solid detectors that work on the ionisation principle because air is a poor stopping material for X-rays. It is also bulky and operates over only a limited range of X-ray energies. However, it is a primary measuring device and all other devices must be calibrated against it.

Ionisation chambers are used to measure the dose rates and doses produced during the routine testing of diagnostic imaging systems. To accommodate the wide variation in field size and intensity, field intensity chambers of several different sizes have to be used.

4.4 The Geiger–Müller Counter

If, when using the equipment shown in Figure 4.1, the X-ray beam intensity was fixed but the potential difference between the plates were gradually increased from zero, the current flowing from the collector plate would vary as shown in Figure 4.3. Initially, all ion pairs recombine and no current is registered. As the potential difference increases (region AB) more and more electrons are drawn to the collector, until, at the first plateau BC, all the ion pairs are being collected. This is the region in which the ionisation chamber operates and its potential difference must be in the range BC.

Beyond C, the current increases again. This is because secondary electrons gain energy from the electric field between the plates and eventually acquire enough energy to cause further ionisations (see Figure 4.4). *Proportional counters* operate in the region of CD. They have the advantage of increased sensitivity, the extra energy having been drawn from the electric field, and, as the name implies, the strength of signal is still proportional to the amount of primary and secondary ionisation. Hence, proportional counters can be used to

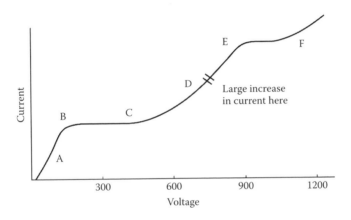

FIGURE 4.3
Variation in current appearing across capacitor plates with applied potential difference for a fixed X-ray beam intensity. (AB) Loss of ions by recombination. (BC) Ionisation plateau. (CD) Proportional counting. (EF) Geiger–Müller region. Beyond F—continuous discharge. The voltage axis shows typical values only.

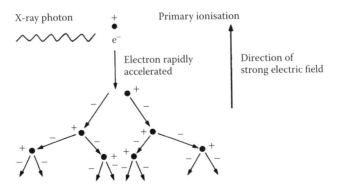

FIGURE 4.4
Amplification of ionisation by the electric field.

measure radiation exposure. However, very precise voltage stabilisation is required since the amplification factor is changing rapidly with small voltage changes. Due to this, very few portable radiation-measuring instruments which are required to measure precisely are designed to work in this region.

Beyond D the amplification increases rapidly until the so-called Geiger–Müller (GM) plateau is reached at EF. Beyond F there is continuous discharge.

4.4.1 The Geiger–Müller Tube

Essential features of a GM tube are shown in Figure 4.5 and the most important details of its design and operation are as follows:

1. In Figure 4.3, the GM plateau was attained by applying a high voltage between parallel plates. However, it is the electric field $E = V/d$, where d is the distance between the plates, that accelerates electrons. High fields can be achieved more readily using a wire anode since near the wire E varies as $1/r$, where r is the radius of the wire. Thus the electric field is very high close to a wire anode even for a working voltage of only 300–400 V. When working on the GM plateau EF the count rate changes only slowly with applied voltage so very precise voltage stabilisation is not necessary.

2. The primary electrons are accelerated to produce an avalanche as in the proportional counter, but in the avalanche discharge excited atoms as well as ions are formed. They lose this excitation energy by emitting X-ray and ultraviolet photons which liberate outer electrons from other gas atoms creating further ion pairs by a process of photoionisation. As these events may occur at some distance from the initial avalanche, the discharge is spread over the whole of the wire. Because of the high electric fields in the GM tube, the positive ions reach the cathode in sufficient

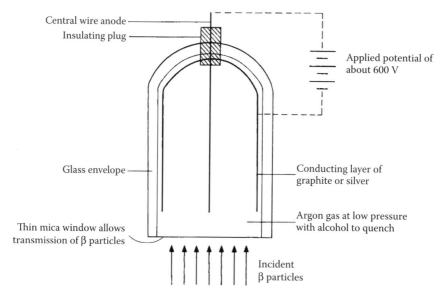

FIGURE 4.5
Essential features of an end-window Geiger–Müller tube suitable for detecting β particles.

numbers and with sufficient energies to eject electrons. These electrons initiate other pulses which recycle in the counter producing a continuous discharge.

3. The continuous discharge must be stopped before another pulse can be detected. This is done by adding a little alcohol or bromine to the counting gas which is either helium or argon at reduced pressure. The alcohol or bromine molecules 'quench' the discharge because their ionisation potentials are substantially less than those of the counting gas. During collisions between the counting gas ions and the quenching gas molecules, the ionisation is transferred to the latter. When these reach the cathode, they are neutralised by electrons extracted by field emission from the cathode. The electron energy is used up in dissociating the molecule instead of causing further ionisation. The alcohol or bromine also has a small effect in quenching some of the ultraviolet photons.

4. The discharge is also quenched because a space charge of positive ions develops round the anode, thereby reducing the force on the electrons.

5. Finally, quenching can be achieved by reducing the external anode voltage using an external resistor and this is triggered by the early part of the discharge.

Once discharge has been initiated, and during the time it is being quenched, any further primary ionisation will not be recorded as a separate count. The instrument is effectively 'dead' until the externally applied voltage is restored to its full value, typically after about 300 μs. This is known as the *dead time*.

Thus the true count is always higher than the real count. The difference is minimal at 10 counts per second but at 1000 counts per second the monitor is dead for $1000 \times 300 \times 10^{-6} = 0.3$ s in every second and losses become appreciable.

4.4.2 Comparison of Ionisation Chambers and Geiger–Müller Counters

Both instruments have important but well defined roles in radiological monitoring so their strengths and weaknesses must be clearly understood.

4.4.2.1 Type of Radiation

Both respond to X- and gamma rays and to fast β particles. By using a thin window the response of the GM tube can be extended to low energy β particles (and even α particles), but not the very low energy β particles from H-3 (18 keV max).

4.4.2.2 Sensitivity

Because of internal amplification, the GM tube is much more sensitive than the ionisation chamber and may be used to *detect* low levels of contamination (but see Section 4.8).

4.4.2.3 Nature of Reading

The ionisation chamber is designed to collect all primary and secondary radiation and hence to give a reading of exposure or exposure rate. With the GM tube there is no proportional relationship between the count rate and the number of primary and secondary ionisations so it is not a radiation *monitor*.

4.4.2.4 Size

The ionisation chamber must be big enough to collect all secondary electron ionisations. Since the GM tube does not have the property of proportionality, there is no point in making it large and it can be much more compact.

4.4.2.5 Robustness and Simplicity

These requirements generally favour the GM tube.

   In conclusion a GM tube is an excellent *detector* of radiation, but it must not be used to measure radiation exposures for which an appropriate *monitor* is required.

Insight

Compensated GM Tubes

A 'Compensated Geiger' can be used as a radiation monitor. This instrument uses a specially designed Geiger tube called an *energy compensated Geiger tube*, in which the pulse rate can be related to the exposure. Dose rate meters based on an energy compensated Geiger tube are, typically, only suitable for measuring photons with an energy between 60 keV and 1.25 MeV. Below 60 keV the sensitivity of the Geiger tube falls quite rapidly and the instrument under records. Therefore it cannot be used for making measurements in diagnostic X-ray departments, even when the primary photon dose is of high energy because the radiation field also contains scattered radiation at a much lower energy.

4.5 Relationship between Exposure and Absorbed Dose

The second aspect of radiation measurement is to obtain the absorbed dose or amount of energy deposited in matter. In respect of the biological damage caused by ionising radiation, this is more relevant than radiation exposure.

   The unit of absorbed dose is the gray (Gy), where $1\ \text{Gy} = 1\ \text{J kg}^{-1}$.

   This has replaced the older unit, called the *rad*. One rad is $100\ \text{erg g}^{-1}$. 1 Joule is equal to 10^7 ergs and thus $1\ \text{Gy} = 100\ \text{rad}$.

   Whereas radiation exposure, by definition, refers to ionisation in air, the dose, or energy absorbed from the radiation, may be expressed in any material. Calculation of the dose in, say, soft tissue, when the radiation exposure in air is known, may be treated as a two stage problem as follows:

   1. Conversion of exposure in air E to dose in air D_A by the equation

$$D_A\,(\text{Gy}) = 34\ E\ (\text{C kg}^{-1})$$

   2. Conversion of dose in air D_A to dose in tissue D_T by the equation

$$\left(\frac{D_T}{D_A}\right) = \frac{(\mu_a / \rho)_T}{(\mu_a / \rho)_A}$$

where $(\mu_a/\rho)_T$ and $(\mu_a/\rho)_A$ are the mass absorption coefficients of tissue and air, respectively.

For derivations of the equations see the 'insight'.

When treated in this way, both parts of the calculation become fairly easy. Note that in future absorbed dose will be simplified to dose unless there is possible confusion.

Insight

Conversion of Exposure in Air to Dose in Air

A term that is being used increasingly in radiation dosimetry is KERMA. This stands for kinetic energy released per unit mass and must specify the material concerned. Note that KERMA places the emphasis on removal of energy from the beam of indirectly ionising particles (X- or gamma photons) to create secondary electrons. Absorbed dose relates to where those electrons deposit their energy in the medium.

There are two reasons why KERMA in air (K_A) may differ from dose in air (D_A). First, some secondary electron energy may be radiated as bremsstrahlung. Second, the point of energy deposition in the medium is not the same as the point of removal of energy from the beam because of the range of secondary electrons. However, at diagnostic energies bremsstrahlung is negligible and the ranges of secondary electrons are so short that $K_A = D_A$ to a very good approximation.

Now the number of ion pairs generated in each kilogram of air multiplied by the energy required to form one ion pair (W) is equal to the energy removed from the beam. But the first term is the definition of radiation exposure, say E, and the third term is the definition of KERMA in air K_A. Thus

$$E(\text{C kg}^{-1}) \times W(\text{J C}^{-1}) = K_A(\text{J kg}^{-1})$$

Or, since $K_A = D_A$, expressing dose at the subject

$$D_A(\text{J kg}^{-1}) = E(\text{C kg}^{-1}) \times W(\text{J C}^{-1})$$

The energy to form one ion pair W is close to 34 J C^{-1} (34 electron volts per ion pair) for all types of radiation of interest to radiologists and, coincidentally, over a wide range of materials of biological importance. By definition 1 Gy = 1 J kg^{-1}. Thus

$$D_A \text{ (Gy)} = 34\ E(\text{C kg}^{-1})$$

Many older textbooks use the unit roentgen for exposure in air and rad for dose. Since 1 R = 2.58 10^{-4} C kg^{-1}, and 1 Gy = 100 rads, D_A (rad) = 100 x 2.58 $\times$ 10^{-4} $\times$ 34 E (roentgen). Hence D_A (rad) = 0.88 E (roentgen).

In this book, dose in air will be used in preference to exposure. 'Skin dose' will be used to describe the dose in air at the surface of the patient.

Conversion of Dose in Air to Dose in Tissue

To convert a dose in air to dose in any other material, recall that for a given incident flux of photons, the energy absorbed per unit mass depends only on the mass absorption coefficient of the medium. Hence

$$\left(\frac{D_T}{D_A}\right) = \frac{(\mu_a/\rho)_T}{(\mu_a/\rho)_A}$$

Note the use of a subscript 'a' to distinguish the mass *absorption* coefficient from the mass *attenuation* coefficient.

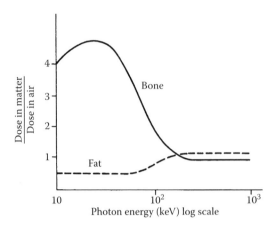

FIGURE 4.6
The ratio (dose in matter/dose in air) plotted as a function of radiation energy for bone and fat.

It follows that only a knowledge of the relative values of μ_a/ρ is required to convert a known dose in air to the corresponding dose in any other material.

The ratio D_T/D_A is plotted as a function of different radiation energies for bone and fat in Figure 4.6. It is left as an exercise to the reader to justify the shapes of these curves from a knowledge of

1. The mean atomic number for each material
2. The relative importance of the Compton and photoelectric effects at each photon energy

Insight

Other Quantities Related to Dose Measurement

In Section 4.1 the terms flux of photons and intensity were introduced without mathematical treatment. Whereas many readers will find the explanations of these terms given there sufficient, readers who wish to study dosimetry more deeply will find it useful to know the mathematical relationships between these and some other related terms.

Consider a situation where there are N photons passing through an area A m^2 in a time t seconds.

Photon fluence, Φ = the number of photons passing through a unit area (N/A photons m^{-2})

Photon flux, ϕ = the number of photons passing through unit area in unit time ($N/A{\cdot}t$ photons m^{-2} s^{-1})

Energy fluence, Ψ = the energy passing through unit area

For a monoenergetic beam $\Psi = \Phi{\cdot}\varepsilon = N{\cdot}\varepsilon/A$ Jm^{-2} where ε is the energy of each photon. (See Section 1.11 for conversion of eV to Joules. Note this is an approximate figure because the charge on an electron has to be measured experimentally.)

Energy flux, or intensity, ψ = the energy passing through unit area in unit time

For a monoenergetic beam $\psi = \varphi{\cdot}\varepsilon = N{\cdot}\varepsilon/A{\cdot}t$ J m^{-2} s^{-1}.

For a beam of mixed energy $\psi = \sum \varphi_i \varepsilon_i$, where i is the ith component of the energy spectrum.

4.6 Practical Radiation Monitors

4.6.1 Secondary Ionisation Chambers

In Section 4.3, one of the problems identified for the free air ionisation chamber was its large volume. Fortunately, there is a technique which, to an acceptable level of accuracy for laboratory instruments, eliminates this problem.

Imagine a large volume of ethylene gas with dimensions much bigger than the range of secondary electrons. Now compress the gas to solid polyethylene, leaving only a small volume of gas at the centre (Figure 4.7). The radiation exposure will be determined by the density of electrons within the gas and if the gas volume is small, the number of secondary electrons either being created in the gas or coming to the end of their range there will be negligible compared to the electron density in the solid.

However, the electron density in the solid is the same as it would be at the centre of the large gas volume. This is because the electron flux across any plane is the product of rate of production per unit thickness and range. The rate of production depends on the number of interactions and is higher in the solid than in the gas by ρ_{solid}/ρ_{gas} (the ratio of the densities). But once the electrons are formed they have a fixed energy and lose that energy more rapidly in the solid due to collisions. Thus the ratio *range of electrons in solid/range of electrons in gas* is in the inverse ratio ρ_{gas}/ρ_{solid}.

Hence the product (rate of production of secondary electrons × range of secondary electrons) is independent of density as shown schematically in Figure 4.8.

An alternative way to view this situation is that, provided the atomic composition is the same, the gas in the cavity does not know if it is surrounded by a big volume of gas or by a much smaller volume of solid resulting from compression of the gas.

The result is precise for polyethylene and ethylene gas because the materials differ only in density. No solid material is exactly like air in terms of its interaction with X-ray photons by photoelectric and Compton processes at all photon energies but good approximations to *air-equivalent walls* have been constructed and a correction can be made for the discrepancy.

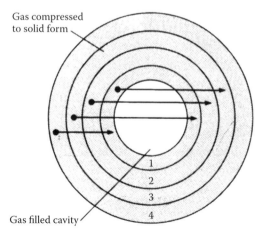

FIGURE 4.7
Secondary electron flux in a gas-filled cavity. Consider the solid as a series of layers starting at the edge of the cavity. Layers 1, 2 and 3 contribute to the ionisation in the cavity but layer 4 does not.

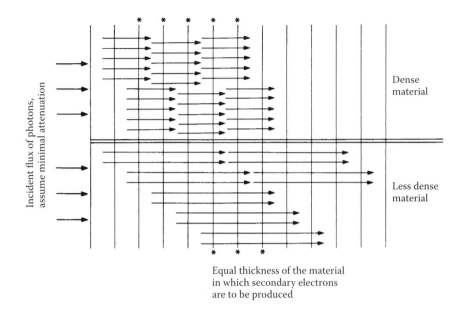

Equal thickness of the material
in which secondary electrons
are to be produced

FIGURE 4.8
A schematic demonstration that the flux of secondary electrons at equilibrium is independent of the density of the stopping material. The upper material has 2.5× the density of the lower. It therefore produces 2.5 times more electrons per slice. However, the range of these electrons is reduced by the same factor. When equilibrium is established, shown by a * for each material, 10 secondary electrons are crossing each vertical slice (the electron density) in each case. Note that equilibrium establishes quicker in the more dense material.

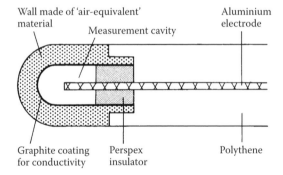

FIGURE 4.9
A simple, compact secondary ionisation chamber that makes use of the 'air equivalent wall' principle.

A simple, compact, secondary instrument, suitable for exposure rate measurements around an X-ray set, is shown in Figure 4.9. The dimensions now only require that the wall thickness should be greater than the range of secondary electrons in the solid medium. Note that a correction must also be made for attenuation of the primary beam in the wall surrounding the measurement cavity.

A modern detector that uses the principle of ionisation in gas is the high pressure xenon gas chamber. Xenon is chosen because it is an inert gas and its high atomic number (Z = 54) ensures a large cross-section for photoelectric interactions. It is sometimes mixed with krypton (Z = 82). The pressure is increased to about 25 atmospheres to improve sensitivity. Although the latter is still poor by comparison with solid detectors (see Sections 4.7 and

4.8), it is possible to pack a large number of xenon gas detectors of very uniform sensitivity into a small space. The use of Xe/Kr high pressure gas detectors in computed tomography is discussed in Section 8.4.1.

4.6.2 Dose Area Product Meters

Dose area product is defined as the absorbed dose to air, averaged over the area of the X-ray beam in a plane perpendicular to the beam axis, multiplied by the area of the beam in the same plane. It is usually measured in Gy cm^2 and radiation back scattered from the patient is excluded.

Provided that the cross-sectional area of the beam lies completely within the detector, it may be shown by simple application of the inverse square law (see Section 1.12) that the reading will not vary with the distance from the tube focus.

Thus the dose area product can be measured at any point between the diaphragm housing the X-ray tube and the patient—but not so close to the patient that there is significant back scattered radiation. Dose area product meters consist of flat, large area parallel plate ionisation chambers connected to suitable electrometers which respond to the total charge collected over the whole area of the chamber. The meter is mounted close to the tube focus where the area of the X-ray beam is relatively small and dose rates are high. It is normally mounted on the diaphragm housing where it does not interfere with the examination and is usually transparent so that when fitted to an overcouch X-ray tube the light beam diaphragm device can still be used. The use of dose area product meters to estimate patient doses is considered in Section 13.2.2.

4.6.3 Pocket Exposure Meters for Personnel Monitoring

Although the secondary ionisation chamber is much more compact than a free air ionisation chamber, it is still too large to be readily portable. However, an individual working in an unknown radiation field clearly may need to have an immediate reading of their accumulated dose or instantaneous dose rate.

An early device for personal monitoring, based on the ionisation principle, was the 'fountain pen' dosimeter. This used the principle of the gold leaf electroscope. When the electroscope was charged the leaves diverged because of electrostatic repulsion. If the gas around the leaves became ionised by X-rays, the instrument was discharged and the leaves collapsed. (These are still available but are now not used much in hospitals, only in industrial situations.)

More widely used nowadays as a practical ionisation instrument is the portable radiation monitor or 'bleeper'. As its name implies, it is light enough to carry around in the pocket and gives both an audible warning of radiation dose rate and displays the dose received. The bleep sounds every 15–30 min on background and the bleep rate increases with dose rate, becoming continuous in high radiation fields. The instrument is quite sensitive, registering doses as low as 1 µGy X-rays and typically giving approximately one bleep every 20 s at 10 µGy h^{-1}.

Note, however

1. Although this instrument appears to be an ionisation chamber since it gives a direct reading of absorbed dose, it works on a modified GM principle.

2. The energy range for this instrument is from 45 keV up to the megavoltage range so it may be unsuitable for use at the lowest diagnostic energies. This poor sensitivity at low energies is a feature of 'a compensated Geiger' (see Section 4.4.2).

4.7 Semi-conductor Detectors

4.7.1 Band Structure of Solids

To understand how semi-conductor detectors work, it is necessary to consider briefly the electron levels in a typical solid. The simple model of discrete energy levels introduced in Chapter 1 and used to explain features of X-ray spectra in Chapter 2 is strictly only true for an isolated atom. In a solid, where the atoms are close together, it applies reasonably well for the innermost K and L shells but for the outer shells the close proximity of other electrons permits each electron to occupy a range or 'band' of energies (see Figure 4.10).

The highest energy level is the conduction band. Here the electrons have sufficient energy to move freely through the crystalline lattice and in particular to conduct electricity. Next is the forbidden band. In pure materials there are no permitted energy levels here and electrons cannot exist in this region. The next highest energy band is the valence band which contains the valence electrons. Depending on the element, there are further bands of even lower energy. These bands carry their full complement of electrons, with forbidden zones and filled bands alternating like layers in a sandwich, but these are of no interest here.

With this simple model, three types of material may be identified. If the forbidden band between the valence and conduction bands is wide there are no electrons in the conduction band and the material is an insulator (point A in Figure 4.10). If the two bands overlap the material is a conductor because electrons can move freely from one atom to the next as the conduction band is continuous between atoms (point B). When the forbidden band is narrow (point C) materials can be made to change from non-conducting to conducting under specific voltage conditions and are termed *semi-conductors*.

However, this is the idealised pattern of a pure crystal. In practice real crystals always contain imperfections and impurities which manifest themselves in various ways. For example, they can cause a small excess of electrons or a deficiency of electrons (holes) in the forbidden band. This model is used to explain the mechanism of action of semi-conductor detectors in the following section. The impurities can also create additional levels or electron traps in the forbidden band. These are the origins of scintillation processes discussed in Section 5.3.

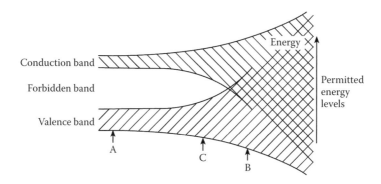

FIGURE 4.10

The band structure of energy levels found in solids. A, B, C show, schematically, the permitted energy levels in insulators, conductors and semi-conductors, respectively.

Impurities are often added deliberately and the manufacture of materials with the requisite properties depends on the production of very pure crystals to which impurities are added under carefully controlled conditions.

4.7.2 Mode of Operation

In a semi-conductor the forbidden band of energy levels is very narrow and therefore only small quantities of energy, sometimes as little as 1–1.5 eV, are required to raise an electron from the filled valence band into the conduction band. An X- or gamma ray photon (with an energy in radiology of thousands of eV) can, therefore, raise thousands of electrons into the conduction band. In these materials electrons can even acquire sufficient energy as thermal energy to make the transition. Semi-conductors are called *n-type* or *p-type* depending on how the current is carried within them. In a n-type the charge is carried by electrons in the conduction band. In a p-type by 'holes' or electron vacancies (which can be considered to have a unit positive charge) in the valence band.

Semi-conductor materials encountered in radiology are silicon (Si), germanium (Ge) and gallium arsenide (GaAs) of which silicon is the most common. Although, in theory, a very pure semi-conductor could be used as a detector, in practice they are generally 'doped' to produce the n- or p-type semi-conductor. Very small quantities of an impurity, say a few parts per million, are deliberately added to the pure silicon. An n-type material would be produced by adding antimony or arsenic which have five valence electrons to the silicon which has four valence electrons. Free electrons will therefore be available. A p-type material would be produced by adding gallium with only three valence electrons to the silicon. There are now 'holes' where spare electrons may reside. If a p-type material and an n-type material are joined together then electrons from the n-type migrate into the p-type and holes from the p-type migrate into the n-type. A region (called the *depletion layer*) is created between the two where, following diffusion, the excess holes are filled by the excess electrons and the electron imbalance has established a potential difference that is sufficient to prevent further flow. The depletion layer is normally of the order of a few tens of micrometres thick.

This is normally called a *p-n junction diode* and has very different electrical properties according to how a voltage supply is connected across it. If connected as in Figure 4.11a

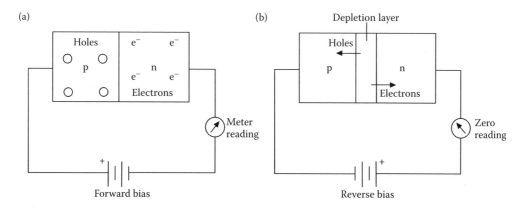

FIGURE 4.11
(a) Voltage applied to p-n diode in forward bias configuration. (b) Voltage applied to p-n diode in reverse bias configuration.

with the positive polarity to the p-type and the negative to the n-type then the device is said to be forward biased and a current will flow. If the polarity is the other way round as in Figure 4.11b then the device is said to be reverse biased and the depletion zone increases in depth as electrons and holes are drawn away from the junction until the internal potential across it is equal and opposite to the applied potential.

The reason this is called a *diode* is that a plot of current flowing against the voltage across the junction has the form of Figure 4.12 similar to a diode. When reverse biased no current flows but when forward biased the current rises very rapidly and exponentially with applied voltage.

If an X- or gamma ray beam interacts with the depletion layer then the electrons produced are attracted to the anode and are proportional to the amount of charge released (i.e. the radiation intensity). Although the depletion layer is very thin (typically 10–30 μm) sensitivity, that is the size of the current, is much higher than might be expected because the detecting medium is a solid (compared to a gas) and the yield of electrons for each interacting X-ray photon (the conversion efficiency) is very high.

The depletion layer provides an excellent radiation detector, behaving very much like a parallel plate ionisation chamber because if any electrons are generated as a result of ionising interactions, they can migrate to the anode and be registered as a current. The thickness of the depletion layer is determined by the magnitude of *V*.

In practice only very thin silicon-based detectors, typically about 200 μm, can be constructed to this design. These detectors also respond very well to visible light and near infra red (Figure 4.13) so they can be used as silicon photodiodes in conjunction with a scintillation crystal (see Section 4.8).

For better detection of X- and gamma radiation, thicker crystals of germanium with lithium diffused into them were originally used to obtain X-ray and particularly gamma ray spectra. They had adequate efficiency but, unlike silicon-based detectors which can operate at room temperature, they must be cooled to liquid nitrogen temperature (–190°C) before they can be used and during the whole of their working life. Very high purity germanium detectors (only about 10^9–10^{11} electrical impurities per cubic centimetre) which only need

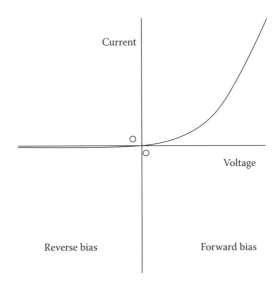

FIGURE 4.12
Variation in current flowing through a diode with voltage across the junction.

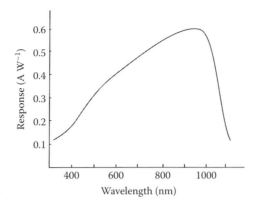

FIGURE 4.13
Spectral response of a silicon photodiode.

to be cooled during operation to reduce noise, are now available and have replaced lithium drifted detectors for most purposes requiring spectral analysis.

Solid-state detectors dispense with the need for relatively bulky photomultiplier tubes (PMTs) (see Section 4.8) and the requirement for stabilised high voltage supplies. As previously intimated, they have a high ionisation yield so the energies of photoelectrons generated by X- or gamma ray absorption can be measured with very high precision.

Insight

More on Sensitivity of Solid-State Conductors

Sensitivity increases with the stopping power of the semi-conductor. Germanium with a Z of 32 is better than silicon with a Z of 14. Volume for volume a silicon-based detector is about 18,000 times more sensitive than an air-filled ionisation chamber. This is partly due to the density as identified above and partly because only 2.5–3 eV are required to release an electron compared to 34 eV in a gas. This produces many more charge carrying pairs. Silicon-based detectors can operate at room temperature but germanium detectors require to be cooled down to the temperature of liquid nitrogen before they can be used.

The production of many more charge carrying pairs means that the statistics of counting are much better. The size of the pulse is proportional to the energy deposited and because the rate of clearance of the charge produced is very rapid, that is the pulse has a short rise time, these solid-state detectors can be used to count the number of photons in high flux radiation fields. This allows high purity germanium detectors to be used for the spectral analysis of X- and gamma photon fields.

4.7.3 Uses of the Silicon Diode

Silicon diode detectors are used extensively in equipment for routine performance testing of diagnostic X-ray systems. By using various filter packs it is possible for the same diode detectors to measure the output kV of the different X-ray units used in diagnostic departments. The same device can also measure the time of an exposure.

They also have an increasingly important role as personal monitors. They have now been developed to the point where they provide both a direct and immediate reading of the dose rate, or cumulative dose and are sufficiently reliable to provide a permanent dose record. The direct reading allows the worker to be aware of their radiation environment and to apply the ALARA principle (Section 14.2.1). The recorded dose over a period of time satisfies legal requirements.

Typical performance characteristics include a range from 1 μGy–16 Gy X-rays, a linear response with dose rate from 5 μGy h^{-1} to 3 Gy h^{-1}, calibration to better than 5%, rapid response time (1 μs), and stability over a wide range of temperature (–20°C to +80°C) and humidity. Since the mean atomic number of silicon is very different from that of air, some variation in detector sensitivity with photon energy would be expected, especially in the photoelectric region (see Section 4.10). However, adequate energy compensation has been applied to give a uniform response (±15%) from 20 keV up to megavoltage energies. They can be affected by non-ionising electromagnetic fields but these problems have been largely overcome. For further information on electronic personal dosimeters see Section 14.7, where traditional methods of personal dosimetry (film and thermoluminescent dosimetry) are discussed.

Insight

CMOS Detectors

The microelectronics which drives the advanced and robust imaging capabilities of camera phones is now an emerging technology for both radiation measurement in radiology and for being trialled and used increasingly in medical X-ray imaging in *single photon emission computed tomography* (SPECT), *positron emission tomography* (PET) and gamma cameras. A detailed description is outwith the scope of this book but an outline is included here as these detectors are likely to become used widely over the next few years.

Complementary metal oxide technology (CMOS) allows the integration of the read out electronics and the radiation sensor onto the same piece of material. CMOS imagers include an array of photo-sensitive diodes which can be sensitive to light or to radiation. Each pixel has one diode. These pixels are active pixels in that each has its own individual amplifier (unlike charge coupled devices, CCDs). In addition, each pixel in a CMOS imager can be read directly on an x-y coordinate system, rather than by the progressive transfer of charge from one detector element to the next—the 'bucket-brigade' process of a CCD (see Section 5.13.4.2). This means that while a CCD pixel always transfers a charge, a CMOS pixel always *detects* a photon directly, converts it to a voltage and transfers the information directly to the output. This fundamental difference in how information is read out of the imager, coupled with the manufacturing process, gives CMOS imagers several advantages over CCDs.

They are physically more robust with high spatial resolution. They are very fast and are capable of detecting high intensity radiation fields without saturation. Because each pixel is read individually there is the potential to carry out pulse size analysis on the photons detected opening up the opportunity for new X-ray imaging contrasts based on tissue spectral properties. They have an inherently low noise and are very radiation tolerant which should give a long useful lifetime. The reading of each pixel individually should mean that the effect of dead or bad pixels is very much reduced compared to other systems where one faulty pixel can have a big effect on obtaining the information from many others. There should also be no latent image 'left over' after readout. At the time of writing the size of these devices is restricted, however, and will have to be increased before they can replace other imaging devices.

4.8 Scintillation Detectors and Photomultiplier Tubes

One of the consequences of the additional energy levels created by impurities in semiconductors is that many of them emit visible light when exposed to X-rays. These light flashes are sometimes referred to as 'scintillations' but the more correct technical term is fluorescence. Fluorescence plays an important role in image formation and, together with the closely related processes of phosphorescence and thermoluminescence, will be discussed in detail in Chapter 5. For the present discussion it is sufficient to record that the light emitted by fluorescence when certain materials are bombarded by X-rays can be used as a basis for radiation monitoring.

One material widely used for such detectors is sodium iodide to which about 0.1% by weight of thallium has been added—NaI (Tl). The energy levels of the traps generated by thallium in the NaI lattice are about 3 eV above the band of valency electrons so the emitted photon is in the visible range. The detector is carefully designed and manufactured to optimise light yield. Note that whereas the *number* of photons emitted is a function of the energy imparted by the X-ray or gamma ray interaction, the *energy* or wavelength of the photons depends only on the positions of the energy levels in the scintillation crystal.

When a scintillation crystal is used as a monitor, its advantages over the detectors discussed so far are as follows:

1. Since it is a high density solid, its efficiency, especially for stopping higher energy gamma photons, is greatly increased. A 2.5 cm thick NaI (Tl) crystal is almost 100% efficient in the diagnostic X-ray energy range. Contamination monitors that must be capable of detecting of the order of 30 counts per second from an area of 1000 mm^2 invariably contain scintillation crystals.

2. It has a rapid response time, in contrast to an ionisation chamber which responds only slowly owing to the need to build up charge on the electrodes (see Figure 4.1).

3. Different scintillation crystals can be constructed that are particularly sensitive to low energy X-rays or even to neutrons. Beta particles and alpha particles can be detected using plastic phosphors.

NaI (Tl) detectors are used extensively in nuclear medicine and the properties that render them particularly appropriate for *in vivo* imaging will be discussed in Chapter 10. Alternative scintillation detectors are caesium iodide doped with thallium and bismuth germanate. Like NaI (Tl), the latter has a high detection efficiency, and is preferable at high counting rates (e.g. for CT) because it has little 'after glow'—persistence of the light associated with the scintillation process. Bismuth germanate detectors also exhibit a good dynamic range and long-term stability.

The light signal produced by a scintillation crystal is too small to be used until it has been amplified and this is achieved by using either a PMT or a photodiode (see Section 4.7.3).

The main features of the PMT coupled to a scintillation crystal (Figure 4.14) are follows:

1. *An evacuated glass envelope, one end of which has an optically flat surface.* Since photon losses must be minimised, the scintillation crystal must either be placed in contact with this surface or if it is impracticable, must be optically coupled using a piece of optically transparent plastic—frequently referred to as a 'light guide' or 'light pipe'.

2. *A layer of photoelectric material such as caesium-antimony.* The characteristic of such materials is that their work function, that is, the energy required to release an electron, is very low. Thus electrons are emitted when visible or ultraviolet photons fall on the photocathode, although the efficiency is low with only one electron emitted for every 10 incident photons.

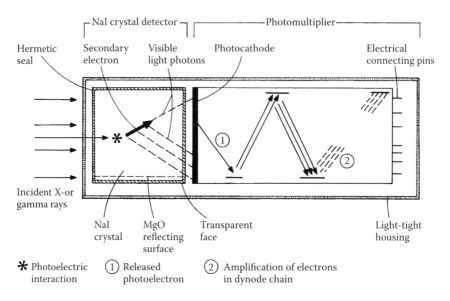

FIGURE 4.14
Main features of a PM tube coupled to a scintillation crystal for radiation detection.

3. *An electrode system to provide further amplification.* This system consists of a set of plates, each maintained at a potential difference of about 100 V with respect to its neighbours and coated with a metal alloy, say of magnesium-silver, designed to release several electrons for every one incident on it. Each plate is known as a dynode and there may be 12 dynodes or more in all, so the potential difference across the PMT will be in the region of at least 1200 V. If each dynode releases four electrons for each incident electron and there are 12 dynodes, the amplification in the PM tube is 4^{12} or about 10^7. Furthermore, this figure will be constant to within about 1% provided the voltage can be stabilised to 0.1%.

Note that during the complete detection process in the crystal and PMT, the signal twice takes the form of photons, once as X-ray photons and once as visible light photons, and twice takes the form of electrons. Since a PMT is an extremely sensitive light detector, great care must be taken to ensure that no stray light enters the system.

4.9 Spectral Distribution of Radiation

Thus far nothing has been said about the third factor in a complete specification of a beam of ionising radiation, namely the energy spectrum of the photons. Under appropriate conditions, a scintillation crystal used in conjunction with a PMT may be used to give such information.

If the crystal is fairly thick, and made of high density material, preferably of high atomic number, most X- or gamma rays that interact with it will be completely stopped within it and each photon will give up all its energy to a single photoelectron. Note that for this discussion it is unimportant whether there is a single photoelectric interaction or a combination of Compton and photoelectric interactions (contrast with the discussion in Section 10.2).

Since the number of visible light photons is proportional to the photoelectron energy, and the amplification by the PMT is constant, the strength of the final signal will be proportional to the energy of the interacting X- or gamma ray photon. A *pulse height analyser* may now be used to determine the proportion of signals in each predetermined range of strengths and if the pulse height analyser is calibrated against a monoenergetic beam of gamma rays of known energy, the results may be converted into a spectrum of incident photon energies (Figure 4.15).

Unfortunately, because of statistical problems, this method of determining the spectral distribution of a beam of X- or gamma rays is not as precise as one would wish. One limitation of the scintillation crystal and PMT combination as a detector of ionising radiation is that the number of electrons entering the PMT per primary X- or gamma ray photon interaction is rather small. There are two reasons:

1. About 30 eV of energy must be dissipated in the crystal for the production of each visible or ultraviolet photon.

2. Even assuming no loss of these photons, only about one photoelectron is produced for every 10 photons on the PMT photocathode.

Thus to generate one electron at the photocathode requires about 300 eV and a 140 keV photon will produce only 400 electrons at the photocathode. This number is subject to considerable statistical fluctuation ($N^{1/2}$ = 20 or 5%). The result is that a monoenergetic beam of gamma rays will produce a range of pulses and will appear to contain a range of energies (Figure 4.16a). This is a particular problem in the gamma camera and will be considered further in Section 10.3.4.5.

As discussed in Section 4.7.2, counting statistics are much better in a semi-conductor detector, so the pulse from a monoenergetic beam is much more precisely defined (Figure 4.16b).

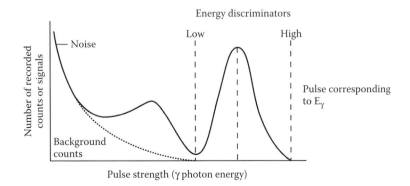

FIGURE 4.15
Use of a pulse height analyser to determine the spectrum of photon energies in an X-ray beam. A typical pulse height spectrum obtained after monochromatic gamma rays, E_γ, have passed through scattering material, and the use of energy discriminators to select the peak are shown. The tail of pulses is due to Compton scattering, which are produced fairly uniformly at all energies but selective absorption at low energies creates the maximum. The sharp rise for very low pulses is due to noise.

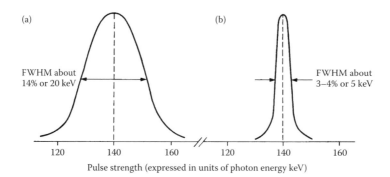

Pulse strength (expressed in units of photon energy keV)

FIGURE 4.16
Typical spread in the strength of signals from a monoenergetic beam of gamma rays when using as the primary detector (a) a NaI (T1) crystal, (b) a solid-state device.

4.10 Variation of Detector Sensitivity with Photon Energy

An important question to consider for any radiation monitor is whether its sensitivity will change with photon energy. The importance of this change can vary depending on the purpose for which it is being used. When used for measuring radiation dose in the body it would ideally have exactly the same response as human tissue. It would therefore need to be constructed of tissue-equivalent material. If used to measure absorbed dose in air then air-equivalent material would need to be used. When not being used for dosimetric purposes, such as described in Section 4.9, changes in sensitivity with photon energy can be accommodated by careful calibration.

Since, in general, the response of a detector is proportional to the amount of energy it absorbs, the question of variation in the response with photon energy can be answered in terms of absorbed energy. The variation of mass absorption coefficient of soft tissue with photon energy is shown in Figure 4.17. For comparison the corresponding graph for a typical scintillator (CsI) is also shown. Now let us suppose one wishes to use a CsI monitor to estimate soft tissue dose. In the vicinity of 1 MeV the measurement would be quite accurate because soft tissue and CsI are absorbing approximately the same energy from the beam. However, at lower photon energies CsI would *over-estimate* the dose to soft tissue. Similar arguments apply if attempting to measure dose in air. For a single photon energy corrections could be made but for a spectrum of energies, such as from an X-ray tube or scattered radiation, correction would be difficult.

Detectors that have a mean atomic number close to that of air or soft tissue, for example, lithium fluoride, show little variation in radiation sensitivity with photon energy, and are sometimes said to be 'tissue equivalent'. Note that lithium fluoride does show some variation in sensitivity at very low energies where even small differences in atomic number can be important.

Insight

Cause of Variation in Mass Absorption Coefficient

In the vicinity of 1 MeV, interactions are by Compton processes. The mass absorption coefficient is therefore independent of atomic number and sensitivity is independent of photon energy.

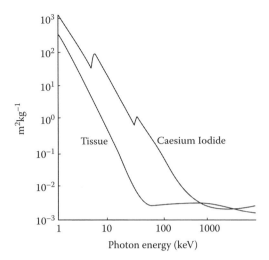

FIGURE 4.17
Variation in mass absorption coefficient with photon energy for soft tissue and caesium iodide.

However, as the photon energy decreases and approaches 100 keV, photoelectric absorption becomes important. The effect is much greater when the detector has a high mean atomic number compared to that in air ($Z = 7.6$) or tissue ($Z = 7.4$). Thus the mass absorption coefficient for CsI increases much more rapidly than that of air and tissue and sensitivity increases.

Note that below about 40 keV the sensitivity does start to decrease. This is because of the effect of the absorption edges. Near an absorption edge, although the incidence of photoelectric interactions is high, conditions are very favourable for the generation of characteristic radiation. Since a material is relatively transparent to its own characteristic radiation, a high proportion of this energy is reradiated and is therefore not available to cause detector response.

4.11 Conclusions

A wide range of properties of ionising radiation is available for radiation measurement. Ionisation in air is taken as the reference standard, partly because it can be related directly to fundamental physical processes and partly because it is very sensitive in terms of the number of ion pairs created per unit energy deposition.

The choice of instrument in a given situation will depend on a variety of factors, including sensitivity, energy, dynamic range (the range of dose over which the monitor will perform reliably), response time, performance at high count rates, variation in response with photon energy, uniformity of response between detectors, long-term stability, size, operating conditions, such as temperature or requirements for stabilised voltage supplies and cost.

It is important to distinguish carefully between radiation measurements with, for example, an ionisation chamber, and radiation detection with a GM counter. The latter is very sensitive and may be very suitable for detecting radiation leakage or contamination from spilled radioactivity. It should not normally be used as a radiation-measuring device unless specially adapted to do so.

There have been big advances in recent years in the use of semi-conductor detectors, not only for radiation measurement but also in the development of a new generation of digital image receptors (see Chapter 5).

Three aspects of radiation measurement have been identified—exposure, absorbed dose and the 'quality', or spectral distribution, of the radiation. If an absolute measure of absorbed dose is required, reference back to the ionisation process and cross-calibration will be required. However, for many applications, for example, in digital radiology and nuclear medicine, there is no need to convert the numerical data into dose routinely so uniformity of response and long-term stability are more important.

One area in which quantitative dose measurement is required is personal monitoring. This subject is deferred until Chapter 14 on 'Practical Radiation Protection', because this is where personal monitoring has most impact. Furthermore, two important mechanisms of personal monitoring, film blackening and thermoluminescence are not explained until Chapter 5.

Further advances in the development of semi-conductor technology, for measurement of X-rays, the development of a new generation of image receptors and to provide information about tissue properties, can be expected.

Further Reading

Greening J R (1985) *Fundamentals of Radiation Dosimetry*, 2nd ed. (Medical Physics Handbooks 15) Adam Hilger Ltd, Bristol, and the Hospital Physicists' Association.

Harrison R M (1997) Ionising radiation safety in diagnostic radiology. Imaging 9, 3–13.

McAlister J M (1979) *Radionuclide Techniques in Medicine*, Cambridge University Press. Cambridge.

Powsner R A and Powsner E R (2006) *Essential Nuclear Medicine Physics*. 2nd ed. Blackwell Publishing. Oxford.

Smith F A (2000) *A Primer in Applied Radiation Physics*, World Scientific Publishing Chapter 5. Singapore.

Exercises

1. Suggest reasons why ionisation in air should be chosen as the basis for radiation measurement.

2. Draw a labelled diagram of a (free air) ionisation chamber and explain the principle of its operation.

3. Explain how the following problems are overcome in an ionisation chamber:

   (a) Recombination of ions

   (b) Definition of the precise volume from which ions are collected.

4. Explain how a (free air) ionisation chamber might be used to measure the exposure at a given point in an X-ray beam. How would you expect the exposure to change if thin aluminium filters were inserted in the beam about half way between the source and the chamber?

5. Outline briefly the important features of an experimental arrangement for measuring exposure in air and discuss the factors which limit the maximum energy of the radiation that can be measured in this way.

6. Explain from first principles why the reading on a free air ionisation chamber will decrease if the temperature increases.

7. Show that under specified conditions the reading on a dose area product meter will not vary with the distance from the tube focus. What conditions must be satisfied?

8. Explain the importance of the concept of electron equilibrium in radiation dosimetry.

9. Describe a small cavity chamber for measurement of radiation exposure. Discuss the choice of material for the chamber wall and its thickness.

10. Describe the operation of a Geiger-Müller tube and explain what is meant by 'dead time'.

11. Define the Gray and show how absorbed dose is related to exposure.

12. Estimate the number of ion pairs created in a cell 10 µm in diameter by a single dose of 10 mGy X-rays.

13. What is meant by the depletion layer in a p-n diode? How does this arise and how can it be used as the basis for a radiation monitor?

14. Explain how a scintillation detector works.

15. Show that the energy of a photon in the visible range is about 3 eV.

16. Describe the sequence of events that leads to a pulse of electrons at the anode of a photomultiplier tube if the tube is directed at a NaI scintillation crystal placed in a beam of photons.

17. Explain how a scintillation detector may be used to measure:

    (a) The energy

    (b) The intensity of a beam of radiation

18. Explain in as much detail as possible the shapes of the curves in Figure 4.17.

19. Discuss the factors which may have to be considered in the choice of a monitor for a specific purpose.

5

The Image Receptor

O W E Morrish and P P Dendy

SUMMARY

- X-rays cannot be seen by the human eye so all X-ray images have to be captured on some form of receptor.
- The essential differences between analogue and digital images are explained.
- The mode of operation of film and film screen combinations as receptors for analogue images is discussed.
- The concepts of optical density, film gamma, film speed and latitude are introduced.
- Receptors which produce digital images in radiography are then considered. They work on quite different principles.
- The use of both analogue and digital receptors in fluoroscopy is considered.
- A brief review of quality control measurements that are necessary to ensure receptors are performing to a high standard completes the chapter.

CONTENTS

5.1 Introduction

Röentgen discovered X-rays when he noticed that a thin layer of barium platinocyanide on a cardboard screen would fluoresce even when the discharge tube (a primitive X-ray tube) was covered by black paper. Simultaneously he had discovered the first X-ray receptor! The receptor is the third essential component, after the X-ray tube which produces the beam and the patient, whose body tissues generate the primary radiation contrast (Chapter 3), required to create high quality radiographic images. In this chapter we shall discuss the various mechanisms by which different receptors produce images and in Chapter 6 concentrate on the important properties that make images 'fit for purpose' in the diagnostic process.

Planar images still comprise the majority of clinical imaging examinations because they produce the highest resolution at lowest cost so this is a good starting point for a planned series of investigations. Furthermore, until about 1990, almost all radiographic images, defined here as static images created with a short pulse of X-rays, would have used specialised radiographic film as the receptor in conjunction with increasingly sophisticated screens to increase sensitivity and reduce patient dose.

The photographic process, whether using X-rays or light in conjunction with film, creates an *analogue image* but in the last 20 years receptors that are capable of producing *digital images* have developed rapidly. The precise difference between analogue and digital images will be explained in the next section because it is fundamental to understanding how different receptors work. However, at this point it is worthwhile explaining why film, having reigned supreme for 80 years, should have been eclipsed so rapidly that in some larger hospitals radiology departments are now almost entirely filmless apart from historical archives.

Digital images have a number of potential advantages over film because the images are collected and stored electronically in such a manner that image acquisition, signal processing, storage and display are virtually four separate stages and can be optimised separately. In particular, post-processing options, especially contrast enhancement (see Section 6.11), can improve visualisation. Also, since digital images are stored in a computer, the ability of the computer to perform routine pre-programmed tasks with a high degree of accuracy means that computer-aided detection and diagnosis may become a useful aide to the radiologist, especially in screening situations where large numbers of 'normal' images have to be examined.

However, these benefits are marginal compared with the advantage that accrues from the fact that digital images are generated in electronic format. There are well-documented problems with film, both at the reporting stage and for archiving images. For example both the radiologist and the film have to be in the same place at the same time, films may be required in theatre, films get lost or may deteriorate with time and there is a big problem with film storage. Thus digital images have a massive advantage in terms of image storage, rapid retrieval and rapid transmission over short and long distances, for example, for remote reporting. Simultaneous viewing by multiple users or students is possible and the productivity and workflow of the radiology department is enhanced. All these factors are considered in the design of Picture Archiving and Communications Systems (PACS) which have been the major driver for digital radiology and will be discussed in Chapter 17, along with issues related to the storage and transmission of the large volumes of data associated with digital images.

5.2 Analogue and Digital Images

As noted in the introduction, it is essential to understand the differences between analogue and digital images, both to appreciate how different receptors work and to understand the discussion in Chapter 6 on the radiological image.

In an *analogue image* there is a direct spatial relationship between the X-ray or visible light photon that interacts with the recording medium and the response of that medium. The result is that the receptor displays a continuously varying function describing the image. Developed photographic film is an example of an analogue image with film blackening varying continuously and smoothly across the image. When the film is a radiograph this format is ideal for visual inspection but it is not easy to extract quantitative data from analogue images.

To extract and manipulate numerical information, the complex distribution pattern of photon interactions must be collected and stored in a computer. In principle the x and y co-ordinates of every X-ray interaction with the receptor could be registered and stored. This is sometimes known as 'list mode' data collection and is used for a few specialised studies in nuclear medicine. With the much larger number of photons in an X-ray image, this approach is impractical and the data have to be 'condensed'. This is done in two ways. First, the image space is sub-divided into a number of compartments, which are normally, but not necessarily, square and of equal size, called *pixels*. In a *digitised image* the X-ray interaction is assigned to the appropriate compartment but the position of the interaction is located no more precisely than this. Thus the image has a discontinuous aspect that is absent from an analogue image. The number and size of compartments is variable. For example in the extreme case of a 2 × 2 matrix illustrated in Figure 5.1a, all interactions would be assigned to just one of four areas. Figure 5.1b illustrates an 8 × 8 matrix.

The data are also condensed in a second way. The electrical signals within a pixel are not allowed to take a continuous set of values but are 'quantised'. Just as photon energies are quantised (see Section 1.11), the signal in each pixel can only take one of a discrete set of values. Each level is known as a pixel value or digital value and the total available number of levels is known as the bit depth. For example, if there were only two allowed pixel values, and these were represented by two grey levels in the final image, all pixels would appear either black or white.

Figure 5.2 illustrates the digitisation process. The image of the wedge on the left shows continuously changing density. The image on the right is discontinuous, with each step representing the pixel or grey level that includes a range of densities in the original image. This example shows the effect of digitising with just eight available grey levels but in normal practice, many more would be available to more accurately capture subject contrast.

(a)   (b)

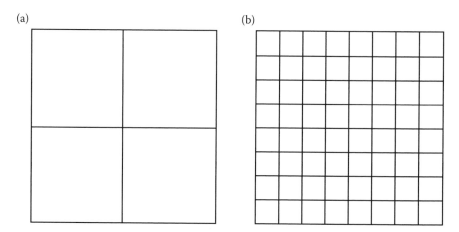

FIGURE 5.1
Examples of coarse pixellation (a) 2 × 2 matrix, (b) 8 × 8 matrix.

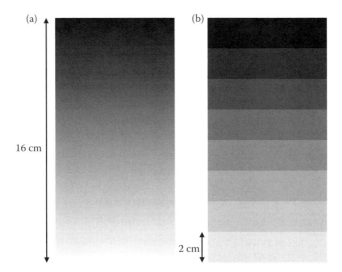

FIGURE 5.2
Images of an aluminium wedge using (a) radiographic film showing a continuous variation of density and (b) the same image digitised with eight grey levels (3-bit) showing a step wedge effect. The effect is analogous to imaging a step wedge which has 8 × 2 cm levels.

Insight

More on Pixels

1. Although digital images are a relatively recent innovation in planar radiology, they may already be familiar to a reader with experience in other modalities since computed tomography (CT) and magnetic resonance imaging (MRI) images, and some nuclear medicine and ultrasound images have been digital since each imaging technique was introduced.

2. Choice of pixel matrix size and number of pixel levels will be discussed in detail later. Suffice to say for the moment that resolution can be no better than the pixel size and the number of pixel levels, (usually chosen on a binary scale—2, 4, 8, 16 etc, to facilitate computer input) must be sufficient to display small contrast differences. For general purpose radiography 3500 × 4300 or 1750 × 2150 pixels (~100–200 μm pixel size for a 35 cm × 43 cm field) are typical and each pixel is assigned to a distinct digital level. The number of available levels typically ranges from 1024 (2^{10}) to 65,536 (2^{16}).

3. There is nothing to be gained by decreasing pixel size below the resolving capability of the imaging system. Thus, whereas a 1024 × 1024 matrix (0.2 mm pixel size) might be fully justified for a high resolution screen used in cine-angiography of the skull, larger pixels are acceptable in nuclear medicine where the system resolution of a gamma camera is no better than 5 mm.

4. As the pixel size becomes smaller, the size of the signal becomes smaller and the ratio of signal to noise gets smaller. When counting photons, the signal size N is subject to Poisson statistical fluctuations (noise) of $N^{1/2}$ (see Section 7.5), so the signal-to-noise ratio is proportional to $N^{1/2}$ and decreases as N decreases. In MRI (see Chapter 16), electrical and other forms of noise cannot be reduced below a certain level and thus the size of the signal, from the hydrogen atoms, is a limiting factor in determining the smallest useful pixel size.

5. Finer pixellation places a burden on the computer in terms of data storage and manipulation. A 1024 × 1024 matrix contains over a million pixels and the data from each one must be stored and examined individually.

6. The properties of the resulting image may not correspond to the optimum imaging capability of the detector if any compression or processing of the data has been applied following acquisition.

5.3 Fluorescence, Phosphorescence, Photostimulation and Thermoluminescence

All these phenomena depend on the fact that within the energy levels in the forbidden band (band structure of solids—see Section 4.7.1) there are imperfections, often referred to as traps (Figure 5.3a). The essential features of *fluorescence* are as follows:

1. Electron traps are normally occupied.

2. An X-ray photon interacts by the photoelectric or Compton process to produce a photoelectron which dissipates energy by exciting other electrons to move from the valence band to the conduction band in which they are free to move.

3. Holes are thus created in the filled valence band.

4. A hole, which has a positive charge numerically equal to that of the electron, moves to a hole trap at a luminescent centre in the forbidden band.

5. When both an electron and a hole are trapped at a luminescent centre, the electron may fall into the hole emitting visible light of characteristic frequency f where $E_1 = hf_1$ (see Figure 5.3b).

6. The electron trap is refilled by an electron that has been excited up into the conduction band.

In fluorescence, the migration of electrons and holes to the fluorescent centres and the emission of a photon of light happens so quickly that it is essentially instantaneous. Not all

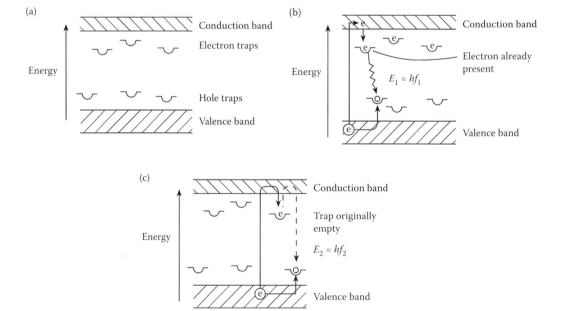

FIGURE 5.3
(a) Schematic representation of electron and hole traps in the 'forbidden' energy band in a solid. (b) The change in energy level resulting in light emission in fluorescence. (c) The change in energy level resulting in light emission in phosphorescence.

transitions of electrons at luminescent centres produce light. The efficiency of the transfer can vary enormously between different materials.

The phenomenon of *phosphorescence* also depends on the presence of traps in the forbidden band but differs in the following respects from fluorescence:

1. The electron traps are now normally empty.

2. X-ray interactions stimulate transitions from the valance band to the conduction band, as in fluorescence, but for reasons too complex to discuss here, these electrons fall into the traps rather than back into the valence band.

3. Furthermore, these electrons must return to the conduction band before they can descend to a lower level.

4. Visible light is, therefore, only emitted when the trapped electron acquires sufficient energy to escape from the trap up into the conduction band.

5. Visible light of a characteristic frequency f_2 (see Figure 5.3c) is now emitted. (As the electron is falling through a larger energy gap than in the fluorescence process f_2 is different from f_1.)

If the electron trap is only a little way below the conduction band, the electron eventually acquires this energy by statistical fluctuations in its own kinetic energy. Thus light is emitted after a time delay (phosphorescence). The time delay that distinguishes phosphorescence from fluorescence is somewhat arbitrary and might range from 10^{-10} s to 10^{-3} s. The two processes can be separately identified by heating the material. Light emission by phosphorescence is facilitated by heating, because the electrons more readily acquire the energy required to escape. Fluorescence is temperature independent. In radiology phosphorescence is sometimes called 'afterglow' and, unless the light is emitted within a very short time when it may contribute to the quantum yield, its presence in a fluorescent screen is detrimental.

If the energy difference between the electron trap and the conduction band is somewhat greater, the chance of the electron acquiring sufficient energy by thermal vibrations at room temperature is negligible and the state is metastable. However, if the electron is given extra kinetic energy, it may be released and then light is emitted. When the energy difference is not too great, the photons in a visible light laser beam may have sufficient energy to cause *photostimulated* phosphorescence. If the energy difference is greater, the extra energy must be supplied as heat and this is essentially the process of *thermoluminescence*. Generally in thermoluminescent material the traps are at several different energy levels so light emissions can occur at quite different temperatures resulting in a characteristic "glow curve."

5.4 Phosphors and Photoluminescent Screens

5.4.1 Properties of Phosphors

The special process which makes photoluminescent materials important in imaging is that more efficient use is made of the X-ray energy. Since an optical photon has only about 2–3 eV of energy, a 40 keV X-ray photon can be converted into more than 1000 visible photons and the imaging process is much more efficient, resulting in better quality images and lower dose to the patient.

A number of factors must be taken into consideration when maximising phosphor sensitivity and performance:

1. Only the radiation that interacts with the detector contributes to the signal so the *quantum efficiency* (QE), which describes the probability of a single quantum of X-rays interacting with the detector, must be high. It is increased by increasing the linear attenuation coefficient, and hence the atomic number and density of the detector and also by increasing the detector thickness. Note, however, that increasing detector thickness will increase unsharpness (see Section 5.8).

2. The packing density of phosphor grains, or the fill factor in some digital detectors (see Section 5.10) also affects the QE, since it alters the area available to stop X-rays.

3. The overall luminescent radiant efficiency, or light output per unit beam intensity of the phosphor, depends on its efficiency for converting X-ray photon energy to light photons and must be high.

4. The spectral output of the phosphor must be matched to the spectral sensitivity of the next stage in the imaging system.

5. Finally, there are practical considerations. The chemical properties of a phosphor can limit how it is used, for example, a hygroscopic phosphor must always be used either encapsulated or inside a vacuum. Phosphors must also be commercially available in known crystal sizes of uniform sensitivity at a reasonable price.

Insight

Quantum Efficiency

If the linear attenuation coefficient of the detector is $\mu(E)$ (E indicates that it is photon energy dependent), and detector thickness d, the transmitted intensity I is

$$I = I_0 e^{\mu(E)d}$$

Thus the X-ray photons stopped are

$$I_0 - I_0 e^{\mu(E)d}$$

And the QE is

$$\frac{(I_0 - I_0 e^{\mu(E)d})}{I_0} = 1 - e^{\mu(E)d}$$

TABLE 5.1

Atomic Numbers and Luminescent Radiant Efficiencies for Some Important Phosphors. The Element behind the : Sign Is the Activator to the Phosphor Salt before the : Sign

Phosphor	Z of Heavy Elements	Luminescent Radiant Efficiency (%)
BaFCl: Eu^{2+}	56	13
BaSO$_4$: Eu^{2+}	56	6
CaWO$_4$	74	3.5
CsBr: Tl	35/55	8
CsI: Na	53/65	10
CsI: Tl	53/55	11
Gd$_2$O$_2$S: Tb	64	15
La$_2$O$_2$S: Tb	57	12
Y$_2$O$_2$S: Tb	39	18
(ZnCd)S: Ag	30/48	18
NaI: Tl	53	10

Photoluminescent screens may be used as the input for three basic imaging systems:

1. Coupled to film in radiography
2. Coupled to a photocathode in an image intensifier or indirect digital detector
3. As the image-storing medium in some digital receptors (computed radiography [CR])

Until the mid 1960s, the most widely used phosphors were calcium tungstate in radiography and zinc-cadmium sulphide in fluoroscopy and image intensifiers. As a direct result of the American space programme, new phosphors were developed which have since been adopted for medical use. They are mainly crystals of salts of the rare earth elements or crystals of barium salts activated by rare earth elements.

The reason why the rare earth phosphors have replaced calcium tungstate for many purposes can be seen in Table 5.1. The Z value of the new rare earth phosphors is slightly lower than that of the heavy elements in calcium tungstate phosphor, and thus slightly less energy may, under some conditions, be absorbed from an X-ray beam. However, this is more than compensated for by the fact that the luminescent radiant efficiency for the rare earth phosphors is at least three times higher. The same light output from the phosphor can thus always be achieved with a much lower X-ray dose. In practice, because of absorption edges (see Section 3.8), at certain photon energies X-ray absorption would also be higher (see Figure 5.4). For some elements in the phosphors (e.g. gadolinium with a K-edge at 50.2 keV) this edge is close to the peak in the intensity spectrum of the X-ray beam after it has been transmitted through a patient (see for example Figure 5.4).

In Table 5.1 figures are also given for comparison for thallium-doped sodium iodide (NaI:Tl) the primary detector used almost universally in gamma cameras in nuclear medicine.

Spectral outputs are shown for some phosphors in Figure 5.5 and receptor sensitivities for some typical receptors in Figure 5.6. It is clear that there can be a major loss in sensitivity if these spectra are not matched.

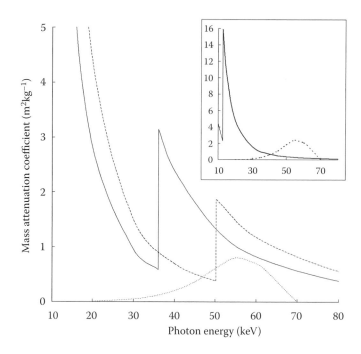

FIGURE 5.4
Mass attenuation coefficients for the majority element in some important receptors showing the positions of absorption edges: solid line showing caesium (Z = 55); long-dashed line showing gadolinium (Z = 64). The inset shows the coefficients for selenium (Z = 34). A typical spectrum for the unscattered component of a 70 kVp beam after heavy filtration in the patient is also shown by the short-dashed line (continuous spectrum shown only). Note the positions of the K-edges in relation to the transmitted spectrum.

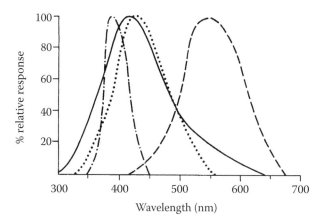

FIGURE 5.5
The spectral output of different phosphors. Solid line CsI:Na. Dashed line (ZnCd)S:Ag. Dot dash line BaFCl:Eu^{2+}. Dotted line CaWO$_4$.

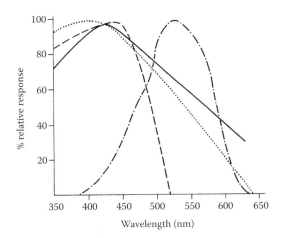

FIGURE 5.6
The spectral response of different light receptors. Solid line—S20 photocathode. Dashed line—X-ray film. Dot dash line—the eye. Dotted line—S11 photocathode.

Insight

Efficiency of the Phosphor Layer

There are three distinct aspects to this efficiency:

1) Absorption efficiency—the ability of the phosphor to absorb X-rays.
2) Conversion efficiency—the inherent ability of the phosphor to convert X-ray energy to light.
3) Emission efficiency—the ability of the light produced to escape the phosphor layer.
   - A thicker phosphor increases absorption and reduces emission slightly.
   - A greater phosphor density (closer crystal packing) increases absorption.
   - Flat-shaped crystals increase absorption due to their greater incident surface area.
   - The addition of dye to the phosphor to reduce light from distant regions reduces emission but improves sharpness.
   - A reflective layer behind the phosphor increases emission but reduces sharpness.
   - Increasing kVp reduces absorption and increases conversion since the higher energy X-ray photons are producing a greater number of light photons—which are of the same energy regardless of X-ray photon energy.
   - Matching the K-edge of the phosphor to the incident X-ray spectrum increases absorption.
   - Rare earth phosphors increase absorption by about 60% compared to calcium tungstate and increase conversion by about 20%.

5.4.2 Production of Photoluminescent Screens

Most photoluminescent screens are produced by laying down a phosphor crystal/binder suspension onto a paper or metallic substrate and then drying it on the substrate. The size of the crystals affects screen performance with larger crystals producing more light for a given screen thickness. If small crystals are used the effective crystal area is small because there is a high proportion of interstitial space. However, larger crystals produce a screen with a much lower resolution. For a given thickness of screen, with an acceptable resolution, the best packing density (ratio of phosphor to interstitial space) of phosphor crystals that can be achieved is approximately 50%. For high resolution screens the ratio is lower.

A very limited number of phosphors based on halide salts, for example, CsI:Na, can be vapour deposited in sufficiently thick layers to enable a useful detector to be made. The packing density is almost 100% thus giving a gain of approximately two over crystal deposited screens with a consequently higher X-ray absorption per unit thickness. For these reasons it is possible to get better resolution and less noise from these phosphors, and this is important in the construction and performance of, for example, an image intensifier (see Section 5.13.1).

5.4.3 Film-Phosphor Combinations in Radiography

For medical applications, radiographic film is almost always used in combination with a fluorescent screen, with only a very few specialised exceptions. However, the properties of the combination are essentially governed by the properties of the film.

Film responds to the light emitted from a fluorescent screen in the same way as it does to X-rays except in one or two minor respects. Thus the many advantages and few disadvantages of using a screen can best be understood by first examining the response of film to X-rays and then considering how the introduction of a screen changes this response.

5.5 X-ray Film

5.5.1 Film Construction

The basic film construction is shown in Figure 5.7. To maintain rigidity and carry the radiation-sensitive emulsion, all radiographic films use a transparent base material of 'polyester' approximately 0.15 mm thick which is completely stable during development. Normally the polyester base has an emulsion bound to both sides (as in Figure 5.7). Using emulsion on both sides of the base material gives a mechanical advantage during processing as it stops the film from curling. However, some special films have emulsion on one side only, for example in mammography a single screen is used on the side of the film remote from the breast (see Section 9.2.5). A single emulsion is used next to the screen to eliminate parallax effects. If transparency film is used in nuclear medicine it is single-sided because by that stage in the imaging process only visible light is being recorded. Both sides of the film are coated with a protective layer to prevent mechanical damage.

The emulsion is finely precipitated silver bromide suspended in gelatine and coated on both sides of a transparent blue-tinted polyester support. The crystals of silver (Ag^+) and bromide (Br^-) ions, which are in a cubic lattice, would be electrically neutral in the pure crystal. However, the presence of impurities distorts the lattice and produces on the surface of the crystal, a spot, called the *sensitivity speck*, which will attract any free electrons produced within the crystal. Visible light and X-rays can release electrons from the bromide ions

$$Br^- + hf \quad \rightarrow \quad Br \quad + \quad e^-$$

$$\text{ion} \quad \text{light} \qquad \text{atom} \quad \text{free electron}$$

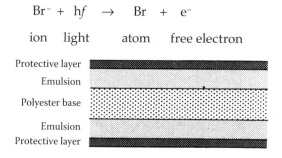

FIGURE 5.7
The basic construction of an X-ray film.

These electrons move through the crystal and are trapped by the sensitivity speck. An electron at a sensitivity speck then attracts a positively charged silver ion to the speck and neutralises it to form a silver atom. This occurs many times and the result is an area of the crystal with a number of neutral silver atoms on the surface. This crystal is then said to constitute a latent image. For the crystal to be developable, between 10 and 80 atoms of silver must be produced. During development with a reducing alkaline agent, crystals with a latent image in them allow the rest of the silver ions present to be reduced and thus form a dark silver grain speck on the film. The film is fixed and hardened at the same time using a weakly acidic solution. The crystals which did not contain a latent image are washed off at the fixation stage leaving a light area on the film.

If the developing agent is too strong it will develop crystals in which no latent image is present. Even in an unexposed film some crystals will be developed to produce a low level of blackening called *fog*. The 'fog' level can be increased by using inappropriate developing conditions, for example, too strong a developer or too high a developing temperature.

5.5.2 Characteristic Curve and Optical Density

The amount of blackening produced on a film by any form of radiation—visible light or X-rays—is measured by its density. Note the use of density here relates to optical density and must not be confused with mass density (ρ) used elsewhere in the book. The preface 'optical' will not be used unless there is some possibility of confusion.

The density of a piece of blackened film is measured by passing visible light through it (Figure 5.8a). Density is defined by the equation

$$D = \log_{10} \frac{I_0}{I}$$

where I_0 is the incident intensity of visible light and I is the transmitted intensity.

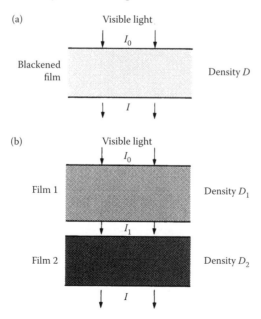

FIGURE 5.8
A simple interpretation of optical density: (a) for a simple film, (b) for two films superimposed.

Basing the definition on the log of the ratio of incident and transmitted intensities has three important advantages:

1. It represents accurately what the eye sees, since the physiological response of the eye is also logarithmic to visible light.
2. A very wide range of ratios can be accommodated and the resulting number for the density is small and manageable (see Table 5.2).
3. The total density of two films superimposed is simply the sum of their individual densities.

From Figure 5.8b,

$$\text{Total } D = \log \frac{I_0}{I} = \log \frac{I_0}{I_1} \cdot \frac{I_1}{I}$$

$$= \log \frac{I_0}{I_1} + \log \frac{I_1}{I}$$

$$= D_1 + D_2$$

When different amounts of light are transmitted through different parts of the film (Figure 5.9), the difference in density between the two parts of the film is called the *contrast*. Hence,

$$C = D_1 - D_2 = \log_{10} \frac{I_0}{I_1} - \log_{10} \frac{I_0}{I_2} = \log_{10} \frac{I_2}{I_1} \tag{5.1}$$

The eye can easily discern differences in density over a range from approximately 0.25–2.5, the minimum discernible difference being about 0.02.

If the density produced on a film is plotted against the log of the radiation exposure producing it, the characteristic curve of the film is generated (Figure 5.10). This is frequently

TABLE 5.2

Relationship between Optical Density and Transmitted Intensity

Transmitted Intensity as a Percentage of I_0 (%)	$OD = \log_{10} I_0/I$
10	$\log_{10} 10 = 1.0$
1	$\log_{10} 100 = 2.0$
0.01	$\log_{10} 10000 = 4.0$

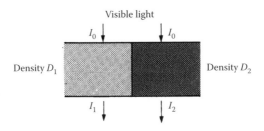

FIGURE 5.9
Representation of contrast between two parts of a blackened film as a difference in transmitted light intensities.

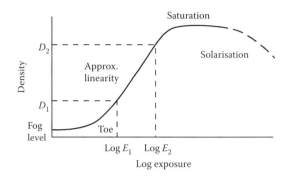

FIGURE 5.10

A typical characteristic curve for an X-ray film. This is often called an H and D curve after Hurter and Driffield who developed it for photographic analysis.

referred to as an 'H and D curve' after Hurter and Driffield who developed it for photographic analysis. The log scale again allows a wide range of exposure to be accommodated. Each type of film has its own characteristic curve although all have the same basic shape.

The finite density at zero exposure is due to a small contribution from 'fogging', that is, the latent images produced during manufacture, by temperature, humidity and other non-radiation means. This can be kept to a minimum but never completely removed. Note that the apparently horizontal initial portion of the curve arises primarily because one logarithmic quantity (D) has been plotted against another (log E). This has the effect of compressing the lower end of the curve. If the data were re-plotted on linear axes, there would be a steady increase in film blackening with exposure from zero dose but when using the characteristic curve, a finite dose is required to be given to the film before densities above the fog level are recorded.

The initial curved part of the graph is referred to as the 'toe' of the characteristic curve and this leads into the approximately linear portion of the graph covering the range of densities and doses over which the film is most useful. Eventually, after passing over the shoulder of the curve, the graph is seen to saturate and further exposure produces no further blackening. This decrease in additional blackening is due to the black spots from developed crystals overlapping until eventually the production of more black silver spots has no further effect on the overall density. At very high exposures (note that the scale is logarithmic), film blackening begins to decrease again, a process known as solarisation.

Insight

Solarisation

The reader may like to satisfy themselves that if a film is exposed to light intensities in the solarisation region where the film gamma is –1, an exact copy can be created. However, this technique is no longer used as copying is simple and straightforward using digital techniques.

5.5.3 Film Gamma and Film Speed

The gamma of the film is the maximum slope of the approximately linear portion of the characteristic curve and from Figure 5.10 is defined as

$$\gamma = \frac{D_2 - D_1}{\log E_2 - \log E_1}$$

If no part of the curve is approximately linear, the average gradient may be calculated between defined points on the steepest part of the curve.

The gamma of a film depends on the type of emulsion present, principally the distribution and size of the silver bromide crystals, and second on how the film is developed. If the crystals are all the same size a very 'contrasty' film is produced with a large gamma. A wide range of crystal sizes will produce a much lower gamma (Figure 5.11). A 'fast' film with large crystals generally also has a wide range of crystal sizes. Finally, increased grain size reduces resolution although unsharpness in the film itself is rarely a limiting factor.

The correct characteristic curve for a film can only be obtained by using the developing procedure recommended by the film manufacturer, including the concentration of developer, the temperature of the developer, the period of development and even the amount of agitation to be applied to the film. An increase in any of these factors will result in the over-development of the film, a decrease will under-develop the film.

Within realistic limits, over-development increases the fog level, the film gamma and the saturation density. Under-development has the opposite effect (Figure 5.12).

The amount of radiation required to produce a given density is an indication of film speed. The speed is usually taken to be the reciprocal of the exposure that causes unit density above fog so a fast film requires less radiation than a slow film. The speed of the film depends on the size of the crystals making up the emulsion and on the energy of the X-rays striking the film. If the crystals are large then fewer X-ray interactions are required to blacken a film and, because of this, fast films are often called 'grainy' films as the crystals when developed give

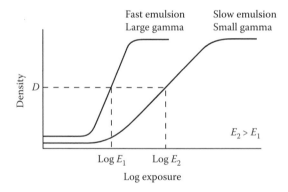

FIGURE 5.11
Characteristic curves for films of different gammas and different speeds.

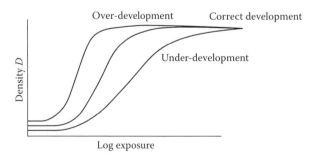

FIGURE 5.12
Variation of the characteristic curve for a film, for different development conditions. Note how the gamma and fog level are affected.

a 'grainy' picture. This is because the energy deposited by a single X-ray photon is sufficient to produce a latent image in a large crystal as well as a small crystal. Fewer large crystals need be developed to obtain a given density. The speed of the film varies with the energy of the X-ray photon, basically because the atomic numbers of the elements in the receptor, and hence their absorption properties, are energy-dependent (see Section 4.10). In practice, due to the wide range of photon energies present in a diagnostic X-ray beam this variation in sensitivity can be neglected during subjective assessment of X-ray films.

The relative speed of two films is dependent on their characteristic curves and the density at which the speed is compared. In the extreme, if the curves cross then there will be places where one film is faster than the other, one place where they have the same speed and other places where the relative speeds are reversed.

5.5.4 Latitude

Two distinct, but related aspects of latitude are important, film latitude and exposure latitude. Consider first film latitude. The optimum range of densities for viewing, using a standard light box, is between 0.25 and 2.5. Between these two limits the eye can see small changes in contrast quite easily. The latitude of the film refers to the range of exposures that can be given to the film such that the density produced is within these limits. The higher the gamma of the film, the smaller the range of exposures it can tolerate and thus the lower the latitude. For general radiography a film with a reasonably high latitude is used. There is an upper limit however, because if the gamma of the film is made too small the contrast produced is too small for reasonable evaluation.

Exposure latitude is related to the object and can be understood by reference to Figure 5.10. If a radiograph is produced in which all film densities are on the linear portion of the curve, (i.e. the object contains a narrow range of contrasts) the exposure may be altered, shifting these densities up or down the linear portion, without change in contrast. (Radiologists often prefer a darker film to a lighter film although there may in fact be no difference in contrast when the densities are measured and the contrast evaluated.) In other words there is 'latitude' or some freedom of choice over exposure. If, on the other hand, the range of densities on a film covers the whole of the linear range, (i.e. the object contains a wide range of contrasts), exposure cannot be altered without either pushing the dark regions into saturation or the light regions into the fog level. There is no 'latitude' on choice of exposure. Exposure latitude can be restored by choosing a film with a lower gamma, (greater latitude), but there is a loss of contrast.

Note that in other imaging applications the ratio of the largest signal that can be handled by the detector without saturation to the weakest signal that can be distinguished from the noise is known at the *dynamic range*. The dynamic range of film is very small.

5.6 Film Used with a Photoluminescent Screen

Most properties of film described to this point have assumed an exposure to X-radiation. Film also responds to light photons but as the quantum energy of one light photon is only 2–3 eV (see Section 1.11) several tens of light photons have to be absorbed to produce one latent image. In contrast, the energy from just one X-ray photon is more than sufficient to produce a latent image.

The advantages of using film in conjunction with fluorescent screens are two-fold. First a much greater number of X-ray photons are absorbed by the screen compared to the number absorbed by film alone. The ratio varies between 20 and 40 depending on the screen composition.

Second, by first converting the X-ray photon energy into light photons, the full blackening potential is realised. If an X-ray photon is absorbed directly in the film, it will sensitise only one or two silver grains. However, each X-ray photon absorbed in an intensifying screen will release at least 400 photons of light—some screens will release several thousand photons. Thus, although tens of light photons are required to produce a latent image, the final result is that the density on the film for a given exposure is between 30 and 300 times blacker (depending on the type of screen) when a screen is used than when a film alone is exposed.

Insight

Intensification Factor

The increase in blackening when using a fluorescent screen is quantified by the *intensification factor*, defined as follows:

$$\frac{\text{The exposure required when a screen is not used}}{\text{The exposure required when a screen is used}}$$

for the same film blackening. A value for the intensification factor is only strictly valid for one density and one kVp. This is because, as shown in Figure 5.13, when used in conjunction with a screen, the characteristic curve of the film is altered. Not only is it moved to the left, as one would expect, but the gamma is also increased.

When using screens, double-sided film has radiological advantages as well as processing advantages since two emulsion layers enable double the contrast to be obtained for a given exposure. This is because the superposition of two densities (one on each side of the film base) produces a density equal to their sum. Although, theoretically, the same contrast could be achieved by doubling the thickness of a single emulsion, this is not possible

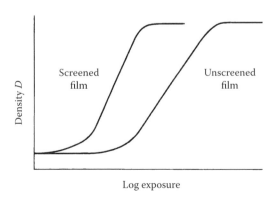

FIGURE 5.13
Change in the characteristic curve of film when using a screen.

in practice due to the limited range in emulsion of the light photons from the screens. Note that, because the range of X-ray photons is not limited in this way, the contrast for unscreened films would be almost the same for a one-sided film of double emulsion thickness. The mechanical advantage of double-sided film would, of course, still remain.

Reference was made in Section 5.5.3 to the speed of the film. In practice it is the speed of the film/screen combination which is important. This is described by a 'speed class'. This system is based on a similar numerical system to the American Standards Association (ASA) film speed system used in photography. The special feature of ASA film speeds 32, 64, 125 is that the logarithms of these numbers 1.5, 1.8, 2.1 show equal increments. In other words speed classes are equally spaced on the 'log exposure' axis of the characteristic curve. The higher the speed class the more sensitive the film so speed classes for film-screen combinations used in radiography are generally higher (Table 5.3).

Very high resolution, high contrast combinations are likely to have a speed class in the region of 100–150, whereas very fast combinations (typically 600 or above) will make sacrifices in terms of resolution and possibly contrast. Note that quantum mottle limits the speed of the system that can usefully be used (see Section 6.10).

A film-screen combination must be used and contained in a light-tight cassette, the film is fogged by ambient light. The front of the cassette is made of a low atomic number material, generally either aluminium or plastic. The back of the cassette is either made of, or lined with, a high atomic number material. This high atomic number material is more likely to absorb totally the X-ray photons passing through both screens and film by the photoelectric effect than to undergo a Compton scatter reaction, which could backscatter photons into the screen.

The two screens are kept in close contact with the film by the felt pad exerting a constant pressure as shown in Figure 5.14. If close contact is not maintained then resolution is lost due to the greater opportunity for the spread of light leaving the screen before reaching the film.

The light emitted from a screen does not increase indefinitely if the screen thickness is increased to absorb more X-ray photons. A point is reached where increasing the thickness produces no more light because internal absorption of light photons in the screen takes place. The absorption of X-ray photons produces an intensity gradient through the screen and significantly fewer leave the screen than enter. The light produced at a given point in the screen is directly proportional to the X-ray intensity at that point.

TABLE 5.3

The Range of Speed Classes Used in Film Screen Radiography and the Corresponding Linear Increases on a Logarithmic Scale

Speed Class	100	200	400	800
Logarithm	2.0	2.3	2.6	2.9

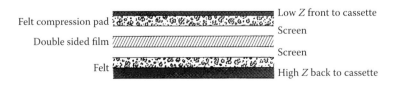

FIGURE 5.14

Cross-section through a 'loaded' cassette containing a 'sandwich' of intensifying screens and double-sided film.

In the case of the back screen, the reduction in intensity through it is of no significance. This screen must be thick enough to ensure that the maximum amount of light for a given X-ray intensity is produced, but once this is achieved it need be no thicker. The front screen thickness is a compromise between achieving the maximum light output through X-ray photon absorption and not reducing the X-ray photon intensity by too much in the area of effective maximum light production, that is, in the layers of the screen closest to the film. The compromise thickness for the front screen, giving maximum light production, is in fact somewhat less than optimum for the back screen. Since unsharpness is less from the back screen, attempts to optimise light production can improve image sharpness. For further discussion on the way in which the screen can affect image quality see Section 5.8.

If the screens are of unequal thickness they must never be reversed but for some modern cassettes there is no difference between the thickness of the front and back screens.

5.7 Reciprocity

Basically the law of reciprocity states that, if two quantities x and y are multiplied together, the same result will be obtained by using $10x$ and $0.1y$ or vice versa. A good example is multiplying current and voltage, $10A \times 0.1V$ and $0.1A \times 10V$ both give one watt of power.

The exposure received by a cassette can be considered to depend on two basic parameters, the intensity of the radiation beam striking the cassette and the time of exposure. The intensity at a fixed kVp is proportional to the milliamps (mA) of the exposure, and the exposure is thus proportional to the milliampseconds (mAs). For unscreened film the same mAs will always give the same blackening of the film regardless of the period of exposure, that is the law of reciprocity is obeyed. When using screened film, however, it is found that for very short exposures and very long exposures, although the same mAs are given, the blackening of the film is less.

For long exposures this effect is known as latent image fading and the amount of fading depends to a large extent on whether the image has been produced by X-ray interactions or by visible light photons. In general, a single X-ray photon will form a developable grain because it deposits so much energy. Hence image fading does not occur. Conversely, many visible light photons are required to produce a latent image and if their rate of arrival is slow the silver halide lattice may revert to its normal state before sensitisation of the speck is completed. This effect is called *failure of the law of reciprocity*.

5.8 Film-Screen Unsharpness

When screens are used there is some loss of resolution. This is because light produced in the screen travels in all directions (see Figure 5.15). Since the screen is of finite thickness, those photons travelling in the direction of the film spread out a little before reaching it.

A major cause of unsharpness in double-sided film is 'crossover', where light from the upper intensifier screen sensitises the lower film layer and vice versa. The development of emulsions containing grains that are flat or tubular in shape, with the flat surface facing

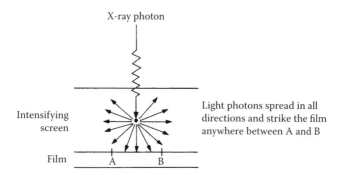

FIGURE 5.15
Schematic representation of image unsharpness created by the use of an intensifying screen.

the X-ray beam, rather than pebble-shaped, has helped to reduce this source of image unsharpness with no loss in sensitivity (see Section 5.13.1).

This unsharpness, or loss of resolution caused by photoluminescent screens, will be discussed again in Section 5.11 since it is a major problem in both analogue and many forms of digital image.

5.9 Introduction to Digital Receptors and Associated Hardware

There are several ways in which digital images can be produced and these will be discussed in the sections which follow. Recall from Section 5.2 that digital images differ from analogue images in two ways, first the image space has to be divided into pixels for sampling purposes and then the signal has to be quantised into one of many discrete levels or pixel levels. There are several ways in which the image can be divided into pixels but basically only one way in which the signal is quantised so this will be considered first.

5.9.1 Analogue to Digital Converters (ADCs)

All digital systems contain an ADC. The electronic configuration of an ADC is beyond the scope of this book but its function is straightforward, namely to decide, by quantisation, the discrete value that should be given to the signal within each pixel. An important property of the ADC is its *accuracy*, which determines how close the assigned digital level is to the analogue level it represents. The number of quantisation levels is determined by the number of bits (binary digits). More bits means more accurate conversion. For example 3 bits ($2^3 = 8$ levels) would be sufficient to describe the step wedge in Figure 5.2b. However, this is not a very accurate representation of the image in Figure 5.2a since a range of values of attenuation are assigned to the same pixel level. ADCs for radiographic images typically have at least 4096 pixel levels (12 bits) so the accuracy is about 1 in 4000 (0.03%).

5.9.2 Pixellating the Image

The second requirement of the digital image is to provide accurate spatial information about the distribution of X-rays. This requires the sub-division of the image space into an array of very small pixels that are no bigger than the final resolution required, say 150 μm.

Basically, there are two ways this can be done. The first is to use one or more very small discrete detectors and, if necessary, to cover the area of interest by suitable detector movement. For example, a linear array of detectors will require linear movement to cover the second dimension. Flat panel imagers (FPIs), which contain a matrix array of small detectors covering the full image space, come into this category, but with no mechanical detector movement required. Their use for imaging is known as *direct digital radiography.*

The alternative approach is to sample, at discrete regular intervals, the analogue image produced at some stage of the imaging process. For example, this may be the X-ray film, the output from an image intensifier or the latent image produced on a photostimulable phosphor plate (computed radiography). Each of these methods will also be discussed, with emphasis on the way in which the required pixel dimensions are achieved.

5.10 Digital Radiography (DR)

In theory there are several ways in which a digital image can be produced directly at the image plane. The possibilities are illustrated in Figure 5.16.

1. A small detector scans across the area occupied by the image (Figure 5.16a). This is a very simple geometry but will require a long time to acquire all the data, even in a relatively coarse 256×256 matrix.

2. A linear array of 200×1.0 mm detectors only has to make linear movements to cover the image space (Figure 5.16b). Data collection is now speeded up but may still take several seconds.

3. A static array of detectors may be used (Figure 5.16c). This will be inherently more rapid since there is no mechanical movement, provided that the electronics is fast enough to handle the rapid collection of large amounts of data.

Insight

Slot Scanning Systems

A few systems have adopted the scanning principle, using a one-dimensional array with linear motion. In an early system proposed for DR of the chest (Tesic et al. 1983) a vertical fan beam of X-rays scanned traversely across the patient in a PA orientation. An entrance slit 0.5 mm wide defined the beam, an exit slit 1.0 mm wide removed much of the scatter. The beam fell on a gadolinium oxysulphide screen backed by a vertical linear detector array consisting of 1024 photodiodes with a 0.5 mm spacing (Figure 5.17). The complete system of X-ray tube, collimators and detectors scanned over the patient in 4.5 s, sampling in 1024 horizontal positions. The estimated entrance dose was 250 μGy.

The advantage of this type of system is that there is no need for an anti-scatter grid. The disadvantage however is that there is a greater load on the X-ray generator as the exposure time is much greater than in conventional radiography.

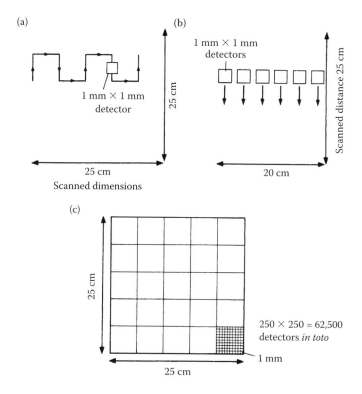

FIGURE 5.16
Detector arrangements for digital radiology (a) scanning detector, (b) linear array of detectors, (c) static array of independent detectors: 250 × 250 = 62,500 detectors *in toto*.

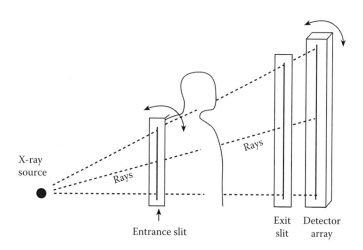

FIGURE 5.17
Diagram of a scanning digital radiography system. The collimated fan beam moves horizontally across the patient. The exit slit and detectors move with the beam.

DR has been revolutionised by the introduction of FPIs using semiconductor technology developed in the production of active matrix liquid crystal displays. There are two essential components to all FPIs:

(a) A large-area, thin layer of X-ray sensitive material which is capable of generating either light (a phosphor) or electrons (a semiconductor)

(b) A two-dimensional array of tiny matrix elements constructed as a thin film transistor (TFT). To achieve the necessary resolution (say 150 µm) across a 30 cm × 45 cm image requires an array of 2000 × 3000 individual, electrically isolated, TFTs

FPIs can work in two slightly different ways (see Figure 5.18). In an indirect detector active matrix FPI the X-ray sensitive material is a phosphor, typically thallium-doped columnar caesium iodide (see Section 5.13.1) or gadolinium oxysulphide. The light emitted from the phosphor is converted by a photodiode, typically amorphous silicon (a-Si) into electrical charges which are stored as an electronic image in the TFTs.

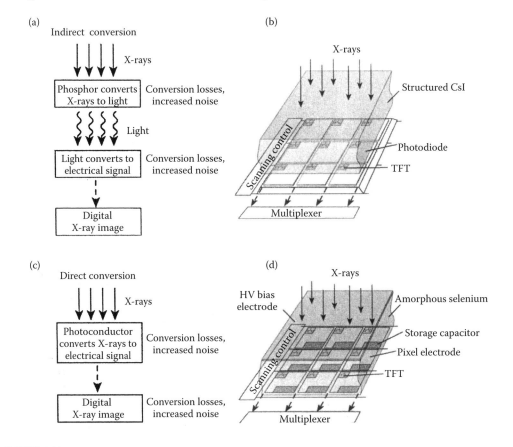

FIGURE 5.18
Illustrating indirect and direct conversion in active matrix flat panel imagers. In indirect conversion (a and b) the X-ray photons are first converted by a phosphor to visible photons which are subsequently converted to an electrical signal by a photodiode. In direct conversion (c and d) a photoconductor is used to convert the X-ray photons directly into an electrical signal. (Reproduced with permission from Zhao W, Andriole K P and Samei E Digital radiography and fluoroscopy in *Advances in medical physics* – 2006. Wolbarst A B, Zamenhof R G and Hendee W R, eds, Medical Physics Publishing, Madison, WI, 1–23, 2006.)

In the direct detector active matrix FPI the X-ray receptor is a layer of amorphous selenium (a-Se) which converts the X-ray pattern into a spatial pattern of electron-hole pairs. An electric field now drives the electrons (or holes depending on field direction) towards the TFT array where they are collected on the pixel electrodes.

With both FPIs, at the end of the exposure the X-ray image has been converted to an array of electrical charges, with the charge at each pixel proportional to the absorbed radiation in the corresponding region of the image. Each TFT acts like a microscopic valve, so for read out a scanning control circuit applies a bias voltage which activates the TFTs one row at a time and the image charges are transferred to charge-sensitive amplifiers. Because all scanning movements are controlled electronically with no mechanical movements, and modern computers can handle the rapid data flow, scanning is very fast. Each row takes about 30 µs to read out so a detector with 2000 × 2000 pixels can be read out in real time at about 15 frames s^{-1}.

Insight (1)

Indirect Detector Active Matrix FPIs

Some further points on indirect detector FPIs affecting overall sensitivity response are as follows:

1. Columnar CsI will allow thicker layers (500 µm) to be used giving better QE without too much loss of resolution.
2. About 25 eV are required to produce a light photon in CsI (i.e. almost 10 times the energy of the photon itself), so a 50 keV X-ray photon produces about 2000 photons.
3. The light output from CsI, which covers a broad band from 400 to 700 nm, is well matched to the sensitivity response of a-Si.
4. The fraction of the pixel area sensed by the photodiode (the fill factor) is not 1.0 because part of each pixel is occupied by electronics associated with the TFT and scanning lines. As the pixel size decreases the fill factor falls because the obstructed area becomes a bigger percentage of the whole. Conversely, the continuing reduction in size of the electronic components of each pixel will increase the fill factor in the future. Fill factors for currently available detectors range from 0.5 to 0.9 depending on pixel size.

Insight (2)

Direct Detector Active Matrix FPIs

Some further points on the direct detector FPIs are as follows:

1. Because of the relatively low atomic number of selenium, the thickness of this layer is quite high (about 1 mm) to give good QE.
2. Since X-rays are converted directly to electrons, which can be collimated by electric fields, the image blurring caused by light spreading in a scintillator is much reduced, even for the relatively thick a-Se layer.
3. The pixel electrode can be built on top of the TFT and scanning lines so the fill factor is close to 1.0.
4. Overall conversion gain at 1000 electrons per 50 keV photon is about the same for direct and indirect FPIs but very small pixels are easier to achieve with the direct system (because there are no photodiodes) and the fill factor is not compromised.

An array of detectors that uses a somewhat different principle to produce digital images is the charge coupled device (CCD) discussed in Section 5.13.4.

5.11 Photostimulable Phosphors—Computed Radiography (CR)

The most prevalent technique for producing digital images by sampling analogue images is currently CR. This approach is based on phosphor screens which have similar properties and construction to intensifying screens, but also have one additional property which has allowed the development of CR. On excitation, only a proportion of the energy is released immediately as light in a luminescent process. This light plays no part in image formation. A small percentage is stored as a latent image and can only be released following stimulation. This latent image can be made to have a life time long enough to be useful as an imaging process and may retain as much as 75% of its signal for up to 8 hours although eventually it will spontaneously decay to a level which is no longer producing an acceptable image. The latent image is read out by using a laser scanning system to stimulate the phosphor into releasing its stored energy. The wavelength of the stimulating laser is in the red or near infrared region and the light emitted has a wavelength in the green, blue or ultraviolet. This process is known as *photostimulable luminescence*. The amount of light emitted is directly proportional to the number of X-ray photons absorbed and is hence completely linear within the normal diagnostic range of exposures.

5.11.1 The Phosphor Screen

This is a re-usable plate coated with crystals of an appropriate photostimulable phosphor. Desirable properties of the phosphor are (a) high emission sensitivity at the wavelength of a readily available laser, for example, He/Ne at 633 nm; (b) light emission in the range 300–500 nm where photomultiplier tubes have a high QE. A class of europium activated barium fluorohalide crystals (BaFX:Eu where X may be Cl/Br/I) appears to be most suitable. The wavelength range for photostimulation is 500–700 nm and light emission is in the region of 400 nm.

The CR cassette has only one phosphor layer, unlike most film screens, and must be placed with this layer uppermost to the X-ray beam. It is not essential for the CR cassette to be as light tight as film-screen cassettes. In fact the phosphor plate can be exposed without any cassette if necessary, as is the case with some dental imaging devices. However the cassette provides protection from damage to the plate as it is handled during radiographic examinations and transported to the reader. In addition, it is desirable to minimise the amount of ambient light incident on the phosphor following exposure, before read out, to ensure that no latent signal is lost to excessive light.

5.11.2 Read Out Process

A difference in wavelength between the stimulating light and the output light is essential to the operation of the phosphor because the intensity of the former is of the order of 10^8 larger. To detect the output, special filters have to be used which only transmit the output wavelength. The He/Ne laser beam defines the size of each pixel and is typically 100–200 μm for general work but may be as small as 50 μm for high resolution work. The stimulated output from each delineated small area on the screen (now identified as a pixel) is collected by a light pipe which has a photomultiplier attached to it and the laser beam scans across the phosphor in a raster pattern sampling about 5–10 pixels per mm (Figure 5.19a). The photomultiplier amplifies the signal which is further amplified by a logarithmic amplifier so that the resultant image can be accurately sampled by an ADC with a limited

(a)

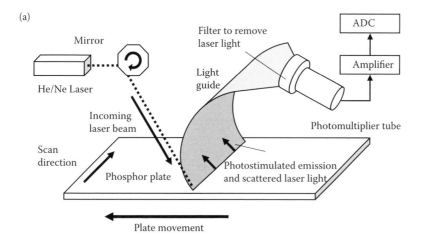

(b)

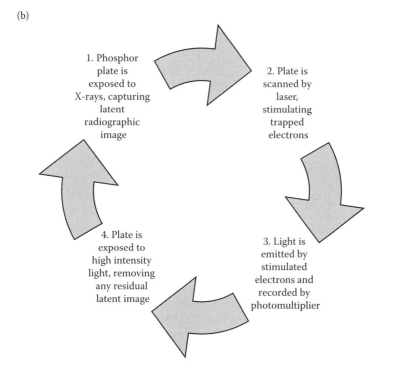

FIGURE 5.19

(a) The read out system for a digital phosphor plate. (b) The life cycle of the plate. (Reproduced with permission from Zhao W, Andriole K P and Samei E Digital radiography and fluoroscopy in *Advances in medical physics – 2006*. Wolbarst A B, Zamenhof R G and Hendee W R, eds, Medical Physics Publishing, Madison, WI, 1–23, 2006.)

bit depth. Ultimately each small area of the phosphor has a light output allocated to it. The image is stored in the computer memory.

To prepare the screen for reuse it has to be exposed to a high intensity flood light for a short period of time. This completely removes any remaining energy of excitation associated with residual trapped electrons from the screen and recreates a uniform response on the screen. The life cycle of the plate is shown in Figure 5.19b.

5.11.3 Properties

As both CR and standard film/screen systems are essentially based on the same image forming device, many of the properties are almost the same. The doses required to produce a usable image are similar with the lower limit on dose reduction essentially being the noise produced by quantum mottle. Spatial resolution for general CR work is slightly inferior to that of film-screen and the causes of loss of resolution are different. In a film/screen system the light spreads out isotropically following the conversion of an X-ray photon to light photons. In the storage phosphor system the main contribution to unsharpness comes from the scattering of the stimulating beam as it enters the phosphor thus creating a spread in the luminescence along its path. The intensity of the stimulating laser light also affects the resolution. A higher intensity beam results in a lower resolution but this is compensated for to some extent by the increase in the amount of stored signal released. The resolution is also limited by the size of the pixels sampled and read out time. If the laser spot has moved to the next point on the phosphor before all of the emitted light from the previous stimulation has been collected, this light will contribute to the signal of the neighbouring pixel. A matrix array of approximately 2000 by 2500 is typical.

The relative intensity of light emitted is proportional to X-ray exposure at the plate over approximately four decades. This 'latitude' allows CR to be used over a wide range of exposure conditions without adjusting the detectors. Image processing frequently permits images of acceptable diagnostic quality to be extracted from over-exposed or under-exposed plates—situations which would necessitate retakes with film screen. Also this wide dynamic range can be exploited in chest radiography using a windowing technique (Section 6.3.3) or in dual energy subtraction (Section 9.5.3).

Insight

CR and DR Compared and Contrasted

1. CR is a cassette system, DR a fixed permanent array of very small detectors.
2. As digital systems, both have all the advantages of systems that can electronically process, modify and display images and enter them into a PACS system.
3. Both have very wide dynamic ranges—thousands of grey levels and much greater than the latitude of film-screen combinations—see Section 5.2 (Insight) and Section 5.5.4.
4. They have comparable matrix sizes (typically 2000 × 2500) and spatial resolutions (3–5 lp mm^{-1}), which is inferior to that of film-screen (5–10 lp mm^{-1} or better sometimes).
5. CR systems require comparable or even higher patient doses than a 400 speed film-screen system. DR has higher detective QE than CR which implies diagnostic quality images at lower doses. However, there is no hard evidence that this is the case, perhaps suggesting that quantum noise is not the noise-limiting factor on image quality in these images.
6. Optimisation of both systems is required to ensure that the most appropriate radiation dose is delivered for the diagnostic outcome (see Section 7.9.1).
7. Digital imaging does not overcome the basic problem of all projection imaging—that the recorded image represents the superposition of attenuations through the whole patient thickness. The data comes from a volume element (voxel) which has a very small cross-sectional area but is very long. This problem is solved in CT (see Chapter 8).
8. Both CR and DR have widespread applications in general radiography, mammography, tomography and subtraction techniques.
9. CR has, until recently, provided more versatility for examination, for example, in the emergency room, intensive care and operating theatre. However, portable DR detectors are now able to provide this facility.

10. In terms of workflow, a DR room can image a greater number of patients in a session than CR if those patients are presenting for a standard examination, for example, chest radiography or mammography.
11. CR is generally available at a much lower initial cost than DR. This is especially true if replacing an existing film-screen system as the same X-ray equipment can be used. When a DR system is installed it is normally as part of a complete radiographic room. Quoted lifetimes of both CR and DR detectors are similar at about 7 years; however, CR plates are normally replaced at much shorter intervals than this due to mechanical damage associated with the handling and readout process of most readers. This provides an additional cost for CR systems.

5.12 Film Digitisation

For the foreseeable future a facility will be required to generate digitised images from plane films produced by departments lacking digital facilities so that they can be entered into a PACS system.

A simple but effective system can use a laser scanner and the steps in the procedure are as follows:

(a) The beam from either a He/Ne gas laser or a solid state diode laser is focussed with lenses and scanned under computer control in a raster pattern onto the film using a beam deflecting device.

(b) The light transmitted through the film at any point depends on the optical density at that point and falls on a photo detector.

(c) The laser and photo detector scan along the line of film producing a continuously varying intensity at the photo detector.

(d) This light is sampled at regular intervals, once per pixel.

(e) The analogue electrical signal is digitised and stored in a matrix of pixel addresses (x, y, co-ordinates and digitised photo detector reading).

The process is illustrated in Figure 5.20.

Insight

Advantages of Laser Beams for Digitising Films

1. The beam can be highly collimated—for example, producing spots as small as 50 µm.
2. This small spot and low light scattering results in high spatial resolution—4000 × 5000 pixels and 10 lp mm^{-1}.
3. Because of the high light intensity of the laser beam there are a large number of pixel levels (12 bits) so the accuracy is high. The system can easily accommodate the full OD range of film.

Flat bed CCD scanners, or less frequently digital cameras or analogue cameras with ADCs, may also be used to digitise films.

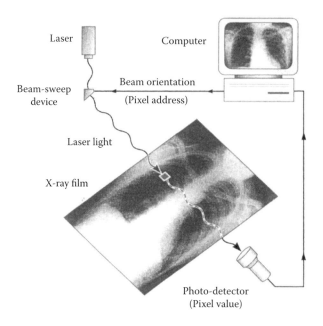

FIGURE 5.20

A simple arrangement for film digitisation—for detail of operation see text. (Reproduced with permission from Zhao W, Andriole K P and Samei E Digital radiography and fluoroscopy in *Advances in medical physics* – 2006. Wolbarst A B, Zamenhof R G and Hendee W R, eds, Medical Physics Publishing, Madison, WI, 1–23, 2006.)

5.13 Receptors Used in Fluoroscopy

In this section the production of analogue and digital images in real time using continuous X-ray exposure, that is fluoroscopy, is discussed. Direct viewing of a fluorescent screen is no longer acceptable because light output is low and the image is of inferior quality. The light intensity emitted from the fluorescent screen can only be increased by increasing the exposure rate from the X-ray tube or by increasing the screen thickness. Neither of these methods is acceptable, as the first increases the dose to the patient and the second reduces the resolution. Any increase in signal strength must, therefore, be introduced after the light has been produced. This is commonly achieved by using an image intensifier which increases the light level to such a degree that cone vision rather than rod vision may be used.

Insight

Eye-Phosphor Combination in Fluoroscopy

Before the invention of the image intensifier in the 1950s, fluoroscopic screens were the primary means by which moving radiographic images were viewed. It is instructive to investigate the reason why direct viewing of the screen results in poor images.

A fluoroscopic screen consists of plastic backing for protection, a zinc-cadmium sulphide (ZnCdS) scintillation screen, and a thick lead glass protective screen. The resolution of the ZnCdS screen varies with crystal size in exactly the same way and for the same reasons as the resolution

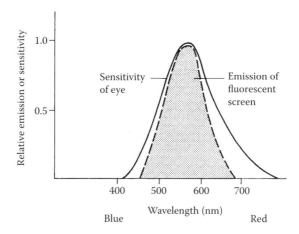

FIGURE 5.21
Comparison of the emission spectrum of visible light from a fluorescent screen with the sensitivity of the eye.

of a radiographic screen. The wavelength of emitted light is of course independent of the X-ray photon energy. When this technique was used historically, it was essential that certain basic preparatory steps were taken. For example, the light output was so low that the room had to be darkened and the eye fully dark-adapted. However, this caused further problems. The spectral output of a cadmium activated zinc sulphide screen closely matches the spectral response of the eye using cone vision at high light intensities (Figure 5.21) but the eye is poorly equipped for detecting changes in light levels or resolution at the low intensities of light given out by a fluoroscopic screen. Furthermore, these levels cannot be increased by increasing the X-ray intensity without delivering an even higher dose to the patient. At the levels of light involved, 10^{-3}–10^{-4} cdm^{-2}, the resolution of the eye is no better than about 3 mm at a normal viewing distance of 25 cm and only changes in light levels of approximately 20% can be detected. This is because vision at these levels is by rod vision alone, cone vision having a threshold of brightness perception of about 0.1 cdm^{-2}. Also the spectral response of rods peaks at a wavelength of 500 nm, where screen output tends to be poor (Figure 5.6). For further discussion see Section 7.3.2.

All of the figures quoted above are for a well dark-adapted eye, so the eye had to be at low light levels for 20–40 min before the fluoroscopic screen was viewed. This allowed the visual purple produced by the eye to build up and sensitise the rods. Alternatively, because the rods are insensitive to red light, red goggles could be worn in ambient light, producing the same effect. Exposure to ambient light for even a fraction of a second completely destroyed the build-up of visual purple.

5.13.1 Image Intensifiers

The basic image intensifier construction is shown in Figure 5.22. The intensifier is partially evacuated and the fluorescent screen is protected by a thin metal housing. This housing also excludes all fluorescent ambient light. The fluorescent screen is laid down on a very thin metal substrate and is now generally CsI. This has two advantages over most other fluorescent materials—see Section 5.4.1. It can be laid down effectively as a solid layer thus allowing a much greater X-ray absorption per unit thickness to be achieved. As described in Section 3.8 the K-absorption edges of caesium and iodine are at 36.0 and 33.2 keV, respectively. They thus occur at the maximum intensity of the X-ray beam used in most fluoroscopy examinations. The screen is also laid down in needle-like crystals (see Figure 5.23). These crystals internally reflect much of the light produced so it does not

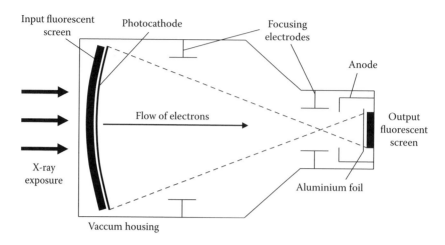

FIGURE 5.22
Construction of a simple image intensifier.

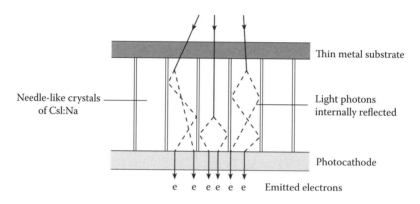

FIGURE 5.23
Greatly enlarged view of the needle-like crystalline structure of a CsI:Na screen.

spread out to cover a large area. Although not as efficient a collimator as a light pipe, light spreading is only about half that with an unstructured screen of the same thickness. CsI can thus be used in thicker layers without any significant loss in resolution although the packing advantage means that a thickness of only 0.1 mm is required.

Intimately attached to the CsI is the photocathode, a photoemissive material comprising antimony and caesium compounds that converts the light photons into electrons. The output of the fluorescent screen must be closely matched to the photo response of the photocathode (see Section 5.4.1). The input phosphor and photocathode are curved to ensure that the electrons emitted have the same distance to travel to the output phosphor. The output of the intensifier is via a second fluorescent screen, often of ZnCdS:Ag, shielded from the internal part of the intensifier by a very thin piece of aluminium. This stops light from the screen entering the image intensifier. This screen is much smaller than the input phosphor. A voltage of approximately 25 kV is maintained between the input and output phosphors.

The mode of action of an image intensifier is as follows. When light from the input fluorescent screen falls on the photocathode it is converted into electrons, and for the present discussion electrons have two important advantages over photons:

1. They can be accelerated.
2. They can be focused.

Thus under the influence of the potential difference of 25–30 kV, the electrons are accelerated and acquire kinetic energy as they travel towards the viewing phosphor. This increase in energy is a form of amplification. Second, by careful focusing so as not to introduce distortion, the resulting image can be minified, thereby further increasing its brightness. Even without the increase in energy of the electrons, the brightness of the output screen would be greater by a factor equal to the ratio of the input and output screen areas. The output phosphor is generally made of small crystals of silver activated zinc-cadmium sulphide which are laid down in a thin layer so as not to affect significantly the resolution of the minified image. The resolution of the image is 3–5 lp mm^{-1} in a new system (see Section 5.13.4).

The amplification resulting from electron acceleration is usually about 50. In other words, for every light photon generated at the input fluorescent screen, approximately 50 light photons are produced at the output screen. The increased brightness resulting from reduction in image size depends of course on the relative areas of the input and output screens. If the input screen is approximately 25 cm (10 inches) in diameter and the output screen is 2.5 cm (1 inch), the increase in brightness is $(25/2.5)^2 = 100$ and the overall gain of the image intensifier about 5000. By changing the field on the focusing electrodes it is possible to magnify part of the image. A direct consequence of this is that the intensification factor due to minification is reduced and the exposure of the patient must be increased to compensate. Magnification always leads to a greater patient dose.

5.13.2 Viewing the Image

The use of image intensifiers has not only produced an important improvement in light intensity, it has also opened up the way for the introduction of further technology into fluoroscopy. For example, direct viewing of the output of the image intensifier is not very convenient and somewhat restrictive, and although it is possible to increase the size of the image by viewing it through carefully designed optical systems, even with this arrangement there may be a considerable loss of light photons and, hence, information. In the worst cases this loss in light gathering by the eye can result in a system not much better than conventional fluoroscopy. A much better and more convenient way to display the image is through a video camera, for example a television camera.

The use of a television viewing system has several advantages. The first is convenience. The amplification available through the intensifier/television system allows a large image to be viewed under ambient lighting conditions thus eliminating the necessity for dark adaptation. The light output from the television monitor is well above the threshold for cone vision and no limit on resolution is imposed because of the limitations of the human eye. It is also a very efficient system allowing good optical coupling which results in little loss of information after the input stage of the image intensifier. The video signal can be recorded allowing a permanent record of the investigation to be kept. These records are available for immediate playback but there may be some loss of information and thus a reduction in image quality during the recording/playback sequence, depending on the equipment used. It is now also possible to use 'frame grabbing' techniques or a last frame hold. The latter allows intermittent screening to be carried out under some circumstances and some systems have fluoroscopy replay functions where the last sequence is stored for a second look if required. Most importantly, the method of operation of a TV camera is essentially a process of image digitisation so further discussion on the exact mode of operation will be deferred until Section 5.13.4.

5.13.3 Cinefluorography and Spot Films

Before the use of digital cameras, permanent records of the analogue image from a TV camera could be obtained by either cine or spot films. Figure 5.24 shows how a half-silvered mirror can be used to deflect part of the signal from the TV camera to a cine camera or a 70 mm/100 mm spot film camera.

Cinefluorography may place stringent demands on the X-ray tube whatever the receptor. For example, although the cine camera only needs to operate at 10 frames s^{-1} to see dynamic movement in the stomach, up to 30–40 frames s^{-1} may be required for coronary angiography. To avoid the X-ray tube rating problems and excessively high patient doses associated with continuous exposures, the X-rays are pulsed with a fixed relationship between the pulse and the frame movement (see Figure 5.25). The X-ray pulses are normally 2–5 ms long but the light output from the image intensifier does not fall to zero immediately after the X-ray pulse is terminated due to after-glow in the tube. The after-glow has, however, decreased to zero before the next frame is taken.

The images produced during cinefluorography may suffer from all the artifacts of conventional images (Section 6.12). If the investigation requires only a small field of view, it may be possible to use a smaller image intensifier with a 6° anode angle on the X-ray tube. This gives a greater output for a smaller effective spot size, thus reducing some of the artifacts as well as the heel effect.

Since cine film has a rather small dynamic range, some form of brightness control is required to compensate for variations in patient thickness. The sensing device, which is analogous to a phototimer, can either measure the current flowing across the image intensifier or the brightness of the output phosphor. This detector then provides feed-back control of either the tube kilovoltage, the tube current, or the effective exposure time per frame (pulse width). Each control mechanism has disadvantages. If the kilovoltage is driven too

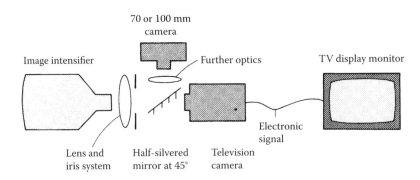

FIGURE 5.24
Method of coupling an image intensifier with a television camera using a half-silvered mirror to deflect part of the image to a spot or cine camera.

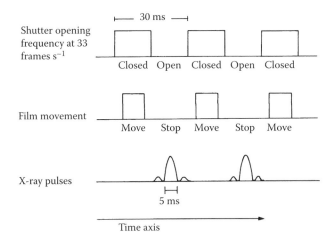

FIGURE 5.25
Synchronisation of the X-ray pulse with film frame exposure during cineflurography.

high, contrast is lost, whereas adjustment of tube current or pulse width has only a limited brightness range. These limitations can be largely overcome with a digital receptor.

Spot films (70 mm or 100 mm) are larger than cine film so the lens on the front of this camera has a longer focal length than that on a cine camera. In comparison with cinefluorography, typical advantages and disadvantages are as follows:

(a) Reduced screening time and patient dose
(b) Constant monitoring of films that can be processed quickly
(c) Film is cheaper and storage easier
(d) Higher doses to medical personnel because staff stay close to the patient
(e) More experience is required to learn the panning technique
(f) Smaller than normal films are not always easy to read

However, as with cinefluorography, a good digital system makes spot film redundant.

5.13.4 Digital Fluoroscopy

5.13.4.1 Image Intensifier-TV Systems

As mentioned briefly in Section 5.13.2, the mode of operation of a television camera, which converts the visual information in the image into electronic form, can be readily adapted for digitising images. The components of a system based on a TV camera coupled to an image intensifier are shown schematically in Figure 5.26. Note that the camera and ADC are shown separately here for convenience. In later digital cameras they would be combined in a single unit.

The two main components of the video camera (Figure 5.27a) are a focused electron gun and a specially constructed light sensitive surface. As shown in Figure 5.27b, the light sensitive surface is actually a double layer, the lower of which is the more important. It consists of a photoconductive material, usually antimony trisulphide, but constructed in such a way that very small regions of the photoconductor are insulated one from another by a matrix of mica.

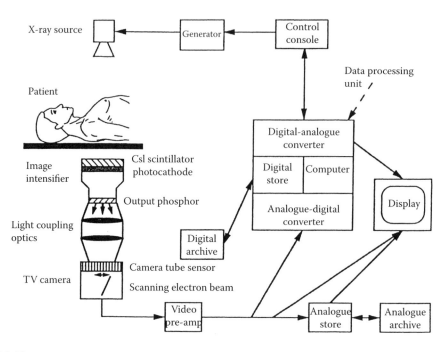

FIGURE 5.26
Block diagram representation of an image intensifier and TV system for digital radiography. (Reproduced with permission from Dendy P P Recent technical developments in medical imaging part 1: Digital radiology and evaluation of medical imaging. *Curr Imag*, 2, 226–236, 1990.)

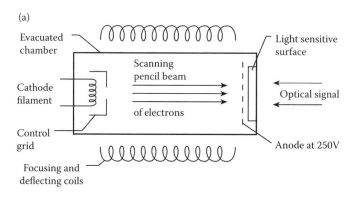

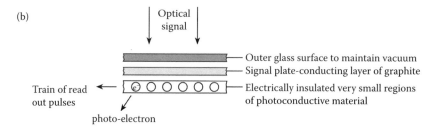

FIGURE 5.27
(a) Basic features of the construction of a video camera, (b) the light sensitive surface shown in detail.

When the camera is directed at visible light, photoelectrons are released from the antimony trisulphide matrix, to be collected by the anode and removed, leaving positive charges trapped there. The amount of charge trapped at any one point is proportional to the light intensity that has fallen on it. Thus the image information has now been encoded in the relative sizes of the positive charges stored at different points in the image plate matrix. These insulated positively charged areas draw a current onto the conductive plate until there is an equal negative charge held there.

The electrons emitted by the cathode are formed into a very narrow beam by the control grid. This beam is attracted towards the fine mesh anode which is at 250 V positive to the cathode. The signal plate has a potential some 225 V less than the anode, allowing the photoelectrons to flow from the signal plate to the anode. The electrons from the cathode pass through the mesh and this reversal in the potential field slows them down until they are almost stationary. They have an energy of only 25 eV when they strike the target plate. The field between the anode and the signal plate also straightens out the path of the electron beam so that it is almost at right angles to the signal plate. The electron beam is scanned across the signal plate by the scanning electrodes so that it only interacts with a few of the insulated areas at a time. Electrons flow from the electron beam to neutralise the positive charge on the target. The reduction in the positive charge releases an equivalent charge of electrons from the conducting layer and the flow of electrons from this layer constitutes the video signal from the camera. The natural line-scanning motion of the beam provides digitisation in one dimension and the data along the scan line can be digitised by registering the accumulated signal in regular, brief intervals of time.

The following additional points should be noted about a television system:

1. Irrespective of the resolving capability of the image intensifier system, the television system will impose its own resolution limit. Vertically the limit is set by the number of scan lines the electron beam executes. Whatever the size of the image, the electron beam only executes a fixed number of scan lines (for a standard TV camera, 625 in the United Kingdom, equivalent to about 313 line pairs). The effective number of line pairs, for the purpose of determining vertical resolution, is somewhat less than this—probably about 200. Now for a small image, say 5 cm in diameter, this represents four line pairs per mm which is comparable to the resolving capability of an image intensifier (see Section 5.13.1). However, any attempt to view a full 230 mm diameter (9 inch) screen would provide only about 0.8 line pairs per mm and would severely limit the resolving capability of the complete system. Therefore 1000 line and 2000 line TV cameras have been developed. Horizontally the limit is determined by the frequency at which the electrons are modulated as the electron beam scans across the screen.

2. Contrast is modified by the use of a TV system. It is reduced by the camera but increased by the television display monitor, the net result being an overall improvement in contrast (for further discussion of contrast see Section 6.3).

3. Rapid changes in brightness seriously affect image quality when using a television system. Thus a photo cell is incorporated between the image intensifier and the television camera with a feed-back loop to the X-ray generator. If there is a sudden change in brightness as a result of moving to image a different part of the patient, the X-ray output is quickly adjusted to compensate.

Insight

Image Reconstruction at Fast Frame Rates

Two approaches to image reconstruction are possible. With continuous low tube current exposure the TV frame can be digitised at 25 frames s^{-1}. Because of the low X-ray output, the noise in each frame is high but this can be reduced by summing several frames, provided there is little patient movement. For certain investigations, for example digital cardiac imaging, acceptably low noise rates are essential at framing rates that can be anything from 12.5 to 50 frames s^{-1}. This requires pulsed operation with a much higher output during the pulse.

5.13.4.2 Charge Coupled Devices

The second method of digitising fluoroscopic images uses the technology of solid state optical imaging scanners such as CCDs or complementary metal oxide semiconductors (CMOSs). The CCD will be described. CMOS are currently not in widespread use in medical imaging and are outside the scope of this book. All modern image intensifiers are coupled to a CCD camera.

A CCD sensor is an amorphous silicon (a-Si) wafer that has been etched to produce an array of elements (pixels) which are insulated from each other. A typical element would have a length of 40–50 μm and matrix size 1024 × 1024 up to 4096 × 4096. By applying a suitable biasing voltage, each element can be made to act as a capacitor storing charge. Thus upon exposure to radiation, electric charge proportional to the beam intensity is released in the a-Si and collected in individual pixels as in the TV camera to produce a charge image across the CCD. Unlike the TV camera, however, methods of read out are available that do not require a scanning beam of electrons. One method is to use a shift register (Figure 5.28). This depends for its action on the fact that each charge sits within a potential well in the matrix (see Section 1.1) but will move to an adjacent pixel if it has a deeper well. Hence by carefully controlling the depth of the potential well in each pixel, a given charge may be systematically moved around the matrix without overlapping or mixing with charge in adjacent pixels. Thus one set of control gates between the elements allows transfer of charges in one row of elements into the shift register line as shown by the vertical arrows. Once in the shift register the charges can be moved, in turn, to the right into the output gate where they can be read. This process is so rapid that read out of the matrix is effectively instantaneous. The analogue signal from each pixel is passed through the

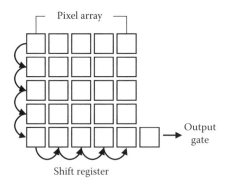

FIGURE 5.28
Schematic representation of charge shifting in a CCD.

signal wire from the shift register to an ADC and the digitised signal is then placed in a pixel array generated in the computer to match the detector element array.

An alternative method of read out allows the charge on each element to be measured by using the fact that simultaneous connections to a vertical line and a horizontal line uniquely define one element where they cross, allowing the charge on that element to be read.

Charge coupled devices are not suitable for recording digitised images directly from X-rays, since the a-Si layer is very thin with poor detection efficiency. In addition the CCD is susceptible to radiation damage at high fluxes. However, a CCD may be used very successfully in conjunction with a luminescent screen or a photoelectric cathode surface (e.g. the output from an image intensifier). The X-ray photons are first converted into visible light photons and then by the a-Si into electrons which are readily captured in the potential wells.

At present CCDs are small area detectors (up to about 60 mm × 60 mm). Ideally the visible light output from, say, the image intensifier should be focussed by an optical lens onto a single large-format CCD. If images obtained from a mosaic of smaller CCDs are stitched together there may be image distortion or loss of data at the seams.

The CCD has been used very successfully in digital fluoroscopy, with the following benefits over a TV camera:

(a) The resolution is fixed by the size and interspacing of the photosensitive elements. It does not vary with field size.

(b) There is no scanning electron beam to cause drifting.

(c) The geometry of the CCD camera is precise, uniform, distortion-free and stable (but see comment above on combining images from several small CCDs).

(d) The sensor is linear over a wide range of illumination.

(e) The CCD has a low read out noise which is helpful when using an optically coupled light detector in which there may be significant loss of light.

Insight

More on CCDs

Since their invention in 1969 CCDs have gradually been introduced into many photographic and video devices. They were largely introduced into diagnostic radiology in the 1990s when they replaced video camera tubes such as the Vidicon and Pumbicon to record the output of image intensifiers.

Since this introduction, CCDs have developed many uses within the diagnostic imaging department demonstrating their versatility. An example can be found in small area digital mammography systems used to localise areas of interest in the breast for biopsy. A further use is in large-area radiography systems where one or more CCDs are used in conjunction with a fibre optic taper or optical lens to record the output of a photoluminescent screen.

5.13.4.3 Flat Panel Detectors

The introduction of direct digital devices in radiography has been followed by the development of flat panel detectors for use in fluoroscopy. To achieve dynamic imaging at high speed, such as that required for cardiac imaging, some significant technical challenges have had to be overcome.

As with the digital receptors used in radiography, two main types are used in fluoroscopy, amorphous silicon or amorphous selenium (see Section 5.10). To achieve high speed imaging of up to 30 frames s^{-1}, the detector must be read very fast. As previously explained, an electric charge accumulates on the TFT array during exposure to X-rays. Each TFT in the array must be activated and read out in turn by a high speed processing unit, the output of which is then amplified and passed through an ADC. This clearly needs data to be processed by the acquisition computer at high bitrates. X-ray sensitivity is also a critical issue for digital detectors used in fluoroscopy, as the diagnostic images are acquired in a limited period of time.

Flat panel detectors cost more than image intensifiers, but their advantages include the fact that images appear uniform and un-distorted due to the lack of focussing electronics. Also they are much smaller and lighter which may result in easier movement and a less confined environment for both patient and operators (for further discussion see Section 9.6.2).

5.14 Quality Control of Image Receptors

Quality control checks of the performance of the image receptor are essential if X-ray imaging systems are to be used optimally, that is to obtain the required diagnostic information for the lowest achievable radiation dose to the patient.

The practicalities of quality control will depend on the type of image receptor used and to some extent, the type of work carried out by the system under examination.

5.14.1 X-ray Film

QC procedures are well-established and documented. They include a check on the condition of cassettes and screens—poor film/screen contact can cause loss of resolution; measurement of the film speed index; checking that a batch of cassettes used in a particular department or area have similar sensitivity (i.e. similar optical density when exposed to the same dose); measurement of gross fog, including suitability of safelight and absence of light leaks in the dark room.

One of the major causes of inferior quality X-ray films is poor processing and strict attention to developing parameters must be paid at all stages of processing. The development temperature must be controlled to better than 0.2 K and the film must be properly agitated to ensure uniform development. All chemicals must be replenished at regular intervals, generally after a given area of film has been processed, and care must be taken to ensure that particulate matter is removed by filtration. Automatic processing units monitor the chemicals continuously and automatically replenish them as required. Thorough washing and careful drying are essential if discolourations, streaks and film distortion are to be avoided. The whole process can be controlled by the preparation, at regular intervals, of test film strips obtained using a suitable *sensitometer*. This sensitometer consists of a graded set of filters of known optical density (Figure 5.29) which is placed over the film and exposed to a known amount of visible light in a darkened room. After processing, the film can be densitometered and a characteristic curve constructed (see Figure 5.10). This may be compared with previous curves and the manufacturer's recommendations. Since the construction of complete characteristic curves is a rather time-consuming process, it is normal only to make three or four spot checks routinely. These can then be plotted on a graph and if they stay within ± 0.1–0.15 of the original base line optical density they can be considered acceptable.

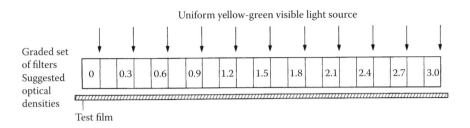

FIGURE 5.29
Use of a sensitometer, consisting of a graded set of filters of known optical density, for quality control of film processing.

5.14.2 CR and DR Receptors

In many ways the introduction of digital imaging has helped to make the quantitative assessment of performance easier. Obtaining measurements from a digital image is facilitated by software tools on the acquisition workstation and the image itself can be extracted for use in image analysis software packages. However, any processing algorithms that are applied to the image can have the effect of inherently changing the data that are being used to make the measurements and an understanding of the system's characteristics is essential, as is access to the unprocessed image.

CR and DR receptors are inherently more complex than film and, due to the relatively short time they have been in use, criteria for assessing their optimum performance are still being determined. Tests are constantly being reviewed in the light of new evidence regarding long term behaviour and relevance to clinical image quality. Some of the quality control tests carried out both CR and DR systems are as follows:

5.14.2.1 Dark Noise

This is the level of signal in a system when no radiation has been used to create an image. Sources for such noise are usually electronic but may be caused by residual signal from previous images.

5.14.2.2 Signal Transfer Property

Exposing the image receptor to a range of known doses and measuring the average pixel value produced in the resulting image gives information about the signal transfer property, the relationship between the input and output of the detector. For most DR systems this relationship would be expected to be linear within the diagnostic range of doses. For CR systems this relationship is normally logarithmic due to the photomultiplier response in the reader. Changes to the pixel value versus dose curve can be indicative of changes to the sensitivity of the imaging system. An assessment can also be made of the variation of noise with dose. Quantum noise will vary with the square root of the dose (see Section 7.5) but other sources of noise (e.g. electronic or structural noise) within the imaging system may vary over time.

5.14.2.3 Erasure Efficiency

Before taking an image, digital systems need to be clear of any residual signal from the preceding image. Any residual, or ghost, image could be mistaken for pathology or make the image diagnostically unacceptable.

5.14.2.4 Detector Uniformity

All flat panel detectors contain pixels that vary in their response to radiation and some will be defective and not respond at all. Software is often used to create a flat-field which presents a 'uniform' response to a uniform radiation field, either by applying a gain to each pixel to standardise its response or by interpolating a value from neighbouring pixels. Regular uniformity checks are essential for a DR system. Note that although the image is uniform, the noise is not because the heel effect (see Section 2.4) reduces the flux over part of the image field. Similarly non-uniformities can present themselves in CR images and these should be checked too.

5.14.2.5 Image Quality

A check of contrast and resolution can be made using test objects. While measurement can be performed visually, with image analysis software image quality can be measured in a number of different ways and will be discussed in Chapters 6 and 7.

When measuring the limiting resolution of a digital system an appreciation of the achievable resolution given by the pixel pitch is important. In CR it is important to measure resolution in both the scan and subscan directions, that is the direction of the laser scan and the direction of travel of the plate, since the resolution in each of these directions is governed by a different property.

5.14.2.6 Detector Dose Indicator (DDI) Calibration

The DDI gives the user an indication of the dose a receptor has received for a particular image. This is a very important indication that the image has been taken at the correct exposure, although it cannot be used to derive the dose to the patient. Each manufacturer currently uses their own methodology to derive DDI values and their relationship to receptor dose. Calibration is done under very specific beam energy and filtration conditions and should be checked as part of a QC program.

5.14.2.7 CR Plate Sensitivity

Batches of CR plates are exposed to a set dose of radiation and the spread of their DDIs is noted in a method similar to that for film-screen combinations.

5.14.3 Fluoroscopic Imaging Devices

Assessment of the performance of fluoroscopy systems introduces a number of problems in addition to those considered in Section 2.7 for standard X-ray sets. Performance of the system is closely related to the image it produces and the reader may wish to return to this section after studying Chapter 6. Among the more important measurements that should be made are as follows:

5.14.3.1 Field Size

A check that the field of view seen on the monitor is as big as that specified by the manufacturer should be made. The area of the patient exposed to radiation must not exceed that required for effective screening.

5.14.3.2 Image Distortion

The heavy dependence on electronic focusing in image intensifiers makes distortion much more likely than in a simple radiograph. Measurements on the image of a rectangular grid permit distortion to be checked.

Insight

Image Distortion in an Image Intensifier

Possible causes of distortion are as follows:

1. Focussing a curved input screen onto a flat output screen causes pincushion distortion.
2. Spatial distortion is caused by stray magnetic fields, including the Earth's magnetic field. Residual magnetisation in the steel of the building can also cause problems.
3. Large field of view modes suffer more distortion than small field of view.
4. The collection efficiency of electron-optical focussing lenses in the image intensifier, and optical lenses coupling the intensifier to cameras and display units is better in the centre. Radial fall-off in brightness is called *vignetting*.

5.14.3.3 Conversion Factor

For an image intensifier this is the light output per unit exposure rate, measured in candela per square metre for each $\mu Gy\ s^{-1}$. In general, this will be markedly lower for older intensifiers, partly due to ageing of the phosphor and partly because better phosphors are being used in the more modern intensifiers where conversion factors are typically 20–30 $cdm^{-2}\mu Gy^{-1}s$. There is little evidence that a poor conversion factor affects image quality directly, but of course to achieve the same light output a bigger radiation dose must be given to the patient. This test is not easily carried out and an alternative indication that a fall in conversion factor is taking place is to measure the input dose rate to the intensifier using an ionisation chamber and a standard, uniform phantom.

5.14.3.4 Contrast Capability and Resolution

As with the CR and DR receptors already mentioned, a number of test objects, for example, the Leeds Test Objects (Hay et al. 1985) have been devised to facilitate such measurements. One of them which is used to estimate noise in the image consists of a set of discs, each approximately 1 cm in diameter with a range of contrasts from 16% to 0.7% (for a definition of contrast in these terms see Section 7.4.2). They should be imaged at a specified kVp and with a specified amount of filtration (typically 70 kVp and 1 mm Cu to simulate the patient). At a dose rate of about 0.3 $\mu Gy\ s^{-1}$, a contrast difference of 2%–4% (certainly better than 5%) should be detectable. It is important to specify the input dose rate because, as shown in Figure 5.30, the minimum perceptible contrast difference is higher for both sub-optimum and supra-optimum dose rates. When the dose rate is too low this occurs because of quantum mottle effects. When it is too high there is loss of contrast because the video output voltage begins to saturate.

The limiting resolution is measured by using a test pattern consisting of line pairs, in groups of three or four, separated by different distances, and viewing the monitor for the minimum resolvable separation. Under high contrast conditions (50 kVp), at least 1.2 line pairs per mm should be resolvable for a 25 cm field of view. Note that the contrast conditions

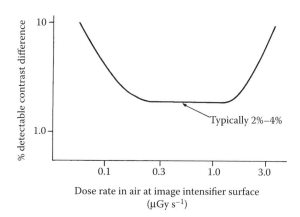

FIGURE 5.30
Variation of minimum perceptible contrast with dose rate for an image intensifier screening system.

must be controlled carefully since, in common with other imaging systems, the finest resolvable detail varies with contrast level for the image intensifier (see Section 7.3.3).

5.14.3.5 Automatic Brightness Control

It is important to ensure that screening procedures do not result in unacceptably high dose rates to the patient. One way to achieve this is by monitoring the output signal from the image receptor, changing the X-ray tube output (kV and/or mA) by means of a feedback loop whenever the attenuation in the X-ray beam changes, for example, with various sized patients. The way in which kVp and mA are changed can affect the dose received by the patient. Increasing the kVp before the mA will produce a lower patient dose but will reduce contrast. There is also a point of diminishing return where the energy of the peak in the X-ray spectrum moves beyond the K-edge of the input phosphor and the attenuation coefficient starts to fall rapidly. A number of kVp-mA curves or relationships may be available within a system for different purposes. Figure 5.31 shows three such curves. Note that each curve is limited by the maximum operating power of the X-ray tube—in this case 400 W—that cannot be exceeded. The input power of the fluoroscopic exposure can be determined by the product of the kVp and the mA. This leads in some cases to a reduction in mA if an increased patient thickness requires an increase in kVp—as seen with curve 3 of Figure 5.31.

Automatic control may mask deterioration in performance somewhere in the system. For example, loss of light output from the image intensifier screen could be compensated by increasing mA. X-ray output should therefore be checked directly.

Most systems should be capable of operating in the range 0.2–1 μGy s^{-1} but dose rates as high as 5 μGy s^{-1} have been reported for incorrectly adjusted equipment! Note that the skin dose to the patient will be, typically 100 times greater than this—about 500 μGy s^{-1} or 30 mGy min^{-1}.

5.14.3.6 Viewing Screen Performance

A calibrated grey-scale step wedge may be used to check that the contrast and brightness settings on the television monitor are correctly adjusted.

For further information on quality control see IPEM (2005).

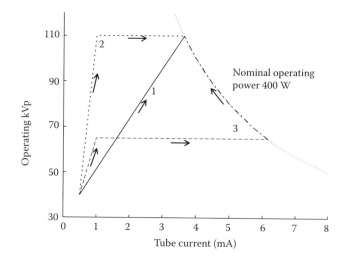

FIGURE 5.31

Different ways in which kV and/or mAs can be increased to give greater output. Curve 1 (solid line) gives a compromise between contrast and dose requirements, curve 2 (short dashes) minimises the dose by raising the kVp as required in paediatric imaging, and curve 3 (long dashes) holds a low kVp value where contrast is essential, for example, to visualise catheters. A curve similar to curve 3 may be used if the maximum keV is to be locked on to a K-shell absorption edge as used in iodine contrast studies.

5.15 Conclusions

A primary photon image cannot be viewed directly by the human eye and for the best part of 100 years photographic film was the only receptor used for simple radiographs in a wide range of applications, including general purpose radiography, mammography, dental work and photofluorography. At an early stage it was recognised that much greater sensitivity could be obtained with a fixed number of X-ray photons if a phosphor was used to convert each X-ray photon into a large number of visible light photons. Fluorescent screens with gradually improving performance have become an integral part of many image receptors resulting in greatly reduced doses to patients. It is important to understand the concepts of optical density, film gamma, film speed and latitude to appreciate how the images can be optimised.

The introduction of digital receptors has made rapid and dramatic inroads into the supremacy of film. Two of the three main imaging modalities, CR and DR indirect detectors with an active panel FPI still use a fluorescent layer as part of the receptor, but DR direct detectors convert the pattern of X-ray photons directly into a pattern of electrons and holes. Digital detectors have been, or are being developed for all applications of radiography.

Digital systems have advantages over film in respect of dynamic range and post-processing, including contrast enhancement (see Section 6.3.3) but their biggest attraction is that they greatly facilitate the management of image data in the X-ray department. When DR and CR are compared and contrasted, although the physical methods by which the images are produced are very different, in general radiography the final images do not show major differences. However, both technologies are developing rapidly so this equilibrium may not hold indefinitely.

Receptors used for imaging in real time with continuous X-ray exposure (fluoroscopy) are undergoing a similar revolutionary change. Many years ago the image intensifier was a major break-through in terms of image brightness and dose reduction and in the past it has been linked to a variety of hardware to display images, including TV monitors and cine cameras. These are now being replaced with technology that is fast enough to produce digital images at up to 30 frames s^{-1} using CCD cameras coupled to the image intensifier, or the image intensifier itself is being replaced by DR FPIs.

Receptors that incorporate optical and electronic imaging systems must be carefully maintained and subjected to quality control. For example with many phosphors it is possible for a slow degradation of the final image to occur. This degradation may only be perceptible if images of a test object taken at regular intervals are carefully compared, preferably using quantitative methods. Digital receptors are inherently more complex than a film-screen cassette and require correspondingly greater checking.

One final point is that each imaging process has its own limits on resolution and contrast and these will be discussed in greater detail in Chapter 6. However, it is important to emphasise here that the inter-relationship between object size, contrast and patient dose is a matter of everyday experience and not a peculiarity of quality control measurements or sophisticated digital techniques. Furthermore, the choice of imaging process cannot always be governed solely by resolution and contrast considerations. The speed at which the image is produced and is made available for display must often be taken into account.

References

Dendy PP *Recent technical developments in medical imaging part 1: digital radiology and evaluation of medical imaging.* Curr Imag 2, 226–236, 1990.

Hay GA, Clark OF, Coleman NJ and Cowen AR *A set of X-ray test objects for quality control in television fluoroscopy.* Br J Radiol 58, 335–344, 1985.

IPEM Report no 91 *Recommended standards for routine performance testing of diagnostic X-ray imaging systems.* Institute of Physics and Engineering in Medicine, York, UK, 2005.

Tesic M M, Mattson R A, and Barnes G T et al. *Digital radiography of the chest; design features and considerations for a prototype unit.* Radiology 148, 259–264, 1983.

Zhao W, Andriole K P and Samei E (2006) *Digital radiography and fluoroscopy in "Advances in medical physics" – 2006.* Wolbarst A B, Zamenhof R G & Hendee W R, eds, Medical Physics Publishing, Madison, WI, 2006, 1–23.

Further Reading

AAPM *Acceptance testing and quality control of photostimulable storage phosphor imaging systems.* Report of AAPM Task group 10, number 93, American Association of Physicists in Medicine, Maryland, US, 2006. (And earlier AAPM publications on QA of receptors).

Allisy-Roberts PJ and Williams J *Farr's Physics for Medical Imaging 2nd ed.* Saunders Elsevier, 2008, 65–102.

Bushberg J T, Seibert J A, Leidholdt E M and Boone J M *The essential physics of medical imaging in 2nd ed.,* chapter 11 *"Digital radiography."* Lippincott Williams & Wilkins Philadelphia, 2002.

Carroll Q B *Practical radiographic imaging 8th ed.* Charles C Thomas Publisher Ltd., Springfield, IL, 2007.

Carter C E and Vealé B L *Digital radiography and PACS* Mosby Elsevier, St Louis, Missouri, 2008.

Spahn M *Flat detectors and their clinical applications (review).* Eur Radiol 15, 1934–1947, 2005.

Exercises

1. What are the differences between digital and analogue images?

2. Explain how the intensification factors of a set of radiography screens might be compared. Summarise and give reasons for the main precautions that must be taken in the use of such screens.

3. Draw on the same axes the characteristic curves for

   (a) A fast film held between a pair of calcium tungstate plates

   (b) The same film with no screen and explain the difference between them.

4. Why is it desirable for the gamma of a radiographic film to be much higher than that of a film used in conventional photography and how is this achieved?

5. Explain what is meant by the speed of an X-ray film and discuss the factors on which the speed depends.

6. A radiograph is found to lack contrast. Under what circumstances would increasing the current on the repeat radiograph increase contrast, and why?

7. Make a labelled diagram of the intensifying screen-film system used in radiology. Discuss the physical processes that occur from the emergence of X-rays at the anode to the production of the final radiograph.

8. Discuss the factors which affect the sensitivity and resolution of a screen-film combination used in radiography and their dependence upon each other.

9. What is meant by the *accuracy* of an analogue to digital converter? How is its accuracy determined?

10. What are the essential differences between digital radiography and computed radiography?

11. Why is it important to retain the capability to digitise analogue images on films? Outline briefly how this can be done.

12. How does the difference in diameter of the input and output screens of an X-ray image intensifier contribute to the performance of the system?

13. Discuss the uses made of the brightness amplification available from a modern image intensifier, paying particular attention to any limitations.

14. Compare and contrast the use of fluorescent screens in radiography and fluoroscopy.

15. Explain how an image intensifier may be used in conjunction with a photoconductive camera to produce an image on a TV screen.

16. What is a charge coupled device and how may it be used in conjunction with an image intensifier to produce digital images?

17. Automatic brightness control indicates that more output is required from the X-ray generator for a screening procedure. Discuss the different ways in which kV and mA can be varied to achieve this and the relative merits of each.

6

The Radiological Image

O W E Morrish and P P Dendy

SUMMARY

- This chapter investigates the various factors that affect the quality of a radiological image.

- Contrast is a difficult quantity to define but since all features of an image depend on contrast this is a good place to start.

- When imaging a patient, tissue overlying and underlying the region of interest generates scatter, so the factors affecting scatter and its reduction are important.

- The sharpness of an image or its resolution is another important criterion of quality.

- Quantum mottle caused by the statistical nature of the interaction of X-ray photons with matter can seriously affect image quality.

- Image processing, made possible with the advent of digital images, is now an important way to enhance the image.

- Finally, some artefacts and distortions caused by the geometrical relationship between the receptor, patient and X-ray source will be considered.

CONTENTS

6.1 Introduction—the Meaning of Image Quality

As shown in Chapter 3, the fraction of the incident X-ray intensity transmitted by different parts of a patient will vary due to variations in thickness, density and mean atomic number of the body. This pattern of transmitted intensities therefore contains the information required about the body and can be thought of as the primary image.

However, this primary image cannot be seen by the eye and must first be converted to a visual image by interaction with a secondary imaging device. This change can be undertaken in several ways, each of which has its own particular features. The definition of quality for the resultant image in practical terms depends on the information required from it. In some instances it is resolution that is primarily required, in others the ability to see small increments in contrast. More generally the image is a compromise combination of the two, with the dominant one often determined by the personal preference of the radiologist. (This preference can change; the 'contrasty' crisp chest radiographs of several years ago are now rejected in favour of lower contrast radiographs which appear much flatter but are claimed to allow more to be seen.)

The quality of the image can depend as much on the display system as on the way it was produced. A good quality image viewed under poor conditions such as inadequate non-uniform lighting may be useless. The quality actually required in an image may also depend on information provided by other diagnostic techniques or previous radiographs.

This chapter extends the concept of contrast introduced in Chapter 5 to the radiological image and then discusses the factors that may influence or degrade the quality of the primary image. Methods available for improving the quality of the information available at this stage are also considered. Other factors affecting image quality in the broader context of the whole imaging process are discussed in Chapter 7.

6.2 The Primary Image

The primary image produced when X-ray photons pass through a body depends on the linear attenuation coefficient (μ) and the thickness of the tissue they traverse. At diagnostic energies μ is dependent on the photoelectric and Compton effects. For soft tissue, fat and muscle the effective atomic number varies from approximately 6 to 7.5. In these materials μ is thus primarily dependent on the Compton effect, which falls relatively slowly with increasing photon energy (see Section 3.4.3). The photoelectric effect is not completely absent, however, and at low photon energies forms a significant part of the attenuation process. In mammography low energy X-ray photons are used to detect malignant soft tissue which has a very similar Z value to breast tissue. The difference in attenuation between the two is due to the higher photoelectric attenuation of the higher Z material. For bone with a Z of approximately 14, most of the attenuation is by the photoelectric effect. This falls rapidly with increasing photon energy.

The Compton effect also decreases with increasing energy above about 50 keV and the resultant fall in attenuation coefficient with photon energy for tissue and for bone is shown in Figure 6.1.

6.3 Contrast

The definition of contrast differs somewhat depending on the way the concept is being applied. For conventional radiography and fluoroscopy, the normal definition is an

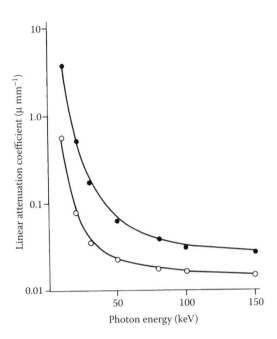

FIGURE 6.1
Variation of linear attenuation coefficient with photon energy for (○) muscle and (●) bone in the diagnostic region.

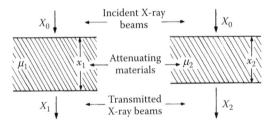

FIGURE 6.2
X-ray transmission through materials that differ in both thickness and linear attenuation coefficient.

extension of the definition introduced in Section 5.5.2. In Section 7.4.2 an alternative approach will be considered based on the signal to noise ratio in the image.

Consider the situation shown in Figure 6.2. This is clearly very similar to that in Figure 5.9. Contrast in the primary image will be due to any difference between X_1 and X_2 and by analogy with Equation 5.1 may be defined as

$$C = \log_{10} \frac{X_2}{X_1}$$

Converting to Naperian logarithms:

$$C = 0.43 \ln \frac{X_2}{X_1} = 0.43(\ln X_2 - \ln X_1)$$

Since, from Chapter 3

$$X_1 = X_0 \exp(-\mu_1 x_1)$$

and

$$X_2 = X_0 \exp(-\mu_2 x_2)$$

thus

$$C = 0.43(\mu_1 x_1 - \mu_2 x_2)$$

If μ_1 and μ_2 were the same, the difference in contrast would be due to differences in thickness. If $x_1 = x_2$ the contrast is due to differences in linear attenuation coefficient. It is conceivable that the product $\mu_1 x_1$ might be exactly equal to $\mu_2 x_2$ but this is unlikely. Note from Figure 6.1 that the difference in μ values decreases on moving to the right, thus contrast between two structures always decreases with increasing kVp.

6.3.1 Contrast on a Photoluminescent Screen

If this primary image is allowed to fall on a photoluminescent screen, the light emitted from those parts of the screen exposed to X_1 and X_2 say L_1 and L_2 will be directly proportional to X_1 and X_2. Hence

$$L_1 = kX_1 \quad \text{and} \quad L_2 = kX_2$$

The contrast

$$C(\text{screen}) = \log\frac{L_2}{L_1} = \log\frac{kX_2}{kX_1} = \log\frac{X_2}{X_1}$$

Hence the contrast on the screen, $C = 0.43(\mu_1 x_1 - \mu_2 x_2)$, is the same as in the primary image. As this is how the eye perceives the image, this is known as the radiation contrast and will be denoted by C_R.

Note that simple amplification, in a fluorescent or intensifying screen, does not alter contrast.

6.3.2 Contrast on Radiographic Film

If the transmitted intensities X_1 and X_2 are converted into an image on radiographic film, the contrast on the film will be different from that in the primary image because of the imaging characteristics of the film.

As shown in Section 5.5.2 the imaging characteristics of film are described by its characteristic curve. By definition

$$\gamma = \frac{D_2 - D_1}{\log E_2 - \log E_1}$$

So for ionising radiation

$$\gamma = \frac{D_2 - D_1}{\log X_2 - \log X_1} \tag{6.1}$$

Now from Section 6.3

$$\log X_2 - \log X_1 = 0.43(\mu_1 x_1 - \mu_2 x_2)$$

and from Section 5.5.2

$$D_2 - D_1 = C$$

Hence substituting in Equation 6.1

$$C = \gamma 0.43(\mu_1 x_1 - \mu_2 x_2)$$

Thus the contrast on film C_F differs from the contrast in the primary image by the factor γ which is usually in the range 3–4. Gamma is often termed the *film contrast*, thus

$$\text{Radiographic contrast} = \text{Radiation contrast} \times \text{Film contrast}$$
$$C_F \qquad\qquad\qquad C_R \qquad\qquad\qquad \gamma$$

Note that contrast is now modified because the characteristic curve relates two logarithmic quantities. Film can be said to be a *'logarithmic amplifier'*. A TV camera can also act as a logarithmic amplifier.

6.3.3 Contrast on a Digital Image

The final output of most direct digital image receptors (DR) is linear with the input exposure whereas the final output signal of a computed radiography phosphor plate (CR) is logarithmic with exposure.

   With both systems the raw data contains a very large number of levels of data (pixel levels). These have to be condensed to the much smaller number of grey levels that can be distinguished by the eye on the displayed black and white image. This is achieved with a look-up table (LUT), a function relating the range of grey level to the range of digital pixel values. Clearly, since this LUT is applied after the image acquisition, the contrast of the digital image may be easily altered to enhance contrast in the range of pixel values containing the greatest diagnostic interest (see Figure 6.3). This is one of the major benefits of digital radiography. Note that manipulation of receptor contrast cannot compensate for limitations on radiation contrast imposed by signal to noise ratio (see Section 7.4.2).

   The LUT is often characterised by the *window level* (the mean of the pixel values across which the grey levels are changing) and the *window width*. For example in Figure 6.3 the window level remains unaltered but on the right the window width has been reduced from 1000 to 200. Window levels and window widths are also important in the display of CT images—see Section 8.3.

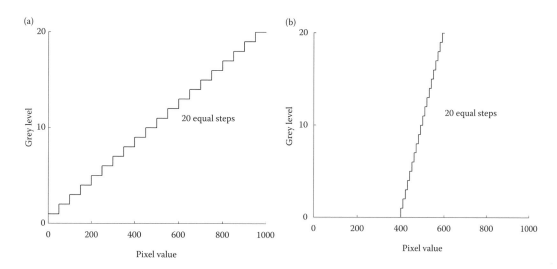

FIGURE 6.3
Examples of two simple LUTs; (a) Equal weight is given to all pixel values. Each grey level represents 50 values; (b) Greater emphasis is given to pixel values in the middle of the range. Each grey level now represents only 10 pixel values, enhancing contrast. All pixel values below 400, or above 600 will show white/black, respectively.

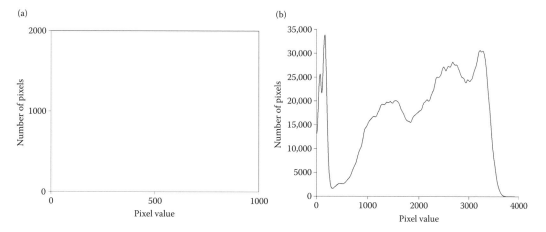

FIGURE 6.4
Histograms showing the frequency of each pixel value; (a) A very simple example—there are 1000 possible pixel values and 2×10^6 pixels in the image. Each pixel value is found in exactly 2000 pixels. Note that the histogram gives no information about the spatial distribution of individual pixel values; (b) A more realistic histogram for a chest X-ray. The majority of the pixel values are in the range 1000–3500 and the LUT can be adjusted so that the grey scale contrast is changing most rapidly over this range, which contains most of the diagnostic information. Note that the large peak to the left of the histogram is due to unattenuated X-rays that have not passed through the patient. Moving to the right, the next three peaks broadly represent the lung, soft tissue and bone areas of the image, respectively.

Application of the LUT to the raw image data is normally done in conjunction with *histogram equalisation*. This is a technique where the signal level of each pixel from the imager's output is represented on a histogram allowing the system to determine the distribution of pixel values in the image (see Figure 6.4), effectively assigning the average grey level to the average pixel value in the image. If necessary the upper and lower limits of the greyscale can be determined, analogous to the latitude of film.

The histogram information may be used to prepare the radiographic data for display such that it is optimised for the observer. An example can be given for a skull radiograph. If a dose in air of 10 mGy is incident on the skull, approximately 10^6 photons mm^{-2} will be transmitted. The fluctuation in this signal due to Poisson statistics is $\pm 10^3$ which is only 0.1% of the signal. Thus changes in signal of this order produced in the skull should be detectable. An alternative way of stating that a change of 0.1% in the signal can be detected is to say that there are 1000 statistically distinguishable pixel levels present in the image. Unfortunately, the eye can only distinguish about 20–30 grey levels and one of the major benefits of digital radiographic techniques over older techniques is that the histogram allows the grey levels to be matched to pixel levels containing clinically useful information.

Insight

Histogram Equalisation

An alternative way to think about histogram equalisation is that it makes maximum use of the available grey shades. If the number of pixel values is 1000 but the maximum pixel value in the histogram is only 500, only half the grey shades are being used (see Figure 6.5a). We wish to shift and stretch this histogram to make better use of the range. This may be done by applying to each pixel value a gain and bias given by the equation

$$\text{Pixel value}_{out} = (\text{Gain} \times \text{Pixel value}_{in}) + \text{Bias}$$

The gain is defined by the user—suppose we choose a value of 1.9—and the bias is obtained from the expression

$$\text{Bias} = \text{Mean}_{out} - (\text{Gain} \times \text{Mean}_{in})$$

Now set the new mean at 600. Then the bias is $600 - 570 = 30$, and our operating equation is

$$\text{Pixel value}_{out} = (1.9 \times \text{Pixel value}_{in}) + 30$$

(see Figure 6.5b).
    New minimum and maximum values are 95 and 980 and we have stretched and shifted the histogram as required (see Figure 6.5c).

6.3.4 Origins of Contrast for Real and Artificial Media

As discussed in Chapter 3 (see Table 3.2), attenuation and hence contrast will be determined by differences in atomic number and density.

Any agent introduced into the tissues, globally or selectively, to modify contrast may be termed a 'contrast enhancing agent'. Contrast may be changed artificially by introducing materials with either a different atomic number or with a different density and enhancement may be positive (more attenuation than other regions) or negative (less attenuation).

The physical principles of positive contrast enhancement have changed little since the earliest days. Iodine ($Z = 53$) is the obvious element to choose for contrast enhancement. Its K shell binding energy is approximately 34 keV so its cross-section for a photoelectric interaction with X-ray photons in the diagnostic energy range is high. Barium compounds

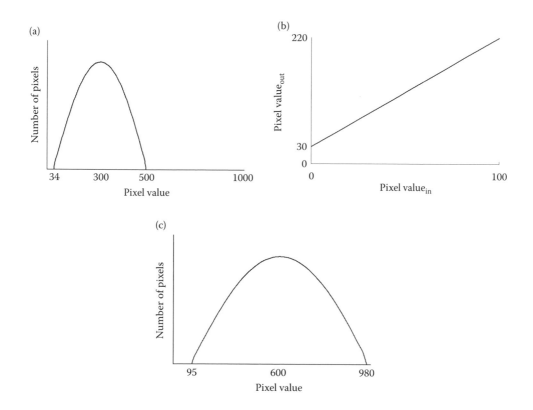

FIGURE 6.5

Stretching the histogram to use all pixel values; (a) Initial histogram—minimum = 34, mean = 300, maximum = 500; (b) The operating equation: Pixel value out = 1.9 × Pixel value in + 30; (c) Stretched histogram—minimum = 95, mean = 600, maximum = 980. Note that since (a) and (c) are histograms, the outline should be a series of steps but these are too small to show here.

($Z = 56$) with a similar K shell energy are used for studies of the alimentary tract and colon.

Negative contrast may be created by the introduction of gas, for example CO_2 in the bowel in double contrast studies. This makes use of the big difference in density between gas and soft tissue. Note that modification of atomic number is very kVp dependent whereas modification of density is not.

An example of the use of a contrast agent is in the study of urinary tract function. This can be investigated by injecting into the circulatory system an iodine compound which will then be filtered out of the blood by the kidneys and passed along the ureters to the bladder. Alternatively catheters can be inserted into the ureters and an iodine compound injected in retrograde fashion up the ureters into the renal pelvis.

The number of 'contrast' materials available is limited by the requirements for such materials. They must have a suitable viscosity and persistence and must be miscible or immiscible with body fluids as the examination requires. Most importantly they must be non-toxic. One contrast material, used for many years in continental Europe, contained thorium which is a naturally radioactive substance. It has been shown in epidemiological studies that patients investigated using this contrast material had an increased chance of contracting cancer or leukaemia.

Iodine-based compounds carry a risk for some individuals and most developments in the past 60 years have been directed towards newer agents with lower toxicity (Dawson and Clauss 1998). These have included achieving the same contrast at reduced osmolarity by increasing the number of iodine atoms per molecule and reducing protein binding capacity by attaching electrophilic side chains.

A recent possibility has been the production of non-ionic dimers which provide an even higher number of iodine atoms per molecule. These rather large molecules impart a high viscosity to the fluid but this can be substantially overcome by warming the fluid to body temperature before injection.

6.4 Effects of Overlying and Underlying Tissue

Under scatter free conditions, and in the absence of beam hardening, it may be demonstrated that a layer of uniformly attenuating material either above or below the region of differential attenuation has no effect on contrast. Consider the situation shown in Figure 6.6 where the two regions are shown separated for clarity.

Radiation contrast,

$$C_R = \log_{10} \frac{X_2'}{X_1'} = 0.43 \ln \frac{X_2'}{X_1'}$$

Now

$$X_1' = X_0' \, e^{-\mu_1 x_1}$$

And

$$X_2' = X_0' \, e^{-\mu_2 x_2}$$

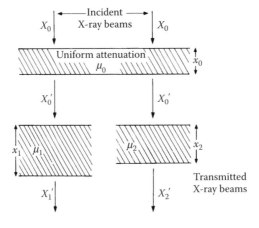

FIGURE 6.6
Diagram showing the effect of an overlying layer of uniformly attenuating material on the X-ray transmitted beam of Figure 6.2.

Hence

$$C_R = 0.43 \ln \frac{X_0' \, e^{-\mu_2 x_2}}{X_0' \, e^{-\mu_1 x_1}}$$

Thus

$$C = 0.43(\mu_1 x_1 - \mu_2 x_2) \text{ as before.}$$

The fact that some attenuation has occurred in overlying tissue and that $X_0' = X_0 e^{-\mu_0 x_0}$ is irrelevant because X_0' cancels. A similar argument may be applied to uniformly attenuating material below the region of interest.

An alternative way to state this result is that under these idealised conditions logarithmic transformation ensures that equal absorber and/or thickness changes will result in approximately equal contrast changes whether in thick or thin parts of the body.

6.5 Reduction of Contrast by Scatter

In practice, contrast is reduced by the presence of overlying and underlying material because of scatter. The scattered photons arise from Compton interactions. They are of reduced energy and travel at various angles to the primary beam.

The effect of this scatter, which is almost isotropic, is to produce a uniform increase in incident exposure across the image receptor. It may be shown, quite simply, that the presence of scattered radiation of uniform intensity invariably reduces the radiation contrast.

If $C_R = \log_{10}(X_2/X_1)$ and a constant X_0 is added to the top and bottom of the equation to represent scatter,

$$C_R' = \log_{10} \frac{X_2 + X_0}{X_1 + X_0}$$

The value of C_R' will be less than the value of C_R for any positive value of X_0.

The presence of scatter will almost invariably reduce contrast in the final image for the reason given above. The only condition under which scatter might increase contrast would be for photographic film if X_1 and X_2 were so small that they were close to the fog level of the characteristic curve. This is a rather artificial situation.

The amount of scattered radiation can be very large relative to the unscattered transmitted beam. This is especially true when there is a large thickness of tissue between the organ or object being imaged and the film. The ratio of scatter to primary beam in the latter situation can be as high as eight to one but is more generally in the range between two and four to one (see Figure 6.7).

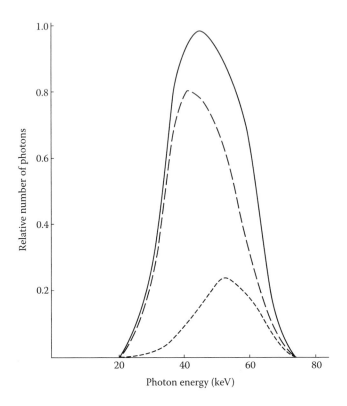

FIGURE 6.7
Typical primary, scatter and total spectra when a body-sized object is radiographed at 75 kVp ——— total; – – – – –
scatter; ------- primary (only the continuous spectrum is shown).

Insight

Effect of Scatter on Contrast

If $C_R = \log_{10}(X_2/X_1)$ and the difference between X_2 and X_1 is small, this may be written as

$$C_R = \log_{10}\left(\frac{X_1 + \Delta X}{X_1}\right) \quad \text{or} \quad \log_{10}\left(1 + \frac{\Delta X}{X_1}\right)$$

Since $\Delta X/X_1$ is small, this simplifies to $C_R = \Delta X/X_1$.

Re-label this equation as $C_0 = \Delta P/P$ where C_0 is the contrast in the absence of scatter and ΔP is the variation in primary beam due to attenuation.

In the presence of added scatter S:

$$C_S = \frac{\Delta P}{P + S}$$

Rearranging

$$C_S\left(1 + \frac{S}{P}\right) = \frac{\Delta P}{P} = C_0$$

$$C_S = C_0 \left(1 + \frac{S}{P} \right)^{-1}$$

Hence when $S = P$, $C_s = C_0/2$, that is, if the scatter contribution is equal to the primary, the contrast is halved.

Note that for a well-filtered beam the effect of beam hardening will be small compared with the effect of scatter.

6.6 Variation in Scatter with Photon Energy

If it is necessary to increase the kVp to compensate for loss of intensity due to lack of penetration or to try to reduce the radiation dose to the patient, the amount of scatter reaching the receptor increases. This is the result of a complex interaction of factors, some of which increase the scatter, others decrease it.

1. The amount of scatter actually produced in the patient is reduced because
   (a) The probability that an individual photon will be scattered decreases as the kVp is increased, although the Compton interaction coefficient only decreases slowly in the diagnostic range.
   (b) A smaller amount of primary radiation is required to produce a given response at the receptor since there is less body attenuation.
2. However, the forward scatter leaving the patient will be increased because
   (a) The fraction of the total scatter produced going in a forward direction increases as the kVp rises.
   (b) The mean energy of the scattered radiation increases and thus less of it is absorbed by the patient.

In practice, as the kVp rises from 50 to 100 kVp, the fall in linear attenuation coefficient of the low energy scattered radiation in tissue is much more rapid than the fall in the scatter-producing Compton cross-section. Thus factor 2(b) is more important than factor 1(a) and this is the prime reason for the increase in scatter reaching the receptor.

The increase in scatter is steep between 50 and 100 kVp but there is little further increase at higher kVp and above 140 kVp the amount of scatter reaching the receptor does start to fall slowly.

6.7 Reduction of Scatter

There are several ways in which scatter can be reduced.

6.7.1 Careful Choice of Beam Parameters

A reduction in the size of the beam to the minimum required to cover the area of interest reduces the volume of tissue available to scatter X-ray photons.

A reduction of kVp will not only increase contrast but will also reduce the scatter reaching the receptor. This reduction is, however, limited by the patient penetration required and, perhaps more importantly, an increase in patient dose due to the increase in mAs required to compensate for the reduction in kVp. (For film, a decrease of 10 kVp would require a doubling of the mAs for the same blackening.)

6.7.2 Orientation of the Patient

The effect of scatter will be particularly bad when there is a large thickness of tissue between the region of interest and the receptor. When the object is close to the receptor it prevents both the primary beam and scatter reaching the receptor. The object stops scatter very effectively since the energy of these photons is lower than those in the primary beam. Thus the region of interest should be as close to the receptor as possible—see also Section 6.9.1 on geometric effects. In practice other requirements of the radiograph generally dictate the patient orientation.

6.7.3 Compression of the Patient

This is a well-known technique that requires some explanation. It is important to appreciate that the process a physicist would call compression, for example a piston compressing a volume of gas, will not reduce scatter. Reference to Figure 6.8 will show that the X-ray photons encounter exactly the same number of molecules in passing through the gas on the right as on the left. Hence there will be the same amount of attenuation and the same amount of scatter.

When a patient is 'compressed', soft tissue is actually forced out of the primary beam, hence there is less scattering material present and contrast is improved.

6.7.4 Use of Grids

This is the most effective method for preventing the scatter leaving the patient from reaching the film and is discussed in the next section.

6.7.5 Air Gap Technique

If the patient is separated from the film, some obliquely scattered rays miss the film. This technique is discussed more fully in Section 9.3.3.

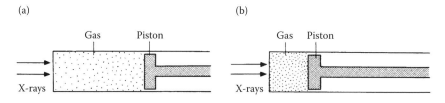

FIGURE 6.8
Demonstration that compression in the physical sense will not alter the attenuating properties of a fixed mass of gas (a) gas occupies a large volume at low density; (b) gas occupies a much smaller volume at a higher density.

6.7.6 Design of Intensifying Screen and Cassette

Since some radiation will pass right through radiographic film, it is important to ensure that no X-ray photons are back-scattered from the cassette. A high atomic number metal backing to a film cassette will ensure that all transmitted photons are totally absorbed by the photoelectric effect at this point. With daylight loading systems and a shift towards a reduction in weight of the cassettes this high Z backing has tended to be reduced or discarded. However, in CR, due to the higher sensitivity of storage phosphor systems to scattered radiation (see Section 5.11), their cassettes incorporate a lead screen to minimise the effects of backscatter.

The slightly greater sensitivity of photoluminescent screens to the higher energy primary photons may be only of marginal benefit in reducing the effect of scatter on the film if the position of the K-edge of the screen phosphor is well below the peak photon energy in the respective intensity spectrum (see Sections 2.2.4 and 3.8).

6.8 Grids

6.8.1 Construction

The simplest grid is an array of long parallel lead strips held an equal distance apart by a material with a very low Z value (an X-ray translucent material). Most of the scattered photons, travelling at an angle to the primary beam, will not be able to pass through the grid but will be intercepted and absorbed as shown in Figure 6.9. Some of the rays travelling at right angles, or nearly right angles, to the grid are also stopped due to the finite thickness of the grid strips, again shown in Figure 6.9. Both the primary beam and the scatter are stopped in this way but the majority of the primary beam passes through the grid along with some scattered radiation. Scatter travelling at an angle of $\theta/2$ or less to the primary beam is able to pass through the grid to point P (Figure 6.10).

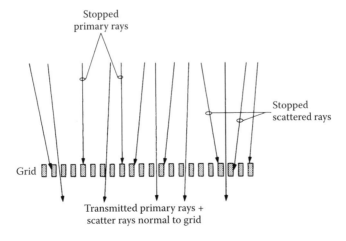

FIGURE 6.9
Use of a simple parallel grid to intercept scattered radiation.

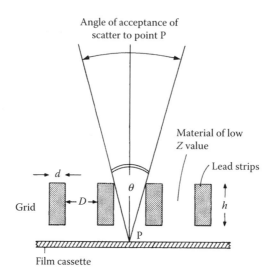

FIGURE 6.10
Grid geometry. Number of strips mm^{-1}, $N = 1/(D + d)$; typically N is about 4 for a good grid. Grid ratio $r = h/D$; typically r is about 10 or 12. Fraction of primary beam removed from the beam is $d/(D + d)$. Since d might be 0.075 mm and $(D + d)$ 0.25 mm, $d/(D + d)$ will be about 0.3. Tan $(\theta/2) = D/2h$.

As grids can remove up to 90% of the scatter there is a large increase in the contrast in radiographs when a grid is used. This increase is expressed in the 'contrast improvement factor' K where

$$K = \frac{\text{X-ray contrast with grid}}{\text{X-ray contrast without grid}}$$

K normally varies between 2 and 3 but can be as high as 4. The higher values of K are normally achieved by increasing the number of grid strips per centimetre. As these are increased, more of the primary beam is removed due to its being stopped by the grid. The proportion stopped is given by

$$\frac{d}{D+d}$$

where d is the thickness of a lead strip and D is the distance between them (Figure 6.10). Reduction of the primary beam intensity means that the exposure must be increased to compensate. The use of a grid therefore increases the radiation dose to the patient and thus there is a limit on the number of strips per centimetre that can be used. In addition, the inter-space material will also absorb some of the primary beam. Crossed grids (see Section 6.8.2) require a greater increase in patient dose than parallel grids.

6.8.2 Use

As grids are designed to stop photons travelling at angles other than approximately normal to them, it is essential that they are always correctly positioned with respect to the central ray of the primary beam. Otherwise, as shown in Figure 6.11, the primary beam will be stopped. The fact that the primary photon beam is not parallel but originates from a point source limits the size of film that can be exposed due to interception of the primary beam by the grid (Figure 6.12). The limiting rays are shown where

$$\tan \psi = \frac{C}{\text{FRD}}$$

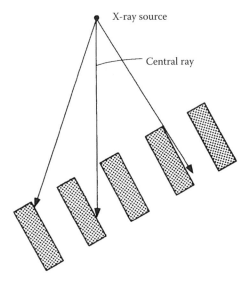

FIGURE 6.11
Diagram showing that a grid which is not orthogonal to the central X-ray axis may obstruct the primary beam.

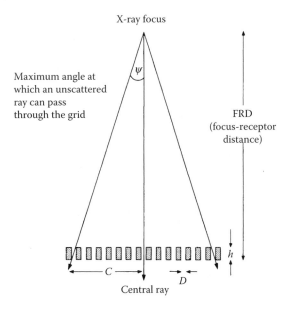

FIGURE 6.12
Demonstration that the field of view is limited when using a simple linear grid.

Tan ψ can be calculated from the grid characteristics. By similar triangles,

$$\tan \psi = \frac{D}{h}$$

where D is the distance between strips and h is the height of a strip.

Grids are identified by two factors. The grid ratio, which is defined as h/D (from Figure 6.10) and the number of strips per centimetre. In practice the two are interdependent. This is because there is an optimum thickness for the strips as they cannot be reduced in thickness to allow more strips per centimetre without reducing their ability to absorb the scattered radiation and thus their efficiency. Decreasing the gap between the strips to increase the number of strips will change the grid ratio unless the height h of the strip is reduced. The grid then becomes too thin to be of any use. The grid ratio varies between 8:1 and 16:1. A ratio of 8:1 is only likely to be used for exposures at a kilovoltage of less than approximately 85 kVp. For exposures at higher kVps the choice is between 10:1 and 12:1 grid. Most departments probably choose the 12:1 grid because although it results in a higher dose to the patient the improvement in contrast for thick sections is thought to be justified. Grid ratios as high as 16:1 should not be used except under exceptional circumstances as the increase in contrast is rarely justified by the increase in dose to the patient.

Insight

Grids and Computed Radiography

The use of grids can create a moiré pattern artefact when used with computed radiography systems if careful selection of grid ratios is not considered. Moiré patterns occur when two grids, or patterns, of a similar spatial frequency are overlaid onto each other with a slight offset, creating areas of interference.

The lines of the anti-scatter grid and the lines created by the scanning laser in a CR reader are two such patterns. Consider an image of a grid taken using a storage phosphor and scanned by a laser with the grid aligned such that the grid lines are parallel to the scan lines. The resultant signal detected by the scan line will either represent a bar, a gap or some combination of both depending on where the scan line falls relative to the grid lines. This would normally create a repeating pattern within the image seen in Figure 6.13a. However, if the frequency of the grid were to be slightly different to the frequency of the scan lines, it can be seen that the repetition of the pattern in the image would be disrupted and replaced by the type of interference seen in Figure 6.13b leading to an under-sampling of the grid lines causing aliasing (see Section 8.4.2).

This effect can be avoided by having a sufficiently high grid frequency and passing the pre-sampled signal through an anti-aliasing filter to remove frequencies above the Nyquist limit of the reader. Some systems provide a grid suppression algorithm such that the appearance of grid lines is reduced. This relies on the grid frequency being within a range that can be detected by the algorithm. When using a grid with CR systems, the grid lines should run perpendicular to the direction of travel of the scanning laser.

The grids shown in previous figures are termed linear grids and should be used with the long axis of the grid parallel to the cathode anode axis of the tube so that angled radiographs can be taken without the primary beam hitting the lead strips.

The simple linear grid has now been largely replaced by the focused linear grid where the lead strips are progressively angled on moving away from the central axis (Figure 6.14).

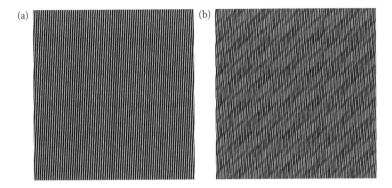

FIGURE 6.13

Illustration of moiré patterns. Image (a) shows the image of a regular series of lines, such as those found with radiographic grids, sampled at a frequency sufficiently high to provide an accurate image. Image (b) demonstrates the moiré patterns generated when the grid in image (a) is under-sampled at a lower frequency.

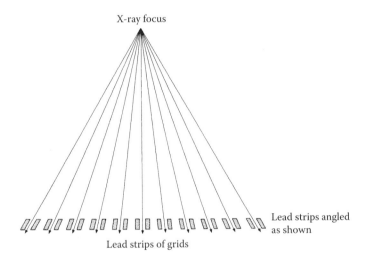

FIGURE 6.14

Construction of a linear focused grid.

This eliminates the problem of cut-off at the periphery of the grid but imposes restrictive conditions on the focus-receptor distances (FRDs) that can be used, the centering of the grid under the focal spot and having the correct side of the grid towards the X-ray tube. If any of these is wrong the primary beam is attenuated. If the grid is upside down then a narrow exposed area will be seen on a test film with very little blackening on either side of it. Decentering tends to produce generally lighter films which get lighter as the amount of decentering is increased. Using the wrong FRD will not affect the central portion of the radiograph but will progressively increase cut-off at the edge of the film as the distance away from the correct FRD is increased. Note that these faults might be difficult to recognise in a digital system since the processing software will tend to compensate for them.

Crossed grids with two sets of strips at right angles to each other are also sometimes used. This combination is very effective for removing scatter but absorbs a lot more of the primary beam and requires a much larger increase in the exposure with consequent increase in patient dose.

6.8.3 Movement

If a stationary grid is used it imposes on the image a radiograph of the grid as a series of lines (due to the absorption of the primary beam). With modern fine grids this effect is reduced but not removed.

The effect can be overcome by moving the grid during exposure so that the image of the grid is blurred out. Movements on modern units are generally oscillatory, often with the speed of movement in the forward direction different from that on the return. Whatever the detailed design, the movement should be such that it starts before the exposure and continues beyond the end of the exposure. Care must also be taken to ensure that, in single phase machines, the grid movement is not synchronous with the pulses of X-rays from the tube. If this occurs, although the grid has moved between X-ray pulses, the movement may be equal to an exact number of lead strips. The lead strips in the grid are thus effectively in the same position as far as the radiograph is concerned. This is an excellent example of the stroboscopic effect. Medium and high frequency machines are, of course, not troubled by this effect. One disadvantage of the focussed over the simple linear grid is that the decentering of the grid during movement results in greater absorption of the primary beam.

6.9 Resolution and Unsharpness

The resolution of a radiological image depends on factors associated with different parts of the imaging system. The most important are geometric unsharpness, patient unsharpness and the resolution of the final imager. For the present discussion, resolution and unsharpness are considered synonymous—a reasonable assumption in most cases. For a more detailed discussion of the relationship between resolution and unsharpness see Section 7.7.

Except on rare occasions in the case of fluoroscopic screens, the imaging devices used in radiography are not the major cause of loss of resolution. There are various other sources of unsharpness of which the more important are geometric unsharpness and inherent patient unsharpness.

6.9.1 Geometric Unsharpness

Geometric unsharpness is produced because the focal spot of an X-ray tube has finite size (Figure 6.15). Although the focal spot has a dimension b on the anode, the apparent size of the focal spot for the central X-ray beam a is much reduced due to the slope on the anode. The dimension normal to the plane of the paper is not altered. If, as shown in Figure 6.15, a sharp X-ray opaque edge is placed directly under the centre of the focal spot, the image of the edge is not produced directly underneath at T but extends from S to U, where S is to the left of T and U to the right of T. By analogy with optics, the shadow of the object to the left of S is termed the *umbra*, the region SU is the *penumbra*. This penumbra, in which on

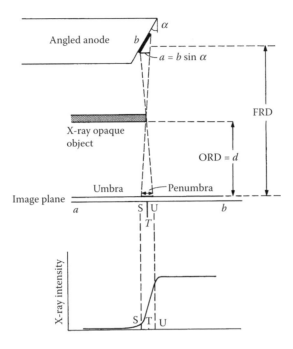

FIGURE 6.15
The effect of a finite X-ray focal spot size in forming a penumbral region. FRD = focus-receptor distance; ORD = object-receptor distance.

moving from S to U the number of X-ray photons rises to that in the unobstructed beam, is termed the *geometric unsharpness*. The magnitude of SU is given by

$$SU = b \sin \alpha \frac{d}{(FRD - d)} \tag{6.2}$$

For target angle $\alpha = 13°$, $b = 1.2$ mm, FRD = 1 m and $d = 10$ cm then SU = 0.06 mm. Since the focal spot size is strictly limited by rating considerations (Section 2.5.3), a certain amount of geometric unsharpness is unavoidable.

Note that

(1) The size of the focal spot places a lower limit on the size of object that can be distinguished. Very small objects, for example, a small calcification on a mammogram or a small vascular embolism may not be visualised if the penumbra from the focal spot is too large.

(2) Careful examination of Figure 6.15 shows that as the penumbra increases the umbra decreases. However, the actual size of the image on the radiograph is only altered significantly when the size of the object to be radiographed approaches or is less than the size of the focal spot so in normal practice the size of the focal spot has little effect on the magnification.

The effect of geometrical factors on unsharpness may be investigated using a line-pair test object. This consists of groups of highly attenuating lead bars separated by gaps of thickness equal to the bars. When imaged by relatively low energy X-rays the lead attenuates most of the beam while the gap transmits most of it resulting in a high contrast 'pair' of lines. Each group of line-pairs is progressively thinner than the last defined by the number that fit with a unit length—line-pairs per millimetre or lp mm^{-1}. A typical range for the template would be 0.5 to 10 lp mm^{-1} corresponding to resolutions of 1 to 0.05 mm.

Figure 6.16a and 6.16b show that for fixed values of FRD and d, resolution increases with decreasing focal spot size. Similarly, Figure 6.17a and 6.17b show that for fixed focal spot size, and fixed d, resolution improves with increasing FRD as predicted by Equation 6.2.

6.9.2 Patient Unsharpness

Other sources of unsharpness can arise when the object being radiographed is not idealised, that is, infinitely thin yet X-ray opaque. When the object is a patient, or part of a

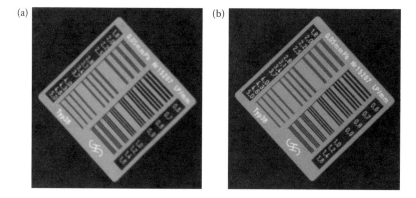

FIGURE 6.16
Effect of focal spot size on resolution. Two images of a line-pair test object taken with fixed values of FRD (100 cm) and d (30 cm); (a) focal spot size = 1.3 mm, limiting resolution 1.7 lp mm^{-1} ; (b) focal spot size = 0.6 mm, limiting resolution 2.2 lp mm^{-1}.

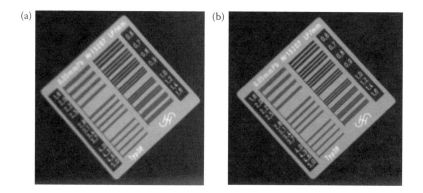

FIGURE 6.17
Effect of FRD on resolution. Two images of a line-pair test object taken with fixed values of focal spot size (1.3 mm) and d (30 cm); (a) FRD = 70 cm, limiting resolution = 1.2 lp mm^{-1}; (b) FRD = 100 cm, limiting resolution = 1.7 lp mm^{-1}.

patient, it has a finite thickness generally with decreasing X-ray attenuation towards the edges. These features can be considered as part of the geometric unsharpness of the resulting image and are in fact often much larger than the geometric unsharpness described in Section 6.9.1. The effect on the number of photons transmitted is shown in Figure 6.18. As can be seen there is a gradual change from transmission to absorption, producing an indistinct edge.

Another source of unsharpness arises from the fact that during a radiograph many organs within the body can move either through involuntary or voluntary motions. This is shown simply in Figure 6.19, where the edge of the organ being radiographed moves from position A to position B during the course of the exposure. Again the result is a gradual transition of radiographic density resulting in an unsharp image of the edge of the organ. The main factors that determine the degree of movement unsharpness are the speed of movement of the region of interest and the time of exposure. Increasing the patient-receptor distance increases the effect of movement unsharpness.

6.9.3 Receptor Unsharpness

An ultimate limit on resolution is provided by the inherent resolution of the image recording system. As discussed in Section 5.8 the resolution of film is much better than even the best film-screen combination which is limited by light spread within the intensifying screen. The resolution of the most up-to-date image intensifier/television systems is about 0.5 mm. In digital imaging systems resolution is limited by the size of the pixels of the captured image.

Receptor unsharpness may be demonstrated using the line-pair test object. If this test object is imaged in contact with the image receptor it is possible to assess the receptor unsharpness in the absence of geometric unsharpness. If direct contact is not possible the

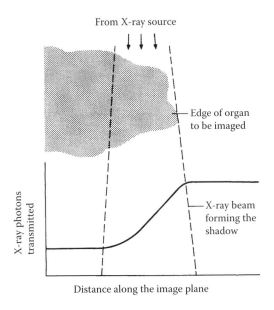

FIGURE 6.18
Contribution to image blurring which results from an irregular edge to the organ of interest.

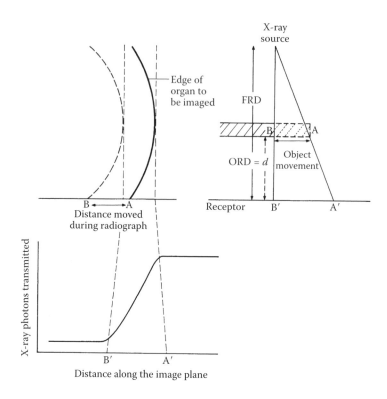

FIGURE 6.19
Effect of movement on radiographic blurring. If the object moves with velocity v during the time of exposure t then AB = vt and A'B' = AB.FRD/(FRD – d).

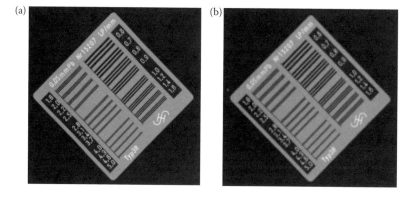

FIGURE 6.20
Examples of receptor unsharpness obtained with the line-pair test object in contact with the receptor; (a) digital image with small pixels (0.16 mm), 4.3 lp mm^{-1}; (b) digital image with large pixels (0.54 mm), 1.3 lp mm^{-1}.

penumbra caused by the focal spot can be minimised by increasing the focus to image distance. Figure 6.20 shows examples in which the test object was in contact with the receptor. Note that as data are passed down the imaging chain, for example, from a CR storage plate to the display device (monitor) and hard copy (film) the resolution can only get worse, never better!

The Star Test Pattern

The line-pair object method described, measures the combined effects of the finite focal spot size and the relative values of FRD and *d*. Assessment of the geometric unsharpness caused by the finite focal spot—and by inference an estimation of the size of focal spot itself—can be made by use of a star test object. Such a test object again consists of lead bars, but this time they are arranged in a radial pattern like the spokes of a wheel with the bars and gaps becoming equally wider with distance from the centre of the test object. This gives a continuous variation of lp mm^{-1}. The detailed theory of operation of the star test pattern can be found in Spiegler and Breckinridge (1972), but in use it is imaged to determine the point at which the resolving capacity of the system fails and causes blurring of the line pairs.

6.9.4 Combining Unsharpnesses

It will be apparent from the preceding discussion that in any radiological image there will be several sources of unsharpness and the overall unsharpness will be the combination of all of them.

Unsharpnesses are combined according to a power law with the power index varying between 2 and 3. The power index 2 is most commonly used and should be applied where the unsharpnesses are all of the same order. The power index 3 should be used if one of the unsharpnesses is very much greater than the rest.

Note that, because of the power law relationship, if one contribution is very large it will dominate the expression. If this unsharpness can be reduced at the expense of the others the minimum overall unsharpness will be when all contributions are approximately equal. For example, if the geometric unsharpness is U_G, the movement unsharpness is U_M, the receptor unsharpness U_R then the combined unsharpness of the three is

$$U = \sqrt{U_G{}^2 + U_M{}^2 + U_R{}^2}$$

If $U_G = 0.5$ mm, $U_M = 1.0$ mm, $U_R = 0.8$ mm, then $U = 1.37$ mm.
If $U_G = 0.7$ mm, $U_M = 0.7$ mm, $U_R = 0.8$ mm, then $U = 1.27$ mm.

6.10 Quantum Mottle

An amplification of the primary image of several thousand is available through a modern X-ray imaging system. However, this does not mean that the dose of radiation delivered to the patient can be reduced indefinitely. Although the average signal or brightness of the image can be restored by electronic means, image quality will be lost.

To understand why this is so, it is important to appreciate that image formation by the interaction of X-ray photons with a receptor is a random process. A useful analogy is rain drops coming through a hole in the roof. When a lot of rain has fallen, the shape of the hole in the roof is clearly outlined by the wet patch on the floor. If only a few drops of rain have fallen it is impossible to decide the shape of the hole in the roof.

(a)    (b)

FIGURE 6.21
One of the authors (PD) suffering from quantum mottle! (a) The image was taken with a digital camera at normal light intensity. (b) The light intensity has been reduced by a factor of 25. The camera can adjust the grey scale but noise is apparent in the image.

Similarly, if sufficient X-ray quanta strike a photoreceptor, they produce enough light photons to provide a detailed image but when fewer X-ray quanta are used the random nature of the process produces a mottled effect which reduces image quality (see Figure 6.21). A very fast rare earth screen may give an acceptable overall density before the mottled effect is completely eliminated. For an example of quantum mottle in a radiograph see Figure 9.16b. Thus the information in the image is related to the number of quanta forming the image.

When there are several stages to the image formation process, as in the image intensifier/CCD system, overall image quality will be determined at the point where the number of quanta is least, the so-called '*quantum sink*'. This will usually be at the point of primary interaction of the X-ray photons with the photoluminescent screen (see Figure 6.22). If insufficient X-ray photons are used to form this image, further amplification is analogous to empty magnification in high power microscopy, being unable to restore to the image detail that has already been lost. It follows that neither electron acceleration nor minification in an image intensifier improves the statistical quality of the image if the number of X-ray photons interacting with the input screen remains the same.

This subject will be considered again under the heading of 'quantum noise' in Section 7.5.

6.11 Image Processing

An important difference between analogue images on film and digital images is that in the latter the processes of image collection and image display are decoupled. This permits

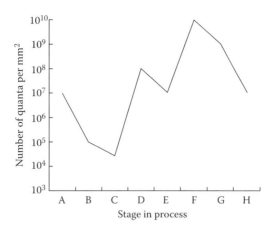

FIGURE 6.22

Quantum accounting diagram for an image intensifier coupled to a CCD camera. The various stages are A—input flux to patient; B—exit flux from patient; C—X-ray photons absorbed (the quantum sink); D—light photons emitted; E—photoelectrons released into the image intensifier (II); F—light photons emitted from the II output; G—light photons entering CCD; H—digital signal from CCD.

a wide variety of image manipulations, or processing, which is not possible with film where collection and display cannot be separated.

Image processing can be considered either in real space or in terms of spatial frequencies. The latter is more amenable to mathematical analysis and is the approach used by manufacturers. It is considered in an insight later. However, the techniques are more readily understood in real space so this is the approach adopted here.

6.11.1 Point Operations

A very simple approach to processing is to calculate a new pixel value for each pixel. This is known as a point operation. Some examples include the following:

(a) Inverting an image—this has the effect of making the pixels which appeared light appear dark, and vice versa. The equivalent with film would be converting a negative into a positive. It can be done quite simply by replacing a pixel value by its 'mirror image' relative to an arbitrary mean. For example, if the mean pixel value is 500, a pixel value of 620 (500 + 120) is replaced by 380 (500 − 120). In more general terms the slope of the LUT has been inverted, thereby inverting the grey scale about its mean.

(b) Enhancing contrast by subtracting a fixed value (n_0) from each pixel (see Figure 6.23). The new contrast is

$$\log\left(\frac{n_1 - n_0}{n_2 - n_0}\right) > \log\left(\frac{n_1}{n_2}\right)$$

(c) Adding two distributions (e.g. two image frames of the heart) to improve the ratio of signal to noise. If the signal is N counts and the noise is assumed to be

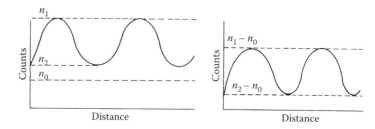

FIGURE 6.23

Background subtraction as a potentially useful method of data manipulation.

Poisson, $\sqrt{N}$, then the ratio of signal to noise is $\sqrt{N}$ and is improved by increasing N. Conversely the ratio is worse if images are subtracted—a point which has to be borne in mind in digital subtraction imaging (see Section 9.5.2).

(d) Contrast stretching, or windowing, discussed in Section 6.3.3 is also an example of a point operation.

6.11.2 Local Operations

This is the process by which the value of a pixel is influenced and modified by the value of nearby pixels. An array of pixel weighting factors, called a *kernel* or *filter*, is applied to the image on a pixel by pixel basis in a process known as convolution. The products of the pixel values covered by the filter and their weighting factors are summed and divided by the sum of the weighting factors (to maintain the overall average pixel value of the image). Some examples are as follows:

(a) *Smoothing filters* to reduce the impact of noise due to poor counting statistics, or if pixellation is coarse (e.g. in nuclear medicine) the intrusive effect of pixellation. With a simple smoothing filter all 9 pixels (i.e. the target pixel and 8 adjacent) are given equal weight—see Table 6.1a.

A template with equal weights leads to rather heavy smoothing so different weights are normally assigned. Consider the application of the filter in Table 6.1b to the array in Table 6.2 where the central pixel is high because of Poisson noise.

The new computed value of the central pixel is

$$\frac{(1\times1)+(2\times2)+(1\times1)+(1\times2)+(9\times4)+(2\times2)+(2\times1)+(2\times3)+(1\times1)}{16} = 3.55$$

TABLE 6.1

Nine Point Arrays of Smoothing Filters; (a) Heavy Smoothing; (b) Lighter Smoothing

1	1	1
1	1	1
1	1	1

(a)

1	2	1
2	4	2
1	2	1

(b)

TABLE 6.2

Hypothetical Set of Raw Pixel Values

1	2	1
1	9	2
2	3	1

TABLE 6.3

Weighting Factors for a Sharpening Filter

−1	−1	−1
−1	4	−1
−1	−1	−1

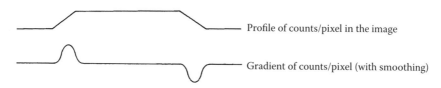

Profile of counts/pixel in the image

Gradient of counts/pixel (with smoothing)

FIGURE 6.24
Edge enhancement or use of gradients to improve visualisation.

Note that the image shrinks slightly as the template cannot be applied to the pixels round the edge—but this is not critical for say a 1024 × 1024 pixel matrix.

(b) *Sharpening filters.* In contrast to smoothing filters, image features become sharper if negative weight is given to surrounding pixels (see Table 6.3).

(c) *Edge enhancement.* A plot of the gradient of counts per pixel is much more pronounced than the change in absolute counts across the edges (see Figure 6.24).

Figure 6.25 shows examples of the effect of different filters on an image. Applications of these processes include, for example, improving the image of a low-contrast boundary, such as a tumour within the body or enhancing the visualisation of the character of the boundary—is it fuzzy or sharp? This is an important distinction when deciding if the tumour is benign or malignant.

6.11.3 Global Operations

These are processes which manipulate the image as a whole. They are not normally amenable to analysis in real space and require spatial frequency domain methods (see Insight). Two applications are mentioned briefly.

(a) *Image Segmentation.* In this approach the image is first divided into several regions of anatomical interest. The mathematics is mostly beyond the scope of this book but a simple example of regional orientation segmentation and thresholding illustrates the principle.

Suppose one wishes to know if a lung nodule is increasing in size with time. Look at the pixel values in the area defined by the nodule and in the surrounding

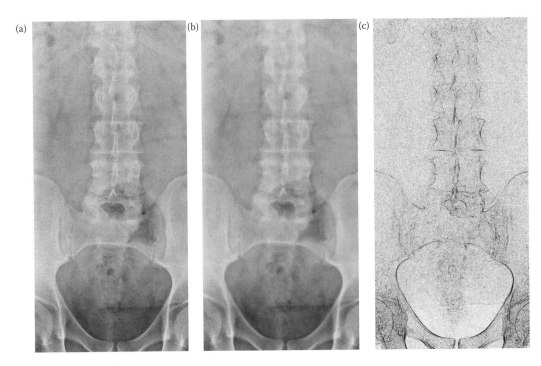

FIGURE 6.25
Three pictures of an X-ray image showing the effect of filters; (a) none; (b) smoothing; (c) edge enhancement.

area, establish a threshold and set all pixel values below the threshold to 0 and all those above to 1. It is then easy to measure the diameter, perimeter and area of the nodule in sequential images.

(b) *Histogram analysis.* The idea of making a histogram of pixel values was introduced in Section 6.3.3. A number of useful operations can be performed using this information. One is to find the region of interest of the image. In many radiographs there are areas within the image where the X-ray beam does not pass through the body. In these cases unattenuated radiation interacts with the receptor producing a signal that is much greater than in the rest of the image. Histogram analysis can identify these areas easily and exclude them from further processing. Systems that use 'automatic collimation' or 'masking' can also use these identified areas to improve the visualisation of the image by automatically assigning a 'black' value to them to reduce glare. A further use of this identification of the region of interest in the image can be to calculate the average receptor dose in that region which is useful in monitoring patient radiation doses.

(c) *Data compression.* This is essential for efficient data storage and will be discussed in Chapter 17.

Insight

Frequency Domain Methods of Image Processing

A full treatment involves a thorough understanding of Fourier transforms, the principle of which is that any image in real space can be expressed in terms of a large number or sinusoidal waves

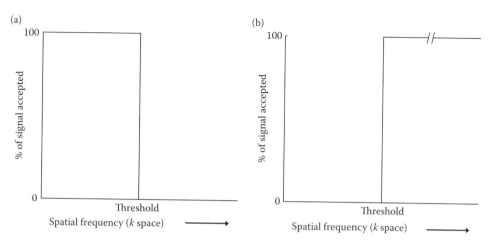

FIGURE 6.26
Properties of filters represented in frequency space; (a) low pass—all spatial frequencies up to a threshold value are transmitted equally; (b) high pass—spatial frequencies above a threshold are transmitted equally.

of different spatial frequencies, expressed in cycles cm⁻¹. Note that the spatial frequency k when dealing with waves in space (k space) is the analogue of temporal frequency $f(\omega/2\pi)$ for waves varying with time.

Thus an image in real space $f(x,y)$ is transformed into $F(u,v)$ where u is a measure of spatial frequencies in the x-direction and v is a measure of spatial frequencies in the y-direction.

The point of this manoeuvre is that applying a processing operation, $H(u,v)$, in frequency space,

$$G(u,v) = H(u,v) \cdot F(u,v)$$

is much easier mathematically than carrying out the same process in real space.

Smoothing filters are also known as *low pass filters* and are very easy to understand in terms of spatial frequencies because they selectively reduce the high frequency components of the image which carry information about edges and sharp detail (see Figure 6.26a). Conversely *high pass filters* have the opposite effect, reducing the low spatial frequencies associated with the slowly varying characteristics of an image, such as overall intensity and contrast, and enhancing the edges and detail (Figure 6.26b). *Band pass filters* accept a pre-selected range of spatial frequencies.

6.12 Geometric Relationship of Receptor, Patient and X-ray Source

The interpretation of radiographs eventually becomes second nature to the radiologist who learns to ignore the geometrical effects which can, and do, produce very distorted images with regard to the size and position of organs in the body. Nevertheless, it is important to appreciate that such distortions occur.

Most of the effects may be easily understood by assuming that the focus is a point source and that X-rays travel in straight lines away from it.

6.12.1 Magnification without Distortion

In the situation shown in Figure 6.27 the images of three objects of equal size lying parallel to the image receptor are shown. The images are not the same size as the object. Assume magnification M_1, M_2 and M_3 for objects 1, 2 and 3, respectively given by

$$M_1 = \frac{AB}{ab} \quad M_2 = \frac{XY}{xy} \quad M_3 = \frac{GH}{gh}$$

Consider triangles *Fxy* and *FXY*. Angles *xFy* and *XFY* are common; *xy* is parallel to *XY*. Triangles and *FXY* are thus similar. Therefore

$$\frac{XY}{xy} = \frac{FRD}{FRD - d} \tag{6.3}$$

By the same considerations triangles *aFb* and *AFB* are similar and triangles *gFh* and *GFH* are similar

$$\frac{AB}{ab} = \frac{FRD}{FRD - d}$$

and

$$\frac{GH}{gh} = \frac{FRD}{FRD - d}$$

Therefore

$$M_1 = M_2 = M_3$$

That is, for objects in the same plane parallel to the receptor, magnification is constant. This is magnification without distortion.

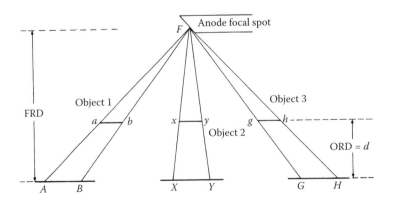

FIGURE 6.27
Demonstration of a situation in which magnification without distortion will occur.

The magnification increased if

1. The FRD is decreased (with d fixed)
2. d is increased (with FRD fixed)

The deliberate use of magnification techniques is discussed in Section 9.4.

6.12.2 Distortion of Shape and/or Position

In general, the rather artificial conditions assumed in Section 6.12.1 do not apply when real objects are radiographed. For example, real objects are not infinitesimally thin, they are not necessarily orientated normal to the principal axis of the X-ray beam and they are not all at the same distance, measured along the principal axis, from the X-ray source. All of these factors introduce distortions into the resulting image. Figure 6.28a shows the distortion that results from twisting a thin object out of the horizontal plane, and Figure 6.28b shows distortion for objects of finite thickness. Although all the spheres are of the same diameter, the cross-sectional area projected parallel to the receptor plane is now greater if the sphere

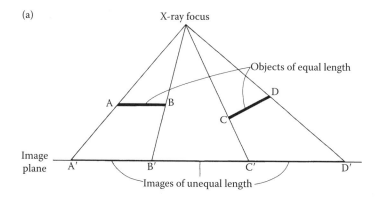

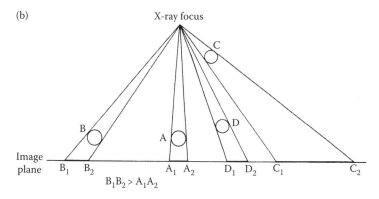

FIGURE 6.28

Demonstration of (a) distortion of shape of an object; (b) distortion of shape when objects are of finite thickness (A,B,C) and of relative position when they are at different depths (C,D). Note that the object B has been placed very wide and the object C has been placed very close to the X-ray focus to exaggerate the geometrical effects. In particular a patient would not be placed as close to the X-ray focus as this because of the high skin dose (see Section 9.6.3).

is off-axis and the image is enlarged more (compare A and B). Note that when the sphere is in a different plane (e.g. C) the distortion may be considerable. Figure 6.28b also shows that when objects are in different planes, distortion of position will occur. Although C is nearer to the central axis than D, its image actually falls further away from the central axis.

A certain amount of distortion of shape and position is unavoidable and the experienced radiologist learns to take such factors into consideration. The effects will be more marked in magnification radiography.

6.13 Review of Factors Affecting the Radiological Image

It is clear from this and the preceding chapter that a large number of factors can affect a radiological image and these will now be summarised.

6.13.1 Choice of Tube Kilovoltage

For a film-screen combination a high kV gives a lower contrast and a large film latitude and vice versa. Digital images can be manipulated to counteract these effects. X-ray output from the tube is approximately proportional to $(kVp)^2$ and for film blackening is proportional to $(kVp)^4$. The relationship is different for digital detectors because the receptor materials have different attenuating properties.

The higher the kV the lower the entrance dose to the patient. A higher kV, however, increases the amount of scatter and if this necessitates use of a grid, the entrance dose may actually be higher.

Increasing the kV allows the tube current or the exposure time to be reduced. At 70 kVp an increase of 10 kVp allows the mAs to be approximately halved.

6.13.2 Exposure Time

In theory, the exposure time should be as short as possible to eliminate movement unsharpness. If movement will not be a problem, exposure time may be increased so that other variables can be optimised.

6.13.3 Focal Spot Size

As the spot size increases so does geometric unsharpness. The minimum size (from a choice of two) should be chosen consistent with the choice of other factors which affect tube rating (kVp, mA, s).

6.13.4 Quality of Anode Surface

As discussed in Section 2.4, damage to the anode surface will result in a non-uniform X-ray intensity distribution and an increased effective spot size. Because of the heel effect there is always some variation in X-ray beam intensity in the direction parallel to the line of the anode-cathode. If a careful comparison of the blackening produced by two structures is required they should be orientated at right angles to the electron flux from the cathode to anode.

6.13.5 Tube Current

In an ideal situation, with film as the receptor all other variables would be chosen first and then a mA would be selected to give optimum film blackening. If this is not possible, for example because of rating limits, the system becomes highly interactive and the final combination of variables is a compromise to give the best end result. For digital images post-processing will also be a major factor.

6.13.6 Beam Size

This should be as small as possible, commensurate with the required field of view, to minimise patient dose and scatter. Note that, strictly speaking, collimation reduces the integral dose, that is the absorbed dose multiplied by the volume irradiated, rather than the absorbed dose itself.

6.13.7 Grids

These must be used if scatter is significantly reducing contrast, for example when irradiating large volumes. Use of grids requires an increased mAs thus increasing patient dose.

6.13.8 Focus-Receptor and Object-Receptor Distance

A large FRD reduces geometric blurring, magnification and distortion, but the X-ray intensity at the patient is reduced because of the inverse square law. The working distance is thus governed eventually by the tube rating. The object-receptor distance cannot normally be controlled by the operator but is kept as small as possible by equipment manufacturers. Movement and geometric blurring and magnification can be influenced by patient orientation either anterior/posterior or posterior/anterior.

6.13.9 Contrast Enhancement

Modification of either the atomic number of an organ, for example by using barium or iodine-containing contrast agents, or of its density, for example by introducing a gas, alters its contrast relative to the surrounding tissue.

6.13.10 Image Receptor

When using film-based imaging systems there is generally only a limited choice. A fast film-screen combination will minimise patient dose, geometric and movement unsharpness. Associated screen unsharpness and quantum mottle may be higher than when a slow film-screen combination is used. If extremely fine detail is required, a non-screen film may be used but this requires a much higher mAs and thus gives a higher patient dose.

A range of digital imaging receptors is available with similar issues of resolution, sensitivity and quantum mottle affecting the choice. Plates with thinner, less sensitive phosphor layers are available for detail work in computed radiography and some systems give the option for a slower readout of the plate to reduce unsharpness arising from a delay in the stimulated output from the previous pixel while reading from the next.

6.13.11 Film Processing

This vital part of image formation must not be overlooked if using radiographic film. Quality control of development is extremely important as it can have a profound effect on the radiograph. Bad technique at this stage can completely negate all the careful thought given to selecting correct exposure and position factors.

6.13.12 Post-processing

For digital images a variety of post-processing techniques is available—see Section 6.11. These will all alter the appearance of the final image.

It will be clear from this lengthy list that the quality of a simple, plain radiograph is affected by many factors, some of which are interactive. Each of them must be carefully controlled if the maximum amount of diagnostic information is to be obtained from the image. For further information see BIR (2001).

References

British Institute of Radiology (BIR). *Assurance of Quality in the Diagnostic X-ray Department.* British Institute of Radiology, London, UK, 2001.
Dawson P. and Clauss W. *Advances in X-ray Contrast. Collected Papers.* Springer, Heidelberg, Germany, 1998.
Spiegler P. and Breckinridge W.C. *Imaging of Focal Spots by Means of the Star Test Pattern,* Radiology 102:679–684, 1972.

Further Reading

Carrol Q.B. *Practical Radiographic Imaging*, 8th edn. Charles C Thomas, Springfield, Illinois, 2007.
Curry T.S., Dowdey J.E. and Murry R.C. Jr, *Christensen's Introduction to the Physics of Diagnostic Radiology*, 4th edn. Lea and Febiger, Philadelphia, 1990.
Fauber T.L. *Radiographic Imaging and Exposure* (Chapter 4; Radiographic image quality pp. 50–99). Mosby Inc, St Louis, 2000.
Oakley J. (ed) *Digital Imaging. A Primer for Radiographers, Radiologists and Health Care Professionals.* Cambridge University Press, Cambridge, UK, 2003.

Exercises

1. What is meant by contrast?
2. Why is contrast reduced by scattered radiation?

3. The definition of contrast used for radiographic film cannot be used for digital images. Discuss the reasons for this.

4. What are the advantages and disadvantages of having a radiographic film with a high gamma?

5. A solid bone 7 mm diameter lies embedded in soft tissue. Ignoring the effects of scatter, calculate the contrast between the bone (centre) and neighbouring soft tissue.

   Film gamma = 3

   Linear attenuation coefficient of bone = 0.5 mm^{-1}

   Linear attenuation coefficient of tissue = 0.04 mm^{-1}

6. Discuss the origins of scattered radiation reaching the receptor.

7. Give a sketch showing how the relative scatter (scattered radiation as a fraction of the unscattered radiation) emerging from a body varies with X-ray tube kV between 30 kVp and 200 kVp and explain the shape of the curve. What measures can be taken to minimise loss of contrast due to scattered radiation?

8. How can X-ray magnification be used to enhance the detail of small anatomical structures? What are its limitations?

9. What are the advantages and disadvantages of an X-ray tube with a very fine focus?

10. List the factors affecting the sharpness of a radiograph. Draw diagrams illustrating these effects.

11. A digital radiograph is taken of a patient's chest. Discuss the principal factors that influence the resultant image.

12. What factors, affecting the resolution of a radiograph, are out of the control of the radiologist? (Assuming the radiographer is performing as required.)

13. A radiograph is found to lack contrast. Discuss the steps that might be taken to improve contrast. Distinguish carefully between film-screen, CR and DR images.

14. Explain the meaning of the terms quantum mottle and quantum sink. What methods are available for reducing the effects of quantum mottle?

15. Explain the difference between global operations and local operations in image processing and give examples of each.

7

Assessment of Image Quality and Optimisation

P P Dendy and O W E Morrish

SUMMARY

- This chapter looks at image quality in the wider context to determine if the image is 'fit for purpose', that is accurate diagnosis.
- Factors affecting image quality are reviewed.
- Image perception is important so the operation of the visual system is explained.
- An alternative definition of contrast is presented and quantum noise is considered in greater detail.
- The importance of modulation transfer function (MTF) as a quantitative measure of imager performance is explained.
- Receiver operator characteristic curves are explained as a versatile method of analysis of both technical and clinical features of images.
- Examples of optimisation of imaging systems and image interpretation are given.
- The chapter concludes with a brief introduction to clinical trials.

CONTENTS

7.1 Introduction

In earlier chapters the principles and practice of X-ray production were considered and the ways in which X-rays are attenuated in different body tissues were discussed. Differences in attenuation create 'contrast' on the image receptor, and differences in 'contrast' provide information about the object. Imaging systems are often described in terms of physical quantities that characterise various aspects of their performance.

However, when an X-ray image, or indeed any other form of diagnostic image, is assessed subjectively, it must be appreciated that the use made of the information is dependent on the observer, in particular the performance of his or her visual response system. Therefore, it is important to consider those aspects of the visual response system that may influence the final diagnostic outcome of an investigation. Quantitative methods for assessing noise and image quality will be considered in greater detail, and the chapter will conclude with examples of ways in which imaging systems and image interpretation may be optimised and a brief introduction to the design of clinical trials.

7.2 Factors Affecting Image Quality

A large number of factors may control or influence image quality. They may be subdivided into three general categories.

7.2.1 Image Parameters

1. The signal to be detected—the factors to consider here will be the size of the abnormality, the shape of the abnormality and the inherent contrast between the suspected abnormality and non-suspect areas.

2. The number and type of possible signals—for example, the number and angular frequency of sampling in computed tomography (CT).

3. The nature and performance characteristics of the image system—spatial resolution (this is important when working with an image intensifier, digital radiography systems and in nuclear medicine), sensitivity, linearity, noise (both the amplitude and character of any unwanted signal such as scatter) and speed, especially in relation to patient motion.

4. The interplay between image quality and dose to the patient.

5. Non-signal structure—interference with the wanted information may arise from grid lines, overlapping structures or artefacts in CT and ultrasound.

7.2.2 Observation Parameters

1. The display system—features that can affect the image appearance include the brightness scale, gain, offset, non-linearity (if any) and the magnification or minification.

2. Viewing conditions—viewing distance and ambient room brightness.

3. Detection requirements.

4. Number of observations.

7.2.3 Psychological Parameters

1. *A priori* information given to the observer.

2. Feed back (if any) given to the observer.

3. Observer experience from other given parameters—this may be divided into clinical and non-clinical factors and includes, for example, familiarity with the signal and with the display, especially with the types of noise artefacts to be expected.

4. Information from other imaging modalities.

Some of these points have been considered in earlier chapters and others will be considered later in this chapter. For further information on display systems, see Sections 10.3.5 and 17.5. It is important to realise that final interpretation of a diagnostic image, especially when it is subjective, depends on far more than a simple consideration of the way in which the radiation interacts with the body.

7.3 Operation of the Visual System

This is a complex subject and it would be inappropriate to attempt a detailed treatment here. However, five aspects are of particular relevance to the assessment and interpretation of diagnostic images and should be considered.

7.3.1 Response to Different Light Intensities

It is a well-established physiological phenomenon that the eye responds logarithmically to changes in light intensity. Thus it can accommodate a large range of densities from very bright to very dark. This fact has already been mentioned in Section 5.5.2 and is one of the reasons for defining contrast in terms of log (intensity).

7.3.2 Rod and Cone Vision

When light intensities are low, the eye transfers from cone vision to rod vision. The latter is much more sensitive, but this increased sensitivity brings a number of disadvantages for radiology.

First, the capability of the eye to detect contrast differences is very dependent on light intensity (Figure 7.1). Whereas, for example, a contrast difference of 2% would probably be detectable at a light intensity or brightness of 300 cd m^{-2} (about 100 millilambert) (typical film viewing conditions), a contrast difference of 20% might be necessary at 3×10^{-3} cd m^{-2} (the light intensity that might be emitted from a bright fluorescent screen).

Second, there is a loss of resolving ability, or visual acuity, at low light intensities. The width of a cone in the macula lutea is about 3 μm, so the maximum visual acuity of the eye is inherently very high. In practice, the minimum detectable separation with cone vision between two objects viewed from about 25 cm is better than 1 mm. Since rods respond as bundles of fibres rather than individually, visual acuity is worse and strongly dependent on light intensity. At 10^{-3} cd m^{-2}, the minimum detectable separation is probably no better than 3 mm.

Loss of contrast perception and visual acuity are the two main reasons why fluoroscopic screens have been abandoned in favour of image intensifiers. Other disadvantages of rod

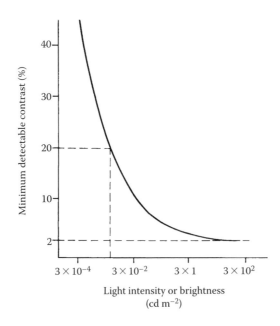

FIGURE 7.1
Variation of minimum detectable contrast with light intensity, or brightness, for the human visual system.

vision are the need for dark adaptation, which may take up to 30 minutes, and a loss of colour sensitivity, although the latter is not a real problem in radiology.

7.3.3 Relationship of Object Size, Contrast and Perception

Even when using cone vision, it is not possible to decide whether an object of a given size will be discernible against the background unless the contrast is specified. As expected, there is a minimum increase in the brightness of an object that will produce a minimum increase in the sensory experience of the eye, but the exact relationship between minimum perceptible contrast difference and object size will depend on a number of factors, including the signal/noise ratio and the precise viewing conditions. Contrast-detail diagrams may be generated using a suitable phantom (Figure 7.2a) in which simple visual signals (e.g. circles) are arranged in a rectangular array, such that the diameter changes monotonically vertically and the contrast changes monotonically horizontally. The observer has to select the lowest contrast signal in each row that is considered detectable. A typical curve is shown in Figure 7.2b and illustrates the general principle that the higher the object contrast, the smaller the detectable object size.

Some typical figures for a modern intravenous angiographic system will illustrate the relationship. Suppose an incident dose in air of 0.3 μGy per image is used. At a mean energy of 60 keV this will correspond to about 10^4 photons mm^{-2}. If the efficiency of the image intensifier for detecting these photons is 30%, about 3×10^3 interactions mm^{-2} are used per image. Assuming a typical signal/noise ratio of 5, then objects as small as 0.4 mm can be detected if the contrast is 30%, but the size increases to 4 mm if the contrast is only 3%.

(a)

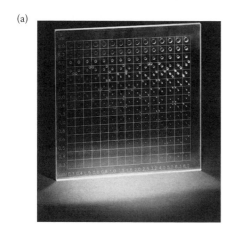

(b)

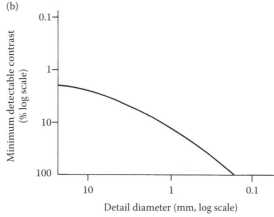

FIGURE 7.2
Relationship between minimum perceptible contrast difference and object size. (a) A suitable phantom for investigating the problem. Holes of different diameter (vertical axis) are drilled in Plexiglas to different depths to simulate different amounts of contrast (horizontal axis). As the holes become smaller the contrast required to visualise them becomes greater. (Photograph kindly supplied by Artinis Medical Systems after a design by Thijssen M A O, Thijssen H O M, Merx J L et al. A definition of image quality: The image quality figure. In *Optimisation of image quality and patient exposure in diagnostic radiology*, B M Moores, B F Wall, H Eriska and H Schibilla eds., BIR Report 20, 1989, 29–34.) (b) Typical result for an image intensifier TV camera screening unit.

Insight

Alternative Display for Contrast-Detail Data

An alternative way to display information on the relationship between threshold contrast and detail size is to plot the threshold detection index, H_T, as a function of the square root of the detail area on a log-log scale.

H_T is defined as

$$H_T = \frac{1}{(C_T \times \sqrt{A})}$$

where C_T is the threshold contrast, expressed as a percentage difference in response of the receptor to the detail and the background, and A is the area of detail. The shape of the C_T versus A curve means that H_T is small when either C_T or A is large, taking its maximum value in the middle of the range. Example threshold contrast-detail detectability (TCDD) curves are shown in Figure 7.3.

This form of plot has some advantages:

1. Increasing sensitivity or lower noise both increase H_T.
2. If quantum noise dominates the image, H_T will increase with increasing air kerma.
3. The maximum provides a well-defined working point as a basis for comparison—for example, between different equipment designs, or to check for overall design improvement with time, or for routine quality assurance checks.

7.3.4 Eye Response to Different Spatial Frequencies

When digitised images are displayed, the matrix of pixels will be imposed on the image, and it is important to ensure that the matrix is not visually intrusive, thereby distracting the observer.

To analyse this problem, it is useful to introduce the concept of *spatial frequency*. Suppose a 128×128 matrix covers a square of side 19 cm. Each matrix element will measure 1.5 mm across, or alternatively expressed, the frequency of elements in space is 1/0.15 or about 7 cm^{-1}. Thus the spatial frequency is 7 cm^{-1}, 700 m^{-1} or more usually expressed in radiology as 0.7 mm^{-1}.

Clearly, the effect of these pixels on the eye will depend on viewing distance. The angle subtended at the eye by pixels having a spatial frequency of 0.7 mm^{-1} will be different,

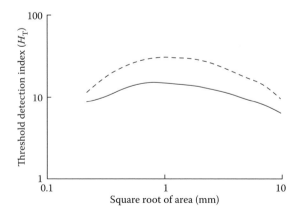

FIGURE 7.3
Typical threshold contrast-detail detectability curves for fluoroscopy (solid line) and digital acquisition (dashed line).

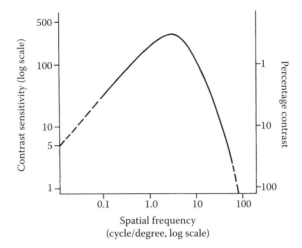

FIGURE 7.4
Contrast sensitivity of the human visual system defined as the reciprocal of threshold contrast measured with a sinusoidal grating plotted against spatial frequency at a brightness of approximately 500 cd m^{-2}. (After Campbell F W. The physics of visual perception, *Phil. Trans. Roy. Soc. Lond.* 290, 5–9, 1980, with permission.)

depending on whether viewed from 1 m or 50 cm. Campbell (1980) has shown that contrast sensitivity of the eye-brain system is very dependent on spatial frequency and demonstrates a well-defined maximum. This is illustrated in Figure 7.4 where a viewing distance of 1.0 m has been assumed. Exactly the same curve would apply to spatial frequencies of twice these values if they were viewed from 50 cm.

Thus it may be seen that the regular pattern of matrix lines is unlikely to be intrusive when using a 1024 × 1024 matrix in digital radiology, but it may be necessary to choose both image size and viewing distance carefully to avoid or minimise this effect when looking at 128 × 128 or 64 × 64 images sometimes used in nuclear medicine.

7.3.5 Temporal Resolution and Movement Threshold

Important information about anatomy and pathology is sometimes obtained from studies of movement or variation in time. Therefore it is necessary to be aware of the response of the human eye to time-related factors.

One important question is, Will a rapid sequence of images give a visual impression of flicker or will the images appear to fuse together? Quantitative information on this question can be obtained by causing a small uniform-luminance source to fluctuate sinusoidally and to observe the effect. Experiments show that, for the image to be flicker-free, the stimulus to the retina from the first pulse must decrease by no more that 10% before the second pulse arrives. A typical frame rate of 16–25 pictures s^{-1} is necessary, but this is very dependent on overall brightness and also on the degree of adaptation and the angle made with the optical axis.

Insight

Temporal Resolution of the Eye

If luminance fluctuates sinusoidally about a mean brightness L, the stimulus will be $L + I \sin(2\pi f t)$ where f is the frequency and the signal varies between $L + I$ and $L - I$. The value of I that causes a

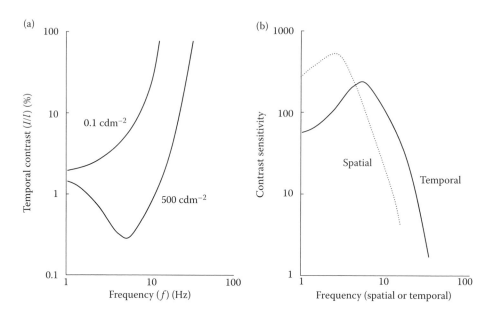

FIGURE 7.5

(a) Temporal contrast for two values of retinal luminescence (0.1 cd m^{-2} and 500 cd m^{-2})—based on de Lange 1958 and Kelly 1961. (b) Temporal contrast sensitivity, or flicker sensitivity $L/I(f)$ at 500 cd m^{-2} and compared with the spatial contrast sensitivity (redrawn from Figure 7.4).

flicker sensation is determined for various values of f, say $I(f)$, and the temporal contrast $I(f)/L$ can be plotted as a function of f (see Figure 7.5a). At 0.1 cd m^{-2}, typical of the darker part of a low-level display, the temporal contrast required for a flicker sensation increases with increasing frequency, as one would expect intuitively. At 500 cd m^{-2}, typical of a cathode ray tube display, $I(f)/L$ actually decreases with increasing low frequency so the required frequency for flicker-free viewing must exceed the value of f at $(I/L)_{min}$.

Figure 7.5b is the 500 cd m^{-2} curve from Figure 7.5a redrawn to show the temporal contrast sensitivity of the eye, or flicker sensitivity, the reciprocal $L/I(f)$. There is a strong similarity with the spatial contrast sensitivity (redrawn from Figure 7.4).

Similar considerations apply to movement threshold. For experience of motion to occur, the stimulation of a cone must have decreased by more than 10% of its maximum value by the time the stimulation of the neighbouring cone reaches its maximum. This response time is about 0.1 s, corresponding to an optimum frequency response of about 10 Hz.

7.4 Objective Definition of Contrast

7.4.1 Limitations of a Subjective Definition of Contrast

In Chapter 5 a definition of contrast based on visual or subjective response was given. However, this definition alone is not sufficient to determine whether a given boundary between two structures will be visually detectable because the eye-brain will be influenced by the type of boundary as well as by the size of the boundary step. There are two reasons for this. First, the perceived contrast will depend on the sharpness of the boundary. Consider, for example, Figure 7.6. Although the intensity change across the boundary is the same in both cases, the boundary illustrated in Figure 7.6a would appear more contrasty on X-ray film because it is sharper. Second, if the boundary is part of the image of a small object in a rather uniform background, contrast perception will depend on the size of the object. A digital system which can artificially increase the density at the edge of a structure will subjectively increase the contrast of the structure.

7.4.2 Signal-to-Noise Ratio and Contrast-to-Noise Ratio

The definition of contrast given in Section 6.3 is often the most appropriate for X-ray film where it is only necessary to measure the optical density of the film. However, it has limitations.

1. Film is a non-linear device
2. Film cannot readily be extended to other imaging systems

An alternative definition starts from the task of perceiving a single object in a background that is uniform apart from the presence of noise and measuring the signal/noise ratio. Imagine, for example, that the object consisted of a single region of lightly attenuating material (Figure 7.7).

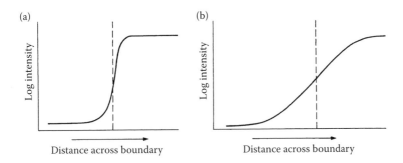

FIGURE 7.6
Curves showing the difference in transmitted intensity across (a) a sharp boundary; (b) a diffuse boundary.

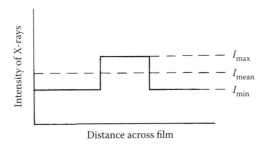

FIGURE 7.7
Appearance of a perfect image of a simple object consisting of a single strip of lightly attenuating material.

The signal may be represented by $I_{max} - I_{min}$ and the noise has an average value I_{min}. Hence the signal/noise ratio is $(I_{max} - I_{min})/I_{min}$ or if the signal is small, to a good approximation $(I_{max} - I_{min})/I_{mean}$. Contrast is often given as $(I_{max} - I_{min})/(I_{max} + I_{min})$. Since $I_{mean} = \frac{1}{2}(I_{max} + I_{min})$ this will give values that differ by a factor of two from those obtained using I_{mean}.

This definition can be easily extended to digital systems. For example, in a digital radiograph the signal will be the number of X-ray photon interactions per pixel and a quantitative expression for the contrast-to-noise ratio will be

$$CNR = \frac{N_1 - N_2}{\text{noise}}$$

where N_1, N_2 are the mean numbers of counts in pixels in the regions of differing contrast. If the only significant source of noise is quantum noise (see next section)

$$CNR = \frac{N_1 - N_2}{\sqrt{N_1 + N_2}}$$

7.5 Quantum Noise

When a radiographic detector is exposed to a uniform intensity distribution of X-rays, the macroscopic density distribution of the response (e.g. optical density of a developed film) fluctuates around some average value. This fluctuation is called *radiographic noise* or *radiographic mottle,* and the ability to detect a signal will depend on the amount of noise in the image. There are many possible sources of noise, for example, noise associated with the imaging device itself, film graininess and screen structure mottle. In digital systems the electronics is a major source of noise. Noise is generally difficult to analyse quantitatively, but the effect of one type of noise, namely quantum noise, can be predicted reliably and quantitatively. Quantum noise is also referred to variously as quantum mottle or photon noise (see Section 6.10).

The interaction of a flux of photons with a detector is a random process, and the number of photons detected is governed by the laws of Poisson statistics. Thus if a uniform intensity of photons were incident on an array of identical detectors as shown in Figure 7.8a, and the mean number of photons recorded per detector were N, most detectors would record either more than N or less than N, as shown in Figure 7.8b. The width of the distribution is proportional to $N^{\frac{1}{2}}$ and thus, as shown in Table 7.1, the variation in counts due to statistical fluctuations is dependent on the value of N, and expressed as a percentage N, that is, a noise/signal ratio, always increases as N decreases. Since the number of photons detected depends on the patient dose, the importance of quantum noise always increases as the patient dose is decreased. Whether a given dose reduction and the consequent increase in quantum noise will significantly affect image quality, and hence the outcome of the examination, can only be determined by more detailed assessment of the problem. For example, a radiographic image might contain 10^6 photons mm^{-2} for an incident skin dose of 3 mGy (see Insight). Thus the noise $N^{\frac{1}{2}} = 10^3$ and the noise-to-signal ratio $N^{\frac{1}{2}}/N = 0.1\%$. In this situation changes in transmission of 1% or more will be little affected by statistical fluctuations. However, a nuclear medicine image comprising 10^5 counts *in toto* may be

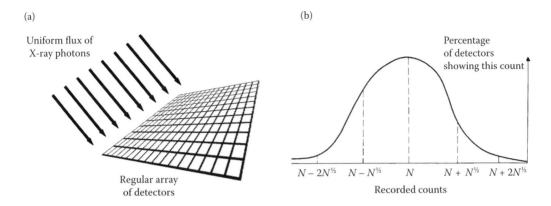

(a)

Uniform flux of
X-ray photons

Regular array
of detectors

(b)

Percentage
of detectors
showing this count

$N - 2N^{\frac{1}{2}}$ $N - N^{\frac{1}{2}}$ N $N + N^{\frac{1}{2}}$ $N + 2N^{\frac{1}{2}}$

Recorded counts

FIGURE 7.8
Demonstration of the statistical variation of detected signal: (a) uniform flux of X-ray photons onto a regular array of detectors; (b) the spread of recorded counts per detector. If the mean count per detector is N, 66% of the readings lie between $N - N^{\frac{1}{2}}$ and $N + N^{\frac{1}{2}}$; 95% of the readings lie between $N - 2N^{\frac{1}{2}}$ and $N + 2N^{\frac{1}{2}}$.

TABLE 7.1

Variation in Counts Due to Statistical Fluctuations as a Function of the Number of Counts Collected (or the Number of Counts per Pixel), N

N	$N^{\frac{1}{2}}$	$N^{\frac{1}{2}}/N \times 100$ (%)
10	3	30
100	10	10
1000	30	3
10000	100	1
100000	300	0.3
1000000	1000	0.1

digitised into a 64×64 matrix (approximately 4000 pixels) giving a mean of 25 counts per pixel. Now $N^{\frac{1}{2}}/N = 20\%$! Quantum noise is always a major source of image degradation in nuclear medicine.

Insight

Relationship between Photon Fluence and Dosimetric Quantities

The entrance surface dose (D) without backscatter is the product of the photon fluence (number of photons per unit area, N/A), the mass energy absorption coefficient for tissue (μ_a/ρ), and the mean photon energy (E).
    Thus,

$$D = \frac{N}{A} \times \frac{\mu_a}{\rho} \times E$$

If $D = 3$ mGy for a skull view (3×10^{-3} J kg^{-1}), $\mu_a/\rho = 0.004$ m^2 kg^{-1}, $E = 50 \times 10^3 \times 1.6 \times 10^{-19}$ J (note, care is required to ensure units are consistent), $N/A \approx 10^{14}$ photons $\cdot$ m^{-2} or 10^8 photons $\cdot$ mm^{-2}.

Assuming transmission through the skull is about 1% (it varies from 10% for a chest to 0.1% for a lateral lumbar spine), and that detector efficiency is 100% (in practice it may be about 70%–80% for a flat panel digital imager), there will be about 10^6 photons mm^{-2} in the image. The number of photons per pixel can be obtained by multiplying by the pixel area and the quantum noise in the pixel can be calculated.

There are now also a number of situations in radiology where quantum noise must be considered as a possible cause for loss of image quality or diagnostic information. These include the following:

1. The use of very fast intensifying screens may reduce the number of photons detected to the level where image quality is affected (Section 6.10).

2. If an image intensifier is used in conjunction with a television camera or the image is converted into digital format using a charge coupled device (CCD) camera, the signal may be amplified substantially by electronic means. However, the amount of information in the image will be determined at an earlier stage in the system, normally by the number of X-ray photon interactions with the input phosphor to the image intensifier. The smallest number of quanta at any stage in the imaging process determines the quantum noise, and this stage is sometimes termed the *quantum sink*.

3. Enlargement of a radiograph decreases the photon density in the image and hence increases the noise.

4. In digital images the smallest detectable contrast over a small area, say 1 mm^2, may be determined by quantum noise (see Section 7.5).

5. In CT the precision with which a CT number can be calculated will be affected. For example, the error on 100,000 counts is 0.3% (see Table 7.1), and the CT number (see Section 8.3) for different pixels in the image of a uniformly attenuating phantom (e.g. water) must vary accordingly.

Insight

Quantum Sinks

In Figure 6.22 the number of photons at each stage in the imaging system for an image intensifier coupled to a digital output was shown.

The critical stage (the quantum sink) was the number of quanta absorbed by the input phosphor. This is one reason why the quantum efficiency of the phosphor was such an important consideration in Section 5.4.1.

The signal-to-noise ratio (SNR) for this sequence will be $3 \times 10^4/(3 \times 10^4)^{1/2}$ or about 120. In this example subsequent amplification (or in the latter stages reduction) in the number of quanta will not affect S/N because signal and noise are amplified equally.

Note, however, that if the number of quanta were subsequently to fall below 3×10^4 during image processing, for example, as a result of digital subtraction imaging, the Poisson noise of the secondary quanta would further degrade S/N causing an even lower, or secondary, quantum sink.

Note that quantum mottle confuses the interpretation of low contrast images. In Section 7.7.1, imager performance will be assessed in terms of MTFs, which provide information on the resolution of small objects with sharp borders and high contrast. Other sources of noise may be the ultimate limiting factor in these circumstances.

7.6 Detective Quantum Efficiency (DQE)

As discussed in the previous section, there will be various sources of noise in a complete imaging system. Noise enters the equation at different stages in the imaging process and, with the exception of quantum noise, is difficult to quantify. This means that SNR, which is a very important quantity in image perception, will vary during the imaging process as more noise is introduced.

DQE provides a measure of these changes and is defined as

$$DQE = \frac{SNR_{out}^2}{SNR_{in}^2}$$

When it is applied to the whole system it compares the output of the device (i.e. SNR in the image signal) to the SNR in the input (incident radiation beam). Alternatively, to monitor how noise levels are changing, the DQE can be calculated for each stage in the imaging process. The DQE for the whole system is then the product of the individual DQEs.

Assuming the only source of noise in the incident beam is quantum noise,

$$SNR_{in}^2 = \frac{N_0^2}{(N_0^2)^{1/2}} = N_0 \text{ (the number of incident quanta)}$$

Thus $DQE = SNR_{out}^2/N_0$ and SNR_{out}^2 is known as the *noise equivalent quanta*, that is, the number of quanta that are equivalent to the total noise in the system.

Note that the concept of *quantum efficiency of a detector* was introduced as an 'insight' in Section 5.4.1. As the name implies, this specifies the efficiency of the detector in registering quanta, or the fraction of quanta incident on the detector that are stopped by it. It must not be confused with DQE, which is a much more general term that can be applied to the complete imaging system and quantifies all sources of noise. If the DQE concept is applied just to the first stage of the imaging process, that is to the detector, then the upper limit of DQE, in which no extraneous noise is introduced, is the quantum efficiency of the detector.

7.7 Assessment of Image Quality

7.7.1 Modulation Transfer Function

It is important that the quality of an image should be assessed in relation to the imaging capability of the device that produces it. For example, it is well known that an X-ray set is capable of producing better anatomical images of the skeleton than the radionuclide 'bone scan' images produced by a gamma camera—although in some instances they may not be as useful diagnostically.

One way to assess performance is in terms of the resolving power of the imaging device—that is, the closest separation of a pair of linear objects at which the images do not merge. Unfortunately, diagnostic imagers are complex devices and many factors contribute to the overall resolution capability. For example, in forming a conventional analogue X-ray image these will include the focal spot size, interaction with the patient, type of film, type of intensifying screen and other sources of unsharpness. Some of these interactions are not readily expressed in terms of resolving power and even if they were there would be no easy way to combine resolving powers.

A practical approach to this problem is to introduce the concept of MTF. This is based on the ideas of Fourier analysis, for detailed consideration of which the reader is referred elsewhere, for example Gonzalez and Woods (2008). Only a very simplified treatment will be given here.

Starting from the object, at any stage in the imaging process all the available information can be expressed in terms of a spectrum of spatial frequencies. The idea of spatial frequency can be understood by considering two ways of describing a simple object consisting of a set of equally spaced parallel lines. The usual convention would be to say the lines were equally spaced 0.2 mm apart. Alternatively, one could say that the lines occur with a frequency in space (spatial frequency) of five per mm.

Fourier analysis provides a mathematical method for relating the description of an object (or image) in real space to its description in frequency space. Two objects and their corresponding spectra of spatial frequencies are shown in Figure 7.9. In general, the finer the detail, that is sharper edges in real space, the greater the intensity of high spatial frequencies in the spatial frequency spectrum (SFS). Thus, fine detail, or high resolution, is associated with high spatial frequencies.

For exact images of these objects to be reproduced, it would be necessary for the imaging device to transmit every spatial frequency in each object with 100% efficiency. However, each component of the imaging device has a MTF which modifies the SFS of the information transmitted by the object.

Each component of the imager can be considered in turn so

$$\begin{array}{ccccccc} \text{SFS out of} & = & \text{SFS into} & \to & \text{Component} & \to & \text{SFS out of} \\ \text{component } M & & \text{component } N & & N & & \text{component } N \end{array}$$

Where

$$(\text{SFS}_{out})_N = (\text{SFS}_{in}) \times \text{MTF}_N$$

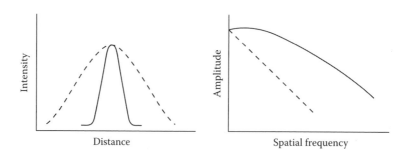

FIGURE 7.9

Two objects and their corresponding spectra of spatial frequencies. Solid line is a sharp object. Dotted line is a diffuse object.

Hence

$$(\text{SFS})_{\text{image}} = (\text{SFS})_{\text{object}} \times (\text{MTF})_{\text{A}} \times (\text{MTF})_{\text{B}} \times (\text{MTF})_{\text{C}} \cdots$$

where A, B, C are the different components of the imager.

No imager has an MTF of unity at all spatial frequencies. In general, the MTF decreases with increasing spatial frequency, the higher the spatial frequency at which this occurs the better the device.

The advantage of representing imaging performance in this way is that the MTF of the system at any spatial frequency v is simply the product of the MTFs of all the components at spatial frequency v. This is conceptually easy to understand and mathematically easy to implement. It is much more difficult to work in real space.

Insight

More on MTF

As indicated in the main text, all objects in real space can be represented by a set of spatial frequencies. This mathematical process is known as a Fourier transform (FT) and simplifies the subsequent mathematics considerably (as previously discussed in an insight in Section 6.11 in connection with image processing in the frequency domain).

It should be noted that rigorous mathematical derivation of MTF assumes a sinusoidally varying object with spatial frequency expressed in cycles mm^{-1}. In the figures presented in this chapter it has been assumed that this sinusoidal wave can be approximated to a square wave and spatial frequencies are given in line pairs mm^{-1} for ease of interpretation by the reader.

The MTF relates the sharpness in the image to the sharpness in the object, expressed in terms of spatial frequencies. Since the FT of a point object is a constant value (normalised to 1.0) at all spatial frequencies, the MTF of an imaging device is the FT of the point spread function when the device images a point source (see Figure 7.10). Of course an ideal imager would reproduce the object faithfully and have an MTF of 1.0 at all values of v.

In its simplest form, MTF only applies strictly to a system in which resolution does not vary with signal strength or location in the image. To achieve this, sampling of the data must occur at a frequency that is appreciably higher than the Nyquist frequency (see Section 8.4.2). This is readily achieved with an analogue detector (e.g. film-screen combination or image intensifier).

However, with digital detectors the inherent information content of the image is higher than the spatial frequency of the sampling. With a sampling distance d of 100 μm, the Nyquist frequency is $1/2d = 5$ cycles mm^{-1}, but the MTF of an aperture the size of a pixel (assumed for simplicity also to be d) contains spatial frequencies much higher than this so the MTF is under-sampled. Thus strict analysis of the MTF of a digital detector is quite complex—for further detail see Zhao et al. 2006.

Some examples of the way in which the MTF concept may be used will now be given. First, it provides a simple pictorial representation of the imaging capability of a device. Figure 7.11a shows MTFs for five image receptors. It is clear that non-screen film transfers higher spatial frequencies, and thus is inherently capable of higher resolution, than screen film. Of course this is because of the unsharpness associated with the use of screens. A similar rationale is made for the greater resolution associated with direct digital detectors over indirect detectors that incorporate a photoluminescent layer (Figure 7.11b). Note that the MTF takes no account of the dose that may have to be given to the patient.

A similar family of curves would be obtained if MTFs were measured for different film-screen combinations. The spatial frequency at which the MTF fell to 0.1 might vary from

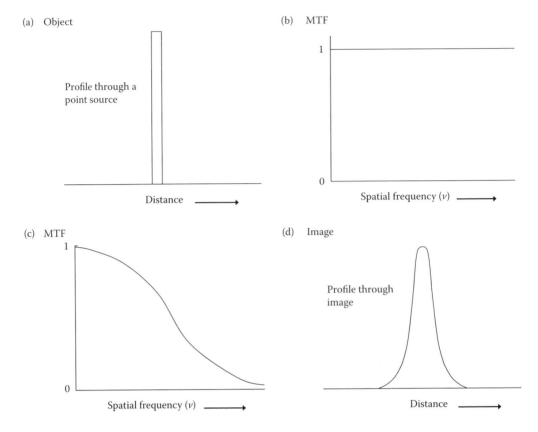

FIGURE 7.10
Diagrams showing the effect of an imaging device on simple objects. To reproduce the profile of the object (a) exactly, all spatial frequencies would have to be transmitted by the imager equally (b). In practice, most imagers transmit high spatial frequencies less well (c), which results in loss of sharpness in the image (d).

10 line pairs per mm for a slow film-screen combination to 2.5 line pairs per mm for a faster film-screen combination. This confirms that slow film-screen combinations are capable of higher resolution than fast film screens.

Since the MTF is a continuous function, an imaging device does not have a 'resolution limit' that is, a spatial frequency above which resolution is not possible, but curves such as those shown in Figure 7.11 allow an estimate to be made of the spatial frequency at which a substantial amount of information in the object will be lost.

Second, by examining the MTFs for each component of the system, it is possible to determine the weak link in the chain, that is, the part of the system where the greatest loss of high spatial frequencies occurs. Figure 7.12 shows MTFs for some of the factors that will degrade image quality when using an image intensifier TV system. Since MTFs are multiplied, the overall MTF is determined by the poorest component—the television camera in this example. Note that movement unsharpness will also degrade a high resolution image substantially.

As a third example, the MTF may be used to analyse the effect of varying the imaging conditions on image quality. Figure 7.13 illustrates the effect of magnification and focal spot size in magnification radiography. Curves B and C show that for a fixed focal spot size, image quality deteriorates with magnification and curves C and D show that for

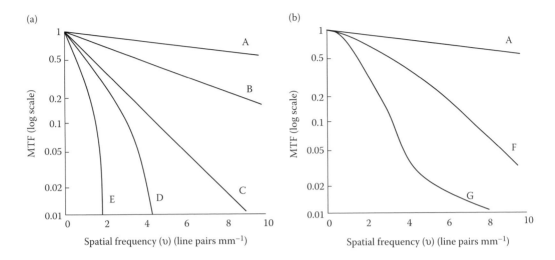

FIGURE 7.11
Some typical MTFs for different imaging receptors. (a) A = non-screen film; B = film used with high defini-tion intensifying screen; C = film with medium speed screens; D = a 150 mm Cs:Na image intensifier; E = the same intensifier with television display. (Adapted from Hay GA. Traditional X-ray imaging. In *Scientific Basis of Medical Imaging*, PNT Wells, ed. Churchill Livingstone, Edinburgh, 1–53, 1982, with permission.) (b) F = a direct digital detector; G = a computed radiography system or indirect digital detector; A is reproduced from Figure 7.11a for reference. Footnote to legend—Note that as a result of technical improvements in resolving capability since the 1980s, the MTFs D and E have moved to the right. However, the resolving capability of an image inten-sifier is still inferior to that of a film screen, and the television camera still causes further degradation.

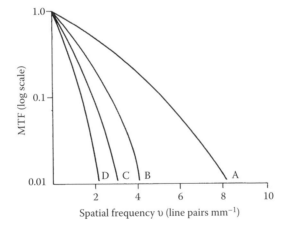

FIGURE 7.12
Typical MTFs for some factors that may degrade image quality in an image intensifier TV system. A = 1 mm focal spot with 1 m focus film distance and small object film distance; B = image intensifier; C = movement unsharpness of 0.1 mm; D = conventional vidicon camera with 800 scan lines.

fixed magnification image quality deteriorates with increased focal spot size. Note that if it were possible to work with $M = 1$, then the focal spot would not affect image quality and an MTF of 1 at all spatial frequencies would be possible (curve A). All these changes could of course have been predicted in a qualitative manner from the discussions of mag-nification radiography in Sections 6.12.1 and 9.4. The point about MTF is that it provides

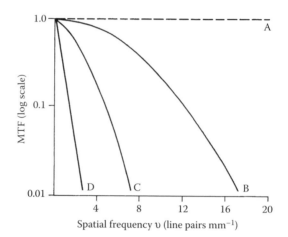

FIGURE 7.13
MTF curves for magnification radiography under different imaging conditions. A = magnification of 1; B = magnification of 1.2 with a 0.3 mm focal spot size; C = magnification of 2.0 with a 0.3 mm focal spot size; D = magnification of 2.0 with a 1 mm focal spot size.

a quantitative measure of these effects, and one that can be extended by simple multiplication to incorporate other factors such as the effect of magnification on the MTF of the receptor, and the effect of movement unsharpness (see for example Curry et al. 1990). Hence it is the starting point for a logical analysis of image quality and the interactive nature of the factors that control it.

Insight

An Unusual MTF

The majority of systems transmit low spatial frequencies better than high spatial frequencies. Xeroradiography (see insert to Section 9.2.5) has an unusual MTF with better data transfer at about one line pair per mm than at lower spatial frequencies. In practice this property means that discontinuities in the image (edges) that contain high spatial frequency information are enhanced. The MTF falls again at even higher spatial frequencies because of loss of resolution caused by other problems (e.g. focal spot size and patient motion).

7.7.2 Physical/Physiological Assessment

Although the MTF provides a useful method of assessing physical performance, its meaning in terms of the interpretation of images is obscure. For this purpose, as already noted in the discussion on contrast, it is necessary to involve the observer and in this context image quality may be defined as a measure of the effectiveness with which the image can be used for its intended purpose.

The simplest perceptual task is to detect a detail of interest, that is the signal, in the presence of noise. Noise includes all those features that are irrelevant to the task of perception—for example, quantum mottle, anatomical noise and visual noise (inconsistencies in observer response). Three techniques that have been used to discriminate between signal and noise will be mentioned briefly (for further details see Sharp et al. 1985 and ICRU Report 54 1996).

It should be noted in passing that the visual thresholds at which objects can be (a) detected, (b) recognised and (c) identified are not the same and this is a serious limitation when attempting to extrapolate from studies on simple test objects to complex diagnostic images.

7.7.2.1 Method of Constant Stimulus

Consider the simple task of detecting a small region of increased attenuation in a background that is uniform overall but shows local fluctuations due to the presence of noise. Because of these fluctuations, the ease with which an abnormality can be seen will vary from one image to another even thought its contrast remains unchanged.

Figure 7.14 shows the probability of generating a given visual stimulus for a set of images of this type in which the object contains very little contrast (C_1). If the observer adopts some visual threshold T, only those images which fall to the right of T will be reported as containing the abnormality (about 25% in this example).

If a set of images is now prepared in which the object has somewhat greater contrast (C_2), the distribution curve will shift to the right (shown dotted) and a much greater proportion of images will be scored positive.

If this experiment is repeated several times using sets of images, each of which contains a different contrast, visual response may be plotted against contrast as shown in Figure 7.15. The contrast resulting in 50% visual response is usually taken as the value at which the signal is detectable.

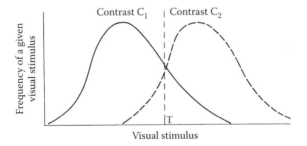

FIGURE 7.14
Curves showing how, for a fixed visual stimulus threshold T, the proportion of stimuli detected will increase if contrast is increased from C_1 to C_2.

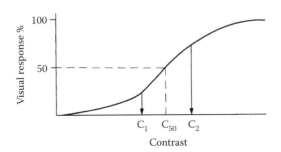

FIGURE 7.15
Visual response (percent positive identification) plotted as a function of contrast on the basis of observations similar to those described in Figure 7.14.

Experiments of this type may be used to demonstrate, for example, that in relation to visual perception, object size and contrast are inter-related.

7.7.2.2 Signal Detection Theory

To apply the method of constant stimulus, contrast must be varied in a controlled manner. Therefore the method cannot be applied to real images but is ideally suited to perceptive studies of 'abnormalities' digitally superimposed on digitised normal images.

In signal detection theory, all images are considered to belong to one of two categories, those which contain a signal plus noise and those which contain noise only. It is further assumed that in any perception study some images of each type will be misclassified. This situation is represented in terms of the probability distribution used in the previous section in Figure 7.16.

There are four possible responses based on the decision made by the observer and its ultimate verification or contradiction with the correct decision: true positive (TP), true negative (TN), false positive (FP) and false negative (FN). These four responses are subject to the following constraints:

$$TP + FN \text{ (all 'signal + noise' images)} = constant$$
$$TN + FP \text{ (all 'noise only' images)} = constant$$

The greater the separation of the distributions, the more readily is the signal detected.

The main problem with both signal detection theory and the method of constant stimulus is that they require the visual threshold level T to remain fixed. In practice this is well nigh impossible to achieve, even for a single observer. If T is allowed to vary, for example if there are several observers, a more sophisticated approach is required (see Section 7.8).

7.7.2.3 Ranking

In this method the observer is presented with a set of images in which some factor thought to influence quality has been varied. The observer is asked to arrange the images in order of preference. This approach relies on the fact that an experienced viewer is frequently able to say if a particular image is of good quality but unable to define the criteria on which this judgement is based.

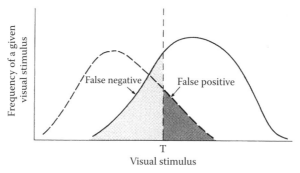

FIGURE 7.16

Curves illustrating the concept of false negative and false positive based on the spread of visual stimuli for an object of fixed contrast. Dotted line = probability distribution of true negatives. Solid line = probability distribution of true positives.

When the ranking order produced by several observers is compared, it is possible to decide if observers agree on what constitutes a good image. Furthermore, if the image set has been produced by varying the imaging conditions in a controlled manner, for example by steadily increasing the amount of scattering medium, it is possible to decide how much scattering medium is required to cause a detectable change in image quality.

The simplest ranking experiment is to compare just two images. If one of these has been taken without scattering material, the other with scattering material, then the percentage of occasions on which the 'better' image is identified can be plotted against the amount of scattering material. The 75% level corresponds to correct identification 3/4 times, the 90% level to correct identification 9/10 times. It might be reasonable to conclude that when the better image was correctly identified 3/4 times, detectable deterioration in image quality had occurred (Figure 7.17).

The ranking approach is particularly useful for quality control and related studies. All equipment will deteriorate slowly and progressively with time and although it may be possible to assess this deterioration in terms of some physical index such as resolution capability with a test object, it is not clear what this means in terms of image quality. If deterioration with time (Figure 7.18a) can be simulated in some controlled manner (Figure 7.18b), then a ranking method applied to the images thus produced may indicate when remedial action should be taken.

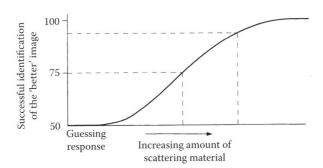

FIGURE 7.17
Curve showing how, the ability to detect the under-graded image might be expected to increase steadily as the amount of degradation was increased.

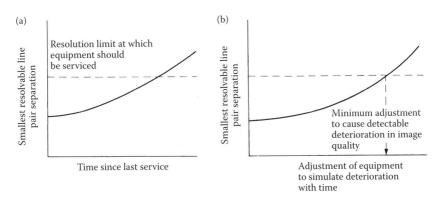

FIGURE 7.18
Curves showing how a logical policy towards the frequency of servicing might be based on measurements of deterioration of imager performance.

7.8 Receiver Operator Characteristic (ROC) Curves

7.8.1 Principles

Images from true abnormal cases will range from very abnormal, a strong visual stimulus, to only suspicious, a weak visual stimulus. Similarly, some images from true normal cases will appear more suspicious than others and in all probability the two distributions will overlap. Note that similar distributions will be produced, perhaps with rather less overlap, if two sets of images are presented in different ways or perhaps after different pre-processing.

The number of TPs identified now depends on the threshold detection level. If a very lax criterion is adopted, point A in Figure 7.19, the TPs = 100% but there is also a high percentage of FPs. If the threshold detection level is higher, or the observer adopts a more stringent criterion say point D, the result is quite different with few FPs but many TPs being missed.

Receiver operator characteristic curves overcome this problem. The observer is now encouraged to change their visual threshold and for each of the working points A to E, the FPs are plotted against the TPs (Figure 7.20). Two extreme distributions can be selected to illustrate the limits of the ROC curve. First, if the distributions are completely separate a range of thresholds may be chosen which first increases the percentage of TPs without any FPs and then gradually increases the FPs until the criterion is so lax that both distributions are accepted as abnormal. In other words, in the case of perfect discrimination, the ROC curve tracks up the vertical axis and then along the horizontal axis. The other extreme is where the distributions overlap completely. Now each relaxation of the detection threshold which allows in a few more TPs allows an equal number of FPs and the ROC curve follows the 45 degree line (shown dotted in Figure 7.20).

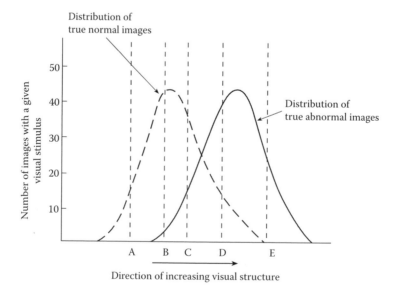

FIGURE 7.19

Use of different visual thresholds with distributions of visual stimuli.

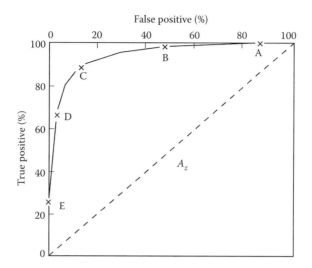

FIGURE 7.20
The ROC curve that would be constructed if the two distributions shown in Figure 7.19 were sampled at five discrete points. (Reproduced with permission from Dendy PP. Recent technical developments in medical imaging Part 1: Digital radiology and evaluation of medical images. *Curr. Imag.* 2, 226–236, 1990.)

A useful numerical parameter is the proportion of the ROC space that lies below the ROC curve, A_Z. In the two extreme distributions just discussed $A_Z = 1.0$ and 0.5, respectively. For intermediate situations the larger the value of A_Z the greater the separation of the distributions. Note that a value of A_Z less than 0.5 indicates that the observer is performing worse than guessing!

A two-dimensional ROC curve can be converted into a three-dimensional ROC curve in which the signal intensity is the third dimension (see Figure 7.21). For example when the signal intensity is high, the ROC curve will be close to the axes. When the signal intensity is zero the ROC curve will be the 45 degree guessing line. Thus a constant value for the imaging parameter produces an ROC curve whereas a profile through the surface at a constant FP fraction yields the response curve that would be produced by the method of constant stimulus.

The simple approach to ROC analysis described here can be applied very successfully when the whole image is being assessed as normal/abnormal. However, there is an obvious weakness if the observer's task is to identify a specific small lesion, for example, a lung nodule or a cluster of microcalcifications on a mammogram. The observer might 'correctly' identify the image as abnormal on the basis of an (erroneously) perceived abnormality in a part of the image that is actually normal. When the observer is required both to identify abnormal images and to specify correctly the position of the abnormality, localisation ROC curves (LROC) may be constructed. A further complication with clinical images is that the radiologist's decision is often based on more than one feature in the image. Clearly, there will be greater confidence in the conclusion when the different features provide corroborative information but the different image features may provide conflicting information. For a review of variants on ROC analysis see Chakraborty (2000).

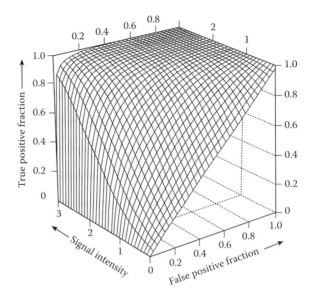

FIGURE 7.21

A 'three-dimensional' graph in which two-dimensional ROC curves are plotted as a function of signal intensity, thereby generating a surface in three dimensions. (Reproduced with permission from ICRU Report 54, Medical imaging—The assessment of image quality. International Commission on Radiation Units & Measurements, Bethesda, USA 1996.)

Insight

Free Response Operating Characteristic (FROC) and Alternative Free Response Receiver Operating Characteristic (AFROC) Curves

These are two approaches developed to overcome some limitations of simple ROC analysis. In the FROC method the observer searches each image for suspicious lesions and indicates both their location and their confidence level, usually from 1 to 4, where 4 is the highest level of confidence. If an indicated lesion is within a predetermined distance of an actual lesion, this is recorded as a TP, otherwise it is a FP. A plot of the fraction of lesions localised on the *y*-axis (range 0–1) against the mean number of FPs per image (on the *x*-axis) is the FROC curve. Low values of the mean number of FPs per image indicate a strict criterion of reporting lesions. There is no limit to the *x*-axis.

Since the mean number of FPs/image is unbounded, there is no equivalence between ROC curves and FROC curves. However, by considering the signal and noise stimuli as overlapping Gaussian variables, somewhat analogous to Figure 7.14, it is possible to convert the *x*-axis into a probability of a FP image, (i.e. one or more FPs). This plot is known as an AFROC curve with limits of 0 and 1.0 on both axes. The area under the curve A_1 (to distinguish from A_z) ranges from 0 for random guessing (cf. $A_z = 0.5$) to 1.0 for a perfect observer.

Note that FROC and AFROC methodology require a set of images where the location of every true lesion is known. This restricts somewhat the use of the methods on clinical images.

Good work can be done on systematic analysis of image quality using analogue images. However, digitised images are preferable because the data are available in numerical form. For example, it is possible to investigate the interaction between, say contrast, resolution and noise for carefully controlled, quantitative changes in the images or to look at the relative importance of structured or unstructured noise. One reason why this is desirable is that the eye-brain system is very perceptive, so there is only a narrow working region in which an observer might be uncertain or different observers disagree. Digital images provide more opportunity for fine tuning than analogue images.

Finally, under computer control it is possible to superimpose known lesions of different size, shape and contrast on normal or apparently normal clinical images that have been digitised. In the subsequent analysis the true abnormals can be unambiguously identified.

7.8.2 ROC Curves in Practice

Receiver operator characteristic analysis has been used to investigate numerous imaging problems. Three examples will be given to illustrate the power of the technique. They are all rather old now but they do illustrate well the value of the technique when used to answer an appropriately well-formulated question. Also they use simple ROC analysis as described here. Recent examples of ROC analysis using more sophisticated approaches are given in Section 7.9 on optimisation.

7.8.2.1 Pixel Size and Image Quality in Digitised Chest Radiography

Several authors have used ROC methods to address the question 'What is the largest pixel size that will generate digitised images that are indistinguishable from analogue chest images?' Clearly there are many features of the image one can examine, including nodules, fine detail, mild interstitial infiltrates and subtle pneumothoraces and MacMahon et al. (1986) were able to show, by digitising images to different pixel sizes that diagnostic accuracy increased significantly as pixel size was reduced, at least to 0.1 mm (Figure 7.22).

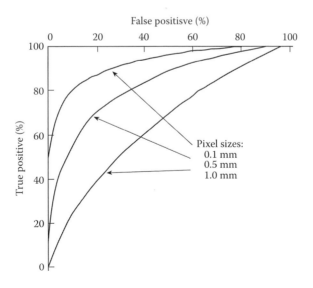

FIGURE 7.22
Use of ROC curves to show how diagnostic accuracy for fine detail in the pneumothorax varies with pixel size in digitised images. (Adapted with permission from MacMahon H, Vyborny CJ, Metz CE et al. Digital radiography of subtle pulmonary abnormalities—An ROC study of the effect of pixel size on observer performance. *Radiology* 158, 21–26, 1986.)

Although a pixel size of 0.1 mm × 0.1 mm may be acceptable for chest radiographs, subsequent work by the same authors (Chan et al. 1987) showed that for mammograms detection accuracy for microcalcifications was still significantly reduced if images were only digitised to this level.

7.8.2.2 Assessment of Competence as Film Readers in Screening Mammography

As a second example consider the question 'Is there a potential to use radiographers as film readers in screening mammography?'

Double reading of screening mammograms has been shown to increase the cancer detection rate (Anderson et al. 1994) but the acute shortage of radiologist resources makes this difficult. Pauli et al. (1996) therefore set out to determine if suitably trained radiographers could play a useful role as second readers in screening mammography. The data are suitable for ROC analysis because there are four possible outcomes to a film report—abnormal mammogram/recall (TP), abnormal mammogram/return to routine screening (false negative), normal mammogram/recall (FP), normal mammogram/return to routine screening (true negative).

A modified version of A_Z, represented here by A_z' was used to allow for the fact that distributions are very skewed because of the large percentage of normals. Results for seven radiographers were compared with those from nine radiologists and an extract from the data is shown in Table 7.2.

Results showed that after suitable training radiographers achieved and maintained the same high standard of performance on training test sets as a group of radiologists.

7.8.2.3 Image Quality Following Data Compression

As a final example consider the question 'Is it possible to compress data without loss of image quality?' Radiology departments have moved rapidly towards digital systems and filmless images. However, this has created formidable data transfer and archiving problems. Consider the memory required to store an 18 cm × 24 cm mammogram digitised at 0.07 mm × 0.07 mm resulting in a matrix size of about 2500 × 3500 (~8.5 million pixels). For a 14-bit image each pixel can hold any one of 16384 (2^{14}) grey levels, so a 4-view study can generate about 60 megabytes of digital data.

One way to reduce the problem is to compress the data. There is clearly redundant information in the complete image. For example, the counts in pixels in the corners will be no higher than background and there will be a high degree of correlation between adjacent

TABLE 7.2

Radiographer and Radiologist Performances in Reading Mammographic Films by Comparison of the Area under an ROC Curve

	A_z' values	
	Radiographers	**Radiologists**
Pre-training	0.77	0.87
Post training	0.88	0.89
After 200 screening mammograms read	0.91	–
After 5000 screening mammograms read	0.88	–

Source: Adapted from Pauli R, Hammond S, Cooke J and Ansell J. Radiographers as film readers in screening radiography: An assessment of competence under test and screening conditions. *Br. J. Radiol.* 69, 10–14, 1996.

TABLE 7.3

Mean Values of A_Z for Five Radiologists Who Read
Uncompressed and Compressed Chest Radiographs

	A_Z	
	Digitised	**Compressed**
Interstitial disease	0.95 ± 0.04	0.95 ± 0.03
Lung nodules	0.87 ± 0.06	0.88 ± 0.05
Mediastinal masses	0.83 ± 0.08	0.80 ± 0.10

Source: Adapted from Aberle DR, Gleeson F, Sayre JW et al.
The effect of irreversible image compression on diagnostic accuracy in thoracic imaging. *Invest. Radiol.* 28,
398–403, 1993.

pixels. Data compression will then have a number of advantages, especially in respect of less archival demands, faster data transfer and faster screen build up.

Compression algorithms are basically of two types. The first is reversible or loss-less compression. Provided that the compression/decompression schemes are known, a true copy of the original image may be recovered. Unfortunately, the maximum compression ratio that can be achieved in this way is about 4:1, often less with noisy data.

Higher compression ratios require the use of irreversible processing which results in some alteration in the information content. Note that this is not a problem *per se*. Most image data undergoes some alteration in information content before it is viewed (e.g. filtering the raw projections during image reconstruction in CT—section 8.4.2). The important requirement is to be able to image a lesion optimally without introducing false readings. Provided that data compression does not influence this process negatively, a certain degree of loss of information can be tolerated.

A full discussion of the relative merits of different compression algorithms is outside the scope of this book but some further detail is given in Section 17.8. It is clear that ROC analysis provides a useful way to select the approach which gives maximum compression with minimum loss of image quality. For example, Aberle et al. (1993) studied 122 PA chest radiographs which had been digitised at 2000 × 2000 pixels and 12 bit resolution and then compressed at an approximate compression ratio of 20:1. Five radiologists read the digitised images and the digitised/compressed images and results were analysed using ROC curves (see Table 7.3).

Although the compression process was irreversible, there was no evidence of any difference in detectability for interstitial disease, lung nodules or mediastinal masses.

7.9 Optimisation of Imaging Systems and Image Interpretation

'Optimisation' is a term that has become very popular in recent years when discussing radiological imaging. Most readers may think they already know what optimisation means, but on closer inspection it is apparent that the concept is rather more complex than it seems.

Central to the idea is the inter-relationship between image quality and patient dose. We have already seen in Section 7.5 that image quality improves as the number of photons used to form the image increases. However, in radiology this means X-ray photons and hence radiation dose. Perceived wisdom is that there is a concomitant extra radiation risk associated with any increase in dose and optimisation in terms of ensuring that all radiation doses to patients are 'as low as reasonably achievable' will be considered in Chapter 13 when doses and risks to patients are reviewed. However, the corollary to this constraint is that *all* other aspects of the imaging process must also be optimised as failure to do so may have an indirect impact on the dose of radiation required to provide the necessary diagnostic information.

These aspects are many and varied, ranging from purely physical features of the imaging process to evaluation of the process of image interpretation. A comprehensive review is beyond the scope of the book but it will be useful to consider a few illustrative examples. The ones chosen have two things in common. First, they are all fairly topical and second there is a quantitative measure of the optimisation process in each of them. The latter point seems entirely appropriate for a book with strong emphasis on physics and highly desirable in ensuring that the overall process of image optimisation gradually improves.

7.9.1 Optimising kVp for Digital Receptors

As discussed in Sections 3.5 and 6.3, X-ray tube potential has a significant effect on image contrast and also on the detection efficiency of the receptor. This latter influence will vary with the attenuating properties of the materials used in the receptor so it is appropriate, especially with the introduction of digital receptors, to try to optimise the tube voltages used for radiographic examinations.

Honey et al. (2005) performed a study that compared the image quality of adult chest examinations using a range of beam energies for both film-screen and computed radiography (CR). A threshold contrast test object (Figure 7.2a), embedded in 9 cm of Perspex to generate realistic attenuation and scatter conditions, was used. The relationship between dose and image quality for the two types of receptor was different, largely due to the different K-edges of each material used (about 50 keV for the Gd_2O_2S used in rare earth screens and about 37 keV for the BaF in the CR phosphor). This means that the efficiency of absorption of the two phosphors depends on the energy of the incident photon in different ways, so exposure factors determined as optimal for film-screen systems are unlikely to be optimal for digital systems. For a fixed effective dose a significant improvement in image quality when using CR was noted when tube potentials were decreased to the range 75–90 kVp from the higher values (~125 kVp) often used with film-screen systems.

Insight

More on Optimisation of kVp

The relationship between image quality and tube potential is a complex function of the primary X-ray spectrum, scatter spectrum and the energy dependence of the absorption efficiency of the receptor. For digital radiography this suggests that an optimisation strategy concerning itself with the energy of the primary beam spectrum can do more than examine the peak tube potential (kVp) alone. Other studies (e.g. Samei et al. 2005) have investigated the influence of the overall shape of the spectrum after alteration by additional tube filtration. This can further reduce patient dose while keeping image contrast at acceptable levels by taking advantage of the wide dynamic range and post-processing options of digital systems—not usually achievable with film-screen systems due to their inflexible contrast properties. Of course, additional filtration does lead to longer exposure times, increasing both wear on the X-ray tube and any movement unsharpness.

Experiments using test objects to quantify improvements or otherwise of criteria of image quality are essential when studying parameters to optimise in diagnostic imaging. However, the transfer of the results of these studies to the clinical situation usually requires verification by clinical trial (see Section 7.10).

7.9.2 Temporal Averaging

In Section 7.3.5 we discussed temporal resolution and movement threshold of the eye-brain system. An important practical application of this aspect of the eye-brain response is that it affects the amount of temporal averaging (i.e. persistence) that should be applied to X-ray fluoroscopy images. In modern digital fluoroscopy systems digital images are collected in 512 × 512 or 1024 × 1024 format at 25 frames s^{-1}. Temporal averaging is user-variable, with time constants typically varying between 0.02 s and 0.3 s, and needs to be optimised for different examinations. Kotre and Guibelaide (2004) used a threshold contrast-detail diameter test object (similar in principle to the one shown in Figure 7.2a but of a different design) and moved it across the image intensifier field of view at controlled speeds. They showed that for the high speeds associated with cardiac motion (75 mm s^{-1} was taken as an indicative value), the averaging provided by the observer's visual system is optimum for small object diameters. For the lower rates of movement typical of the abdomen (about 10 mm s^{-1}) a persistence time constant of 0.15 s is optimum for image perception. The optimum time averaging will vary with exact conditions but these values are broadly in line with the known temporal response of the eye and the general conclusions should apply to all image-intensifier-based digital fluoroscopy systems.

Note that in this type of work the intensifier input dose rate (0.5 μGy s^{-1} here) must be specified so that the quantum noise can be estimated and imaging conditions must be sufficiently realistic for all other sources of noise to be included.

7.9.3 Viewing Conditions

Another important aspect of the imaging chain is the way in which the observer assesses and evaluates the image. Since the observer is influenced by the conditions under which the image is viewed, it is important that satisfactory viewing conditions for the task are established and maintained so that as much diagnostically relevant information as possible can be extracted from the image.

As early as 1946 Blackwell showed that the ability of a human observer to detect the presence of a circular target against a contrasting background depends on the degree of contrast, the angular size of the target and the level of illumination. Since that time a generally held consensus has developed that viewing conditions will affect image interpretation. Other factors that may be important are the luminance of the image, uniformity, glare and reflection from the image. The Commission of the European Communities (CEC 1997) has specified minimum or maximum values for general radiography.

Most studies on the effect of viewing conditions on perception of contrast and detail have looked at the above-mentioned factors as a group but Robson (2008) has pointed out that for the purposes of optimisation they should be examined individually. Images of a contrast-detail test object (such as the one shown in Figure 7.2a), viewed under strictly controlled conditions are suitable for this purpose and the authors showed that, in the absence of ambient light, glare and reflection, the threshold luminance contrast remained the same for details of different diameter over a wide range of viewing luminance box settings (45 cd m^{-2} to 50,000 cd m^{-2}) (Figure 7.23). Similar results were found for ambient

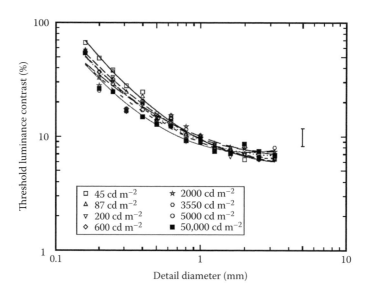

FIGURE 7.23
Contrast-detail curves showing threshold luminance contrast for different viewing box luminance settings. (Reproduced with permission from Robson K J. An investigation into the effects of sub-optimal viewing conditions in screen-film mammography. *Br. J. Radiol.* 81, 219, 2008.)

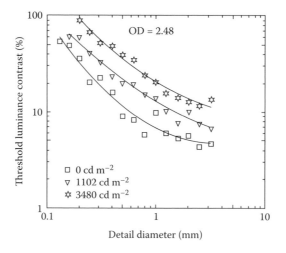

FIGURE 7.24
Curves showing the contrast-detail performance for films with an OD of 2.48 under different levels of uniform reflection. (Reproduced with permission from Robson K J. An investigation into the effects of sub-optimal viewing conditions in screen-film mammography. *Br. J. Radiol.* 81, 219, 2008.)

light and glare but perception was very sensitive to reflected light, with the threshold detail diameter increasing with the amount of reflected light, especially when reflection was non-uniform (Figure 7.24). This work was done within the context of mammography because viewing conditions are especially critical in the detection of subtle abnormalities in mammograms. However, the conclusions are likely to apply to the viewing of clinical images generally and may lead to a reappraisal of the optimum viewing conditions.

Insight

More on Viewing Conditions

The findings of this work are at variance with those of Blackwell, especially regarding the effect of the level of illumination. A possible explanation is that Blackwell used noiseless signals so the only noise contribution was from the visual system σ_{vis}. If there is a significant external noise contribution σ_{ext}, for example, from the film or in a digital system from the electronics, the total noise σ_{tot} will be $(\sigma^2_{vis} + \sigma^2_{ext})^{1/2}$. σ_{vis} may be affected by light intensity but if $\sigma_{ext} > \sigma_{vis}$ this effect may be obscured. Evidence that the total noise in the recent work was much higher than in Blackwell's work comes from the fact that the threshold contrast was higher.

7.9.4 Optimising Perception

There is an intuitive assumption that optimising the conspicuity of image features will improve the ability of the observer to make the right decision and hence improve the diagnostic process. Thus a great deal of attention has been given to defining and improving objective measures such as contrast, resolution and SNR.

However, there are also a number of factors that influence the ability of the observer to *interpret* the information. Some of these are image dependent and relate to the visual conspicuity of features relevant to the clinical problem, others are image independent and relate to the cognitive processes of the observer.

We know very little about the cognitive factors influencing perception but there is increasing evidence that in some circumstances they may be more important than previously thought. This is a difficult research area and a full treatment is beyond the scope of the book but a set of experiments reported by Manning et al. (2004) illustrates some of the issues. The authors investigated the hypothesis that in a complex imaging task with a search component and multiple targets, there may be too many confounding factors for conspicuity of image features always to be the dominant criterion. Nodule detection on PA chest radiographs as evidence of early lung cancer was chosen as the model system because it is well reported that up to 30% of such nodules may be missed by radiologists.

A group of radiologists, selected as expert observers, was asked to report on 120 digital chest images containing 81 pulmonary nodules of varying visibility and location. Eye tracking procedures were used to analyse the dwell-time on each suspicious area and AFROC techniques were used to record both the conspicuity and locations of the lesions. Quantitative methods were used to measure lesion size, grey level contrast and sharpness of boundary for each lesion and these were combined, with equal weights, in a conspicuity index (CI).

Correlation between the CI and AFROC score was poor, suggesting that the radiologists were not heavily influenced by conspicuity. More pertinent is an analysis of dwell times (see Figure 7.25), that is, the length of time radiologists spent looking at a particular region of the image. The upper limit for fixations that are associated with detection without recognition is thought to be about 1 s. Longer dwell times are associated with the cognitive processes of recognition and identification. In Figure 7.25, 90% of the true negatives have gaze durations of less than 1 s, consistent with no recognition or identification being involved. As expected, most of the positives, whether true or false, have gaze times of longer than 1 s. The pertinent finding is that 65% of the false negatives had dwell times of more than 1 s, suggesting that the failure is one of interpretation, not detection.

More studies are required, for example, using a wide variety of artificial lesions of varying size, contrast and location, superimposed digitally on normal images. However, the

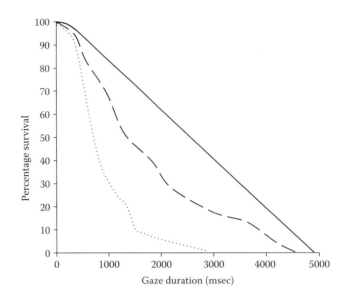

FIGURE 7.25
The percentage of nodules holding visual attention over a 15 s time interval for the four possible decision outcomes of true positive (TP) and false positive (FP), which cannot be distinguished (solid line), true negative (TN) (dotted line), and false negative (FN) (dashed line). (Redrawn with permission from Manning D J, Ethell S C and Donovan T. Detection or decision errors? Missed lung cancer from the anterior chest radiograph. *Br. J. Radiol.* 77, 231–235, 2004.)

work is important because, if confirmed, it may add a new dimension to the way image interpretation is taught and could influence the way in which artificial intelligence is used to assist the radiologist.

7.9.5 Computer-Aided Diagnosis (CAD)

As a final topic in this brief survey we can consider CAD as an example of optimisation of the whole concept of image interpretation. CAD, which can be defined as diagnosis made by a physician who takes into consideration a computer output based on quantitative analysis of radiological images, must be distinguished clearly from the earlier concept of automated computer diagnosis. This attempted to replace radiologists by computers but had little success.

The numerical format of the digitised image and the speed of modern computers are a powerful combination for extracting quantitative data about the image and in CAD these data can be used in two different ways: (a) locating a lesion (detection), (b) making an estimate of the probability of disease (differential diagnosis).

The concept of CAD can be applied to all imaging modalities, images of all parts of the body and all kinds of examinations, for example, skeletal and soft tissue imaging, functional imaging (see Section 10.4.1) and angiography, so the scope is considerable and efforts to date have concentrated on examinations where CAD is likely to be most beneficial. One of these is asymptomatic screening in mammography because large numbers of images, most of which are normal, have to be examined and CAD is ideally suited to carrying highly repetitive tasks with no lowering of standards due to lack of experience, lack of concentration or tiredness. Other diseases where screening has potential advantages, for example, colorectal cancer, have also been targeted. Large numbers of images from the same patient are often generated in 3-D imaging techniques such as helical CT and magnetic resonance imaging (MRI) so some of these studies are being investigated. The other general areas in which CAD has been applied are those where current practice leads to a high percentage of false negative reports.

The basic technologies involved in CAD schemes are (Doi 2005)

1. Image processing for detection and extraction of abnormalities (see Section 6.11)
2. Quantitation of image features for candidates of abnormalities
3. Data processing for classification of image features between normals and abnormals
4. Quantitative evaluation and retrieval of images similar to those of unknown lesions
5. Quantitative studies of observer performance

Most CAD systems work on the basis of 'prompts' to the observer identifying suspicious locations and allocating a 'probability of abnormality'. However, the value of this information is critically dependent on the quality of the algorithm used to generate the prompt. In theory there are a large number of ways in which the same images might be processed and many variations in which the 'prompt level' might be decided. Exact details of algorithms are often commercially secret but examples of the image processing that may be used are well documented.

1. Apply an algorithm to a single image to identify regions with characteristics associated with the abnormality—for example, groups of bright blobs on a mammogram that could correspond to microcalcification clusters.
2. Generate a further image in which background structures are suppressed and potential abnormalities enhanced.
3. Threshold the image based on features such as size of and degree of enhancement at suspect locations.
4. Compare suspect locations with locations in a library of normal images (or in mammography with the contralateral breast).

In the final assessment of a CAD system, ROC analysis (see Section 7.8) is frequently a useful quantitative tool. Consider, for example, the work of Doi and colleagues on detection and classification of lung nodules on digital chest radiographs. As noted in the previous section the literature shows that radiologists may miss about 30% of lung nodules on chest radiographs and even when a nodule is found, classification into benign and malignant is difficult. Image processing techniques were developed into algorithms for module detection and shown both to assist many radiologists and to reduce variation in detection accuracy due to variation in radiologists' experience.

Further algorithms were developed to provide a second opinion on the likelihood of malignancy based on image features such as the shape and size of the nodule and the distribution of pixel values inside and outside the nodule. Linear discrimination analysis is a statistical method that determines the differences between two or more types of object based on an assessment of their features and uses that information to place the object into predetermined categories—such as benign or malignant. This analysis was used to assign a likelihood measure of malignancy. Observer performance studies were

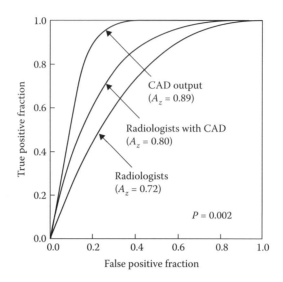

FIGURE 7.26
ROC curves used to evaluate the distinction between malignant and benign nodules on chest radiographs with and without computer-aided diagnosis. (With permission from Doi K. Review Article: Current status and future potential of computer-aided diagnosis in medical imaging. In *Computer Aided Diagnosis*, Gilbert F J and Lemke H, eds., *Br. J. Radiol.* 78, Special Issue, S3–19, 2005.)

analysed by ROC analysis and showed that radiologists could better distinguish benign and malignant nodules when they had the benefit of computer prompts (Figure 7.26). In fact the highest A_z value was achieved by computer analysis alone so perhaps automated computer diagnosis is not dead after all!

7.10 Design of Clinical Trials

By concentrating on the physics of the imaging process, it is easy to overlook the fact that *in vivo* imaging is not an end in itself but only a means to an end. Ultimately the service will be judged by the quality of the diagnostic information it produces and critical, objective evaluation of each type of examination is essential for three reasons. First it may be used to assess diagnostic reliability and this is important when the results of two examinations conflict. Second, objective information allows areas of weakness to be analysed and provides a basis for comparison between different imaging centres. Finally, medical ethics is not independent of economics and increasingly in the future the provision of sophisticated and expensive diagnostic facilities will have to be justified in economic terms.

It is easy to understand why good clinical trials are so difficult to carry out when the constraints are listed:

1. The evaluation must be prospective with images assessed within the reference frame of the normal routine work of the unit—not in some academic ivory tower.

2. A typical cross-section of normal images and abnormal images from different disease categories must be sampled since prevalence affects the predictive value of a positive test.

3. A sufficient number of cases must enter the trial to ensure adequate statistics.

4. Equivalent technologies must be compared. It is meaningless to compare the results obtained using a 1985 ultrasound scanner with those obtained using a 2005 CT whole body scanner or vice versa.

5. Evaluations must be designed so that the skill and experience of the reporting teams do not influence the final result.

6. Finally, and generally the most difficult to achieve, there must be adequate independent evidence on each case as to whether it should be classified as a true normal or a true abnormal.

If these constraints can be met, methods are readily available for representing the results. For example, if images are simply classified as normal or abnormal, the four possible outcomes to an investigation can be expressed as a 2 × 2 decision matrix (Table 7.4).

$$\text{Overall diagnostic accuracy} = \frac{\text{No. of correct investigations}}{\text{Total investigations}} = \frac{a + d}{a + b + c + d}$$

Other indices sometimes quoted are

$$\text{Sensitivity} = \frac{\text{Abnormals detected}}{\text{Total abnormals}} = \frac{a}{a + b}$$

$$\text{Specificity} = \frac{\text{Normals detected}}{\text{Total normals}} = \frac{d}{c + d}$$

and

$$\text{Predictive value of a positive test} = \frac{\text{Abnormals correctly identified}}{\text{Total abnormal reports}} = \frac{a}{a + c}$$

It is well known that prevalence has an important effect on the predictive value of a positive test and to accommodate possible variations in prevalence of the disease, Bayes Theorem may be used to calculate the posterior probability of a particular condition, given the test results and assuming different *a priori* probabilities.

TABLE 7.4

A 2 × 2 Decision Matrix for Image Classification

	Abnormal Images	Normal Images
True abnormal	a	b
True normal	c	d
Totals	$a + c$	$b + d$

TABLE 7.5

Examples of the Confidence Ratings Used
to Produce an ROC Curve

Rating	Description
5	Abnormality definitely present
4	Abnormality almost certainly present
3	Abnormality possibly present
2	Abnormality probably not present
1	Abnormality not present

If the prior probability of disease, or prevalence, is $P(D_+)$ then it may be shown that the posterior probability of disease when the test is positive (T_+) is given by

$$P(D_+ / T_+) = \frac{[a/(a+b)]\, P(D_+)}{[a/(a+b)]\, P(D_+) + [c/(c+b)]\, P(D_-)}$$

Similarly the posterior probability of disease when the test is negative (T_-) is given by

$$P(D_+ / T_-) = \frac{[b/(a+b)]\, P(D_+)}{[b/(a+b)]\, P(D_+) + [d/(c+d)]\, P(D_-)}$$

For a full account of Bayes Theorem see Shea (1978).

As an alternative to simple classification of the images as normal or abnormal, the data can be prepared for ROC analysis if a confidence rating is assigned to each positive response (Table 7.5).

These ratings reflect a progressively less stringent criterion of abnormality and correspond to points E to A, respectively in Figure 7.20. Thus by comparing the ROC curves constructed from such rating data, the power of two diagnostic procedures may be compared.

Much work has been done in recent years on the design of trials for critical clinical evaluation of diagnostic tests and it is now possible to assess better the contribution of each examination within the larger framework of patient health care.

7.11 Conclusions

The emphasis in this chapter has been on the idea that there is far more information in a radiographic image than it is possible to extract by subjective methods. Furthermore, many factors contribute to the quality of the final image and for physiological reasons, the eye can easily be misled over what it thinks it sees. Thus there is a strong case for introducing numerical or digital methods into diagnostic imaging. Such methods allow greater manipulation of the data, more objective control of the image quality and greatly facilitate attempts to evaluate both imager performance and the overall diagnostic value of the information that has been obtained.

References and Further Reading

Aberle D R, Gleeson F, Sayre J W et al., The effect of irreversible image compression on diagnostic accuracy in thoracic imaging. *Invest Radiol. 28*, 398–403, 1993.

Anderson E D C, Muir B B, Walsh J S and Kirkpatrick A E, The efficacy of double reading mammograms in breast screening. *Clin Radiol. 49*, 248–251, 1994.

Blackwell H R, Contrast thresholds of the human eye. *J Opt Soc Am. 36*, 624–643, 1946.

Campbell F W, The physics of visual perception. *Phil Trans Roy Soc Lond. 290*, 5–9, 1980

CEC Commission of the European Communities, *Criteria for the acceptability of radiological (including radiotherapy) and nuclear installations*, Radiation protection no. 91, Luxembourg, CEC, 1997.

Chakraborty D P, The FROC, AFROC and DROC variants of ROC analysis, in *Handbook of Medical Imaging*, J Beatel, HL Kundel and RL van Mettler WA Bellingham, eds, Society of Photo-optical Instrumentation Engineers, 2000, 771–796, 2000.

Chan H P, Vyborny C J, MacMahon H et al., Digital mammography ROC studies of the effects of pixel size and unsharp mask filtering on the detection of subtle microcalcifications. *Invest Radiol. 22*, 581–589, 1987.

Curry T S, Downey J E and Murry R C, *Introduction to the Physics of Diagnostic Radiology*, 4th edn. Lea & Febiger, Philadelphia, 1990.

De Lange D Z N H, Research into the dynamic nature of the human fovea-cortex systems with intermittent and modulated light I. Attenuation characteristics with white and coloured light. *J Opt Soc Am. 48*, 777–784, 1958.

Dendy P P, Recent technical developments in medical imaging Part 1: Digital radiology and evaluation of medical images. *Curr Imaging. 2*, 226–236, 1990.

Doi K, Review Article: Current status and future potential of computer-aided diagnosis in medical imaging, in *Computer Aided Diagnosis*, F J Gilbert and H Lemke, eds, *Br J Radiol. 78* Special Issue pp S3–19, 2005.

Gonzalez R C and Woods R E, *Digital Image Processing*, 3rd edn, Prentice Hall, New Jersey, 2008.

Hay G A, Traditional X-ray imaging, in *Scientific Basis of Medical Imaging*, P N T Wells, ed, Churchill Livingstone, Edinburgh, 1–53, 1982.

Honey I D, MacKenzie A and Evans D S, Investigation of optimum energies for chest imaging using film-screen and computed radiography. *Br J Radiol. 77*, 422–427, 2005.

ICRU Report 54, *Medical Imaging—the Assessment of Image Quality*. International Commission on Radiation Units & Measurements, Bethesda, 1996.

Kelly D H, Flicker fusion and harmonic analysis. *J Opt Soc Am. 50*, 1115–1116, 1961.

Kotre C J and Guibelalde E, Optimisation of variable temporal averaging in digital fluoroscopy. *Br J Radiol. 77*, 675–678, 2004.

MacMahon H, Vyborny C J, Metz C E et al., Digital radiography of subtle pulmonary abnormalities—An ROC study of the effect of pixel size on observer performance. *Radiology. 158*, 21–26, 1986.

Manning D J, Ethell S C and Donovan T, Detection or decision errors? Missed lung cancer from the anterior chest radiograph. *Br J Radiol. 77*, 231–235, 2004.

Pauli R, Hammond S, Cooke J and Ansell J, Radiographers as film readers in screening radiography: An assessment of competence under test and screening conditions. *Br J Radiol. 69*, 10–14, 1996.

Robson K J, An investigation into the effects of sub-optimal viewing conditions in screen-film mammography. *Br J Radiol. 81*, 219, 2008.

Samei E, Dobbins III J T, Lo J Y and Tornai M P A, Framework for optimising the radiographic technique in digital x-ray imaging. *Radiat Prot Dosim. 114*, 220–229, 2005.

Sharp P F, Dendy P P and Keyes W I, *Radionuclide Imaging Techniques*, Academic Press, London, 1985.

Shea G, An analysis of the Bayes procedure for diagnosing multistage disease. *Comput Biomed Res. 11*, 65–75, 1978.

Thijssen MAO, Thijssen HOM, Merx JL et al., A definition of image quality: The image quality fig-
ure, in *Optimisation of Image Quality and Patient Exposure in Diagnostic Radiology*, B M Moores,
B F Wall, H Eriska and H Schibilla, eds, BIR Report 20, 29–34, 1989.
Zaho W, Andriole K P and Samu E, Digital radiography and fluoroscopy, in *Advances in Medical
Physics*, A B Wolbarst, R G Zamenhof and W R Hendee, eds, Medical Physics Publishing,
Madison, Wisconsin, 1–23, 2006.

Exercises

1. What factors affect (a) the sharpness (b) the contrast of clinical radionuclide images?

2. List the factors affecting the sharpness of a radiograph. Draw diagrams to illustrate these effects.

3. What do you understand by the 'quality' of a radiograph? What factors affect the quality?

4. Explain the terms subjective and objective definition, latitude and contrast when used in radiology.

5. Explain what is meant by the term 'perception' in the context of diagnostic imaging. Explain how perception studies may be used to assess the quality of diagnostic images.

6. Why does the MTF of an intensifying screen improve if the magnification of the system is increased?

7. Show that enlargement of an image such that the photons are spread over a larger area increases quantum noise by $(N/m^2)^{1/2}$ where N is the original number of photons mm^{-2} and m is the magnification. Is quantum noise increased by magnification radiography?

8. Explain how quantum mottle may limit the smallest detectable contrast over a small 1 mm^2 area in a digitised radiograph.

9. What is an ROC curve and how is it constructed? Give examples of the use of ROC curves in the assessment of image quality.

10. Discuss the concept of optimisation in radiological imaging and give examples.

8

Tomographic Imaging with X-Rays

S J Yates and P P Dendy

SUMMARY

- Tomographic imaging partially or completely removes the problem of overlying or underlying tissues by viewing the plane of interest from different angles.

- Longitudinal tomography, in which the angles can be up to $20°$ from the normal to the plane interest, will be used to introduce the principle.

- In computed tomography (CT) the angle equals, or is close to, $90°$ and the fundamentals of CT will be explained.

- For single-slice CT the essential features of data collection and data reconstruction will be discussed.

- Spiral CT and multi-slice CT will be introduced. These are developments by which faster scanning times, and in the case of multi-slice CT, isotropic resolution, can be achieved.

- Assessment of image quality and the causes of image artefacts will be addressed.

- CT is generally a high dose technique so methods of dose reduction and good quality assurance programmes will be important topics.

- Finally, some specialist applications of CT and future developments will be considered briefly.

CONTENTS

8.1 Introduction

Three fundamental limitations can be identified that apply to planar imaging with both X-rays and gamma rays from radionuclides (see Chapter 10).

(i) Superimposition

The final image is a two-dimensional representation of an inhomogeneous three-dimensional object with many planes superimposed. In X-ray imaging the image relates to the distribution of attenuation coefficients in the different planes through which the X-ray beam passes.

The confusion of overlapping planes, and in particular the generation of scatter (see Sections 6.4 and 6.5) results in a marked loss of contrast making detection of subtle anomalies difficult or impossible. A simplified example of the consequence of superimposition is shown in Figure 8.1.

(ii) Geometrical Effects

The viewer can be confused about the shapes and relative positions of various structures displayed in a conventional X-ray image and care must be taken before drawing conclusions about the spatial distribution of objects, even when an orthogonal pair of radiographs is available.

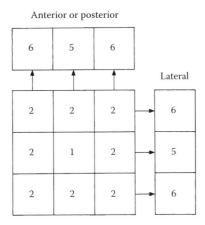

FIGURE 8.1
Illustrating how superposition of signals from different planes reduces contrast. If the numbers in the nine squares represent units of attenuation on a logarithmic scale in nine compartments, the theoretical contrast is 2:1. If, however, the attenuation is viewed along any of the projection lines shown, the contrast is only 6:5.

(iii) Attenuation Effects

When the intensity of X-rays striking an image receptor is described by the equation $I = I_0 e^{-\mu x}$ it must be appreciated that beam attenuation I/I_0 depends on both μ and x. This can cause ambiguity since an observed difference in attenuation can be due to changes in thickness alone (x), in composition alone (μ) or a combination of these factors.

Two methods have been used to overcome the effects of superimposition of information in different planes.

(i) Longitudinal or Blurring Tomography

This relies on the principle that if an object is viewed from different angles, such that the plane of interest is always in the same orientation relative to the detector, the information in this plane will superimpose when the images are superimposed, but information in other planes will be blurred.

In longitudinal tomography, the angle between the normal to the plane of interest and the axis of the X-ray beam ϕ can be anything from 1° to 20°. Since the object space is not completely sampled, the in-focus plane is not completely separated from its neighbouring planes, but the tomographic effect increases with increasing ϕ.

Longitudinal tomography is now not widely used routinely, but it is considered briefly in Section 8.2 because it illustrates how viewing an object from different angles helps to remove the effect of superimposition. It also forms the basis of the relatively new technique of *tomosynthesis* using digital detectors.

(ii) Transverse Tomography

This can be considered as the limiting form of longitudinal tomography with $\phi = 90°$, that is with the axis of the X-ray beam parallel to the plane of interest. The object is considered as a series of thin slices and each slice is examined from many different angles. Isolation of the plane of interest from other planes is now complete and is determined by the width of the interrogating X-ray beam, detector design and the amount of scatter. Transverse tomography provides an answer to geometrical and attenuation problems as well as that of superimposition. The majority of this chapter is devoted to consideration of the many physical factors that contribute to the production of high quality transverse tomographic images using X-rays.

8.2 Longitudinal Tomography

The simplest form of longitudinal tomography is the linear tomograph and historically this is a good way to introduce the principles. As shown in Figure 8.2, the X-ray tube is constrained to move along the line SS′ so that the beam is always directed at the region of interest P in the subject. As the X-ray source moves in the direction S → S′, shadows of an object P will move in the direction F′ → F. Hence if an image receptor is placed in the plane FF′, and is arranged to move parallel to that plane at such a rate that when S_1 has moved to S_2, the point P_1 on the receptor has moved to P_2, all the images of the object at P will be superimposed.

Consider, however, the behaviour of a point in a different plane. As S_1 moves to S_2, the image receptor will move a distance P_1P_2 as before, so the point F_1, which coincides with X_1, will move to F_2. However, the shadow of X will have moved to X_2. Hence the shadows cast by X for all source positions between S_1 and S_2 will be blurred out on the receptor between F_2 and X_2, in a linear fashion. A similar argument will apply to points below P. Thus only the image of the point P, about which the tube focus and receptor have pivoted, will remain sharp.

Consideration by similar triangles of other points in the same horizontal plane as P, for example Q, shows that they also remain sharply imaged during movement. The horizontal plane through P is known as the *plane of cut* or sometimes the *focal plane*. Although

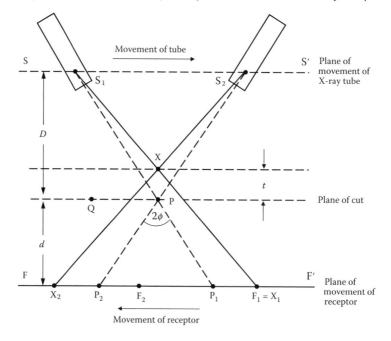

FIGURE 8.2
Diagram illustrating the movement of the X-ray tube and the image receptor in linear longitudinal tomography and the blurring that occurs for points that are not in the plane of cut.

the plane of cut is not completely separated from adjacent planes, there is a minimum amount of blurring that the eye can discern and this effectively determines the 'thickness' of the plane that is in focus. The thickness of cut increases if the value assumed for minimum detectable blurring increases, but decreases if the angle of swing increases. It also decreases if the ratio focus-film distance/focus-plane of cut distance increases.

Insight

It is instructive to calculate the thickness of the plane in terms of the minimum amount of blurring, B say, and the geometry of the system. Referring again to Figure 8.2, suppose the point X suffers the minimum amount of blurring that is just detectable. From the previous discussion:

$$\text{Film blurring} = X_2F_2$$

By geometry this is equal to $X_2X_1 - X_1F_2 = X_2X_1 - P_2P_1$
   This must be set equal to B
Now

$$P_1P_2 = S \cdot \frac{d}{D} \quad (\text{where } S = S_2 - S_1)$$

and

$$X_1X_2 = S \cdot \frac{(d+t)}{(D-t)}$$

Hence

$$B = S\left[\frac{(d+t)}{(D-t)} - \frac{d}{D}\right] \approx S\frac{(D+d)}{D^2}t \quad (\text{since } t \ll D)$$

Since

$$\frac{S}{D} \approx 2\phi$$

$$B = 2t\left(\frac{D+d}{D}\right)\phi$$

The same argument can be applied to points below P so the thickness of cut,

$$2t = B \cdot \frac{1}{\phi} \cdot \frac{D}{(D+d)} \tag{8.1}$$

Assuming typical values:

Angle of swing $2\phi = 10° = 10 \times 2\pi/360$ radians
Focus-film distance $(D + d) = 90$ cm
Focus-plane of cut $D = 75$ cm
Taking $B = 0.7$ mm (a reasonable figure for the blurring that will not detract significantly from the interpretation of most images)
Thickness of cut $2t = 6.6$ mm

Whilst it may be desirable for the slice thickness (t) to be small, the contrast between two structures will be proportional to $(\mu_2 - \mu_1) \cdot t$. Consequently, large angle tomography giving thin slices may be used when there are large differences in atomic number or density between structures of interest, but if $(\mu_2 - \mu_1)$ is small, a somewhat thicker slice may be required to give sufficient contrast.

Note the following additional points:

1. The plane of cut may be changed by altering the level of the pivot.

2. Since the tube focus and film move in parallel planes, the magnification of any structure that remains unblurred throughout the movement remains constant. This is important because the relatively large distance between the plane of cut and the receptor means that the image is magnified.

3. For two objects that are equidistant from the plane of cut, one above and one below, the blurring is greater for the one that is further from the receptor. This may influence patient orientation if it is more important to blur one object than the other.

4. Care is required to ensure that the whole exposure takes place whilst the tube and image receptor are moving.

5. The tilt of the tube head must change during motion—this minimises reduction in exposure rate at the ends of the swings due to obliquity factors but there is still an inverse square law effect.

8.2.1 Digital Tomosynthesis

One of the disadvantages of conventional longitudinal tomography is that the technique only provides information relating to one plane of interest. Digital tomosynthesis overcomes this by replacing the single protracted X-ray exposure with a sequence of projection X-rays, each acquired with the tube in a different position. A conventional tomographic image can be obtained by summing the individual projections. However, by shifting the data in each image, the plane which is in focus can be selected retrospectively. Consequently, a set of images, each in focus in a different plane, can be reconstructed from a single series of exposures. Algorithms can also be applied to remove some of the blurring from structures outside of the plane of interest, further improving on the conventional technique. Digital tomosynthesis thus overcomes many of the limitations of conventional planar radiography, as outlined in Section 8.1, but at a lower radiation dose than that associated with X-ray CT. The technique is increasingly being applied in X-ray mammography, and there is significant interest in a range of other fields, including chest, dental and orthopaedic imaging.

8.3 Principles of X-ray Computed Tomography

Transverse tomography is now used in every major imaging department in the form of X-ray CT. This title is a shortened version of the full terminology—'X-ray *Computed Axial Transmission Tomography*'—which is rarely used nowadays. It is, however, important not to overlook the words omitted since they give a more accurate description of the technique.

As discussed in Section 7.3.3, there is an inverse relationship between the contrast of an object and its diameter at threshold perceptibility. Thus the smaller the object, the larger the contrast required for its perception, and loss of contrast resulting from superimposition of different planes has an adverse effect on detectability. Hence, although the average film radiograph is capable of displaying objects with dimensions as small as 0.05 mm, when due allowance is made for film resolution, geometric movement blur and scattered radiation, a contrast difference of about 10% is required to achieve this.

The objective of single-slice CT is to take a large number of one-dimensional views of a two-dimensional transverse axial slice from many different directions and reconstruct the detail within the slice. Multi-slice CT has developed into a volume imaging technique and is less inherently based on transverse axial slices. However, the basic principles are unchanged.

Conventionally, the x and y axes are in the transverse plane and the z-axis is along the length of the patient. The object plane can be considered as a slice of variable width subdivided into a matrix of attenuating elements with linear attenuation coefficients $\mu(x,y)$ (Figure 8.3). One thousand projections are required for a typical matrix size of 512×512, corresponding to a pixel size of 0.5 mm across a 25 cm field of view. In single-slice CT, resolution in the longitudinal direction is relatively poor, with the slice thickness varying from 1 mm to 10 mm. However, in multi-slice CT, isotropic resolution of around 0.5 mm or better can be achieved (see Section 8.6).

When data are collected at one particular angle, θ, the intensity of the transmitted beam I_θ will be related to the incident intensity I_0 by

$$P_\theta = \ln\left(\frac{I_0}{I_\theta}\right)$$

where P_θ is the projection of all the attenuation coefficients along the line at angle θ.

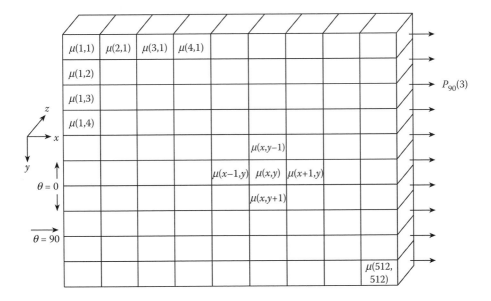

FIGURE 8.3
A slice through the patient considered as an array of attenuation coefficients. If $\theta = 0$ is chosen arbitrarily to be along the y-axis, the arrows show the P_{90} projections.

For example, for $\theta = 90°$, *one* of the values of P_{90}, say $P_{90}(3)$ to indicate that it is the projection through the third pixel in the y direction, is given by

$$P_{90}(3) = \mu(1,3) + \mu(2,3) + \mu(3,3) \ldots \ldots \ldots \ldots \mu(512,3)$$

Note that, strictly, each μ value should be multiplied by the x dimension of the pixel in the direction of X-ray travel. All values of x are the same in this projection. However, pixels contribute different amounts to different projections.

The problem is now to obtain sufficient values of P to be able to solve the equation for the $(512)^2$, about 260,000, values of $\mu(x,y)$. The importance of computer technology now becomes apparent, since correlating and analysing all this information is beyond the capability of the human brain but is ideal for a computer, especially since it is a highly repetitive numerical exercise.

For each pixel in the reconstructed image, a CT number is calculated which relates the linear attenuation coefficient for that pixel $\mu(x,y)$ to the linear attenuation coefficient for water μ_w according to the following equation:

$$\text{CT number} = K \cdot \frac{\mu(x,y) - \mu_w}{\mu_w}$$

where K is a constant equal to 1000 on the Hounsfield scale. Note that since the image space is divided into pixels and the data must take discrete values (CT numbers), CT is a true digital technique. Some typical values for CT numbers are given in Table 8.1.

Contrast may now be expressed in terms of the difference in CT numbers between adjacent pixels. Since a change of 1 in 1000 is 0.1%, contrast changes of about 0.2%–0.3% can be detected. Compare this with 2% visible contrast for fairly large objects on a good planar radiograph. Note that at the high tube potentials used in CT, the Compton effect dominates. As discussed in Section 3.4.3, the Compton effect depends on free electrons and is fairly constant for biological elements when expressed per unit mass. However, because of density differences, *the number of electrons per unit volume* varies and it is this variation we are measuring in CT.

An interactive display is normally used so that full use can be made of the wide range of CT numbers. Since only about 30 grey levels are distinguishable by the eye on a good black and white monitor, adjustment of both the mean CT number (the *window level*), and the range of CT numbers covered by the grey levels (the *window width*) must be possible quickly and easily (see Figure 8.4 and Section 6.3.3). A wide window is used when comparing structures widely differing in μ, but a narrow window must be used when variations

TABLE 8.1

Typical CT Numbers for Different Biological Tissues

Tissue	Range (Hounsfield Units)
Air	−1000
Lung	−200 to −500
Fat	−50 to −200
Water	0
Muscle	+25 to +40
Bone	+200 to +1000

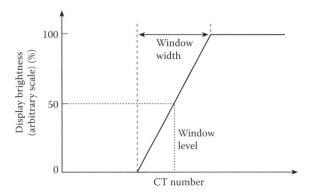

FIGURE 8.4
Illustrating the relationship between window width and window level when manipulating CT numbers.

in μ are small. Consider, for example, a 1% change in μ, which represents a range of about 10 CT numbers. With a window from +1000 to –1000 this may show as the same shade of grey. However, if a narrow window, ranging from –50 to +50 is selected, 10 CT numbers represents a significant proportion of the range and several shades of grey will be displayed.

8.4 Single-Slice CT

8.4.1 Data Collection

The process of data collection can best be understood in terms of a simplified system. Referring to Figure 8.5 which shows a single source of X-rays and a single detector, all the projections $P_0(y)$ that is, from $P_0(1)$ to $P_0(512)$, can be obtained by traversing both the X-ray source and detector in unison across the section. Both source and detector now rotate through a small angle $\delta\theta$ and the linear traverse is repeated. The whole process of rotate and traverse is repeated many times such that the total rotation is at least 180°. If $\delta\theta = 1°$ and there is a 360° rotation, 360 projections will be formed.

Although this procedure is easy to understand and was the basis of the original or first generation systems, it is slow to execute, requires many moving parts and requires a scanning time of several minutes. Thus a major technological effort has been to collect data faster and hence reduce scan time, thereby reducing patient motion artefacts. Historically, scan times were reduced in second generation scanners by using several pencil beams and a line of detectors.

Third generation scanners achieved faster scan times using an arc of detectors and a fan-shaped beam, wide enough to cover the whole body section (Figure 8.6a). One of the initial problems with this generation of scanners was that detector instabilities led to ring shaped artefacts within images. Fourth generation scanners used a complete ring of detectors and only the X-ray source rotated (Figure 8.6b). This design did not result in faster scan times than third generation scanners. However, it offered advantages in terms of detector calibration, and consequently the ring artefact issue was largely overcome. At this stage, scan rotation times were typically 1–2 s.

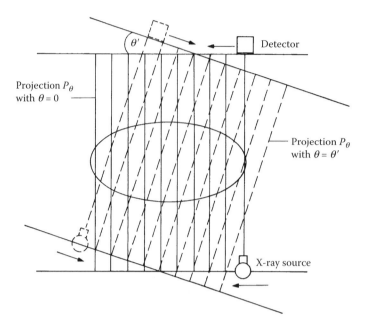

FIGURE 8.5
Illustration of the collection of all the projections using a single X-ray tube and detector and a translate-rotate movement (a *first generation* scanner).

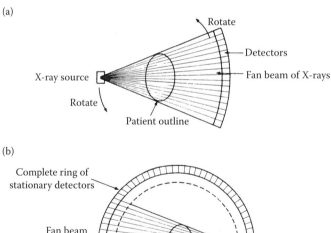

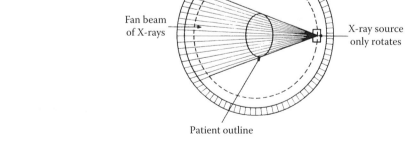

FIGURE 8.6
Design of different generations of CT scanners. (a) A rotate-rotate (third generation) scanner. (b) A fourth generation scanner, in which only the tube rotates.

The fourth generation design has not proved popular, however, and modern multi-slice systems are all based on third generation designs. Improvements in the stability of solid-state detectors mean that ring artefacts are no longer a significant issue with this design. These systems have achieved further increases in overall scan speed using an array of multiple rows of detectors to acquire more than one image in a single rotation (see Section 8.6). Modern equipment contains tens of thousands of detectors and rotation times of around 0.3 s can be achieved.

Even faster scan times can be achieved using 'scanners' with no moving parts. In one approach, termed *electron beam CT*, there is a semicircle of tungsten targets below the patient and a complementary semicircle of detectors above the patient. Note that reconstruction is possible from profiles collected over angles of a little more than 180°. Collection over a full 360° is not necessary. The beam of electrons is deflected electromagnetically so that the tungsten targets are bombarded in sequence. Very high tube currents can be used since each part of the anode is only used very briefly and scan times as short as 50 ms have been achieved.

In spiral CT (see Section 8.5) continuous X-ray output may be required for up to 60 s. This places a number of new demands on the design and construction of X-ray tubes and generators, especially with respect to heat dissipation. For example, 200 mAs per rotation might be required to achieve the necessary signal-to-noise ratio in the image. At an effective energy of 75 keV, 15 kJ of heat is generated per rotation, or for 100 rotations 1.5 MJ of heat is released in the anode. To improve heat storage capacity and tube rating, larger, faster rotating ceramic-mounted anodes have been developed (see Section 2.3.3). Anode angles tend to be smaller than for normal X-ray tubes and to achieve more uniform heat transfer and dissipation the anode is flat with the X-ray beam angled onto it.

Table 8.2 shows typical specifications for an X-ray tube used for CT. Note that rotation times are not likely to get significantly shorter, given the size and weight requirements of the anode for heat storage. A rotation time of 0.4 s imposes a mechanical rotational force of 15 g on the anode (where g is the force experienced by unit mass as the result of gravity). For a rotation time of 0.2 s this force would rise to 60 g.

Other constraints ensure optimum image quality. Voltage fluctuations must be less than 1% and this requires high frequency power supplies with the voltage controlled by a dedicated microprocessor. Generator output must also be controlled to better than 1% so the tube operates in fast pulse mode (a few ms) using grid control (see Sections 2.3.1 and 2.3.6). High resolution requires a small focal spot size, typically of the order of 0.5–1.0 mm and the tube must be arranged with its long axis perpendicular to the fan beam to avoid heel effect asymmetry.

TABLE 8.2

Typical Specifications for an X-ray Tube Used for CT

	Typical Performance
kV	80–140
Maximum mA	500–800
Generator power (kW)	60–100
Anode heat capacity (MJ)	6
Anode angle (°)	7
Focal spot size (mm)	0.5–1.0
Rotation time (s)	0.3–1.0

Choice of X-ray energy is a compromise between high detection efficiency and good image contrast on the one hand, low patient dose and high tube output on the other. Calculation of a unique matrix of linear attenuation coefficients assumes a monoenergetic beam. Heavy filtration is required to approximate this condition—for example, 0.4 mm Cu might be used with a 120 kV beam to give an effective energy of about 75 keV. When corrections are made for differences in mass attenuation coefficient and density, 0.4 mm Cu is approximately equivalent to 6.5 mm Al. Note, however, that the characteristic radiation that would be emitted from copper (Z = 29, K shell energy = 9.0 keV) as a result of photoelectric interactions would be sufficiently energetic to reach the patient and increase the skin dose. Thus the copper filter is backed by aluminium to remove this component.

Modern scanners will also correct for the effects of beam hardening by software in the reconstruction programme. Failure to correct adequately for beam hardening will cause image artefacts (see Section 8.9).

In single-slice CT, beam size is restricted by adjustable slit-like collimators near the tube. Collimators at the detector can also be used to define the slice thickness more precisely and to control Compton scatter, especially at the higher energies. Normally only a few percent of scattered radiation is detected. Note that scattered radiation is more of a problem with a fan beam than with a pencil beam. The reader may find the discussion of broad beam and narrow beam attenuation in Section 3.6 helpful in understanding the reason. Note also that scatter rejection is better in third generation designs than fourth. In third generation CT, the tube and detectors are in fixed orientations with respect to each other. This allows the use of detector collimation which is focussed on the X-ray source. Conversely, in fourth generation CT, the tube moves relative to the detectors, and the degree of detector collimation must be reduced.

The requirements of radiation detectors for CT are a high detection efficiency (i.e. high atomic number and density), wide dynamic range, good linearity, fast response and short afterglow to allow fast gantry rotation speeds (e.g. for ECG gated cardiac studies), stability, reliability, small physical size, high packing density, high geometric efficiency and reasonable cost.

This has been an area of intensive commercial development. Originally thallium-doped sodium iodide NaI (Tl) crystals and photomultiplier tubes (PMTs) were used. One problem with NaI (TI) crystals is afterglow—the emission of light after exposure to X-rays has ceased. This effect is particularly bad when the beam has passed through the edge of the patient and suffered little attenuation because the flux of X-ray photons incident on the detector is high. Detectors are now radiation-sensitive solid-state material (e.g. cadmium sulphate, gadolinium oxysulphide, gadolinium orthosilicate) and PMTs have been replaced by high purity temperature stabilised silicon photodiodes. Modern solid-state detectors have an 80%–90% X-ray quantum detection efficiency, depending on kVp and object, and 80%–90% geometric efficiency. They have low electronic noise thus ensuring that quantum noise is the limiting noise contribution. They respond rapidly and afterglow signals drop by several orders of magnitude in a few ms, so they are suitable for a CT scanner with very high sampling rates. They have good anti-scatter collimation.

The detector element spacing can be as low as 0.5 mm and some designs incorporate a quarter detector shift in the x-y plane to double the spatial sampling density. The current from the silicon photodiode can be amplified and converted into a digital signal in an ADC. The dynamic range is better than 16 bits.

In spite of their inherently low sensitivity, gas-filled ionisation chambers were initially quite widely used as detectors in third generation scanners. High atomic number gases such as xenon and krypton, at pressures of up to 30 atmospheres, in ionisation chambers several cm deep give a quantum detection efficiency of about 50% and the detectors can be closely spaced. There is little variation in sensitivity and, since the detectors work on the ionisation plateau (see Figure 4.3) the current produced is independent of the voltage across the ion chamber so they have excellent stability. This property made them particularly valuable in overcoming the ring artefact issues associated with detector instability in third generation scanners. Unfortunately the low sensitivity is a major drawback and gas-filled detectors have now become obsolete because their design is unsuitable for multi-slice systems (see Section 8.6).

8.4.2 Data Reconstruction

The problem of reconstructing two-dimensional sections of an object from a set of one-dimensional projections is not unique to radiology and has been solved more or less independently in a number of different branches of science. As stated in the introduction, in diagnostic radiology the input is a large number of projections, or values of the transmitted radiation intensity as a function of the incident intensity, and the solution is a two-dimensional map of X-ray linear attenuation coefficients.

Mathematical methods for solving problems are known as *algorithms* and a wide variety of algorithms has been proposed for solving this problem. Many are variants on the same theme so only two, which are both fundamental and are quite different in concept, will be presented. The first is filtered back projection (FBP), sometimes called *convolution* and *back projection*, and the second is an iterative method.

8.4.2.1 Filtered Back Projection

Figure 8.7 shows several projections passing through the point (x,y) which is at some distance l from the centre of the slice. The philosophy of back projection is that the attenuation coefficient $\mu(x,y)$ associated with this element contributes to all these projections, whereas other elements contribute to very few. Hence if the attenuation in any one direction is assumed to be the result of equal attenuation in all pixels along that projection, then the total of all the projections should be due to the element that is common to all of them.

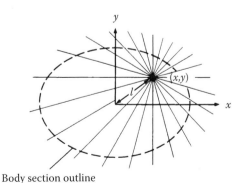

Body section outline

FIGURE 8.7
The projections that will contribute to the calculation of linear attenuation coefficient at some arbitrary point (x,y), a distance l from the (x,y) origin.

Hence one might expect an equation of the form

$$\mu(x,y) = \sum_{\substack{\text{over all} \\ \text{projections}}} P_\theta(l)$$

where $P_\theta(l)$ is the projection at angle θ through the point (x,y) and $l = x \cos\theta + y \sin\theta$, to give an estimate of the required value of $\mu(x,y)$.

An alternative way to look at this approach, which may be easier to understand (and explains how the name arose), is to assume that the attenuation of each profile is 'back projected' along the profile, with each pixel making an equal contribution to the total attenuation. Note that because the beam has finite width, the pixel (x,y) may contribute fully to some projections (Figure 8.8a) but only partially to others (Figure 8.8b). Due allowance must be made for this in the mathematical algorithm. Similar corrections must be applied if a fan beam geometry is being used.

If this procedure is applied to a uniform object that has a single element of higher linear attenuation coefficient at the centre of the slice, it can be shown to be inadequate. For such an object each projection will be a top-hat function, with constant value except over one pixel width (Figure 8.9a). These functions will back project into a series of strips as shown in Figure 8.9b. Thus simple back projection creates an image corresponding to the object but it also creates a spurious image. For the special case of an infinite number of projections, the spurious image density is inversely proportional to radial distance r from the point under consideration.

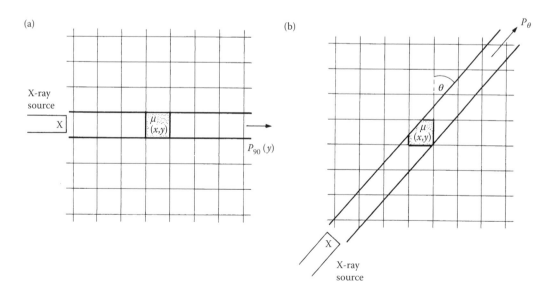

FIGURE 8.8

Illustration of the difference between the matrix size and the size and shape of the scanning beam for most angles. Assuming, for convenience, that the beam width is equal to the pixel width, then for the P_{90} projection (Figure 8.8a) it can be arranged for the beam to match the pixel exactly. However, this is not possible for other projections. Only a fraction of the pixel will contribute to the projection P_θ (Figure 8.8b) and allowance must be made for this during mathematical reconstruction. Any variation in μ across a pixel, or since the slice also has depth, the volume element (voxel), will be averaged out by this process.

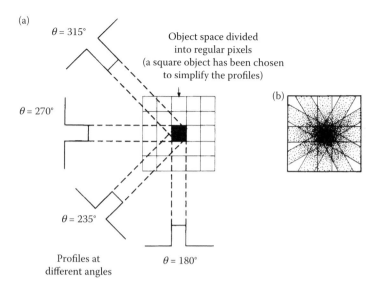

FIGURE 8.9
(a) The projections for a uniform object with a single element of higher linear attenuation coefficient at the centre of the field of view. (b) Back projection shows maximum attenuation at the centre of the field of view but also a star-shaped artefact.

Further rigorous mathematical treatment is beyond the scope of this book and the reader is referred to one of the numerous texts on the subject (e.g. Gordon et al. 1975). Suffice to say that the back-projected image in fact represents a blurring of the true image with a known function which tends to $1/r$ in the limiting case of an infinite number of profiles. Furthermore, correction for this effect can best be understood by transforming the data into spatial frequencies (see Sections 6.11 and 16.8). In terms of the discussion given there, the blurred image can be thought of as the result of poor transmission of high spatial frequencies and enhancement of low spatial frequencies. Any process that tends to reverse this effect helps to sharpen the image since good resolution is associated with high spatial frequencies. It may be shown that in frequency space correction is achieved by multiplying the data by a function that increases linearity with spatial frequency.

An important factor determining the upper limit to the spatial frequency (v_m) is the amount of noise that can be accepted in the image. This upper limit is also related to the size of the detector aperture, the effective spot size of the X-ray tube and the frequency of sampling. The value of v_m imposes a constraint on the system resolution since for an object of diameter D reconstructed from N profiles, the cut-off frequency v_m should be of the order of $N/(\pi D)$ and ensures spatial resolution of $1/2v_m$. For example, if $N = 720$ and $D = 40$ cm, the limit placed on spatial resolution by finite sampling is about 1 mm. Note that reference to Figure 8.9 shows that sampling is higher close to the centre of rotation than at the periphery. Thus if only the centre of the field of view is to be reconstructed, higher spatial frequencies may contribute and better resolution may be achieved.

Finally, the correction function to be applied to the back-projection data cannot rise linearly to v_m and then stop suddenly because this will introduce another source of image artefacts. Figure 8.10 shows how a sharp discontinuity in spatial frequency translates into a loss of sharpness in real space—the effect is very similar to the diffraction of light at a straight edge.

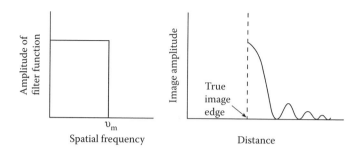

FIGURE 8.10

The effect of a sharp discontinuity in a filter function that has a constant amplitude up to v_m (left) on the sharpness of the edge of an image (right).

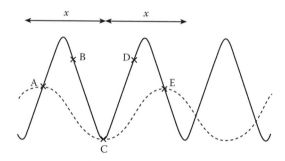

FIGURE 8.11

An example of aliasing. If the sine wave is sampled at A B C D E, it is uniquely determined. If it is only sampled at A C E, the values may be fitted by a lower frequency curve shown dotted.

Insight

Aliasing

Many texts state that the generation of spurious spatial frequencies resulting in multiple low intensity images owing to inadequate sampling of data is known as aliasing. However, relatively few give a detailed explanation in simple terms.

A good starting point is the approach of Cherry et al. (2003). Scan profiles are not continuous functions but collections of discrete point by point samples of the scan projection profile (in digital and CT work, the pixel). The distance between these points is the linear sampling distance. In addition, profiles are obtained in CT scans only at a finite number of angular sampling intervals around the object. The choice of linear and angular sampling intervals and the maximum spatial frequency of the cut-off filter (the cut-off frequency) in conjunction with the detector resolution, determine the reconstructed image resolution.

The sampling theorem (Oppenheim and Wilsky 1983) states that to recover spatial frequencies in a signal up to a maximum frequency v_m, the Nyquist frequency, requires a linear sampling distance d given by $d \leq 1/2v_m$. Alternatively, if the value of d is fixed, there is a limit on v_m of $1/2d$. If frequencies higher than v_m are transmitted (e.g. by the MTF of the detector) or amplified by the filter function, aliasing will result.

To understand how under-sampling can introduce spurious spatial frequencies, consider Figure 8.11 which shows a simple sinusoid with repeat distance x. If it is sampled at 5 equally spaced points A B C D E, the curve will be uniquely defined. No other sinusoid will have the same amplitude values at these 5 points. Note that the data have been sampled five times in the distance $2x$ so the sampling theorem is satisfied.

If the data are only sampled at A C and E, an alternative sinusoid (rising as it passes through A and E, of lower spatial frequency may also be fitted.

Figure 8.12 summarises the position with respect to filter functions. On the left, curve A shows the effect of simple back projection on the amplitudes of different spatial frequencies, curve B shows the (theoretically) ideal correction function. However, this function will cause amplification of high spatial frequency noise. On the right, filter C will give good resolution but will cause artefacts because of the sharp edge in frequency space. Filter D will cause loss of resolution because high spatial frequencies near to but less than υ_m are inadequately amplified. Filter E will give good resolution without too much noise amplification but there will be some aliasing.

Clearly there is no ideal filter function and the final choice must be a compromise. However, the exact shape of the filter function has a marked effect on image quality in radiology and manufacturers generally offer a range of filter options. The performance of a filter function with respect to resolution can be well represented by its modulation transfer function. MTFs for a standard and a high resolution filter are shown in Figure 8.13. These would correspond to spatial resolution limits of about 0.7 mm and 0.4 mm, respectively.

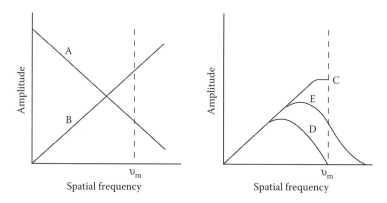

FIGURE 8.12
Typical functions associated with image reconstruction by filtered back projection. For explanation, see text.

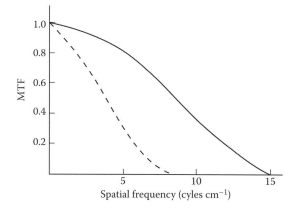

FIGURE 8.13
Modulation transfer functions for two CT filter functions; dashed line—standard, solid line—high resolution.

For a standard body scan one would normally use short scan times because of body movement, with a wide slice and standard convolution filter. For a high resolution scan, say of the inner ear, a sharper filter, that is one which falls to zero more steeply, and thinner slices will be selected. Both of these changes will increase the relative importance of noise. If this reaches an unacceptable level it can be counter-acted by increasing the mAs and hence the dose.

8.4.2.2 Iterative Methods

An iterative method makes repeated use of a mathematical formula to provide a closer approximation to the solution of a problem when an approximate solution is substituted in the formula. A series of approximations that is successively closer to the required solution is obtained.

A simple example of iteration to construct a map of attenuation values in CT, with an array of 9 pixels, is shown in Figure 8.14. The starting point is to guess a distribution of values for $\mu(x,y)$. For example, in the first estimate shown, the true projections P_1 have been assumed to contribute equally to all pixels in the image (ii) (a). One projection to be expected from the assumed values of μ is now compared with the actual value and the difference between observed and expected projections is found (ii) (b). Values of μ along this projection are then adjusted so that the discrepancy is attributed equally to all pixels (iii) (a). This process is repeated for another projection (iii) (b), a revised set of assumed projections is calculated and the whole process is repeated. Further detail on iteration is given in the 'insight'.

Insight

More on Iterative Methods

1. Note the iteration does not always converge, that is the estimate becomes progressively worse, not better. To check for convergence, calculate the root mean square deviation:

$$\sqrt{\frac{\Sigma(x_E - x_0)^2}{n-1}}$$

where the summation is made over all pixels. For the fourth estimate, this is

$$\sqrt{\frac{0+0+1/9+0+1/9+0+4/9+0+1}{8}} = \sqrt{\frac{15}{72}} = 0.46$$

It is left as an exercise for the student to show that the standard errors of previous estimates are higher.

2. In practice it is better to model the input data (μ_{xy} values) and the imaging process (e.g. the shape of the X-ray beam and spectrum) to various degrees of sophistication to predict the image. Unfortunately, the stochastic process of quantum noise prevents an exact quantification of the model parameters from the measured projections. Therefore it is necessary to estimate the most likely parameter combination. For example, the maximum likelihood expectation algorithm establishes a model that is most likely to give the same projection set as the measured object. The deviations between the modelled and measured projections may be used to improve the modelled parameters in an iterative process. Clearly this

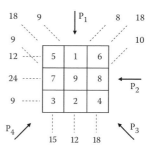

(i) <u>Object</u>
    The object consists of nine elements
    with true attenuations as shown.

(ii) <u>First estimate</u>
    (a) Assume all pixels have contributed
        equally to P_1.
    (b) Compare estimated P_2 projections
        with true P_2 projections. The
        uppermost projection is 3 too high,
        etc.

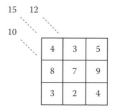

(iii) <u>Second estimate</u>
    (a) Share the discrepancies equally
        among the pixels on that projection.
        E.g. Each pixel on the top line is
        decreased by 1.
    (b) Compare estimated P_3 projections
        with true P_3 projections. The middle
        projection is 3 too low, etc.

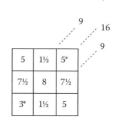

(iv) <u>Third estimate</u>
    (a) Share the discrepancies equally
        among the pixels on that projection.
        Note that pixels marked * remain
        unaltered.
    (b) Compare estimated P_4 projections
        with true P_4 projections.

(v) <u>Fourth estimate</u>
    (a) Share the discrepancies equally
        among the pixels on that projection.
        Note that two other pixels, marked *,
        remain unaltered this time.
    (b) This is now a reasonable
        approximation to the object. The
        process can be repeated for other
        projections, to obtain progressively
        better approximations.

FIGURE 8.14
A simple example of the iterative process.

sequence can be repeated indefinitely but there is a high cost in computing time and a procedure that requires large numbers of iterations is probably unsatisfactory. Also care is necessary to ensure that the iterative process converges to a unique solution because the iterative process tends to model image variations that are due to random (stochastic) changes resulting in increasingly noisy images. If μ values are changed too much, the revised image may be less similar to the object than the original image.

No one reconstruction method holds absolute supremacy over all others and it is essential to assess the efficiency of a particular algorithm for a particular application. A major disadvantage of iterative methods for fast CT imaging is that all data collection must be completed before reconstruction can begin.

8.5 Spiral CT

Until the early 1990s, a major constraint on the whole imaging process was that after each rotation the direction of rotation had to be reversed. This returned the tube and detectors to their original positions to untangle the wires. Since that time, the development of slip-ring technology has permitted continuous rotation of the X-ray tube and detector system for third generation scanners, or continuous rotation of the X-ray tube for fourth generation scanners. Rotation may be in a fixed plane to obtain information about the way in which contrast medium reaches an organ or fills blood vessels over short period of time. Alternatively, the patient may be moved through the gantry aperture during such continuous data acquisition. This provides a spiral or helical data set and is known variously as *spiral CT, helical CT or volumetric CT.*

Clearly, for this mode of data collection it is necessary for the X-ray tube to emit radiation for a long period of time. An initial restriction on spiral CT was set by the heat rating of the tube, limiting the method to low tube currents and poor counting statistics. The developments in anode design discussed in Section 2.3.3 have been important. A modern scanner will easily cover the full length of the patient at standard tube currents.

A second early limitation was that image quality was not comparable with axial acquisition. This was partly due to the noise associated with the low counts, and partly to the fact that data collected at different angles are not in the same orthogonal plane relative to the patient (Figure 8.15). Some degradation of the slice sensitivity profile is inevitable and in single-slice CT this is greater the higher the rate of table feed. One approach to this problem is to use linear interpolation between two adjacent data points obtained at the same angular position—for a simple account see Miller (1996). In Figure 8.15 measured values at A and B for the same arbitrary angle θ may be used to deduce the actual value in the plane of interest. However, B is rather a long way from this plane so an improved algorithm effectively shifts the phase of this sinusoid by 180° and then interpolates between the measured values A and B' for a more accurate estimate of the required value.

An important term in relation to spiral CT is the *pitch*. This relates the volume traced out by the spiral to the nominal beam thickness. It is defined as

$$\text{Pitch} \ = \ \frac{\text{Table feed per rotation}}{\text{Nominal beam thickness}}$$

The pitch is 1.0 if the distance travelled by the couch during one full rotation is equal to the nominal beam thickness (e.g. nominal beam thickness = 5 mm, duration of 360°

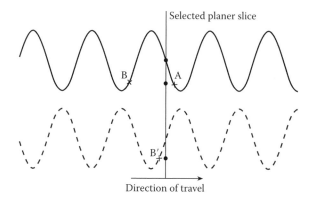

FIGURE 8.15

Graph showing that because the patient moves during data acquisition the X-ray focus traces out a spiral relative to the patient. To reconstruct a plane of interest, data not actually collected in that plane must be used. Interpolation between two positions separated by 180° (AB′) produces a better result than interpolation between positions separated by 360° (AB).

rotation 1s, rate of movement = 5 mm s⁻¹). If the rate of movement is doubled, the table pitch becomes 2.0. Slices of nominal thickness 5 mm can still be reconstructed but there will be degradation of the slice sensitivity profile.

There are a number of advantages of spiral CT:

- Slice thickness, slice interval and slice starting point may be chosen retrospectively rather than prospectively. The simple example of a single 5 mm lesion being detected with 5 mm slices illustrates this advantage. In axial CT, unless the slice starts exactly at the edge of the lesion, it will contribute to two slices with corresponding loss of contrast.

- Reconstruction is not limited to transverse sections. They can also be coronal or sagittal, pathology can be viewed from any angle and vessel tracking techniques may be used to follow tortuous vessels as they pass through the body.

- The faster scan permits a full study during a single breath hold. This is important to eliminate respiration where motion artefacts can be troublesome—for example, when visualising small lesions in the thorax. High levels of contrast can also be maintained in vessels for the duration of the study and CT angiography becomes possible.

8.6 Multi-Slice CT

8.6.1 Data Collection

Rapid acquisition of data, both for improved volume coverage and to minimise patient movement, is a major goal in CT and, as a means to this end, faster rotation times and spiral CT have already been discussed. The development of scanners in the mid 1990s which allowed the simultaneous acquisition of more than one slice presented a further significant advance in CT technology.

The principle of multi-slice CT is relatively straightforward. In a third generation single-slice design, up to 900 detector elements were arranged in an arc that was concentric with the z-axis but the detectors were only one element deep (typically 10 mm) in the z direction. To achieve slices less than 10 mm thick the beam width was restricted using physical collimators, often both between the X-ray tube and the patient and between the patient and the detectors. In multi-slice systems the detectors are physically and electronically separated along the z-axis and thus form a matrix of elements (Figure 8.16).

Several designs of detector array have been developed by different manufacturers. A full review of the options is beyond the scope of this book but they all achieve a flexible combination of the number of simultaneous contiguous slices and slice thickness along the z-axis. Two basic designs have been used—fixed array detectors, in which all elements have the same width, and adaptive array detectors, in which the outer detectors are wider than those nearer the centre. An example of each is shown in Figure 8.16. Figure 8.16a shows a fixed array detector with 64 detector rows and a collimated detector of 0.625 mm giving a maximum coverage of 40 mm along the z-axis. Thicker z values may be obtained by combining the detectors in groups. Figure 8.16b shows an adaptive array detector with 24 detector rows. In the middle there are 16 detectors with a width of 0.75 at the centre of rotation. These are flanked by eight outer detector rows 1.5 mm wide. This array may operate as a 16×0.75 mm array covering 12 mm or as a 16×1.5 mm array covering 24 mm.

The development of multi-slice CT allowed imaging times to be significantly reduced. For example, an early multi-slice scanner offering four 5 mm slices per rotation allowed a volume of data to be acquired in a quarter of the time that a single-slice scanner would have taken to acquire 5 mm images through the same volume. However, the scanner could alternatively be used to acquire thinner slices, and hence achieve better z-axis resolution. A major aim in the development of CT has been to acquire images with isotropic resolution, that is, where the resolution in the z direction matches that in the x-y plane. Once this is achieved, high quality images can be reconstructed in the coronal and sagittal planes, and various 3-D visualisation techniques can be applied. Resolution in the x-y plane is typically of the order of 10 lp cm^{-1} corresponding to resolution of objects of around 0.5 mm in size. The early multi-slice scanners did not achieve sub-millimetre slice widths.

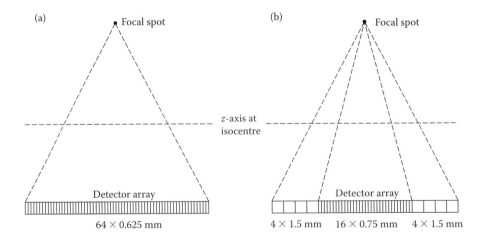

FIGURE 8.16
Examples of multi-detector arrangements; (a) A fixed array detector with 64×0.625 mm detectors, (b) an adaptive array with 16×0.75 mm detectors at the centre and 4×1.5 mm detectors on each side.

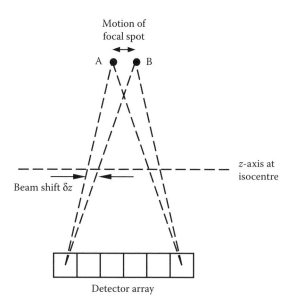

FIGURE 8.17
Principle of double z-sampling; As a result of periodic motion of the focal spot position in the z direction, the projections falling on each detector from spot A are slightly different from those from spot B. The shift at the isocentre δz can be made equal to half the detector width.

However, most scanners offering 16 slices or above offer slice widths of around 0.6 mm or thinner and practically isotropic resolution is achievable.

A further improvement in z-axis resolution has been implemented by one manufacturer in a technique known as double-z sampling. In this technique, the focal spot of the X-ray tube is shifted on a periodic basis between two locations on the anode surface. The principle is illustrated in Figure 8.17. Because each detector 'sees' the two focal spot positions along slightly different directions, the two beams are attenuated along slightly different projections through the patient. If, for example, the detector assembly allows 32 collimated 0.6 mm projections per rotation of the gantry assembly, two consecutive 32 channel projections, created with the focal spot in different positions, can be interleaved to produce 64 overlapping projections. Each projection remains 0.6 mm thick, but the z-axis is now sampled every 0.3 mm at the isocentre. Note that the overall resolution is slightly inferior to this, because there is some loss of resolution in other parts of the field of view.

The latest scanners have now extended the multi-slice concept to detectors giving 16 cm of coverage in the z direction, achieved through 320 rows of 0.5 mm detectors. These scanners offer the potential for imaging entire organs in a single rotation, and offer particular advantages in the field of CT angiography.

Despite the development in the number of rows of detectors, there are no major differences in the basic principle of operation of a multi-slice array of detectors and the single arc of CT detectors discussed in Section 8.4. Each element consists of a fluorescent solid-state material that converts absorbed X-rays into visible light photons, and a silicon photodiode.

8.6.2 Data Reconstruction and Storage

In third and fourth generation single-slice CT, the radiation forms a fan in the x-y plane, giving what is termed a *fan beam geometry* (Figure 8.6). In early multi-slice scanners, offering up to four images per rotation, image reconstruction is carried out by assuming that the data are collected as four parallel fan beams, each one being perpendicular to the z-axis. However, this is an approximation. Whilst the central images are acquired in a plane which is almost perpendicular to the z-axis, towards the edge of the bank of detectors there is increasing angulation away from this ideal. As a result, data are now being collected using a cone of radiation, in what is termed a *cone-beam geometry*. Ignoring the cone-beam geometry and treating the data as being acquired from parallel fan beams can lead to artefacts which result from inconsistencies in the reconstruction process. These arise because the apparent longitudinal position of an off-axis object varies as the tube and detectors rotate (Figure 8.18). However, in up to four-slice multi-slice CT, these cone-beam artefacts can normally be tolerated, and the reconstruction is still essentially 2-D in nature.

For spiral multi-slice CT, data are being acquired in the form of multiple interleaved helices. If the cone-beam geometry is ignored, data reconstruction can still be based on linear interpolation between the two nearest available datasets (see Section 8.5). These data may have come from any of the multiple detector rows, or from the same detector but on a previous or subsequent rotation. Alternatively, the concept of a filter width in the z direction has been introduced. Using this technique, for any particular projection, reconstruction is based on all relevant datasets within a fixed 'filter width' in the z direction. In multi-slice spiral CT, the effect of changes in pitch on image noise, patient dose and z-axis resolution depends critically on whether image reconstruction is based on a two-point linear interpolation or a filter width technique.

For multi-slice CT with more than eight slices, the cone-beam nature of the acquisition process must be properly accounted for (*cone-beam CT*). A number of techniques have been developed. One is to extend the algorithms for filtered back projection (see Section 8.4.2), so that, rather than back projecting into a 2-D plane, data are back projected into a 3-D volume. However, this process requires significant computing power. Alternatively, 'Advanced Single-Slice Rebinning' (ASSR) techniques may be used. These techniques are based around the reconstruction of large numbers of non-parallel overlapping tilted images in planes which correspond to the plane of irradiation of a row of detectors. Subsequently,

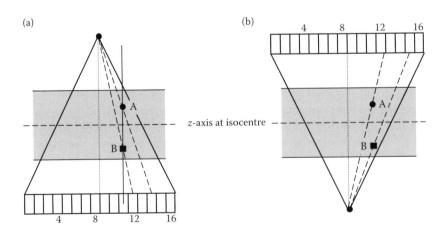

FIGURE 8.18
The effect of cone-beam geometry in multi-slice CT; (a) In single-slice CT, the two objects A and B would be consistently imaged in the same transverse plane, corresponding to the location of detector 11. In multi-slice CT, when the tube is above the patient, objects A and B are imaged by detectors 14 and 12, respectively. (b) When the tube has rotated to beneath the patient, objects A and B are now imaged by detectors 12 and 15, respectively.

parallel axial images, or indeed images in any other plane, are obtained by interpolation between the non-parallel tilted images. Further detail is beyond the scope of this book but a good review is given by Flohr et al. (2005).

There is potential for a large amount of data to be created from a long multi-slice acquisition with narrow detector collimation, and due consideration must be given to data storage and retrieval issues. For a single 512 by 512 pixel image (about 260,000 pixels) stored with a 12 bit accuracy, each image requires about 400 kbytes of storage (1 byte = 8 bits). Furthermore, with multiple reconstructions in different planes or with different reconstruction filters, it is common to create in excess of a thousand images from a routine examination. It may also be necessary to store the original projections if several reconstruction algorithms are available and the radiologist is unsure which one will give the best image. Issues relating to the storage and retrieval of images are discussed further in Chapter 17.

8.7 Image Quality

For CT it is difficult to define image quality. However, as with conventional radiology, image quality, noise and patient exposure or radiation dose are all inter-related. Two forms of resolution can be considered (a) the ability to display two objects that are very close as being separate, (b) the ability to display two areas that differ in contrast by a small amount as being distinct.

For a high contrast object the system resolution is determined by the focal spot size, the size of the detector, the separation of measurement points or data sampling frequency, and the displayed pixel size. The focal spot can affect resolution as in any other form of X-ray imaging, although in practice with the sub-millimeter focal spots now used in CT this is not a problem. However, the width of the detector aperture is important and if resolution is considered in terms of the closest line pairs that can be resolved, simple ray optics show that the line pair separation at the detector must be greater than the detector aperture. However, because the patient is at the centre of rotation, not close to the detectors, there is always some magnification in CT, typically a factor of 1.5–2.0. Consequently, the resolution in the patient will be better than the resolution at the detectors.

The pixel size must match the resolving capability of the remainder of the system. For example, a 512 × 512 matrix across a 50 cm field of view corresponds to a pixel size of just less than 1 mm. This would be capable of displaying an image in which the resolution was 5 lp cm^{-1}. However, it could not display adequately an image in which the resolution was 10 lp cm^{-1}. Note that by restricting the field of view to the central 25 cm, this higher resolution image can be adequately displayed on a 512 × 512 matrix.

The effect of sampling frequency on spatial resolution was discussed in Section 8.4.2.

At low contrast, resolution becomes a function of contrast and dose. When a CT scanner is used in the normal 'fast scan' mode, a typical relationship between resolution and contrast will be as shown in Figure 8.19. The minimum detectable contrast decreases as the detail diameter increases. Note however, that for objects as small as 1 mm, the minimum detectable contrast is about 1%. This compares very favourably with a minimum detectable contrast of about 10% for small objects in conventional radiology and illustrates well one of the principles of tomographic imaging discussed in the introduction to this chapter.

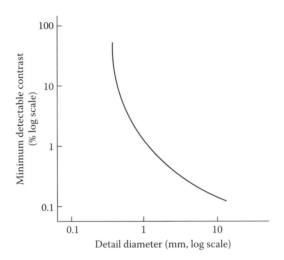

FIGURE 8.19
Relationship between minimum detectable contrast and detail size for a CT scanner.

At low contrast, *quantum mottle* may become a problem. If a uniform water phantom were imaged, then even assuming perfect scanner performance, not all pixels would show the same CT number. This is because of the statistical nature of the X-ray detection process which gives rise to quantum mottle as discussed in Section 6.10. In fact, CT operates close to the limit set by quantum mottle. A 0.5% change in linear attenuation coefficient, corresponding to a change in CT number of about 5, can only be detected if the statistical fluctuation on the number of counts collected ($\sqrt{n}$) is less than 0.5%. Hence,

$$\frac{\sqrt{n}}{n} < \frac{0.5}{100}$$

which gives a value for n of about 4×10^4 photons. This is close to the collection figure for a single detector in one projection.

Any attempt to reduce the dose will increase the standard deviation in the attenuation coefficient due to this statistical noise. When measurements are photon limited, statistical noise increases if the patient attenuation increases. However, it decreases if the slice thickness increases or the pixel width increases because more photons contribute to a given attenuation value.

Quantum mottle is still a considerable problem in the very obese patient especially for a central anatomical site such as the spine.

8.8 Dose Optimisation

Computed tomography is an excellent technique when properly used and, particularly because of the technical advances discussed in this chapter, there has been a dramatic increase in its use—see Table 8.3. Because of this dramatic increase, it is essential to be

TABLE 8.3

Approximate Numbers of CT Scans/
Year (Millions) in the United Kingdom
and United States

Year	UK	US
1985	0.3	8.0
1995	1.0	20
2005	2.7	60

Source: Adapted from Hall E J and Brenner D
J. Cancer risks from diagnostic X-rays.
Br. J. Radiol. 81, 362–378, 2008.

sure that CT is being used optimally in terms of benefit versus risk and use of expensive resources.

Furthermore, as discussed in detail in Section 13.7.1, CT is a relatively high dose technique and in Council Directive 97/43/Euratom of 30 June 1997 on 'Health protection of individuals against the dangers of ionising radiation to medical exposure' (CEC 1997), CT has been given the status of a 'Special Practice' requiring special attention to be given to patient doses. This is especially important in paediatrics where risks are proportionately higher.

There are a number of technical factors which may increase the dose. With the faster scanning capabilities now available there is a tendency to scan more slices or a larger volume than is necessary to answer the diagnostic question. This is thoroughly bad technique and must be avoided at all costs. A more subtle problem is that with narrower slices, higher mAs values are required to maintain acceptable noise levels, but at the cost of increased dose.

Insight

Dose Control

Recall from Section 8.3 that because CT is a digital modality, there is nothing analogous to the saturation effect of film blackening to alert the operator that doses are rising.

The dose is directly proportional to mAs, other factors remaining constant, but variation with kV is more complex. Work with iodine-filled phantoms shows that, at constant dose the signal-to-noise ratio increases with reduced kV and reduced phantom diameter. However, at 80 kVp tube output may not be high enough for, say, CT angiography with larger patients. In paediatrics, where the benefits of low kV (say 80 kVp) should be greatest, beam hardening artefacts may be more of a problem.

For multi-detector arrays, narrow beam widths are less dose efficient. The X-ray intensity (dose) has to be maintained at a fixed level over all the active detectors to maintain a constant noise level. There is a penumbra to the useful part of the beam, which contributes proportionally more to a narrow beam than to a wide beam (Figure 8.20). Consequently, the widest beam collimation consistent with the required slice thickness should be chosen. Note that when a linear array of, say, 64 × 1 mm detectors is covered by a fan beam, the useful beam is so wide that the edge effect is negligible.

Since dose is directly proportional to mAs, a 'one size fits all' approach to mAs is bad technique and protocols which incorporate tube current modulation (see next section) should be used to fit the mAs to the patient. Attenuation through the patient will

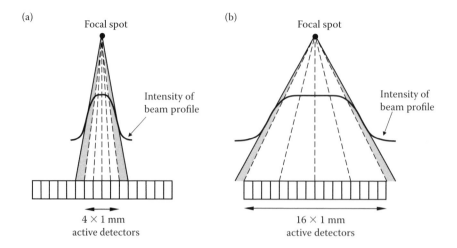

FIGURE 8.20
Diagrams showing that a narrow beam (a) is less dose efficient than a broad beam (b). The edge effects from a narrow beam make a higher proportional contribution to the dose. For further explanation, see text.

determine the final noise level and hence the input mAs for a given image quality, so the cross-sectional diameter of the patient is an important measurement. For a well-filtered 120 kVp narrow beam the intensity should fall by about 50% every 4 cm, but in practice, due to scatter and other factors, attenuation is rather less than this and typically 50% every 10 cm (McCollough et al. 2006).

Note that children show much larger variations in cross-sectional attenuation than adults and are more sensitive to X-rays. For small children, protocols optimised for paediatric work must be used, *not* the corresponding adult protocols (Paterson et al. 2001).

8.8.1 Tube Current Modulation

Traditionally, CT was carried out with a fixed tube current. However, in the late 1990s and early 2000s, manufacturers put a significant effort into developing automatic exposure controls (AEC) for CT, allowing the tube current to be modulated to match attenuation characteristics.

Attenuation in the patient clearly varies with patient thickness and composition, both for different projections in the x-y plane and different positions along the z-axis. To maintain a constant level of noise or image quality, lower tube currents can be used for small patients or z locations in which the patient is less attenuating. Furthermore, the quantum noise in any single image is dominated by the most heavily attenuated projection, where there are least photons reaching the detector. Consequently, tube currents can also be modulated during a rotation to match the attenuation profile, without necessarily degrading the overall image quality achieved. One approach to achieving this is for a CT localisation radiograph (a low-dose procedure—see Section 13.7.1) to be used by the system, together with assumptions about typical patient shapes, to estimate attenuation in the AP and lateral directions. The tube current can also be adapted along the longitudinal z-axis with the aim of keeping a constant noise level for the whole trunk. An alternative approach is to modulate the tube current in real time by constant feed back of the attenuation profile measured over the previous rotation.

It should be noted that for a narrow beam, μ will have a fairly precise value, but wider beams cannot be modulated to the same degree because there will be a spread of attenuation coefficients in the z direction through which the wider beam is passing and an average value of μ must be used. This is of particular significance for the latest generation of scanners offering up to 16 cm of coverage in a single rotation.

These AEC systems for CT have a great potential but it is important to recognise that there is a complex relationship between contrast to noise ratio, a specified level of spatial resolution and minimum dose. In isotropic CT, noise, and therefore dose for a comparable image quality, vary inversely with the fourth power of the voxel dimension. To make full and proper use of AEC, the image quality, or noise levels, acceptable for a given diagnostic task must be understood and defined.

8.9 Artefacts

Computed tomography systems are inherently more prone to artefacts than conventional radiography and these will be summarised briefly.

8.9.1 Mechanical Misalignment and Patient Movement

These will generate streak artefacts, intense straight lines, not always parallel and bright or dark in the direction of motion. Examples of mechanical imperfections include a non-rigid gantry, misalignment and X-ray tube wobble. In each case the X-ray beam position will deviate from that assumed by the reconstruction algorithm. Patient movement is more serious than in conventional imaging because the superimposed profiles have not passed through the same part of the body. Movement may be either voluntary (respiration) or involuntary (peristalsis, cardiac).

8.9.2 X-ray Output Variation and Detector Non-Uniformities

These cause ring artefacts which result from errors in projection over an extended range of views. Variations in X-ray output or detector sensitivity of less than 1% can cause problems. Since rings in image space are equivalent to vertical lines in projection space, the latter can be searched for such lines which may then be removed. Unfortunately these lines have much lower contrast than the rings they reconstruct.

For third generation scanners, previous problems with variations in detector sensitivity have largely been overcome with modern solid-state detectors. In fourth generation scanners there is continuous real-time calibration of the detectors using the leading edge of the rotating fan beam.

8.9.3 Partial Volume Effects

These effects occur when, for example, a long thin voxel has one end in bone, the other in soft tissue. An average value of μ will then be computed, resulting in a loss in contrast. Furthermore, with a diverging beam, an object may be in the scanning plane for one projection but not for another. This leads to inconsistencies in the reconstruction process, resulting in streaking artefacts. Partial volume effects are best overcome by using thinner slices, in which there is less averaging between different tissues.

8.9.4 Beam Hardening

Even if the X-ray beam is heavily filtered, it is still not truly monochromatic. A certain amount of beam hardening will still occur on passing through the patient, owing to selective attenuation of the lower energy X-rays. This will affect the measured linear attenuation coefficients. For example, there may be a false reduction in density at the centre of a uniform object, it appears to have a lower value of μ, and false detail near bone-soft tissue interfaces. Although beam hardening can cause streaks, it is more likely to cause shading artefacts, bright or dark unpredictable CT number shifts, often near objects of high density. For example, apparent areas of low attenuation can develop within hepatic parenchyma adjacent to ribs and these 'rib artefacts' can simulate lesions.

8.9.5 Aliasing

This phenomenon, whereby high frequency noise generated at sharp, high contrast boundaries appears as low frequency detail in the image, was discussed in Section 8.4.2 in connection with the shape of the filter function. It is caused by under-sampling the highest spatial frequencies. The effect is less marked when iterative techniques are used but cannot be eliminated entirely.

8.9.6 Noise

If the photon flux reaching the detectors is severely reduced, for example by heavy attenuation in the patient, there will be statistical fluctuations (quantum noise) resulting in severe streaks.

8.9.7 Scatter

The logarithmic step in the reconstruction process makes the effect of scatter significantly non-linear and the distribution becomes very object dependent. Shading artefacts result. Both detector collimation and software algorithms can be used to reduce the effect of scatter.

8.9.8 Cone-Beam Artefacts

These were discussed in Section 8.6.2. They occur when the cone-beam geometry of a multi-slice acquisition is not properly accounted for. In early multi-slice scanners, treating the data as if it were generated by multiple parallel beams is an example.

8.10 Quality Assurance

As with other forms of X-ray equipment, specification of performance characteristics and their measurement for CT scanners can be considered under three headings (a) type testing, which is a once-only measurement of all relevant parameters associated with the function and use of a particular model of equipment, (b) commissioning tests when a CT scanner is first installed, and (c) routine quality assurance measurements which check that performance is maintained at an acceptable level throughout the scanner's lifetime.

A wide range of in-depth tests are carried out as part of type testing or commissioning. However, these will normally be made by specialists and are outside the scope of this book. For further information see IPEM Report 32 (2003).

There are, however, a number of tests of performance that should be done on a regular basis. These tests should be rapid, so that they take minimal scanner time; uncomplicated, so that staff are capable of performing them; and quantitative, so that as many tests as possible produce objective, numerical answers. A number of basic measurements give a great deal of information about the performance of the scanner:

(i) *Image noise*—an image is taken of a water-filled phantom and the standard deviation in CT number measured in a large region of interest. As explained in Section 8.7, the CT number will not be constant throughout the image. The actual variation in CT numbers will depend on the radiation dose and hence the mAs, but should be constant for fixed exposure parameters. For multi-slice scanners this test can be repeated for multiple slices obtained in a single axial acquisition, to assess inter-slice variation.

(ii) *CT number constancy*—this may be checked by imaging a test object containing inserts to simulate various tissues. Mean CT numbers in a region of interest in each material should be compared with values obtained at commissioning.

If the exact composition and density of the materials is known, these measurements may also be used to check that the CT number is varying linearly with the linear attenuation coefficient. Note, however, that for this measurement to be successful it is necessary to know the effective energy of the X-ray beam at all points in the phantom because of beam hardening.

(iii) *CT number uniformity*—this may be assessed using the same image as that used for the image noise test. Smaller regions of interest are positioned at the centre and around the edge of the image of the phantom, to assess the uniformity of the mean CT number across the phantom.

(iv) *High contrast resolution*—high contrast bar patterns, with decreasing spacing can be used to estimate the limiting high contrast spatial resolution (see Section 6.9.1). Alternatively, a more complex mathematical analysis of the image of a fine bead or wire can be used to establish the modulation transfer function of the system (see Section 7.7.1).

(v) *Axial dose*—assessment of the variation of dose, normally expressed as the Computed Tomographic Dose Index (see Section 13.7.1), with a range of exposure settings should be carried out.

For all these features, careful measurement at commissioning will establish a baseline and subsequent deterioration during routine testing should be investigated.

8.11 Special Applications

8.11.1 Four-Dimensional CT

One of the advantages of faster CT scanning is that it is now possible to introduce time as a fourth dimension in the imaging process. This has many advantages, especially the ability to remove artefacts caused by, for example, respiratory or cardiac motion.

For an example of the use of 4-D imaging to remove respiratory artefacts, see Vedam et al. (2003). The paper explains how an ordinary spiral CT scanner can be used to acquire artefact-free images of the lungs in the presence of respiratory motion. The method involves taking an over-sampled CT scan and then assigning each slice to one of 8 bins, corresponding to different phases of the breathing cycle, as determined by an external respiratory signal. This technique not only gives artefact-free images but also contains respiratory motion information not available in 3-D CT images. Recent work has used this information to investigate breathing irregularities.

All major vendors now offer 4-D CT, which is also becoming increasingly widely used in radiotherapy, for example to make appropriate corrections for respiration in treatment planning.

The latest generation of scanners, with up to 16 cm of longitudinal coverage in a single rotation, introduce even greater possibilities for time-based studies. Such systems can provide complete organ coverage in a single sub-second rotation, opening up new possibilities for functional imaging and CT perfusion studies.

8.11.2 Cardiac CT

The challenge of cardiac CT is to obtain very high resolution images in a very short time, so that cardiac motion does not significantly degrade the quality of the images obtained. The development of multi-slice CT, with increasing numbers of slices and decreasing rotation speeds, has resulted in rapid developments in cardiac imaging within the last few years.

In conventional CT, reconstruction is often based on data collected over 360°. However, as has already been noted, data collected over 180° is sufficient for the reconstruction process, and for a third generation scanner, this can be collected in a little over half the tube rotation time. (Note that to collect 180° worth of data, the tube must rotate through 180° plus the angle subtended by the detector bank in the axial plane.) Consequently, for a tube rotation time of 0.4 s, a temporal resolution of around 200 ms can be achieved in cardiac CT. However, except at very slow heart rates, this is still insufficient for imaging in the 50–200 ms period in mid-late diastole when there is least change in the left ventricular volume. Improvements in temporal resolution can be made using cardiac-gating and a multi-sector reconstruction technique. Rather than acquiring data from a 200 ms portion of a single cardiac cycle, data from two heart beats can be used, each contributing information from the same 100 ms portion of the cardiac cycle. This process can be extended to three or four cardiac cycles. However, if the heart rate isn't perfectly stable there is an increasing chance of mismatch between successive phases in the cardiac cycle. To minimise the number of sectors required, it is common practice to use β-blockers to slow the heart rate, and hence reduce the requirement for such a high temporal resolution.

As an alternative to the multi-sector approach, one vendor has introduced a scanner with two tubes and two detector banks, mounted at 90° to each other. This system allows data from 180° to be collected in only one-quarter of a rotation. Together with a decrease in tube rotation time to 0.33 s, this allows a single-sector cardiac image to be obtained in only 83 ms.

Radiation dose is a significant issue in cardiac CT. High resolution reconstructions inherently result in higher doses. Furthermore, the acquisition process can be very dose-inefficient if large portions of the data, acquired during periods of significant cardiac motion, are never used for image reconstruction. One solution is to use sequential axial scanning, triggered at an appropriate point in the cardiac cycle. This is very dose efficient, but in most scanners is too slow for contrast-based studies. Scanners offering complete cardiac coverage in a single rotation can, however, operate in this mode. An alternative solution to the dose issue is to perform a spiral scan at low pitch, but to modulate the tube current, such that the current is reduced to typically 20% of its full value during periods of significant cardiac motion, when the reconstruction data is unlikely to be used.

8.11.3 Dual Energy and Spectral CT

Conventional CT provides limited information about the attenuating properties of a tissue, averaged across the range of energies in the X-ray spectrum. In recent years there has been considerable interest in developing techniques to obtain information on the energy dependence of the measured attenuation. At its simplest, dual energy imaging allows for the subtraction of contrast-filled blood vessels or bone from the resultant image. However, there is further potential to use the nature of the energy dependence of the attenuation to characterise the tissues being studied, for example, in classifying tumour types and in the characterisation of plaques and different body fluids.

Dual energy imaging can be achieved by a number of different methods. At its simplest, successive high and low energy images, for example, at 140 kVp and 80 kVp can be obtained, and subsequently subtracted or otherwise manipulated. A dual source scanner operating at two different kVps offers the advantage of being able to obtain these two datasets concurrently, avoiding problems associated with the mis-registration of successive images. As an alternative, rapid kV-switching can be implemented in a single tube system to obtain similar results. Dual energy effects can, in principle, also be achieved through the use of variable filtration of the X-ray beam.

An alternative approach to spectral CT is to introduce detectors which offer some form of energy discrimination. At its simplest, layered detectors have been introduced, in which the first layer of the detector is optimised for the detection of low energy photons, the second layer for high energy photons. These methods are similar to those discussed in Section 9.5.3. Looking further to the future, photon counting detector systems with the ability to discriminate and analyse a wide range of photon energies are currently undergoing rapid development.

8.12 Conclusions

A radiographic image is a two-dimensional display of a three-dimensional structure and in a conventional image the required detail is always partially obscured by the superimposition of information from underlying and overlying planes. The overall result is a marked loss of contrast.

Tomographic imaging provides a method for eliminating, either partially or totally, contributions from adjacent planes. Longitudinal tomography essentially relies on the blurring of structures in planes above and below the region of interest. It is a well-established technique and the main consideration is choice of thickness of the plane of cut. If the focus-plane of cut distance and focus-image receptor distance are fixed, the thickness is determined by the angle of swing, decreasing with increasing angle. Large angle tomography may be used when there are large differences in atomic number and/or density between the structures of interest, but if differences in attenuation are small, somewhat larger values of slice thickness may be desirable. With the development of digital detector technology, there are new opportunities for simultaneously collecting a set of images by longitudinal tomography that are in focus in different planes (tomosynthesis).

In single-slice CT imaging, a large number of views are taken of a transverse slice through the patient from different angles. Several generations of CT scanner have been introduced, each designed primarily to give a faster scan time, which is now in the region of 0.4 s per slice. Even faster scanning probably requires 'scanners' with no moving parts and electron beam CT is a potential area for development.

X-ray output is required for several seconds and this has placed new demands on the design and construction of X-ray tubes and generators, especially to absorb and dissipate the large amount of heat released in the anode.

Radiation detectors for CT have also been an area of rapid development to meet the stringent requirements, especially for high detection efficiency, good linearity, wide dynamic range and fast response. Thallium-doped sodium iodide crystals with PMTs and ionisation chambers filled with high atomic number inert gases at high pressure have both been used but the industry standard is now a radiation-sensitive solid-state detector, for example gadolinium oxysulphide, coupled to a high purity, temperature stabilised silicon photodiode.

Two major developments that have greatly expanded the applications of CT are spiral CT and multi-slice CT. In spiral CT, slip-ring technology permits continuous rotation of the X-ray tube with faster scanning and coverage of the whole trunk of the patient. There are numerous advantages—(a) slice thickness, slice interval and slice starting point may be chosen retrospectively, (b) reconstruction is not limited to transverse sections, (c) a study can be completed during a single breath hold, (d) high levels of contrast can be maintained in vessels for the study duration.

The introduction of arrays of detectors of different sizes has permitted multi-slice CT in which slice thickness can be varied. This has resulted in even faster scanning, isotropic resolution and, with increasing coverage in the z direction (e.g. 320 rows of 0.5 mm detectors can cover 16 cm) the potential for imaging entire organs in a single rotation.

Data reconstruction has been a challenging problem for many years. The principal methods developed for single-slice CT were filtered back projection (FBP) and iteration. Both have their strengths and weaknesses—for example, FBP is easy to implement but iteration provides better opportunities for modelling the problem. In FBP data processing can start on the projections, but all projections have to be collected before iteration can begin. Additional problems with data construction in multi-slice imaging result from the requirement to combine data from non-parallel beams (the cone-beam effect).

Image quality is difficult to define in CT but factors that need to be considered include radiation dose, quantum mottle, noise and resolution. Also a number of artefacts can appear on the images arising from one or more of the following causes—mechanical misalignment, patient movement, detector non-uniformities, partial volume effects, beam hardening, aliasing.

Computed tomography is generally a high dose technique and the frequency of CT examinations has increased rapidly in recent years. Both simple dose reduction strategies, for example protocol optimisation, taking the minimum number of slices, using the lowest mAs that will give diagnostically useful images, and more sophisticated methods, for example, tube current modulation, should be considered. A careful quality assurance programme will also help to ensure that image quality is optimised.

Applications of CT continue to expand and diversify, including four-dimensional CT, cardiac CT and spectral CT.

References

CEC, European Communities Council Directive 97/43/Euratom of 30 June 1997, Health protection of individuals against the dangers of ionising radiation in relation to medical exposure and repealing Directive 84/466/Euratom, *Off J Eur Commun* L 180, 1997.

Cherry S R, Sorenson J A & Phelps M E, *Physics in nuclear medicine*, 3rd ed. Saunders/Elsevier, 2003, 486–488.

Flohr T, Schaller S, Stierstorfer K et al., Multi-detector row CT systems and image-reconstruction techniques, *Radiology* 235, 756–773, 2005.

Gordon R, Herman G T & Johnson S A, Image reconstruction from projections, *Sci Am* 233, 56–68, 1975.

Hall E J & Brenner D J, Cancer risks from diagnostic X-rays, *Br J Radiol* 81, 362–378, 2008.

IPEM, *Measurement of the performance characteristics of diagnostic X-ray systems used in medicine. Part III: computed tomography X-ray scanners (IPEM Report 32)* 2nd ed. Institute of Physics and Engineering in Medicine, York, 2003.

Miller D, Principles of spiral CT—practical and theoretical considerations, *Rad Mag* Feb 28–29, 1996.

McCollough C H, Bruesewitz M R & Kofler J M Jr, CT dose reduction and dose management tools: overview of available options, *Radiographics*, 26, 503–512, 2006.

Oppenheim A V & Wilsky A S, *Signals and systems*. Prentice Hall Inc. Eaglewood Cliffs N J, ch 8, 1983.

Paterson A, Frush D P & Donnelly L F, Are settings adjusted for paediatric patients? *AJR* 176, 297–301, 2001.

Vedam S S, Keall P J, Kini V R et al., Acquiring a four-dimensional computed radiography data set using an external respiratory signal, *Phys Med Biol* 48, 45–62, 2003.

Further Reading

Cody D D, Stevens D M & Rong J, CT quality control, in *Advances in medical physics 2008*. Eds Wolbarst A B, Zamenhof R G and Hendee W R, Medical Physics Publishing, Madison, 2008, 47–60.

Flohr T G, Cody D D & McCollough C H, Computed tomography, in *Advances in medical physics 2006*. Eds Wolbarst A B, Zamenhof R G & Hendee W R, Medical Physics Publishing, Madison, 2006, ch 3.

Hsieh J, *Computed tomography: principles, design, artefacts and recent advances*. SPIE Press, Washington, 2003.

ICRP, Managing patient dose in multi-detector computed tomography *(ICRP Publication 102)*. *Ann ICRP* 37, 1, 2007.

Kalender W, *Computed tomography: fundamentals, system technology, image quality, applications*, 2nd ed. Wiley-VCH, Weinheim, 2005.

Pullan B R, The scientific basis of computerised tomography, in *Recent advances in radiology and medical imaging, vol 6*, Eds Lodge T & Steiner R E, Churchill Livingstone, Edinburgh, 1979.

Seeram E, *Computed tomography: physical principles, clinical applications, and quality control*, 3rd ed. Saunders, Philadelphia, 2008.

Swindell W and Webb S, X-ray transmission computed tomography, in *The physics of medical imaging*, Ed Webb S, Adam Hilger Bristol & Philadelphia, 1988, 98–127.

Exercises

1. Explain briefly, with the aid of a diagram, why an X-ray tomographic cut is in focus.

2. List the factors that determine the thickness of cut of a longitudinal X-ray tomograph and explain how the thickness will change as each factor is varied.

3. Explain the meaning of the terms pixel and CT number, and discuss the factors that will cause a variation in CT numbers between pixels when a uniform water phantom is imaged.

4. Explain why the use of a fan beam geometry in CT without collimators in front of the detector would produce an underestimate of the μ values for each pixel.

5. Discuss the factors that would make a radiation detector ideal for CT imaging and indicate briefly the extent to which actual detectors match this ideal.

6. Explain why the technique of transverse tomography can eliminate shadows cast by overlying structures. Suggest reasons why the dose to parts of the patient might be appreciably higher than in many other radiographic examinations.

7. Figures 7.2b and 8.19 show the relationship between contrast and resolution for an image intensifier and CT scanner, respectively. Explain the differences.

8. Describe and explain how the CT number for a tissue might be expected to change with kVp.

9. Explain the design principles of detector arrays for multi-slice CT with particular reference to the way in which different slice thicknesses can be achieved.

10. Discuss the causes of different types of artefact in CT.

9

Special Radiographic Techniques

P P Dendy and B Heaton

SUMMARY

- This chapter highlights some physical principles that are specific to, or especially important in particular radiographic techniques.
- Mammography is a low voltage technique developed to enhance the low contrast in breast tissue.
- High voltage radiography may be useful when increased X-ray output or better penetration is required. There will be some loss of contrast.
- There is some magnification in all images and the reasons are explained. Magnification is generally undesirable but is useful in a few situations.
- Subtraction techniques are used to eliminate unwanted information from an image, thereby making diagnostically important information easier to visualise. Digital images greatly facilitate the use of subtraction methods.
- Interventional radiology (IR) is a wide-ranging term for situations in which X-ray imaging is an aide to other clinical procedures. Since it is not a diagnostic procedure, high quality images may be less important than other aspects of the intervention.
- The final two sections highlight some important points when imaging children (paediatric radiology) and in dental radiology.

CONTENTS

9.1 Introduction

In Chapter 2 the basic principles of X-ray production were presented and Chapter 3 dealt with the origin of radiographic images in terms of the fundamental interaction processes between X-rays and the body. Chapters 5 and 6 showed how the radiographic image is converted into a form suitable for visual interpretation.

On the basis of the information already presented, it is possible to understand the physics of most simple radiological procedures. However, a number of more specialised techniques are also used in radiology and these will be drawn together in this chapter. These techniques provide an excellent opportunity to illustrate the application to specific problems of principles already introduced and appropriate references will be made to the relevant sections in earlier chapters.

9.2 Mammography—Low Voltage Radiography

9.2.1 Introduction

X-ray mammography is used almost exclusively for the early detection of female breast cancer. Breast cancer in men is not unknown but rare.

There are several difficulties associated with imaging the breast to determine whether a carcinoma or pre-cancerous condition exists. First, an important pointer is the relative amounts, distributions and variants in fibroglandular and adipose tissues. However, there is very little difference between the two tissues in terms of the properties which create radiological contrast. Their densities are both close to 1.0×10^3 kg m^{-3}, differing by less than 10%. Adipose tissue has a slightly lower mean atomic number (Z about 5.9) than fibroglandular tissue (Z about 7.4) owing to its higher fat content. To exploit this difference the photoelectric effect must be enhanced (recall that it is very Z-dependent) by imaging at low kV.

Second, one of the prime objectives of mammography is to identify areas of microcalcification, even as small as 0.1 mm diameter. These are very important diagnostically. To achieve the necessary geometric resolution a small focal spot size is required and problems of X-ray tube output and rating must be considered. Also the resolution limit of the image receptor should, ideally, be higher than in normal radiography.

Third, there is a wide variation in the X-ray intensity leaving the breast, being high near the skin and much lower in thicker and denser regions. Therefore the ideal receptor will have a wide latitude (film) or dynamic range (digital receptor).

Finally, breast tissue is very sensitive to the induction of breast cancer by ionising radiation (especially for women between the ages of 14 years and menopause). Therefore in symptomatic mammography the normal requirement that clinical benefit must outweigh the risk applies. However, mammography is also used for asymptomatic screening where there is benefit to only a small percentage of exposed women and limitations on dose are even more stringent. High regard must therefore be paid to the radiation dose received during mammographic examinations whilst ensuring that the dose is high enough to avoid intrusive quantum mottle arising from the inherently low contrast.

There have been numerous historical reviews of the progress made over the years in the physics and technological aspects of mammographic imaging—for a brief account see Brateman and Karellas (2006). We shall concentrate on the 'state of the art' in technical developments that have improved image quality and at the same time reduced patient dose, with an explanation of the underlying physics.

9.2.2 Molybdenum Anode Tubes

As stated in the previous section, in order that maximum contrast may be achieved, a low kV must be used since both linear attenuation coefficients themselves and their difference (e.g. $\mu_{\text{soft tissue}} - \mu_{\text{fat}}$) decrease with increasing energy (see Section 6.2). The choice of kV is, however, a compromise. Too low a kV results in insufficient penetration and a high radiation dose to the breast. X-ray photons below 12–15 keV contribute very little to the radiograph and must, if possible, be excluded. Optimum contrast for typical breast thicknesses of 3–5 cm is achieved with X-rays in a relatively narrow energy range, 17–25 keV.

A typical spectrum for a molybdenum (Mo) (Z = 42) anode tube operating at 35 kV constant potential and using a 50 μm Mo filter was shown in Figure 3.16. A significant proportion of the X-ray photons are in fact K_α (17.4 keV) and K_β (19.6 keV) characteristic X-rays from Mo. These lines lie just below the K shell energy (absorption edge) of Mo at 20.0 keV. Figure 9.1 shows a similar spectrum at 28 kVp for ease of reference. An X-ray tube window of low atomic number (1 mm beryllium with Z = 4) is used so that wanted X-rays are not attenuated. The total permanent filtration on such a tube should be equivalent to at least 0.3 mm of aluminium to remove low energy radiation.

For reasons of geometric resolution discussed in the general introduction to mammography, a small focal spot size must be used. For contact work the source to image receptor

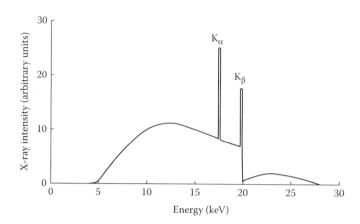

FIGURE 9.1
Spectral output of a molybdenum anode X-ray tube operating at 28 kVp with a 0.05 mm molybdenum filter.

distance is typically 60–65 cm and the object-receptor distance is about 6 cm giving a magnification of about 1.1 (see Section 9.4). This relatively small magnification allows the necessary resolution of at least 12 lp mm^{-1} to be achieved with a focal spot size of about 0.3 mm. For magnification studies, for example to examine small regions of concern in diagnostic work, an even smaller spot (0.1–0.15 mm) is required. This choice of focal spot is achieved either by shaping the anode to provide two target angles or by dual cathode filaments in a focussing cup. Note that the low kV causes a pronounced space charge effect so the relationship between filament and tube currents is highly non-linear.

For a source-receptor distance of 65 cm the effective anode angle must be at least 20° to avoid field cut-off for typical field sizes (18 cm × 24 cm and 24 cm × 30 cm). To achieve the best compromise of field coverage, heel effect and effective focal spot size the entire tube is tilted relative to the image receptor plane. In Section 2.4 the anode angle was defined as the angle between the anode surface and the X-ray beam axis. Thus if the anode angle is 16° and the cathode-anode axis is tilted at 6° to the horizontal, an effective anode angle of 22° can be achieved. Note that the heel effect can now be used, with correct orientation of the anode and cathode, to offset the effect of variation in breast thickness and the central axis of the beam is close to the centre of the breast (see Figure 9.2).

Another factor that can degrade resolution is extra-focal radiation (see Section 2.4) since this effectively enlarges the focal spot. To minimise this effect the anode is grounded so that many of the rebounding electrons are attracted to the housing, not accelerated back to the anode.

The quality of the X-ray beam depends to some extent on the ripple on the kV wave form. The evolution from single phase to three phase and recently high frequency generators reduced the ripple from 100% (one phase) to 6% (three phase) and typically 1%–4% (high frequency). Most commercially available mammography units use rotating anodes and high frequency generators, which have the same advantages as in general radiography (see Section 2.3.4). Tube currents are typically 100 mA on broad focus and, because of rating considerations, 30 mA on fine focus. Exposure times can be long (1–4 s) because the efficiency of output of the low atomic number molybdenum anode is poor and this is long enough for reciprocity failure (see Section 5.7) still to be of concern in film-screen mammography.

Although the characteristic lines in the molybdenum spectrum are ideal for imaging the normal breast, it has been shown that the higher energy photons in the continuous spectrum still make a significant contribution to the image. Thus the kVp has to be controlled

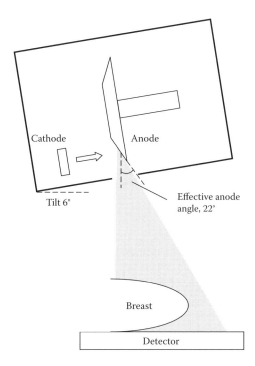

FIGURE 9.2
Effect on the X-ray beam of angling the anode-cathode axis.

rather more precisely than would be necessary if the characteristic lines (which are independent of kVp) totally dominated the spectrum. Also there is some merit in increasing the kVp for thicker breasts (perhaps from 28 to 30 kVp for breasts greater than 7 cm) to reduce dose although this results in some loss of radiographic contrast.

Other features which ensure that patient doses are kept to a minimum include a carbon fibre table and automatic exposure control (AEC). Note that at 30 kVp the attenuation of the table will be much greater than at, say 100 kVp. Furthermore, the table must have very uniform attenuation properties. Any irregularity in table attenuation can appear as an artefact on the image. AEC is discussed in Section 9.2.5.

9.2.3 Rhodium and Tungsten Anode Tubes

Theoretical work several years ago (Jennings et al. 1981) based on signal to noise ratios (S/N) (see Section 7.4.2) indicated that for thicker breasts the optimum energy is somewhat higher than 17–22 keV, perhaps 22–25 keV and subsequent experimental work has confirmed this.

Since the molybdenum spectrum is 'locked in' to the K shell energies of molybdenum, a higher energy spectrum requires the anode material to be changed. To retain the benefits of a characteristic line spectrum, one way to do this is to use an anode material with a slightly higher atomic number, for example, rhodium (Rh) (Z = 45). The characteristic K_α and K_β lines now shift to energies about 3 keV higher than those for molybdenum (20.2 and 22.7 keV). To retain the selective absorption of the high energy X-rays produced by the K-edge, the Mo filter (30–50 μm) must now be replaced by a Rh filter (25 μm). Note that a Rh filter can be used with a Mo target but a Mo filter must *not* be used with a Rh target since the Mo attenuates heavily at the energies of the Rh characteristic lines (see Figure 9.3).

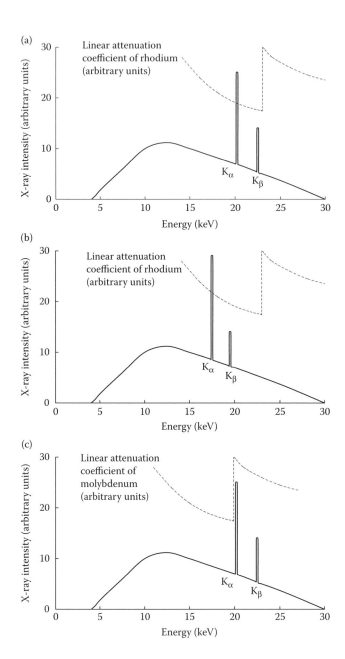

FIGURE 9.3
Attenuation curves for filter materials superimposed on different spectra—(a) rhodium spectrum + rhodium filter; (b) molybdenum spectrum + rhodium filter; (c) rhodium spectrum + molybdenum filter.

Alternatively, for very thick breasts a tungsten (W) anode may be used. The effect of a 50 μm rhodium filter on the output spectrum of a tungsten anode is shown in Figure 9.4. Of course with Z = 74 there are no useful characteristic radiations but the greater efficiency of X-ray output compensates for the extra attenuation and exposure times are short. If a Rh filter is used some of the benefit to the spectrum of the Rh absorption edge is retained. If an aluminium filter is used, it behaves as in general radiographic examinations.

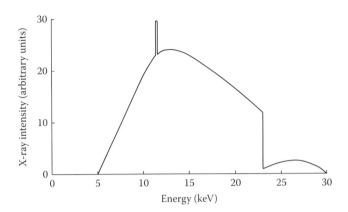

FIGURE 9.4
Spectral output of a tungsten anode X-ray tube operating at 30 kVp with a 50 μm rhodium filter.

Insight

L Shell Characteristic Lines

Note the spike just above 10 keV in the spectrum in Figure 9.4. This is due to the production of L shell characteristic lines by the tungsten anode. This is the only spectrum in the book where evidence of these lines is shown. They will be produced in any tube used for general radiography whenever a tungsten anode is used, but their contribution to the spectrum is normally negligible.

Since a tungsten anode gives a good output, a small focal spot (0.2 mm) can be used and this can be reduced to 0.1 mm by mounting the tube at an angle of 5°. Note that greater care is required in setting and checking the generator kilovoltage since there are no characteristic lines in the spectrum as in the case of molybdenum and rhodium.

Summarising these last two sections, modern mammography tubes may therefore have four selectable focal spots, large and small for both a Mo target and for either a Rh or W target. At higher kV there is slight loss of contrast but better penetration.

9.2.4 Scatter

Scatter is a serious problem in mammography, increasing with both breast thickness and breast area. As shown in Figure 9.5, the scatter to primary ratio increases sharply at high field sizes and can exceed 1.0 for an 8 cm thick breast. Using the analysis in Section 6.5, this corresponds to a 50% reduction in contrast. Over the relatively narrow kVp range used in mammography (25–35 kVp), scatter stays fairly constant. Three methods can be used to reduce the effects of scatter in mammography.

(a) Compression

Compression of the breast is essential if good quality images are to be obtained. As discussed in Section 6.7, compression in the radiological sense actually forces soft tissue out of the direct path of the X-ray beam. In mammography this allows the spread of different anatomical structures with less superposition and is achieved by means of a compression paddle situated between the tube and the breast.

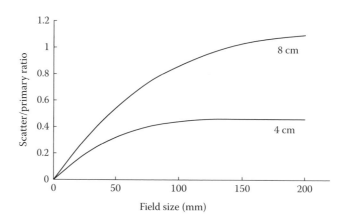

FIGURE 9.5
Typical trends showing the effect of breast thickness and field size on scatter to primary ratio for a molybdenum spectrum with a molybdenum filter; values for both 4 cm and 8 cm thick breasts are shown.

Because there is less breast tissue in the beam, there are three important effects, (i) less scatter is generated, improving primary radiation contrast, (ii) beam attenuation is reduced so exposures are shorter and motion artefacts are reduced, (iii) the integral dose to glandular tissue is reduced and hence there is a lower risk of radiation-induced cancer.

The compression force should be displayed and automatically removed after exposure for patient comfort.

(b) *Grids and Air Gaps*

Grids were discussed in Section 6.8 and the air gap technique is discussed in Section 9.3.3. These general methods of scatter reduction are also used in mammography when appropriate. Inevitably, the use of a grid will increase the patient dose because some of the useful beam that has passed through the patient is stopped, or partially stopped by the strips of lead foil or by the inter-space material. However, by using a low grid ratio (typically 4:1), a low line density (typically 3 lines mm^{-1}) and a carbon fibre inter-space for low attenuation, the dose increase (Bucky factor = exposure with grid/exposure without grid) is limited to about a factor of two. Image contrast can be improved by about 40% but the grid movement must be such that no grid lines are visible on the image.

For small field sizes the air gap technique may be a satisfactory method of scatter reduction. The smallest available focal spot must be used to maintain resolution. Since the focus-image receptor distance is fixed in a mammography unit, the focus-skin distance must be reduced. This will increase the skin dose but the sensitive volume of breast tissue may be reduced.

9.2.5 Image Receptors

(a) *Film-Screen Combinations*

Many different films and intensifying screens are offered by manufacturers exclusively for mammography. The films are single sided, thus eliminating parallax, and are used with a single screen which is often much thinner than standard screens. The screen is positioned behind the film as this causes less loss of resolution and is also more efficient for blackening the film. The explanation for these effects is shown

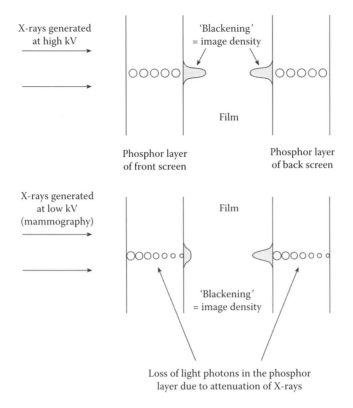

FIGURE 9.6
Schematic diagrams illustrating how high energy X-rays produce light fairly uniformly throughout the intensifying screen (upper diagram) but, because of significant attenuation, low energy X-rays produce most light where they first enter the screen.

in Figure 9.6. In the upper diagram, using high energy X-rays, the loss of X-ray intensity in the screen is small, so the amount of light produced in successive layers of the screen is similar. Both the amount of light reaching the film (which determines blackening) and the mean distance of the light source from the film (which affects resolution) will also be similar. For mammography X-rays, in the lower diagram, the situation is different. Now there is appreciable attenuation of the X-rays so it is important that as much light as possible is produced close to the film. Clearly this occurs when the screen is behind the film not in front of it.

The screens are invariably rare earth, usually gadolinium oxysulphide, and screen and film are pulled into very close contact, for example by using a flexible plastic cassette that can be vacuum evacuated. The screens are only about one-tenth as fast as ordinary screens to ensure that very high resolution can be achieved—typically 20–22 lp mm^{-1}, whilst quantum mottle is also eliminated. Further development of needle-shaped fibre-optic-like crystals (see Section 5.13.1) of small cross-section may improve performance.

To enhance contrast differences, the film has a high gamma and hence a small film latitude—see Section 5.5.4. Therefore AEC is used routinely to provide consistent film density, typically centred on an optical density of about 1.7, over the clinically useful range of breast thicknesses and X-ray tube potentials. The AEC is usually under the cassette to keep magnification as small as possible and consists of an ionisation chamber or an array of semiconductor diodes. Charge released in the sensor is accumulated in a capacitor until a preset voltage is reached and the exposure is terminated. Modern AEC systems incorporate a number of software corrections for variables such as beam quality, screen spectral sensitivity and film reciprocity law failure. Note that if the AEC is not properly set then film density may decrease with increasing breast thickness, beam hardening being the main cause.

Since there is a large variation in attenuation across the breast, the small film latitude is a major limitation. This can be eliminated by using digital receptors—see below.

(b) *Digital Receptors*

With the emergence of full field systems, digital mammography is rapidly replacing film-screen techniques, much as in general radiography.

The three basic systems discussed in Sections 5.10 and 5.11 for collecting digitised images, namely computed radiography (CR) with photostimulable phosphors and digital radiography (DR) with either indirect conversion flat panel imagers or direct flat panel imagers, are all applicable to mammography.

A CR system can be fitted retrospectively to existing screen-film mammography units, DR needs an entirely new unit. Most dedicated digital mammography units currently use direct conversion flat panel imagers. Since high resolution is essential, pixel size is an important consideration. CR may achieve resolution of about 50 μm, equivalent to a spatial frequency of about 10 lp mm^{-1}. For DR pixel size is slightly smaller for direct conversion (75–80 μm) than for indirect conversion (100 μm) and the detector is more efficient.

A slot scanning system using a geometry similar to that in Figure 5.17 can also be used for mammography. At any one time the breast is only exposed to the thin, fan-shaped beam passing through the entrance slit, while the exit slit collimates to the active area of the detector. The two slits scan together across the breast and precise synchronisation of the mechanical scanning system and mechanical read-out electronics is essential to avoid blurring artefacts. Because of the heavy collimation, only a small portion of the X-ray beam is used at any one time so a high output anode must be used to minimise heating problems during the relatively long scan time. Note that the collimation also removes much of the scatter so no grid is necessary.

Insight

Slot Scanning—Practical Detail

A commercial model marketed briefly by Fischer Imaging (Senoscan) illustrates the approach. A tungsten-rhenium anode was used and the total scan time was 5–6 s, although each narrow slice was only irradiated for 200 ms. The digital detector used indirect conversion technology with a 10 mm × 220 mm CsI:Tl scintillation crystal coupled to a linear array of four slit-shaped charge-coupled devices (CCDs). Each CCD was made up of a matrix of 25 μm^2 pixels and each pixel was fed by 25 × 5 μm optical fibres. Spatial resolutions of 54 μm for a large field image (21 cm × 29 cm) and 27 μm for a smaller field of view (11 cm × 15 cm) were attainable.

The technique of tomosynthesis, discussed in Section 8.2, is finding application in mammography because of the inherently low contrast and further contrast reduction caused by overlapping planes. A moving X-ray source may be used with a fixed flat panel digital detector. Objects located at different distances above the detector are shifted relative to each other in the resulting images. Individual slices can be reconstructed from the digital data.

Automatic exposure control is different in a digital system since the image receptor itself, or a sub-component of it, can be used to control the exposure. Therefore adequate exposure for all types of breast can be achieved without the need for critical placement of the AEC sensor. With such an AEC system, a brief pre-exposure may also permit automatic selection of the best anode target and filter.

As for general radiology comparison with screen-film mammography explains why digital is proving popular.

(1) *Ease of use*—dedicated DR mammography units can display an image immediately for preliminary assessment. CR units still require processing but the work involved and time delays associated with film loading/unloading are avoided.

(2) *Post-processing*—this is not possible with a film image. A digital image is an interactive display allowing brightness, contrast, window level, window width, noise and sharpness all to be adjusted to optimise the image, or small regions of it (e.g. suspicious of microcalcification) *after* data collection.

(3) *Radiology service management*—image transfer, multiple viewing, archive and retrieval are all much easier with digital systems.

Insight

Resolution in Mammography

Great emphasis has been placed on resolution in mammography and the reader may have noted that the pixel sizes given for digital systems do not permit the resolution obtainable with the best film-screen combinations. Digital systems will be limited to somewhere between 6 and 10 lp mm^{-1}, compared to 20–22 lp mm^{-1} for film screen. Notwithstanding, for a modern digital system at comparable dose, image quality seems to be very similar to that of film-screen images.

The reasons are as yet unclear but the discussion on factors affecting image quality in Chapter 7 is relevant. In respect of detector performance, MTF, noise characteristics, dose and DQE, as well as resolution all affect image quality. Furthermore, the ability to enhance image features, especially contrast, by post-processing digital images may also offset the benefits of inherently better resolution with film screen.

For a good review of the performance of digital detectors in mammography see Noel and Thibault (2004).

Insight

Xeroradiography

Older books on the physics of mammography would have discussed this approach to breast imaging. It is of considerable interest from the theoretical view point because the imaging technique is rather novel.

Briefly, it depends on the way X-rays interact with a uniformly charged photoconductor (a special form of semiconductor). Where many interactions occur, corresponding to low attenuation in the breast, most of the charge leaks away. Pockets of charge remain where attenuation has been high. When the imaging plate is covered in powder and exposed to a strong electric field, the field lines are distorted to charge edges and highlighted by the powder. Thus xeroradiography produces good spatial resolution and edge-enhanced images—a feature which causes its MTF to be better at high spatial frequencies than at low spatial frequencies. However, the relatively poor contrast sensitivity and relatively high radiation dose, combined with progressive improvement in film-screens and, more recently, digital detectors, have rendered the technique non-competitive.

For more detail see for example, Dendy and Heaton, 2nd edition (1999).

TABLE 9.1

Breast Cancers Induced per Million Women per mGy.

Age	Cancers per 10^6 Women per mGy
25	18–30
35	15–25
45	7–17
55	3–12
65	1–8

Note: The ranges of values reflect the difficulty of making accurate estimates.

Source: Data are from several studies tabulated by Law J, Faulkner K and Young KC, Risk factors for induction of breast cancer by X-rays and their implications for breast screening, *Br. J. Radiol.* 80, 261–266, 2007.

TABLE 9.2

Results of 1994 and 2008 Surveys of Mammography Screening Units in East Anglia UK Using a Standard Breast Phantom (4 cm)

Year of Survey	Number of Units	Average MGD mGy	Range of MGD mGy	Range of Optical Density
1994	23	1.17	0.72–1.68	1.30–1.69
2008	30	1.42	1.20–1.88	1.56–1.90

Source: Courtesy Mr D Goodman and Mr O Morrish.

9.2.6 Quality Control and Patient Doses

Asymptomatic mammography screening is an excellent example of the principle that the medical benefit, that is the number of unsuspected breast lesions detected, must clearly outweigh the number of radiation-induced cancers caused. This criterion must be applied both to the whole population screened and to individual women within the population. Such an analysis requires a knowledge of the excess lifetime risk of mortality in females from breast cancer, a precise knowledge of the mean glandular dose (MGD), accurate information on the extra cancers detected and increasingly a knowledge of any familial disposition to breast cancer. Both the risk of radiation-induced breast cancer (Table 9.1) and the pick up rate are very age dependent.

Skin doses must be kept very low because the breast is so radiosensitive. 30–40 years ago mammograms required a MGD of tens of mGy per exposure, with best current techniques and good quality control (see below) the MGD has fallen to about 1 mGy. Using film screen as the receptor, it has been fairly stable for the past 10 years (see Table 9.2). A contribution to the small increase in MGD comes from the fact that films are now generally darker.

As shown in Figure 9.7, in clinical practice MGD is very dependent on breast thickness.

The patient dose using a tungsten anode tube and a K-edge filter is less than that using a molybdenum anode. At operating voltages of about 30 kVp the dose reduction factor for a very thin breast is approximately two. This dose advantage is partially offset by a slight reduction in image quality since the molybdenum spectrum is better than the tungsten

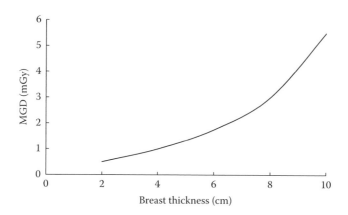

FIGURE 9.7
Typical increase in MGD with breast thickness using a molybdenum anode at 30 kVp and molybdenum filter for a mean optical density of 1.5.

spectrum for radiographing thin breasts. For thicker breasts the dose reduction factor can rise to as much as 5 and contrast enhancement of digital mammograms gives better quality images with higher energy beams.

Factors that can affect MGD are kept under critical review. These include the following:

a. X-ray beam spectrum (kVp, anode, filter)

b. Compression force and compressed breast thickness—the MGD increases rapidly with breast thickness

c. The effect of the age of women at exposure on tissue density in the compressed breast

d. Use of grids

e. Magnification techniques

f. Film-screen combinations/processing/optical density

g. Number of exposures and total mAs

For further comment on the impact of digital receptors on doses in mammography see Section 13.7.2.

Strict attention to image quality is also required to ensure a high pick up rate. Optical density must be checked—a low dose mammogram that produces too light a film is diagnostically useless. Test objects are available to check spatial resolution, threshold resolution and granularity and focal spot size must be checked carefully. Note that because of the line focus principle this may be different in different directions. Typical tolerance limits on some of the more important variables are shown in Table 9.3.

Many of the QC checks for digital mammography are similar to those for screen-film systems; for example, generator performance (accuracy of kV, constancy of output); X-ray collimation and alignment; beam quality, focal spot resolution, the compression device, because they are independent of the receptor.

Clearly established checking procedures for digital receptors, analogous to those for screen film, are still evolving. However, some aspects where careful checks will be necessary are detector uniformity, linearity of signal with dose, noise versus dose, resolution, artefacts and image display—see Section 5.14.2.

TABLE 9.3

Typical Tolerance Limits on Some Important Variables in Mammography

Measure	Tolerance Limit
Resolution	12 lp mm^{-1}
Threshold contrast (large detail)	1.2%
MGD	2.5 mGy with grid
kV	± 1 kV of intended value
Filtration	0.3 mm A1
Output consistency/kV dependence/ variation with the tube current and focus	5%
Film density (measured with 4 cm perspex block under AEC)	± 0.2 of target value and in OD range 1.5–1.9
AEC breast thickness compensation	± 0.2 in OD
AEC variation with the tube kV, tube current	10% of mean

Source: IPEM Report 89, The commissioning and routine testing of mammographic X-ray systems, Institute of Physics and Engineering in Medicine, York, U.K., 2005.

9.3 High Voltage Radiography

9.3.1 Principles

Increasing the generator voltage to an X-ray tube has as number of effects, some of which are desirable with respect to the resulting radiograph, some are undesirable. Among the desirable features are increased X-ray output per mAs, more efficient patient penetration, reduced dose to the patient and, if film is used as the receptor, more efficient blackening. The last of these effects arises because a higher proportion of the X-ray photons have energies above the absorption edges of silver (atomic number 47, K shell energy 26 keV) and bromine (atomic number 35, K shell energy 13 keV). More scattered radiation reaches the film and this is clearly a disadvantage. Finally, the fact that radiographic contrast falls with increasing tube kilovoltage is generally a disadvantage, except when it is necessary to accommodate a wide patient contrast range. These effects may all be illustrated by considering the technique used for chest radiography.

An operating kilovoltage of 60–70 kVp is a sound choice for small or medium-sized patients. It gives a good balance of subject contrast, good bone definition and a sharp, clean appearance to the pulmonary vascularity. However, for larger patients, say in excess of 25 cm anterior-posterior diameter, both attenuation and scatter become appreciable. Tube current can only be increased up to the rating limit, then longer exposure times are required. Also if a scatter-reducing grid is used, the dose to the patient is increased appreciably. Finally, use of a grid in this low kV region may enhance contrast excessively, resulting in areas near the chest wall and in the mediastinum being very light and central regions being too dark.

The problems encountered when a high voltage technique (say 125–150 kVp) is adopted are different. Tube output and patient penetration are good—recall that the heat released in the anode = mAs × kV but as the kV is increased output increases approximately as kV^2 and when this is combined with better penetration of the patient the mAs can be reduced proportionately more than the kV is increased. A change from 75 kVp to 90 kVp (20% increase) can halve the mAs. Thus operating conditions of, for example, 140 kVp and 200–500 mA allow short exposures to be used, typically 1–1.5 ms. Short exposures are an advantage whenever movement may be a problem, for example movement artefact due to the heart and great vessels in chest radiography, radiography of the gastro-intestinal tract, digital subtraction angiography and some computed tomography (CT). Because of the short exposures, the X-ray tube must be designed for high heat input to the anode in the short term. The anode may complete less than one full rotation during the exposure.

Since less heat is produced overall, the tube has a longer lifetime. Note, however, that the X-ray tube has to tolerate consistent use at high voltage so the manufacturer should be informed at the time of installation if the tube is to be used in this way. The kV of the generator output should be checked regularly since the tube will be operating near to its electrical rating limit. High tension cables may develop problems more frequently than at low kV.

9.3.2 The Image Receptor

If film is the receptor, the image will be of markedly lower contrast and may have an over-all grey appearance. If a patient's anterior-posterior diameter is markedly different over the upper and lower chest, an aluminium wedge filter may be used to compensate for differences in tissue absorption.

One consequence of using a higher kV is that, for any object, the primary radiation contrast, or subject contrast, will be less. This is because there are less photoelectric interactions and differences between linear attenuation coefficients, for example, $\mu_{bone} - \mu_{soft\ tissue}$ are less. One advantage is that the ribs interfere less with the lung field. As shown in Figure 9.8, the effect is to compress the spread in transmitted intensity through body parts differing in attenuation. This has important consequences when film screen is the receptor. On the left (Figure 9.8a), there is no exposure latitude (see Section 5.5.4) at low kV since a shift either way along the characteristic curve will result in loss of contrast at either toe or shoulder. At higher kV (Figure 9.8b) the narrow range of exposures AB may be shifted to A_1B_1 or to A_2B_2 without loss of contrast. Furthermore, body parts which cause more

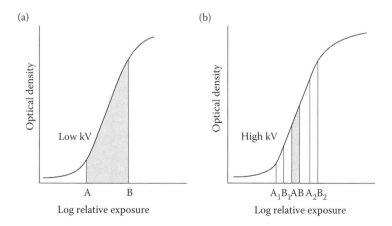

FIGURE 9.8
Graphs showing how, for a given object and a given film-screen combination, there may be no exposure latitude at low kV (a), but there is at high kV (b).

attenuation than A (e.g. the diaphragm) or less attenuation than B would show little or no contrast at low kV. At high kV such tissues are more likely to fall on the linear part of the characteristic curve.

With digital receptors detector latitude is not really a problem because the dynamic range is large so by suitable selection of window level and window width, virtually all signals can be displayed, irrespective of body attenuation. Note, however, that although the algorithms of a computer-based system can easily adjust an image that has received too much radiation at high kV, if there is too little signal because of high attenuation at low kV or because the dose has been reduced too much at high kV, increased noise (quantum mottle) may degrade the image and computer software cannot compensate for this effect.

Other consequences of moving to higher voltage, namely increased scatter, reduced dose, lower heat rating, higher electrical rating, will apply irrespective of the detector. Furthermore, the physical principles on which kVp values and concomitant mAs values are based are largely independent of the receptor and should only be altered when moving to a digital system if careful observation and measurement show that imaging will be optimised as a result.

9.3.3 Scattered Radiation

When using a high kV technique it is not uncommon for five times as many scattered photons as primary photons to reach the film so some form of scatter-reducing technique must be used. If this is by means of a grid a 10:1 grid ratio is probably a good compromise. A higher grid ratio may improve image quality even more for a few large patients but unless the grid is changed between patients, all patients will receive a lot more radiation. Using a grid will increase the entrance skin dose (ESD).

Insight

Does High kV Technique Reduce ESD?

In Section 9.3.1 we stated that an advantage of increasing kV was reduced patient dose. The standard explanation is quite straightforward. As the kV increases, body attenuation is less so to achieve the same dose *at the receptor* a lower entrance dose is required.

However, a practical study of ESD values for patients undergoing chest examination in a number of radiology departments in East Anglia UK some years ago showed that average ESDs were actually higher at high kV (Wade et al. 1995). Further follow-up showed that several of the departments were using grids for the high kV work and this negated the benefits of better penetration. The take-home message is that for patient dose studies it is important to compare practical situations as well as investigating the impact of a single variable.

An alternative method to reduce scatter reaching the film is the *air gap technique*, the principle of which is illustrated in Figure 9.9. Imagine there is a small scattering centre near the point where the X-rays leave the patient. At diagnostic energies, Compton scattered X-rays will travel almost equally in all directions (see Section 3.4.3). Referring to the diagram, as the cassette is progressively moved away from the patient, the number of scattered X-ray photons intercepting it decreases.

It is also clear from the diagram that the first 20–30 mm of gap will be the most important. Since there will be scattering centres throughout the body one might think that the technique would not be very effective because there is a large 'gap' between most of the

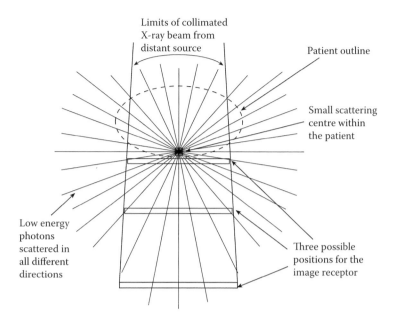

Limits of collimated
X-ray beam from
distant source

Patient outline

Small scattering
centre within
the patient

Low energy
photons
scattered in
all different
directions

Three possible
positions for the
image receptor

FIGURE 9.9

Diagram showing schematically how the number of scattered photons intercepting the image receptor will decrease as it is moved away from the position of contact with the patient.

scattering points and the image receptor even when the latter is in contact with the patient. However, low energy scattered photons produced near the entrance surface are heavily absorbed within the patient, thus it is the scattered photons that originate close to the exit surface which cause most of the problem.

Note also

1. Whereas attenuation of scattered X-rays within the patient is an important factor, attenuation of scattered X-rays within the air gap itself is negligible.
2. On the scale of Figure 9.9, the X-ray source is distant so the margins of the collimated primary beam will be almost parallel—thus the effect of the inverse square law on the primary beam as a result of introducing the gap will be small. Since the source of scatter is much nearer the receptor, the effect of the inverse square law will be much greater (it is left as an exercise for the reader to confirm this for the figures given below).

Insight

More on Scattered Radiation

Figure 9.9 showing almost uniform distribution of scattered radiation is very much a first approximation. Closer study of the problem reveals minor but important variations in the pattern of scatter. First, reference back to the polar diagrams in Figure 3.6 shows that slightly more radiation is scattered in the forward direction than in the sideways directions. Furthermore, calculations such as those in Section 3.4.3 show that photons scattered through small angles retain most energy. Finally, the most energetic photons are attenuated least by body tissues. The result of combining all these factors is that as kV increases, scatter leaving the patient is markedly more in the forward direction. One consequence is that when grids are used the grid ratio will be much higher for high kV work (typically 20:1) than for low kV work (typically 4 or 5:1 for mammography). However, there is still sufficient side scatter in high kV work for the air gap technique to be useful in some circumstances.

A number of other features of the air gap technique—for example, the effect of the finite focal spot size on image sharpness, the effect of the penumbra and patient movement on sharpness and the effect on patient dose are identical to those encountered in macroradiography and will be considered in the next section.

For a high kV chest technique, in the range say 125–150 kVp, a typical air gap might be 20 cm with a focus-receptor distance of 3 m. This would give comparable contrast to a 10:1 grid ratio. Both techniques result in an increased dose to the patient but the increase due to the inverse square law as a result of the air gap is generally less than that required to compensate for the grid.

Although the air gap technique would appear to have a number of advantages, it is not widely used, perhaps because the position of the image receptor relative to the couch has to be changed.

Two final comments will be made on high voltage radiography. First, kV has no effect on resolution, magnification or distortion because it causes no change in beam projection geometry. Second, this discussion has been presented in terms of low voltage (60–70 kVp) versus high voltage (125–150 kVp) techniques but intermediate voltages can of course be used, with the consequent mix of advantages and disadvantages.

9.4 Magnification Radiography

Magnification can be thought of as a form of distortion and in most radiographic applications it is kept to a minimum. For example, chest radiographs are routinely checked for enlargement of the heart and poor technique resulting in magnification could result in misdiagnosis. However, magnification can sometimes be very useful, for example, when confirming microcalcification in mammograms, looking at small bones in the extremities, or some angiographic procedures.

As discussed in Section 6.12.1, when an object is not in contact with the receptor the image is magnified in the ratio focus-receptor distance/focus object distance. This geometry is reproduced in Figure 9.10a for ease of reference but with the relevant distances relabelled. Thus the focus-object distance is d_1, the object-receptor distance is d_2 and $M = (d_1 + d_2)/d_1$. Note that magnification $M = 1$ only if the object is in contact with the film and very thin, that is, when $d_2 = 0$.

For most routine X-ray examinations, M is kept as small as possible. This is because, as discussed in Section 6.9.1, if the focal spot is of finite size, which is always the case in practice, a penumbra proportional to d_2/d_1 is formed, so the penumbra is also absent only if $d_2 = 0$. Comparison of Figure 9.10b and 9.10c, shows that for fixed values of d_1 and d_2, the size of the penumbra depends on the size of the focal spot. Note that magnification implies an increased size of the umbra. If the penumbra increases but there is no change in the umbra, there is no magnification only a loss of sharpness.

In practice, structures of interest are never actually in contact with the image receptor. Taking a typical value of $d_2 = 10$ cm, then for a focus receptor distance FRD($d_1 + d_2$) of 100 cm,

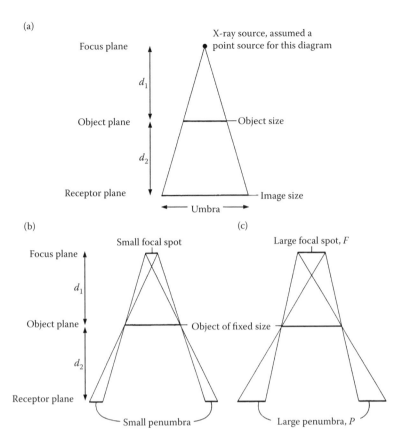

FIGURE 9.10
Geometrical arrangements for magnification radiography. (a) Assuming a point source of X-rays, then by similar triangles the magnification $M = (d_1 + d_2)/d_1$, (b) and (c) demonstrate that for an object of fixed size and fixed magnification, the size of the penumbra increases with the size of the focal spot.

$M \simeq 1.1$. One way to achieve a magnified image is to magnify, using optical methods, a standard radiograph. However, this approach is often not satisfactory as it produces a very grainy image with increased quantum noise (see Sections 6.10 and 7.5). The alternative is to increase the value of M and this can be done in two ways: (a) keep FRD($d_1 + d_2$) constant, but increase d_2, reducing d_1; (b) keep d_1 constant and increase d_2. Unless otherwise stated, it will be assumed that d_2 is increased keeping FRD constant—this is the norm in mammography. Although increasing d_2 achieves the desired result, this has a number of other consequences as far as the radiographic process is concerned and these will now be considered.

(i) *Focal Spot Size*
As shown in Figure 9.10b and c, the size of the penumbra depends on focal spot size. By similar triangles, if the penumbra on one side is P, $P = F(d_2/d_1) = F(M - 1)$, where F is the focal spot size. If it is assumed that the penumbra is part of the magnified image, it is left as an exercise for the reader to show that the true magnification is equal to

$$M + 2(M-1)\frac{F}{xy}$$

where xy is the size of the object. Thus when M is large and F is of the order of xy, the penumbra contributes significantly to the image.

If we think of xy as the smallest resolvable object distance in the image plane, large values of F are clearly a problem. This is illustrated with three worked examples.

(a) *High kV Chest*—note that FRD is increased

$$\text{FRD}(d_1 + d_2) = 2 \text{ m} \text{ or } 2000 \text{ mm}; \quad d_2 = 12 \text{ cm} \text{ or } 120 \text{ mm}$$

$$M = 1 + \frac{d_2}{d_1} = 1.06; \quad P = F\left(\frac{d_2}{d_1}\right) = F\left(\frac{120}{1880}\right)$$

Assuming the finest detail to be resolved is about 200 μm (0.2 mm) and that P should be no greater than half this value gives $F \sim 1.6$ mm.

(b) *Mammography with Minimal Magnification*

$$\text{FRD} = 65 \text{ cm} (650 \text{ mm}); \quad d_2 = 6 \text{ cm} (60 \text{ mm})$$

$M = 650/590 = 1.1$. For a 0.3 mm spot, $P = F(M - 1) = 30$ μm which will permit resolution of 100 μm calcifications.

(c) *Mammography with Magnification*

$$\text{FRD} = 65 \text{ cm} (650 \text{ mm}); \quad d_2 = 22 \text{ cm} (220 \text{ mm})$$

$M = 650/430 = 1.5$. If $F = 0.3$ mm, $P = 0.3 \times 0.5 = 150$ μm which is too big. F must be reduced to 0.1 mm to give an acceptable penumbra of 50 μm.

When the FRD is short, spots larger than 0.3 mm are little use and 0.1 mm is preferable. This may impose rating constraints.

A focal spot of 0.3 mm or less is not easy to measure accurately and its size may vary with the tube current by as much as 50% of the expected value. A pin hole may be used to measure the spot size (Section 2.7) but to estimate the resolution it is best to use a star test pattern (Section 6.9.3). The performance of a tube used for magnification radiography is very dependent on a good focal spot and careful, regular quality control checks must be carried out. It can be difficult to maintain a uniform X-ray intensity across the X-ray field using a very small focal spot. The intensity distribution may be either greater at the edge than in the centre or, conversely, higher in the middle than at the edge. Such irregularities can cause difficulties in obtaining correct exposure factors.

(ii) *Receptor Unsharpness*

Unsharpness caused by the limiting resolution of a film-screen combination is reduced by magnification. To understand the reason for this, consider image formation for a test object that consists of eight line pairs per mm. If the object is in contact with the screen ($M = 1$) the screen must be able to resolve eight line pairs per mm, which is beyond the capability of fast screens. Now suppose the object is moved to a point midway between the focal spot and screen ($d_2 = \text{FRD}/2$ and $M = 2$). The object is now magnified at the screen to four line pairs per mm, thereby making the imaging task easier.

A similar argument applies to digital detectors which have a fixed pixel size—the pixel 'projects back' to smaller dimensions in the object plane. For systems which have a fixed matrix size, the effect would be different if the matrix were expanded to cover a larger field of view, since pixel size would increase. According to the inverse square law, the pixel would project back to the same size as in the unmagnified state so resolution would be unaltered.

Notwithstanding this finding that magnification generally improves receptor resolution, readers may recall that in Figure 7.13 modulation transfer function fell more rapidly as the magnification increased (curves C and D) implying inferior resolution or more unsharpness. This is because focal spot size increases unsharpness and is normally the dominant effect. For more detail see the 'insight'.

Insight

Effect of Magnification on Resolution

From the discussion in the main text, if R is the intrinsic receptor unsharpness for an infinitesimally thin object in contact with the receptor ($M = 1$), for any other value of M, U_R (the receptor unsharpness) $= R/M$.

But U_R is only one contributor to the overall unsharpness. A more general expression for U is $(U_G^2 + U_M^2 + U_R^2)^{1/2}$, see Section 6.9.4.

Referring to Figure 9.10, the penumbra $P = F(d_2/d_1) = F(M - 1)$. If we divide by M, effectively reducing blurring to its value in the object plane, then

U_G (geometric unsharpness) $= F(1 - 1/M)$. Movement unsharpness will be ignored for this argument.

Therefore, expressed in terms of constants other than M, and relative to the object plane, $U_R = R/M$ and $U_G = F(1 - 1/M)$. If $U_R \gg U_G$ then unsharpness is proportional to $1/M$ and *decreases* with increasing M. If $U_G \gg U_R$ then unsharpness is proportional to $(1 - 1/M)$ and *increases* with increasing M.

The critical value is R/F, the ratio of the intrinsic receptor resolution to the focal spot size. If R/F is greater than about 0.5, then magnification radiography will reduce unsharpness, a value less than 0.5 and magnification will increase unsharpness (Dance 1988). In practice for general radiography $R/F \ll 0.5$ (intrinsic receptor resolution ~0.1 mm, focal spot size ~1 mm) so resolution decreases (blurring increases) with increasing magnification.

When U_R and U_G are comparable, unsharpness may decrease for small values of M and subsequently increase for larger values. These conditions can be achieved in magnification mammography using a very small focal spot (~0.1 mm) and a magnification of 1.5. The small focal spot limits tube current to about 25 mA, so if exposures greater than 100 mAs are necessary exposure times will exceed 4 s.

(iii) Movement Unsharpness

One further important source of image degradation in magnification radiography is movement unsharpness (see Section 6.9.2). If an object is moving at 5 mm s^{-1} and the exposure is 0.02 s, then the object moves 0.1 mm during the exposure. If the object is in contact with the film ($M = 1$) and the required resolution is four line pairs per mm, corresponding to a separation of 0.25 mm, then movement of this magnitude will not seriously affect image quality. However, as shown in Figure 6.19, the effect of object movement at the receptor depends on d_2. If $d_2 = $ FRD/2, that is, $M = 2$, the shadow of the object at the receptor will move 0.2 mm and this may cause significant degradation of a system attempting a resolution of 0.25 mm. Note, however, that the size of the penumbra has remained the same *fraction* of the object size.

A further significant contribution to movement unsharpness may come from the increased exposure time resulting from lower tube output with the smaller spot size and increased focus film distance.

(iv) *Quantum Mottle*

This is determined by the number of photons per square mm in the image (see Sections 6.10 and 7.5) which in turn is governed by the required film blackening or digital receptor sensitivity. As the image is being viewed under normal viewing conditions, the number of photons striking the receptor per square mm is exactly the same as on a normal radiograph and the quantum mottle is also the same.

(v) *Patient Dose*

If the FRD is fixed, then exposure factors are unaltered, but if the patient is positioned closer to the focal spot to increase magnification then the entrance dose to the patient is increased. Two factors compensate partially for this increased entrance dose. First, the irradiated area on the patient can, and must be reduced. This will require careful collimation of the X-ray beam and accurate alignment of the part of the patient to be exposed. Second, an 'air gap' has in effect been introduced (see Section 9.3.3) so it may be possible to dispense with the use of a grid.

Some increase in the FRD may also be necessary because if the mid-plane of the patient is positioned 50 cm from the focal spot (to give $M = 2$ for a FRD of 100 cm) the upper skin surface of the patient will be very close to the focal spot and the inverse square law may result in an unacceptably high dose.

Note: (1) That if the FRD is increased, exposure factors will have to be adjusted and a higher kVp may be necessary to satisfy rating requirements.

(2) That the explanations given in both (iv) and (v) assume a geometry in which FRD is kept constant. If magnification is achieved by keeping d_1 fixed and by increasing d_2 the position is a little more complicated. Because of the inverse square law, either the entrance surface dose must be increased or quantum noise in the image will increase.

The way in which the effect on image quality of some of these factors may be analysed more quantitatively was discussed in Section 7.7. Suffice to conclude here that resolution falls off rapidly with magnification and for a 0.3 mm focal spot the maximum usable magnification is approximately 2.0 at the object or about 1.6–1.8 at the skin surface nearest to the tube.

9.5 Subtraction Techniques

9.5.1 Introduction

A radiological image is a map of variations in the X-ray intensity transmitted through different parts of the body. If two images of the same part of the body are taken, and they differ in some way, then, provided body thicknesses remain constant, these differences must be due to changes in linear attenuation coefficient, μ, in parts of the object field. When one image is subtracted from the other, regions in which μ has changed will be much more visible. Therefore subtraction eliminates unwanted information from an image, thereby making the diagnostically important detail easier to visualise. It is particularly useful where sequential images differ in a small amount of detail. A typical example would be during angiography where one wishes to highlight the position and amount of contrast medium in blood vessels between two images separated by a short time interval.

Subtraction methods can be used with analogue images on film. Although no longer the method of choice, a brief description helps to understand the principles involved and some of the problems. The basic principle of the technique is quite simple. A radiograph is a negative of the object data. If a negative of this negative is prepared (a positive of the object data) and positive and negative are then superimposed, the transmitted light will be of uniform intensity. This is because regions that were black on the original negative are white on the positive and vice versa, the two compensating exactly. The positive of the original image is often called the 'mask'. When a second radiograph is taken of the patient, with one or two details slightly different, for example, following the injection of contrast medium, superimposition of the mask and the second radiograph will result in all the unchanged areas transmitting uniformly, but the parts where the first and second radiograph differ will be visualised.

For the technique to be successful, the two images must superimpose exactly and no patient movement must take place between exposures. The exposure factors for the radiograph and the tube output must remain the same to ensure an exact match of optical density which should be in the range 0.3–1.7. The copy film making the mask must have a gamma equal to –1.0. When the films are viewed together the combined optical density is approximately 2.0 and a special viewing box with a high illumination is required.

In practice, digitised images are now used for all subtraction applications. This is because it is relatively straightforward to subtract the counts/pixel in the mask image from the counts/pixel in the modified image on a pixel by pixel basis to obtain the subtracted image. In the remainder of this section it will be assumed that all images are digitised.

9.5.2 Digital Subtraction Angiography

A series of individual X-ray exposures is made, typically at a rate of greater than one per second, before, during and after injection either intravenously or intra-arterially of X-ray contrast material containing a high atomic number element such as iodine. The image collected during each X-ray exposure is stored in digital format.

All the technical details considered in Chapters 5 and 6 for good digital images are relevant. For example, to achieve high resolution, small focal spot sizes (0.5 mm) and detectors with small pixel sizes are necessary. The output from the X-ray generator must be very uniform (otherwise the counts/pixel in unaltered regions will not cancel). Feed-back control from the digital image processor helps to achieve this. Coupled with repeat pulsed exposures of short duration, especially in cardiac work, this places severe demands on X-ray tube rating—high tube currents (~1000 mA), grid controlled output and high anode heat storage capacity (2 MJ).

The digital detector may be either an image intensifier linked via a fibre optic to a CCD device, or a flat panel DR detector. Typical pixel sizes are 150 μm × 150 μm with a resolution of 2–2.5 lp mm^{-1}. The requirement for a modern angiographic system to be capable of recording 30 or more images per second with a 1024 × 1024 matrix and 12–14 bit depth causes major problems for data collection, manipulation, storage and display.

Ideally the signal in the subtracted image should reflect the concentration of contrast agent in the vessel (i.e. the value of μ) and should be independent of the thickness of overlying and underlying tissue. To a first approximation this can be achieved if *logarithmic subtraction* is used—the logarithm of the numerical count in each pixel is calculated before subtraction giving a simple relationship for the signal from the vessel involving only the difference in μ values for the vessel and tissue, and the thickness of the vessel (see Insight).

Insight

Logarithmic Subtraction

In Section 6.4 we showed that, in the absence of scatter, radiation contrast was unaffected by a layer of uniformly attenuating material above (or below) the region of interest. This result arises because contrast is defined as the logarithm of a ratio of intensities, to be consistent with the logarithmic response of film.

A similar argument applies when digitised images are subtracted. Referring to Figure 9.11a, before contrast is added for simplicity assume that μ_t in the vessel is the same as in the tissue, that C_1 is the count in a specified pixel and k is a constant.

$$C_1 = kI_1 = kI_0 \exp(-\mu_t x)$$

After contrast is added (Figure 9.11b)

$$C_2 = kI_2 = kI_0 \exp[-\mu_t(x - y)] \exp(-\mu_c y)$$

where μ_c is the linear attenuation coefficient in the contrast-filled vessel.

$$C_1 - C_2 = kI_0 \exp(-\mu_t x) - kI_0 \exp[-\mu_t(x - y)] \exp(-\mu_c y)$$

This expression cannot be simplified any further. Hence $C_1 - C_2$ depends on both I_0 and on x, the thickness of surrounding tissue.

On the other hand, taking Naperian logarithms ($\log_{10}$ simply introduces a numerical factor which cancels out)

$$\ln C_1 = \ln kI_0 - \mu_t x$$
$$\ln C_2 = \ln kI_0 - \mu_t (x - y) - \mu_c y$$

Hence $\ln C_1 - \ln C_2 = (\mu_c - \mu_t) y$ which is independent of x as required, and of I_0.

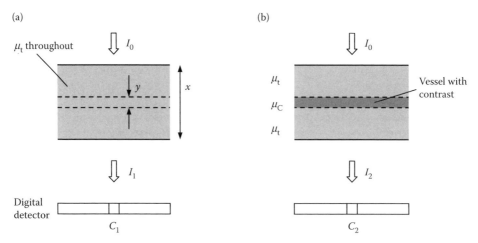

FIGURE 9.11
Simple models for calculating the attenuating effect of a narrow blood vessel, (a) before; (b) after addition of contrast. For explanation see 'insight'.

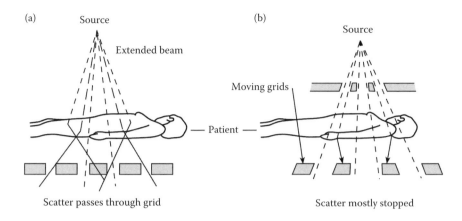

FIGURE 9.12
Use of grids to remove scatter (a) a single grid; (b) grids both in front of the patient and in between patient and detector moved in synchrony. Two grids remove more scattered radiation than a single grid.

Unfortunately, this relationship only applies in scatter-free conditions so for quantitative work, and to optimise contrast, scatter must be greatly reduced or eliminated completely. Several rather complex procedures, beyond the scope of this book, for achieving this have been suggested. One possibility is that grids are placed both in front of and behind the patient and moved in synchrony (Figure 9.12). This removes more scatter than a single grid. Slot scanning systems, see for example, Figure 5.17, are generally more effective than grids for removing scatter.

9.5.2.1 Image Noise

Many of the advantages of digital imaging can now be exploited. For example, any pair of frames may be subtracted, one from the other, to form a new image. The wide dynamic range of the digitised image and the techniques of image processing in Section 6.11 may be used to bring out features of interest. Because of the method of formation, subtracted images are rather noisy. Imagine, for example, that each frame contains on average 10,000 photons per mm^2. The Poisson error ($n^{1/2}$) is 100 photons and this represents the noise. If a small iodine-filled vessel attenuates 10% of the photons, that is, 1000 photons the S/N is 10. However, at least two frames are required to form the subtracted image so the noise is now $(20,000)^{1/2}$ whilst the signal has remained unchanged. This begs the question 'If S/N is worse in the subtracted image than in the contrast image, why do the subtraction?' Recall, however, that the purpose of subtraction is to suppress anatomical detail. This can be thought of as a form of noise, but it is *systematic* noise. Therefore, unlike quantum noise, which is *random* it is greatly reduced by the subtraction process.

Digital techniques may be used to sum frames retrospectively—sometimes known as frame integration—and it is left as an exercise to the reader to show that if, say, four pre-injection frames and four post-injection frames are summed the S/N ratio in the final image will be improved. Note that frame integration reduces the effects of both quantum noise and electronic noise, as a fraction of the signal, whereas increasing the dose only reduces the effect of quantum noise.

Finally, when using contrast note that the ratio S/N (considering only Poisson quantum noise) is $C \times D/D^{1/2} = C \times D^{1/2}$, where C is the concentration of contrast and D is the radiation dose. Thus to improve S/N it is preferable to increase the concentration of contrast rather than the dose (always bearing in mind possible side effects of the contrast medium).

9.5.2.2 Roadmapping

This is an excellent illustration of the power of subtraction imaging and is used to assist the advancement of guide wires and catheters in the cardiovascular tree using as little contrast as possible. The problem is that both the contrast agent and the wire/catheter are radio-opaque and thus are difficult to distinguish.

Exact details vary between systems but the principles are as follows:

1. Acquire a short native scene using fluoroscopy. Use the last image as a mask for subtraction

2. Inject contrast medium and create subtracted images outlining the vascular tree—dark against the background

3. Continue until contrast is maximum. This ensures the vascular tree is optimally visualised

4. Invert the unsubtracted maximum contrast image so that the vascular tree appears radiotranslucent

5. Now advance the wire catheter under fluoroscopy and use the inverted image as a new mask for 'subtraction' (strictly speaking, addition because the contrast image has been inverted). The wire/catheter appears black advancing in a white vessel, thus making visualisation much easier (see Figure 9.13)

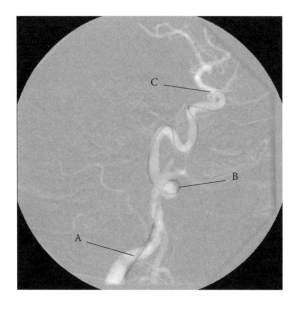

FIGURE 9.13
Illustration of roadmapping. The lighter structure is the internal carotid artery. In the lower part of the image the dark guide catheter (A) is fairly straight because it is sufficiently rigid to alter the shape of the flexible artery. Above the microcatheter delivery system (B), the microcatheter (also darker) follows the contours of the finer arteries up to the aneurysm (C). (Image courtesy of Dr J Higgins and Ms H Szutowicz.)

9.5.3 Dual Energy Subtraction

In some examinations it may be desirable to enhance either soft tissue characteristics or the bony characteristics of the image at the expense of the other. There are two basic approaches but both exploit the difference in mean atomic number of bone (~13.8) and soft tissue (~7.6). Because the photoelectric effect is very dependent on atomic number, low energy radiation must be used to generate soft tissue contrast. Bone will show reasonable contrast even at high kV.

In one approach two digital images are collected in quick succession, one at high energy and one at low energy. In the established technique the beams are spectra generated by a conventional X-ray tube—see Figure 9.14. However, there are problems with the overlapping spectra, beam hardening, quantum noise and scatter. Attempts are being made to use narrower energy windows. It is preferable, but currently only possible in a few research centres, to use intense, highly collimated, narrow line spectra from a synchrotron (see Insight). The images are subjected to weighted logarithmic subtraction, $\ln(C_1) - R \ln(C_2)$ where C_1, C_2 are the counts in a pixel at two energies and R is a weighting factor. The value of R can be varied so that bone is either enhanced or suppressed relative to soft tissue.

Insight

Synchrotron Radiation

A full treatment is beyond the scope of this book but a brief summary may be helpful. Radiation from a synchrotron with energies in the X-ray range can be generated by accelerating electrons almost to the speed of light in a closed, circular trajectory, storage ring. They are then caused to oscillate in a transverse direction and behave like tiny transmitting antennae. The emitted X-rays have many of the properties of laser beams—highly collimated, very intense (anything from 5 to 10 orders of magnitude stronger than a conventional beam) and can be very monochromatic. Techniques exist to produce either a single wavelength or a spectrum. Because the spectrum is so intense, a monochromator can be used to create several narrow spectral lines (typically 30 eV wide) of useable intensity.

In a hospital environment there are, unfortunately, major logistic problems in both installing a synchrotron, which is neither simple nor inexpensive, and in using it routinely.

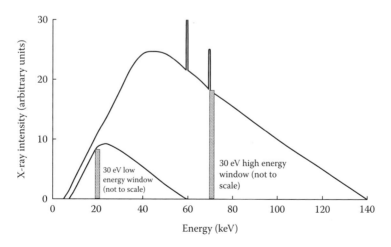

FIGURE 9.14
Illustrating the use of wide energy windows with conventional spectra from a tungsten anode for dual energy subtraction. The spectrum on the left is generated at 60 kVp and a 30 eV window is centred on 20 keV. The spectrum on the right is generated at 140 kVp with a window centred on 70 keV.

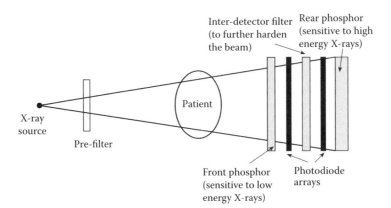

FIGURE 9.15
Schematic diagram of a single exposure dual energy system.

An alternative approach is to use a conventional high energy spectrum with a dual energy cassette (see Figure 9.15). The first part of this detector comprises a yttrium oxysulphide phosphor screen coupled to a photodiode array. Since the K absorption edge for yttrium is at 17 keV, a signal corresponding to the detection of low energy photons, that is, a bone + soft tissue image, is generated. The X-ray beam now passes through a copper filter for further beam hardening before falling on a thick gadolinium oxysulphide phosphor screen. The K-edge for gadolinium is at 50 keV so a signal corresponding to the detection of high energy photons is generated. If the image recorded in the second screen (Figure 9.16b) is suitably weighted and then subtracted from the image obtained in the first screen (Figure 9.16a), an image virtually devoid of bony structures is produced (Figure 9.16c). Note: (1) there is no movement blurring because the images are produced simultaneously; (2) there is an excellent example of quantum mottle in the abdominal region of the high energy image, arising because very few photons have penetrated the abdomen and the added copper filtration.

9.5.4 Movement Artefact

This can be a major problem in digital subtraction imaging since there is normally a time interval between images. In DSA the interval may be several seconds and Figure 9.17 shows how the movement of a bone-tissue interface close to a blood vessel could generate an artefact comparable to the signal from the vessel itself.

The time interval can be very variable. One application of subtraction imaging not yet mentioned is to assess serial changes of a condition or disease status over the course of days or weeks. Consider two chest images of a patient with a suspected lung tumour taken a week apart. If the images are subtracted digitally, subtle changes that might be difficult to spot if the images were compared visually, are more apparent. Because of the long time interval, special measures such as anatomical markers and rotation of the digital images may be necessary to achieve superposition.

Dual energy subtraction may be used in conjunction with contrast agent to reduce the time interval in DSA but if spontaneous movement is to be eliminated completely (e.g. of the heart) the images must be obtained simultaneously. Images obtained with a dual cassette are simultaneous but the technique is not robust enough for this application.

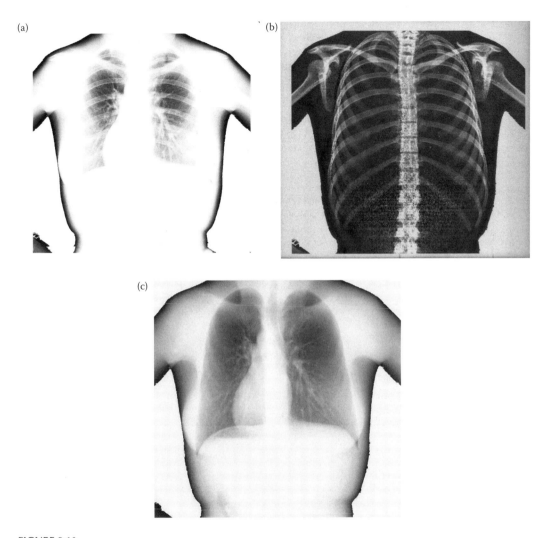

FIGURE 9.16
Images obtained with a prototype dual energy chest unit; (a) the low kV image, (b) the high kV image, (c) the subtracted image. (Images courtesy of Professor Gary T Barnes.)

To overcome this problem *dichromography* is being developed. After administration of iodine contrast, a dual beam monochromator is used to convert the X-ray spectrum from a synchrotron into two monochromatic beams with energies just above and just below the K absorption edge of iodine (33.17 ± 0.300 keV). Working with wavelengths that are so close to the K shell absorption edge increases the sensitivity to iodine by some two orders of magnitude allowing significant reduction in contrast agent administered. The beams pass through the heart at slightly different angles and then fall on two separate detectors creating simultaneous images. To image the whole heart the beams are scanned vertically.

9.5.4.1 Time Interval Differencing

This application of digital subtraction techniques uses a property of digital images that has already been discussed, namely that any pair of frames can be subtracted, retrospectively,

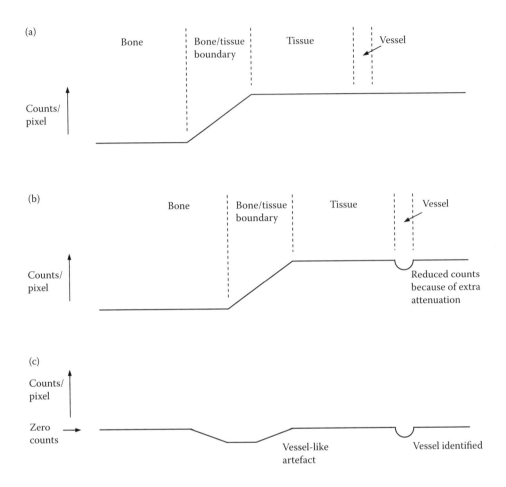

FIGURE 9.17
Creation of an artefact due to movement. The images show, schematically, counts/pixel in (a) the mask frame; (b) the contrast-enhanced frame; (c) the subtracted frame (b–a). A shift to the right in the bone/tissue boundary between frames (a) and (b) appears as an artefact in frame (c).

one from the other. Thus it is not necessary to use the same frame as the mask for each subtraction.

Imagine that 30 frames s^{-1} have been collected for several seconds. In time interval differencing a new series of subtracted images is constructed by using frame 1 as the mask for, say, frame 11, frame 2 as the mask for frame 12, frame 3 as the mask for frame 13 and so on. Thus each subtracted image is the difference between images separated by some fixed interval of time—one-third of a second in this example. Such processing can be effective in situations where an organ is undergoing regular cyclical patterns of behaviour, for example, in cardiology. Note that the statistics on a single frame may be poor but the technique is still applicable to a fixed time difference between two groups of frames. Both the time difference and number of frames can now be adjusted to give the best images.

9.6 Interventional Radiology

9.6.1 Introduction

Interventional radiology is an all-embracing term for situations in which X-ray imaging is used as an aide to other clinical procedures. The first interventional procedure is generally attributed to Dotter and Judkins (1964) who used dilators to relieve arterial stenosis. The use of X-ray images to guide small instruments such as catheters through blood vessels to identify and correct focal narrowing of blood vessels remains an important application of IR in the cardiac catheterisation laboratory.

Since its first introduction, IR has expanded and diversified dramatically. By year 2000, the compound annual rate of increase of IR in the UK across all service settings was ~10% and an entire journal (Seminars in Interventional Radiology) is now devoted to reviews of IR in different clinical specialties.

Advances in technology and demographic changes will be drivers for further developments. For example, the ageing population and increasing obesity are both leading to increased vascular disease and cancers, resulting in more peripheral angioplasty and stents, more biliary ureteric stents and more radiofrequency ablation. Increasing cirrhosis of the liver caused by elevated levels of alcoholic intake and hepatitis C have resulted in more chemoembolisation processes and transjugular intra-hepatic portosystemic shunts (TIPS).

Because of the diversity of procedures, it is beyond the scope of this book to discuss optimum techniques for all possible procedures. However, issues relating to radiation dose are common to all IR because procedures can be very long and staff have to be in the room close to the patient so both groups are potentially exposed to high doses of radiation. Therefore this section will concentrate on the physical principles, techniques and precautions applicable to IR for ensuring that 'the least amount of radiation is used to get the job done' (Dendy 2008).

9.6.2 Equipment Factors

Pulsed Fluoroscopy—This is an important method of dose reduction in IR. Recall that to obtain a good pulse profile, a grid controlled X-ray tube in which the high tension is kept constant and a third grid controls the flow of electrons to the anode must be used (see Section 2.3.1). Both fluoroscopy pulse width and entrance exposure/pulse must be optimised during calibration. The manufacturer's default setting for pulse rate, often about 12–15 frames/s, should be challenged to find out if adequate images can be obtained with a lower rate. For some paediatric studies, rates can be reduced to as little as 3.75 pulses/s without reduction of contrast or spatial resolution but with a substantial reduction in dose.

Filtration—Unwanted low energy radiation must be removed from the beam. 0.1 mm of copper can reduce skin dose rate by 40% but increases tube loading and reduces contrast. (It is left as an exercise for the reader, after studying Chapter 13, to explain why the reduction in effective dose is rather less). A range of filters should be available for different programmes and the filtration used for a particular procedure should always be recorded to assist with dose reconstruction and optimisation if required.

Organ Curves—As discussed in Section 5.14.3 the X-ray set can respond to demand for more output by increasing kV, mA or a combination of both. Several options may be available so it is important to know the impact of each on dose and contrast.

Field Size—Several field sizes are possible, depending on the application, typically from 12 cm for cardiac work to 48 cm for body angiography (diagonal distance across a flat panel detector [FPD]), and it is important to use the smallest that covers the region of interest. Too large a field size will increase the volume of patient in the direct beam and increase scatter both to other parts of the patient and to other staff in the room.

Robotics—This subject is outside the scope of this book but it is important to note the potential of robotic systems in IR, especially where the robot can permit the interventionalist to stay further from the couch.

Image receptor—Digital techniques are having a big impact on IR as in other branches of radiology. Cinefluorography and television monitors have been largely replaced by CCD cameras and there is now increasing interest in replacing image intensifiers with either direct conversion or indirect conversion FPDs (see Section 5.10). The FPDs have a better dynamic range because there is no television camera and inherently better detective quantum efficiency. Both of these properties might reduce patient dose. FPDs also have better spatial resolution, negligible contrast loss and no inherent vignetting.

However, Davies et al. (2007) have recently reported on a careful comparison of three commercial systems, all used for similar coronary angiography procedures. One uses a state-of-the-art image intensifier (II) in which problems of geometric distortion, glare and vignetting have been largely eliminated, coupled to a CCD camera with low noise digital recording. The other two use indirect conversion FPDs. All the systems have high power X-ray tubes, automatic dose control, programmable filters and similar frame rates.

Phantom dose measurements and clinical assessments of both patient dose and image quality were made. Minor differences were noted, and these may be indicative of an important contribution from processing algorithms. However, overall there was no clear evidence that replacing IIs with FPDs produces any great improvement in either dose reduction or image quality. As the cost of FPDs comes down, they will replace IIs because of all the benefits of a totally digitised system (see Section 5.1). It is noteworthy that image quality and dose may be only relatively minor considerations.

A typical specification for a state-of-the-art system for IR is shown in Table 9.4.

TABLE 9.4

Typical Specification for a State-of-the-Art IR System

---

1. X-ray tube mounted on a C-arm which can be rotated round 3 axes to obtain any desired orientation

2. Variable source-detector distance (80–120 cm)

3. *Tube assembly*—microprocessor controlled high frequency X-ray generator, 60–120 kV, 0.5–1000 mA, switching times 0.5–800 ms, pulse frequencies up to 50 frames s^{-1}, 0.4–1.0 mm focal spot size, maximum power 100 kW, continuous power 3 kW, heat capacity 1.5 MJ

4. Collimator to adjust field of view

5. 2–8 mm copper filtration

6. Carbon fibre patient table

7. AEC pre-selecting kV, mA, pulse width and pre-filtration

8. *Typical dose rates to image receptor*—30 nGy per pulse for fluoroscopy (advancement of guide wires and catheters, placement of stents, coils and closure devices). Higher doses of 300 nGy per pulse for digital cine mode to produce low noise images of diagnostic quality

9. 40 cm × 30 cm FPD comprising a-Si diode matrix coupled to CsI scintillators, 3 k × 2 k matrix, 150 μm pixels, 3.25 lp mm^{-1} resolution; *or* distortion-free image intensifier coupled to state-of-the-art CCD camera

9.6.3 Doses to Patients

Patients may be subject to both *stochastic* and *tissue-specific (deterministic)* effects of IR. Readers who are not familiar with these terms may wish to defer reading the remainder of this section until they have read Chapter 12.

Dose area product (DAP), which is a measure of the total energy imparted to the patient as ionising radiation, correlates reasonably well with effective dose, which is a surrogate for the stochastic risks of induced cancer probability and hereditary effects. Since IR procedures are so varied, it is no surprise that DAP and effective dose values also vary widely. Some average figures for specified procedures are given in Section 13.3.3. At the upper end of these values the risk is by no means trivial.

The range of reported values for a particular procedure can vary by even more than in conventional AP/PA or lateral views—in extreme cases a range of 100 or more has been reported. There are several possible causes for this variability, including (a) equipment-related factors such as those discussed in the previous section; (b) patient factors, including size and complexity of the procedure; (c) level of training of the interventionalist; (d) level of experience of the interventionalist; (e) high/low volume case load; (f) poor technique.

Tissue-specific injuries in IR are primarily to the skin (see Section 12.5.1). For the present discussion their main features are that they have well-defined thresholds (from 2–10 Gy) and the time of onset is very variable (from a few hours to more than a year). For several decades many physicians, and some radiologists, were of the opinion that fluoroscopy was a low dose procedure with no possibility of reaching any threshold for tissue effects. However, the increased use of lengthy interventional procedures, combined with a realisation that long-term follow-up may be necessary to detect tissue-specific injuries, have provided clear evidence that this is not correct.

Balter (2001) gives a good review of case reports of skin injuries in patients caused by IR up to about year 2000. The various causes of high ESD are fairly predictable—(a) long irradiation times, sometimes several hours, for example, TIPS, angioplasties, interventional neuroradiology; (b) examination of thick body parts or very obese patients causing severe beam attenuation; (c) body parts close to the X-ray port resulting in very short source to skin distances.

A significant feature of these case reports is that although several of them estimated fluoroscopy time, very few estimated the radiation dose so the reports throw little light on dose thresholds in IR procedures—see insight.

Insight

Dose Thresholds in IR Procedures

There are reasons why thresholds in IR might differ from those quoted in the literature, where thresholds are based mainly on radiotherapy and accidents.

1. Unlike stochastic effects, which are thought to be strictly additive, tissue-specific effects can be repaired so fractionated or low dose rate effects may be less severe.
2. At low dose rates very little is known about the effects of exposed area on thresholds.
3. There is increasing evidence that some patients are more radiation sensitive than others, especially to skin injury. Possible explanations are enhanced genetic susceptibility to radiation or a compromised healing process.

For many procedures, DAP is a poor measure of the maximum skin dose and risk of tissue effects especially in cardiology where there are partially overlapping, or

TABLE 9.5

Good Practice for Investigating a Recorded Tissue Effect after a High Exposure

1. Monitor and record all entrance skin doses and DAP readings
2. Check monitor calibrations and, if possible set tighter tolerances
3. Match the entrance field to the position of the injury—especially for erythema
4. Seek advice from the Radiation Protection Adviser and radiotherapy colleagues
5. Check radiation protection reports for evidence of variation in performance of equipment
6. Review all irradiation programmes with a view to lower dose rates
7. Introduce routine post-procedure patient visits when the estimated skin dose exceeds a specified limit—say 2 Gy

non-overlapping fields. Therefore it is important to record the cumulative dose or cumulative air KERMA at the interventional reference point, chosen to coincide approximately with the skin surface. In recent years greater care has been taken over gathering evidence that will be useful in assessing tissue effects and improving techniques—see Table 9.5.

9.6.4 Staff Doses

All the staff in the interventional room will be exposed to scattered radiation and must have 'appropriate' monitoring. This begs the question 'What is appropriate?'

- Is one personal dose meter sufficient and should it be worn above or below the apron?
- Are two dose meters required, one above and one below the apron?
- Is extremity monitoring—eyes, hands, legs necessary?

Some general points can be made

(1) Actions which reduce the dose to the patient will also reduce the dose to staff.

(2) The pattern of scattered radiation around the patient is *not* uniform. Readers are advised to refer to Figure 3.7 again. This shows that the highest dose rates are in the back-scattered direction. In the forward direction the patient is an effective attenuator and in a practical situation further attenuation will be provided by the image receptor. The implications for the interventionalist are clear. For lateral and oblique views it is much better to work on the image receptor side of the patient than on the X-ray tube side. If the tube axis is vertical, work with an under-couch tube if the clinical procedure allows so that the dose to the upper body and eyes is reduced.

(3) Doses to the hands are very dependent on their position relative to the beam and the hand movements required for a particular examination. These will vary from one examination to another.

(4) The monitoring must be both relevant and proportionate to the risk.

Proper use of protection measures, for example, lead apron, thyroid collar, fixed and moveable lead screens can be very effective. For example, a 0.35 mm lead apron and thyroid shield will reduce the risk by at least a factor of 30, maybe more for some procedures.

Other personal shielding, for example, lead eye glasses and leaded gloves (giving up to 20% beam attenuation) may be necessary but they are inconvenient to wear and leaded gloves severely inhibit manual dexterity.

Because of all the uncertainties, for a new procedure the best advice is 'If in doubt, monitor'. If readings are acceptably low, monitoring may be relaxed. Acceptability should be judged in terms of work load and dose limits (see Section 14.3.3). For whole body monitoring a single dose meter worn under the apron should suffice for annual effective doses less than annual background radiation; an additional monitor should be worn over the apron if doses are higher. Additional extremity monitoring can often be relaxed after initial measurements but should be continued if estimated annual doses exceed one-tenth of the relevant dose limit.

There are also logistic problems with personal dosimetry in IR (e.g. staff forget to wear their badges), often caused because staff are focussed on the clinical procedure, not on their personal dose. Perhaps greater efforts should be made to relieve the operator of the responsibility for their own monitoring. With fully digital systems one way to do this might be by mathematical modelling using the exposure and projection data from the acquisition runs stored by the equipment, generic data on scatter patterns around the X-ray tube, couch and patient, and images of staff positions. This is a formidable project but, if successful, it would have the added benefit of providing precise data on doses to patients.

9.7 Paediatric Radiology

The most recent reports on the long-term effects of radiation (see Section 12.5.2) have confirmed that the risk from X-rays to children is greater than the risk to adults, perhaps by a factor of 2 or 3. This is partly because the radiation risk itself depends strongly on the age at which exposure occurs, and partly because the opportunity for expression, that is, the life expectancy at the time of exposure, is greater. Thus the numerous factors that contribute to the complex relationship between image quality and dose must be re-examined to ensure that, for the imaging process as a whole, the principle of optimisation still applies.

There are many reasons why imaging criteria applicable to the standard adult may not be appropriate for paediatric imaging—the smaller body size, age-dependent body composition, lack of co-operation and functional differences (e.g. higher heart rate, faster respiration). These problems are especially important in the special care baby unit where neonates, particularly those born prematurely—perhaps as early as 24 weeks and with a birth weight of only 600 g—often suffer from a variety of serious life-threatening complications. Early diagnosis and prompt therapy of diseases in the respiratory and cardiovascular system is essential.

9.7.1 Review of Technical Criteria

In this brief review, which will be restricted to factors directly related to the physics of radiology, attention will focus on those aspects where the optimum technique for children may differ from that for adults. Taken together they make a strong case for customised facilities, if not dedicated facilities for paediatric radiology.

Exposure times—these will generally be short, and indeed should be short to minimise movement problems. However, many generators are incapable of delivering the exposures in the 1–4 ms range that are frequently required. There may be problems associated with the rise time to maximum output, which in older units can be longer than the required exposure, and if the cable between the transformer and the tube has a significant capacitance, there may be a post-exposure surge of radiation.

To achieve a rectangular configuration of kV with minimal ripple and hence a uniform X-ray output over a very short exposure time (say 1 ms) requires a grid controlled 12-pulse, multipurpose or constant potential generator. The shorter the exposure, the more stringent the requirements. This leads to the rather paradoxical conclusion that the smallest patients require the most powerful machines.

Note that it is generally not acceptable to achieve more 'convenient' that is, longer, exposure times by reducing the kV. This will increase exposure times, by the converse of the processes discussed in Section 2.5.4, and will also increase contrast but the penalty in terms of increased dose is likely to be high.

Added filtration—an optimised radiograph of a child cannot be obtained simply by reducing the exposure time in proportion to body thickness. This can be appreciated from Figure 9.18. At the depth in an adult corresponding to the entrance surface of the child there will be not only beam attenuation but also beam hardening. This beam hardening must be introduced externally for paediatric investigations. Additional filtration of 1 mm aluminium plus 0.1–0.2 mm copper (equivalent to about 3–6 mm aluminium at standard diagnostic radiographic voltages) is appropriate. Note that in general-purpose equipment it may not be easy to change the filtration frequently, nor to avoid the risk of using the wrong filtration for a given patient.

Attempts have been made to use filter materials with K shell absorption edges (see Section 3.8 and Table 1.1) at specific wavelengths, for example, rhodium and erbium, with K-edges at 23.2 and 57.5 keV, respectively. Considerable dose reduction can be achieved with these filters.

Low attenuation materials—in view of the increased radiation sensitivity of the patient, the use of carbon fibre or some of the newer plastics in materials for table tops, grids (if used), front plates of film changers and cassettes is strongly recommended. In the voltage range used for paediatric patients, dose reduction of up to 40% may be achieved.

Field size and collimation—inappropriate field size is a frequent fault in paediatric radiology. Too small a field may result in important anatomical detail being missed. Too large

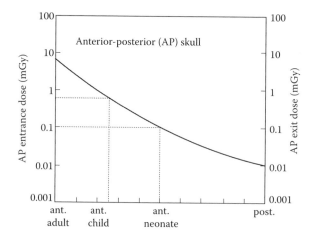

FIGURE 9.18

Schematic representation of the difference in entrance dose for an adult, a child and a neonate. The initial attenuation in the adult must be introduced by added filtration for paediatric radiology.

a field will not only impair image contrast and resolution by increasing the amount of scattered radiation but will also result in unnecessary ionising radiation energy being deposited in the body outside the region of interest. In the neonatal period the tolerance for maximum field size should be no more than 1 cm greater than the minimum value. After the neonatal period this may be relaxed to 2 cm.

Effective immobilisation—incorrect positioning is another frequent cause of inadequate image quality and no exposure should be allowed unless there is a high probability that the exact positioning will be maintained. In paediatric work appropriate immobilisation devices may be required to ensure that (i) the patient does not move, (ii) the beam can be centred correctly, (iii) the image is obtained in the correct projection, (iv) accurate collimation limits the beam to the required area, (v) shielding of the remainder of the body is possible.

In many situations physical restraint will be a poor substitute for a properly designed device.

Use of grids—there is much less scattered radiation from an infant than from an adult and grids are frequently unnecessary especially for fluoroscopic examinations and spot films. Therefore equipment in which the grid can easily be removed should be used.

If a grid is necessary, the grid ratio should not exceed 1:8 and the line number should not exceed 40 per cm. Moving grids may cause problems at very short exposure times.

Image receptor—Table 9.6 shows some typical values for the dose required at the image receptor for film-screen combinations and the lower limit of visual resolution for different speed classes.

Note that these figures, which were obtained at 80 kVp with a suitable phantom, should be used only as a guide. Exact values will vary with kV and filtration, especially that caused by the object.

Reference to Table 9.6 shows that a high speed film-screen combination gives a big dose saving. Furthermore it results in a shorter exposure which minimises motion unsharpness, the most important cause of blurring in paediatric imaging. The slight loss in resolution is rarely a problem although it may be a reason for choosing a lower speed class for selected examinations.

Many of the advantages of digital image receptors and digital image handling are important in paediatric radiology, not least the potential for dose reduction. However, the underlying physics is not basically different from that discussed in Chapters 5 and 6.

Automatic exposure control—There are a number of potential problems with using AEC in paediatric work (i) the system may not be able to compensate for the very large variation in patient size; (ii) the detector may be too big for the critical region of interest; (iii) it may

TABLE 9.6

Relationship between Speed Class, Radiation Dose and Resolution for Different Film-Screen Combinations

Speed Class	Dose Requirement at Image Receptor (μGy)	Lower Limit of Resolution (lp mm^{-1})
25	40	4.8
50	20	4.0
100	10	3.4
200	5	2.8
400	2.5	2.4
800	1.25	2.0

not be possible to move the detector to the critical region; (iv) the usual ionisation chamber of an AEC is built behind the grid which is frequently not necessary; (v) a variety of film-screen combinations may be required for different examinations and the AEC system will have to be calibrated for each of them; (vi) the AEC may require a longer exposure time than the radiographic examination.

An AEC specifically designed for paediatric work will have a small mobile detector that can be positioned very precisely, mounted behind a lead-free cassette and must respond to an absorbed dose that is considerably less than 1 μGy. The technical problems are such that it may be preferable to use predetermined exposure charts based on the infant's size and weight.

9.7.2 Patient Dose and Quality Criteria for Images

Paediatric radiology provides a good example of the essential requirement for patient doses and image quality to be studied conjointly, not independently.

In the early 1990s, a group of paediatric radiologists and medical physicists reviewed the status of quality assurance in paediatric radiology in Europe. They used one set of criteria for radiographic techniques that would result in low doses and a second set of fairly simple criteria of acceptable image quality. As shown in Figure 9.19, many centres were not complying with all the technical criteria and there was a clear inverse correlation between criteria fulfilled and mean surface entrance dose. Conversely, there was no correlation between image quality and radiation dose, indicating a clear potential for dose reduction.

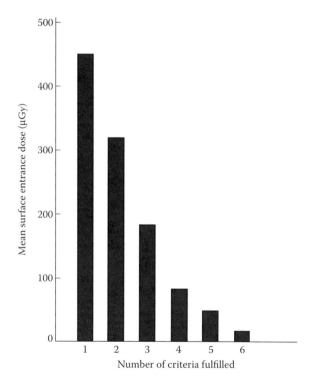

FIGURE 9.19
Correlation between fulfilment of X-ray technique criteria and dose for chest radiography in 10-month old infants. (Adapted from Schneider K, Evaluation of quality assurance in paediatric radiology. *Rad. Protec. Dosimet.* 57, 119–123, 1995, with permission.)

TABLE 9.7

Good Radiographic Technique for an AP Projection of the Chest of a Neonate

Patient Position	Supine
Nominal focal spot value	Less than 1.3 mm
Additional filtration	1 mm Al + 0.2 mm Cu
Grid	No
Screen-film system	Nominal speed class 400
Focus-receptor distance	100 cm
Radiographic voltage	65 kVp
AEC	No
Exposure time	<4 ms
Protective shielding	Lead-rubber masking of abdomen

Source: Adapted from CEC European Guidelines on quality criteria for diagnostic radiographic images in paediatrics. Report EUR 16261 Luxembourg, Office for Official Publications of the European Communities, 1996.

TABLE 9.8

Suggested Image Quality Criteria for AP Neonatal Chest Examinations

Reproduction of the vascular pattern in the central half of the lungs

Visually sharp reproduction of the trachea and proximal bronchi

Visually sharp reproduction of the diaphragm and costophrenic angles

Reproduction of the spine and paraspinal structures

Visualisation of retrocardiac lung and mediastinum

Source: Adapted from CEC European Guidelines on quality criteria for diagnostic radiographic images in paediatrics. Report EUR 16261 Luxembourg, Office for Official Publications of the European Communities, 1996.

More or less as a direct consequence of this work, in 1996 the European Commission published 'European Guidelines on Quality Criteria for Diagnostic Radiographic Images in Paediatrics'. An example of good radiographic technique for an AP projection of the chest of a neonate is shown in Table 9.7.

The suggested guideline skin entrance dose that should not normally be exceeded was set at 80 µGy although this figure has subsequently been reduced to 50 µGy and even 30 µGy should suffice.

A list of image criteria is suggested in Table 9.8 and images may be compared using evaluation methods such as those discussed in Chapter 7 especially receiver operator characteristic (ROC) curves.

For a recent paper on dose and image quality optimisation in neonatal radiology see Dougeni et al. (2007).

9.8 Dental Radiology

In dentistry, as in other specialties, accurate diagnosis is a fundamental pre-requisite for correct treatment and dental radiographic examinations represent one of the most frequently undertaken radiological investigations. In 2001 there were about 10 million intra-oral radiographs each year in the UK for a population of about 60 million. Furthermore, the annual expenditure in the UK on dental radiology is comparable with that on many other aspects of dental care (e.g. extractions). Thus the emphasis given to dental radiology in education and training and the standards expected in dental radiology practice should be at least comparable with those in other areas of dental care.

9.8.1 Technical Detail

9.8.1.1 Intra-Oral Radiography

Historically, a major source of unnecessary dose to the patient was too low an operating potential. Many older sets operate at 45–50 kVp and whereas they give excellent contrast, the dose is high. In newer equipment the operating potential is higher—in the range 60–70 kVp. This still gives adequate contrast and the dose is greatly reduced. Total beam filtration should be at least 1.5 mm aluminium up to 70 kVp and 2.5 mm aluminium above 70 kVp. The added filtration should not be so thick that exposure times are greater than 1 s.

X-ray units using high frequency generators and giving an effectively constant potential output have a number of advantages over X-ray units with self-rectified or half-wave rectified voltage profiles. For example, the higher X-ray output rate permits techniques involving longer focus to skin distances whilst maintaining short exposure times; accurate, very short exposure times are possible for digital intra-oral radiographic systems (see later).

Exposure timers must provide increments that are small enough for accurate selection of the required exposure. Timers must be carefully checked, errors in excess of 0.2 s are unacceptable.

Another problem that is gradually being overcome is that, traditionally, the image receptor, a rectangular film, was not the same shape as the collimated X-ray beam, which was circular. Where such systems are still in use, to minimise unnecessary exposure the beam diameter should not exceed 60 mm at the patient end of the cone. However, this is still a major source of inefficiency since either the X-ray beam will be greater than its useful area, or the field of view will be impractically small. Therefore rectangular collimation has been incorporated on some new equipment and the beam size at the end of the collimator is, ideally, no more than 35 mm × 45 mm, that is, no more than 2.5 mm greater than the standard dental film size (30 mm × 40 mm) on any edge. Combined with beam aiming devices and suitable film holders, rectangular collimation reduces doses and, by improving consistency of set-up, improves the diagnostic quality of radiographs and reduces the proportion of rejected films.

When the output of the X-ray set is low, another consequence of low kV, it may be necessary to work at short focus to skin distance to keep the exposure time short. However, this results in high ESDs because of the inverse square law effect. For dental radiology a minimum focal skin distance of 200 mm is recommended.

9.8.1.2 Panoramic Radiography

Panoramic radiographs are obtained by collimating the X-ray output with a slit and rotating the source and slit in a horizontal plane. A complementary slit in front of the curved cassette holder collimates the transmitted beam onto the receptor. The beam size should not exceed 5 mm × 150 mm and the area of the receiving slit should not exceed that of the transmitting slit by more than 20%. It is important to check that these slits remain aligned as they rotate. Tube kilovoltage should be in the range 60–90 kVp and there should be a range of mA settings to make full use of the sensitivity of modern image receptors. Note that narrower beams and shorter cycle times require an effectively constant output. Any significant fluctuation, for example caused by the kV dropping with the mains frequency, would cause banding on the film. If AEC is used there must be a back-up timing circuit to prevent excessive over-exposure.

Many panoramic radiographs are unacceptable because of poor patient positioning and alignment. Equipment must be provided with effective positioning aids, for example the light beam type. There should be sufficient variation on exposure times to take advantage of the fastest receptors.

9.8.1.3 Image Receptors

Film is still widely used as the receptor in dental radiology and the fastest film consistent with satisfactory clinical results should be used. There is no justification for using slower than ISO (International Organisation for Standardisation) speed group E film and current opinion is that good diagnostic films can be obtained with F speed films if care is taken over processing.

Increased fog levels on dental films due to poor storage conditions before use is a common problem. Films should be stored in a cool, dry place and must be far enough away from, or adequately shielded from scattered X-rays. They should not be used after their expiry date.

Fast films and rare earth intensifying screens with a film-screen speed of at least 400 should be used for panoramic, extra oral and vertex occlusal views where practicable. It is important to ensure that the light sensitivity of the film is matched to the wavelength range emitted by the screen.

CR with photostimulable phosphors is being introduced in intra-oral dental radiology although perhaps not as quickly as elsewhere. An alternative approach that can be used for both horizontal and bitewing images uses caesium iodide as the initial scintillator. Its columnar structure combined with a fibre-like plate maintains good resolution (see Section 5.13.1) and the light is incident on a CCD as the first stage in the digitisation process.

Digital systems for panoramic imaging also use either photostimulable phosphor plates or a CCD device. One advantage of the CCD is that the read-out can be synchronised with the rotation of the X-ray beam. This improves the sensitivity of the system and hence reduces patient dose.

Cone beam CT (see Section 8.6.2) is also being developed for dental applications to provide three-dimensional images of the jaw, but specialised aspects of this technique are currently beyond the scope of this book.

Other points of note in digital dental imaging are as follows:

1. It is important to use appropriate sensor sizes—that is, comparable with film.
2. As with other radiological techniques, the indications are that doses can be lower with digital receptors. Therefore an X-ray set with near-constant operating potential and the ability to select sufficiently low exposure settings must be used to allow the full savings of dose to be realised.
3. Exposure settings and dose must not be reduced so low that image quality is compromised.
4. Retakes are very easy with digital systems—sometimes too easy! All retakes must be properly justified, recorded and included in quality assurance statistics.

9.8.2 Protection and Quality Assurance

Careful attention to techniques that limit the entrance surface dose is the best method of patient protection. In particular there is no case for using a lead apron, except perhaps for the infrequent vertical occlusal projection. The apron will not protect against radiation scattered internally in the body, and direct measurements have shown that under optimal conditions external scatter results in a dose of no more than 0.01 µGy per exposure to the gonads. If the thyroid gland is in the primary beam, there may be a case for using a thyroid collar. The introduction of rectangular collimation and longer focal skin distances may eliminate this requirement almost entirely.

Staff will be adequately protected by distance if they are at least two metres from the set. If this condition can be satisfied there is no need to leave the room but staff should stand behind the protective panel if one is provided. Note that the beam is not fully stopped by the intra-oral film, nor is it fully absorbed by the patient. Therefore it should be considered as extending beyond the patient until attenuated by distance or stopped by suitable shielding for example, a solid wall.

Quality assurance procedures in dental radiology show no essential differences from those used in conventional radiology. Checks should be made on the equipment, films, processing and dark room—a light leak or the wrong safe light are not unknown sources of fog. The dose at the cone tip should not exceed 1.3 mGy for a 70 kVp set when using E speed film, and about 20% less for F speed film. Tube leakage should not exceed 0.25 mGy h^{-1} at 1 m for intra-oral work, or 1 mGy h^{-1} at 1 m for other work.

A number of publications (e.g. NRPB 2001) have drawn attention to the poor quality of many dental radiographs, even reaching the stage where many have no practical value. It is very important to avoid exposures that have no merit.

Image quality is a good test of the entire quality assurance programme. Routine films can be checked quickly and simply by comparison with good quality reference radiographs obtained under carefully controlled conditions. To obtain such films a suitable phantom consisting of human teeth and facial bones embedded in plastic or other tissue equivalent material can be useful. For absolute evaluation a phantom containing calibrated step wedges and wired meshes to test resolution will be required. Image quality should be monitored on a regular basis—say 6 monthly. Using a simple 3-point scale, 1 = excellent, 2 = just diagnostically acceptable, 3 = unacceptable; 70% minimum in category 1, 20% maximum in category 2 and 10% maximum in category 3 would be reasonable targets.

For readers who are already familiar with the concepts of effective dose and collective effective dose (see Sections 12.6 and 13.5), the insight gives an idea of the potential dose savings in dental radiology

Insight

Dose Savings in Dental Radiology

Table 9.9 shows how easily the effective dose can escalate due to poor technique. Fortunately, many of these improvements have now been implemented but it is salutary to realise that with a ball-park figure of 10 million intra-oral radiographs/annum a saving of 10 µSv/film is equivalent to a collective effective dose saving of 100 manSv.

TABLE 9.9

Typical Effective Doses for Dental Examinations under Different Conditions

Examination and Conditions	Effective Dose (mSv)
Two dental bitewings, 70 kVp set, 200 mm FSD, rectangular collimation E speed film	0.002
Two dental bitewings 70 kVp set 200 mm FSD, round collimation E speed film	0.004
Two dental bite wings 50–60 kVp set 100 mm FSD round collimation E speed film	0.008
Two dental bite wings 50–60 kVp set 100 mm FSD round collimation D speed film	0.016

Source: Adapted from NRPB Guidelines on radiology standards for primary dental care. Report by the Royal College of Radiologists and the National Radiological Protection Board. Documents of the NRPB Vol 5 No 3 NRPB Chilton, 1994.

References

Balter S. *Interventional fluoroscopy—physics, technology and safety.* Wiley-Liss, 2001, 170–180.

Brateman L and Karellas A. Mammography and other breast imaging techniques. In *Advances in medical physics.* Eds, Wolbarst A B, Zamenhof R G, Hadlee W R, Medical Physics Publishing, Madison, WI, 2006.

CEC. *European Guidelines on quality criteria for diagnostic radiographic images in paediatrics.* Report EUR 16261 Luxembourg, Office for Official Publications of the European Communities, 1996.

Dance D R. Diagnostic radiology with X-rays. In *The physics of medical imaging.* Ed., S Webb, IOP Publishing Ltd, Bristol, UK, 1988, 38–40.

Davies A G, Cowan A R, Kengyelics S M, et al. Do flat panel detector cardiac X-ray systems convey advantages over image intensifier based systems? Study comparing X-ray dose and image quality. *Eur Radiol,* 17, 1787–1794, 2007.

Dendy P P. Commentary—Radiation risks in interventional radiology. *Br J Radiol,* 81, 1–7, 2008.

Dendy P P and Heaton B. *Physics for diagnostic radiology* (2nd ed.). IOP Publishing, Bristol, UK, 1999, 238–241.

Dotter C T and Judkins M P. Transluminal treatment of arterosclerotic obstruction. Description of a new technic (sic) and a preliminary report of its application. *Circulation,* 30, 654–70, 1964.

Dougeni E D, Delis H B, Karatzu A A, et al. Dose and image quality optimisation in neonatal radiography. *Br J Radiol,* 80, 207–815, 2007.

IPEM Report 89, *The commissioning and routine testing of mammographic X-ray systems.* Institute of Physics and Engineering in Medicine, York, UK, 2005.

Jennings R J, Eastgate R J, Siedband M P and Ergun D L. Optimal X-ray spectra for screen film mammography *Med Phys,* 8, 629–639, 1981.

Law J, Faulkner K and Young KC. Risk factors for induction of breast cancer by X-rays and their implications for breast screening *Br J Radiol,* 80, 261–266, 2007.

Noel A and Thibault F. Digital detectors for mammography – the technical challenges. *Eur Radiol,* 14, 1990–1998, 2004.

NRPB. *Guidelines on radiology standards for primary dental care.* Report by the Royal College of Radiologists and the National Radiological Protection Board. Documents of the NRPB, Vol 5, No 3, NRPB Chilton, 1994.

NRPB. *Guidance for dental practitioners on the safe use of X-ray equipment.* National Radiological Protection Board, Chilton, Didcot, 2001.

Schneider K. Evaluation of quality assurance in paediatric radiology. *Radiat Protect Dosimet,* 57, 119–123, 1995.

Wade J P, Goldstone K E, Dendy P P. Patient dose measurements and dose reduction in East Anglia UK. *Radiat Protect Dosimet,* 57, 445–448, 1995.

Further Reading

Bushberg J T, Seibert J A, Leidholdt E M and Boone J M. *The essential physics of medical imaging* (2nd ed.). Lippincott, Williams and Wilkins, Philadelphia. Mammography, 2002, 191–229.

Dowsett D J, Kenny P A and Johnston R E. *The physics of diagnostic imaging* (2nd ed.). Hodder Arnold, chapter 9, 2006, 227–251.

Iannucci J M and Howerton L J. *Dental radiography – principles and techniques* (3rd ed.). Saunders, Elsevier, St Louis, 2006.

Pisano E D, Yaffe M J and Kuzimiak C M. *Digital mammography.* Lippincott, Williams and Wilkins, Philadelphia, 2004, 4–26.

Exercises

1. What are the basic requirements of an X-ray tube that is to be used for mammography? How can these be achieved with

   (a) a molybdenum anode tube; (b) a tungsten anode tube

2. Compare and contrast the use of film-screen combinations and digital receptors in mammography.

3. Since a low kVp is required for high contrast, why should the use of a high kVp sometimes be advantageous?

4. Outline the principles of magnification radiography and indicate its limitations.

5. Discuss the relative advantages and disadvantages of magnification mammography.

6. Suggest some situations in which a subtraction technique might be used and indicate how the required information would be obtained.

7. What is the effect of digital image subtraction on (a) quantum noise; (b) structure noise? How can the effect of quantum noise be reduced in digital subtraction angiography?

8. Outline the ways in which dual energy subtraction imaging may be used to highlight bony or soft tissue structures.

9. What are the important features of state-of-the-art equipment for interventional radiology?

10. What are the most likely tissue injuries in interventional radiology? What arrangements should be made to minimise such effects and to investigate them when they occur?

11. Discuss the need for effective immobilisation in paediatric radiology.

12. Explain the need for good quality assurance procedures in dental radiology and discuss the potential for reducing the collective dose to the population.

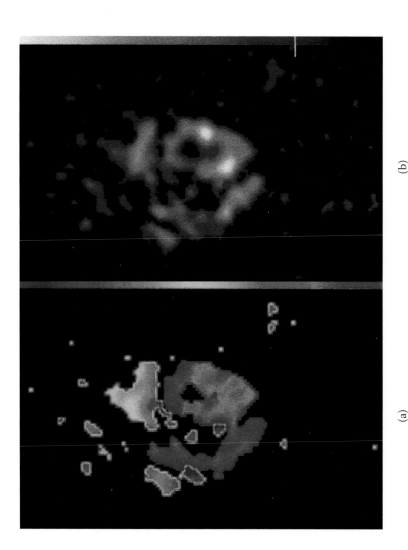

(a)

(b)

FIGURE 10.19

Functional image of a normal MUGA study. (a) The phase image represents the phase (or timing) of the contraction of the heart chambers. (b) The amplitude image represents the amplitude of the contraction. In this case it can be seen that the largest contraction occurs apically and along the lateral wall. Both of these parameters are derived from the Fourier fitted curve (Figure 10.18).

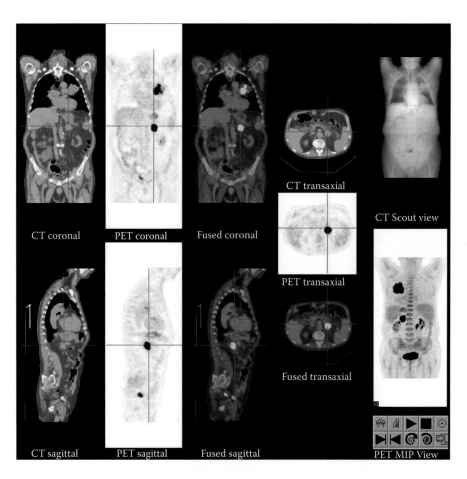

CT coronal   PET coronal   Fused coronal

CT transaxial

CT Scout view

PET transaxial

CT sagittal   PET sagittal   Fused sagittal   Fused transaxial

PET MIP View

FIGURE 11.1

Typical PET/CT review screen showing CT, PET and fused image data sets. The bottom right hand image shows a rotating maximum intensity reprojection image.

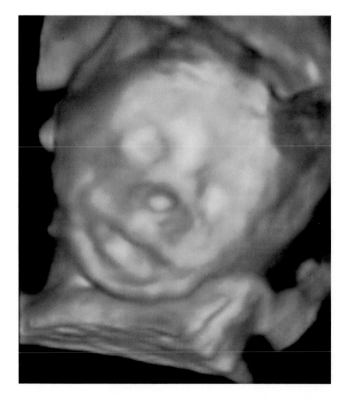

FIGURE 15.38
A 3D ultrasound rendering of a foetal face.

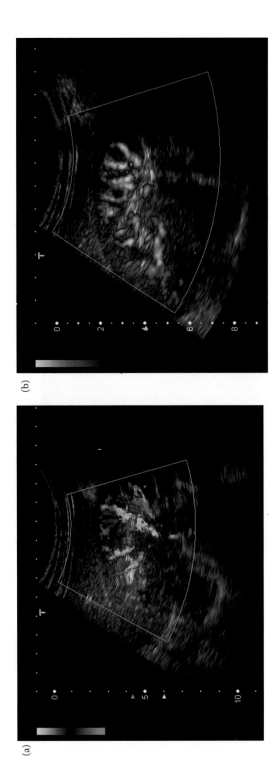

FIGURE 15.44

Doppler ultrasound images showing vasculature in the kidney. There are two main ways of mapping Doppler shifts detected within the 'colour box' on the B-mode image: (a) A colour flow map (CFM) shows mean Doppler shifts at each point; (b) A power Doppler map shows the strength of Doppler-shifted echoes at each point. Notice the reference colour bars to the left of each image.

10

Diagnostic Imaging with Radioactive Materials

F I McKiddie

SUMMARY

This chapter covers the following aspects of imaging with radioactive materials:

- Requirements of imaging systems and techniques for obtaining accurate data
- Principles of operation of the gamma camera
- Additional features of modern gamma camera systems
- Parameters influencing image quality
- Gamma camera performance
- Data display and storage
- Methods of data acquisition
- Quality control of the gamma camera and other aspects of nuclear medicine

CONTENTS

10.1 Introduction

Nuclear medicine is popularly understood to be the use of radioactive materials to produce diagnostic images of biochemical processes within the body. Although the wider term includes all applications of radioactivity in diagnosis and treatment, excluding sealed source radiotherapy, the general perception is taken to mean diagnostic imaging *in vivo*.

However, this does not mean that the *in vitro*, or non-imaging techniques are insignificant. These involve the measurement of samples taken from the patient and are around 7% of the workload in a typical UK department (Hart and Wall 2005). The samples can be blood, breath, urine or faeces and are labelled with both gamma and beta emitting radionuclides. The requirement for accurate mathematical models of the processes under investigation in many *in vitro* tests ensures that the results are absolute measures of physiological processes such as glomerular filtration rate. For further details of the range of *in vitro* tests see Elliott and Hilditch (2005).

The primary requirement in *in vivo* diagnostic imaging is the ability to obtain information concerning the spatial distribution of activity within the patient. This chapter deals with the physical principles involved in obtaining diagnostic quality images after a small quantity of radioactive material has been administered to the patient in a suitable form.

The basic requirements of a good imaging system are as follows:

1. A device that is able to use the radiation emitted from the body to produce high resolution images, supported by electronics, computing facilities and displays that

will permit the resulting image to be presented to the clinician in the manner most suitable for interpretation.

2. A radionuclide that can be administered to the patient at sufficiently high activity to give an acceptable number of counts in the image without delivering an unacceptably high dose of radiation to the patient.

3. A radiopharmaceutical, that is a radionuclide firmly attached to a pharmaceutical, that shows high specificity for the organ or region of interest in the body.

It is important to recognise that, when detecting *in vivo* radioactivity, sensitivity and spatial resolution are mutually exclusive (see Figure 10.1). The arrangement on the left (Figure 10.1a) has high sensitivity because a large amount of radioactivity is in the field of view of the detector, but poor resolution. The arrangement on the right (Figure 10.1b) has better resolution but correspondingly lower sensitivity. Since gamma rays are emitted in all directions, the collimator ensures that the image is only made up of those events travelling perpendicular to the detector. This preserves the relationship between the position within the patient from which the gamma ray was emitted, and its position of interaction in the detector.

In diagnostic imaging spatial resolution is important and sensitivity must be sacrificed. A modern gamma camera (see Section 10.2.1) records no more than 1 in 10^4 of the gamma rays emitted from that part of the patient within the field of view of the camera. Furthermore, any additional loss of counts in the complete system will result in an image of inferior quality unless the imaging time is extended to compensate. Therefore this chapter also considers the factors that limit image quality and the precautions that must be taken to optimise the images obtained using strictly controlled amounts of administered activity and realistic imaging times.

10.2 Principles of Imaging

Medium energy gamma rays in the range 100–200 keV are most suitable for *in vivo* imaging. Lower energy gamma rays are stopped in the body resulting in an undesirable patient dose, whilst higher energy gamma rays are difficult to stop in the detector. This will be discussed further in Section 10.3 where factors affecting the quality of radionuclide images are considered.

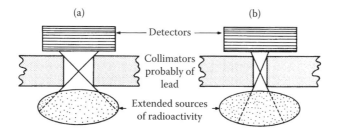

FIGURE 10.1

Collimator design showing conflicting requirements of sensitivity and resolution. Arrangement (a) where the detector has a wide acceptance angle will have high sensitivity but poor resolution, whereas arrangement (b) will have much better resolution but greatly reduced sensitivity.

In all commercial equipment currently available, the radiation detector is a scintillation crystal of sodium iodide doped with about 0.1% by weight of thallium-NaI (Tl). The fundamental interaction process in a scintillation detector is fluorescence which was discussed in Section 5.3. The sodium iodide has a high density (3.7×10^3 kg m^{-3}) and since iodine has a high atomic number ($Z = 53$) the material has a high stopping efficiency for gamma rays. Furthermore, provided the gamma ray energy is not too high, most of the interactions are by the photoelectric effect (see Section 3.4.2) and result in a light pulse proportional to the gamma ray energy. This is important for discriminating against scatter (see Section 10.3.4). The thallium increases the light output from the scintillant, because the traps generated by thallium in the NaI lattice are about 3 eV above the band of valence electrons so the emitted photon is in the visible range and about 10% of the gamma ray energy is converted into light. This yields about 4000 light photons at a wavelength of 410 nm from a 140 keV gamma ray. Note that whereas the number of photons emitted is a function of the energy imparted by the interaction, the energy or wavelength of the photons depends only on the positions of the energy levels in the scintillation crystal. Finally, the light flashes have a short decay time, of the order of 0.2 μs. Thus the crystal has only a short dead time and can be used for quite high counting rates. One disadvantage of the NaI (Tl) detector is that it is hygroscopic and thus must be placed in a hermetically sealed container. Also the large crystals in gamma cameras are easily damaged by thermal or physical shocks.

Alternative scintillation detectors are caesium iodide doped with thallium, and bismuth germanate. Like NaI (Tl), the latter has a high detection efficiency, and is the commonest detector in positron emission tomography (PET) systems. It has the higher stopping power required for the high energy gamma rays and it has a short decay time, allowing it to cope with the high count rates encountered in the absence of a collimator. Bismuth germanate detectors also exhibit a good dynamic range and long-term stability.

The light signal produced by a scintillation crystal is too small to be used until it has been amplified and this is almost invariably achieved by using a photomultiplier tube (PMT). The main features of the PMT coupled to a scintillation crystal were discussed in Section 4.8.

To isolate the output pulses from the PMT corresponding to the photopeak energy of the radionuclide being imaged, the technique of *pulse height analysis* is used (see Section 4.9). For a radionuclide emitting monoenergetic gamma rays, pulse height analysis should, in principle, discriminate completely between scattered and unscattered rays. When a 140 keV gamma ray from technetium-99m interacts with an NaI (Tl) crystal, it does so primarily by the photoelectric effect. This produces a number of visible photons and, hence, a final signal that is proportional to the gamma ray energy. Any photon that has been scattered in the patient by the Compton effect will be of lower energy and will produce a smaller pulse that can be identified and rejected. If an incident pulse is accepted by the pulse height analyser a signal is passed to the computer system and a 'count' is registered. Note that there is further discussion on this point in Section 10.3.4.5.

10.2.1 The Gamma Camera

Modern gamma camera systems consist of one or two collimated detectors mounted on a gantry connected to a desktop-computer (PC) based acquisition and processing terminal. The gantry is also intricately linked to the patient couch and the combination is designed to allow the detectors to manoeuvre freely around the patient. This allows the detectors to obtain static images of any part of the body, or to track over the entire length of the patient's body to obtain what are known as whole body images. The commonest gantry design is the ring gantry which was developed from the slip-ring technology introduced in computed tomography (CT). This also allows the detectors to be rotated around the patient in up to a 540° arc to obtain tomographic image data.

The collimation of the detectors allows the spatial relationship between the point of emission of a gamma ray in the patient and the point at which it strikes the crystal to be established (see Figure 10.2). Note that unlike a grid in conventional radiology, the collimator in radionuclide imaging has no role in discriminating against scatter within the patient. The function is purely to ensure that all photons incident on the crystal are travelling perpendicular to the crystal (or nearly so) when they interact.

The detectors on modern gamma cameras are generally rectangular with a crystal of approximately 400 mm × 500 mm. Up to 100 PMTs will be arranged in a close packed hexagonal array behind the crystal to improve spatial resolution. As shown in Figure 10.3, the number of photons reaching each PMT, and hence the strength of the signal, will be determined by the solid angle subtended by the event at that PMT. Hence, by analysing all the PMT signals, it is possible to determine the position of the gamma ray interaction in the crystal. Essential features of the gamma camera may be considered under five headings.

10.2.1.1 The Detector System

Components of the detector system are shown in Figure 10.4. In the gamma camera, crystal thickness must be a compromise. A very thin crystal reduces sensitivity whereas a very thick crystal degrades resolution (see Figure 10.5). A camera crystal is typically 6–12 mm thick, with most manufacturers now choosing a 9 mm thickness as optimal.

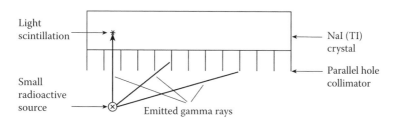

FIGURE 10.2
Use of a collimator to encode spatial information. In the absence of the collimator radiation from the source may strike any point in the crystal.

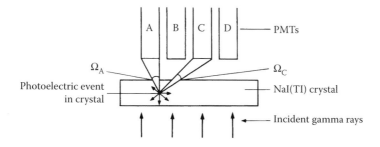

FIGURE 10.3
Use of an array of PMTs to obtain spatial information about an event in an NaI (TI) crystal. Light photons spread out in all directions from an interaction and the signal from each PMT is proportional to the solid angle subtended by the PMT at the event. The signal from PMT A is proportional to Ω_A and much greater for the event shown than the signal from PMT C which is proportional to Ω_C.

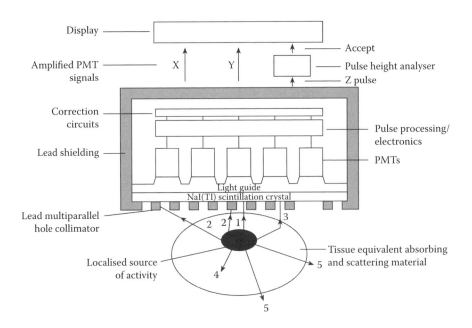

FIGURE 10.4
Basic components of a gamma camera detector system. The fates of photons emitted from the source may be classified as follows: (1) useful photon, (2) oblique photon removed by collimator, (3) scattered photon removed by pulse height analyser, (4) absorbed photon contributing to patient dose but giving no information, (5) wasted photons emitted in the wrong direction.

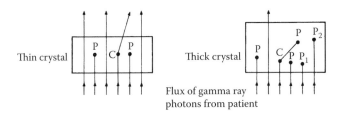

FIGURE 10.5
Interactions of gamma rays with thin and thick NaI (TI) crystals. P = photoelectric absorption. C = Compton scattering. With a thin crystal, many photons may pass through undetected, thereby reducing sensitivity. With a thick crystal the image is degraded for two reasons. First, the distribution of light photons to the PMTs for an event at the front of the crystal such as P_1 will be different from the distribution for an event at the rear of the crystal such as P_2. Second, scatter in the crystal degrades image quality since the electronics will position 'the event' somewhere between the two points of interaction in the crystal.

As shown in Table 10.1 a 12.5 mm crystal stops most of the 140 keV photons from technetium-99m (Tc-99m), the most widely used radionuclide in nuclear medicine (see Section 10.3.2). However, it can also be seen that these crystals are less well suited to higher energies. The detector system is protected by lead shielding to stop stray radiation.

10.2.1.2 The Collimator

The most common type of collimator, which has parallel holes, is shown in Figure 10.6a. It consists of a thick lead plate in which a series of small holes has been microcast or

TABLE 10.1

Stopping Capability of a 12.5 mm Thick NaI (Tl) Crystal for Photons of Different Energy

Photon Energy keV	Interactions %
80	100
140	89
200	60
350	23
500	15

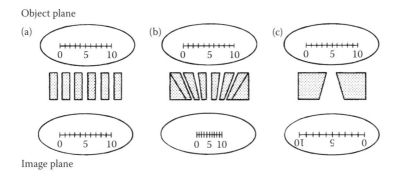

FIGURE 10.6

The effect of different collimator designs on image appearance. (a) The parallel hole collimator produces the most faithful reproduction of the object. (b) The diverging collimator produces a minified image but is useful when the required field of view is bigger than the detector area. (c) The pinhole collimator produces an enlarged inverted image and is useful for very small fields of view.

constructed from stacks of corrugated foil. The axes of the holes are perpendicular to the face of the collimator and parallel to each other.

Performance of the collimator will be determined primarily by its resolution and sensitivity. As shown in Figure 10.7 long narrow holes will produce high resolution but low sensitivity so these two variables work against each other. A typical low energy general purpose collimator will have a resolution of 6 mm and a sensitivity of around 150 cps per megabecquerel. However, a typical low energy high resolution collimator will have a resolution of 5 mm and a sensitivity of around 100 cps per megabecquerel. This emphasises the non-linear relation between resolution and sensitivity in parallel hole collimators. The general purpose and high resolution collimator pairs are the most widely used in routine diagnostic imaging. The 'low energy' in their name refers to the fact that the thickness of the septa and the size of the holes are optimised for gamma rays in the 120–140 keV range.

As the object is moved away from the face of a parallel hole collimator, resolution deteriorates markedly so all imaging should be done with the relevant part of the patient as close as possible to the collimator face. Sensitivity is relatively independent of distance from the collimator face, only decreasing if additional attenuating material is interposed.

Figure 10.7 also illustrates another problem. Higher energy gamma rays may be able to penetrate the septa and this will cause serious image degradation. Thicker septa are now required and for adequate sensitivity this also means larger holes and correspondingly poorer resolution.

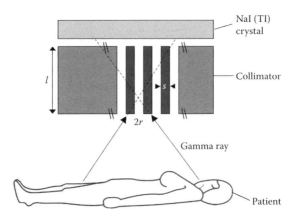

FIGURE 10.7
Diagram showing that oblique gamma rays will pass through many lead strips, or septa, before reaching the detector. Typical dimensions for a low energy collimator are l = 25 mm, $2r$ = 3 mm, s = 0.2 mm. The number of holes will be approximately 15,000.

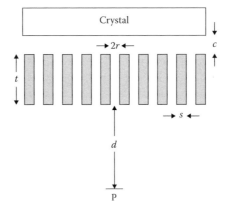

FIGURE 10.8
Diagram showing the physical proportions and geometry of a parallel hole collimator with a point source positioned at P.

Insight

Resolution and Sensitivity of a Collimator

The spatial resolution of a parallel hole collimator depends on the geometry of the holes, corrected for any septal penetration. If the resolution R_P of the image of a point source at P (see Figure 10.8) is measured by its full width at half maximum height (FWHM) then

$$R_P = \frac{2r(t_e + d + c)}{t_e}$$

where r is the hole radius and t_e the effective collimator thickness after septal penetration has been accounted for.

$$t_e = t - \left(\frac{2}{\mu}\right)$$

where μ is the linear attenuation coefficient for gamma rays in the collimator material. The sensitivity (or geometric efficiency) of the collimator is given by

$$\text{Sens} = \left[\frac{Kr^2}{t_e(2r+s)} \right]$$

where K is a factor dependent on the shape and pattern of the holes.

These equations demonstrate that with increasing distance d from the collimator face the resolution deteriorates, that is, R_p increases, but the sensitivity is unaffected (assuming no attenuation).

Other collimator designs are used for special purposes. A converging collimator will magnify the image of a small organ (Figure 10.6b). A variation sometimes used to image the brain is a cone-beam collimator. This gives improved sensitivity and resolution. However, these collimators introduce distortion because the magnification factor depends on the distance from the object plane to the collimator and is therefore different for activity in different planes in the object. There are also variations in resolution and sensitivity across the field of view as the hole geometry varies from being almost parallel at the centre to highly angled near the edge.

To image small objects a pinhole collimator which functions in a manner analogous to the pinhole camera may be useful (Figure 10.6c). The pinhole is a few millimetres in diameter and effectively limits the gamma rays to those passing through a point. The ratio of the size of image to the size of object will depend on the ratio of the distance of the image plane from the hole to the distance of the object plane from the hole. The latter distance must be small if reasonable magnification is to be achieved. The thyroid gland is the organ most frequently imaged in this way. Note that the pinhole collimator suffers from the same distortions as converging collimators.

10.2.1.3 Pulse Processing

Pulse arithmetic circuits convert the outputs from the PMTs into three signals, two of which give the spatial co-ordinates of the scintillation, usually denoted by X and Y, and the third the energy of the event Z (see Figure 10.4).

Each PMT has two weighting factors applied to its output signal, one producing its contribution to the X co-ordinate, the other to the Y co-ordinate. Several different mathematical expressions have been suggested for the shape of the weighting factors. Those which give the greatest weight to PMTs nearest to the event are to be preferred since they will be the largest signals and hence least susceptible to statistical fluctuations due to noise (for fuller discussion see Sharp et al. 1985). The final X and Y signals are obtained by summing the contributions from all tubes.

Insight

Positional Signal Calculation

A simple method of demonstrating the positional calculation is shown in Figure 10.9. This assumes that the field of view of each PMT is triangular, dropping to zero at the centre of each adjacent tube (see Figure 10.9a). If the signals from all the tubes are simply summed, this produces the output shown in Figure 10.9b. This is the energy signal Z. To obtain useful positional information, the output must vary linearly with x. Therefore, the weighting factors ω_j are used. In the case shown in Figure 10.9c the weighting factors are $\omega_1 = 2$, $\omega_2 = 1$, $\omega_3 = 0$, $\omega_4 = -1$, $\omega_5 = -2$.

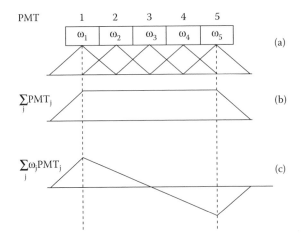

FIGURE 10.9
The use of weighting factors in a positional signal calculation. This example shows a calculation for the *x*-axis. A similar calculation would be carried out for the *y*-axis. (a) A linear array of 5 PMTs; (b) Simply summing the signals produces an output which is independent of *x*, except at the edges of the array; (c) The sum of the weighted signals produces an output which varies linearly with *x*.

As the weighting factors are energy dependent, allowance must be made for this by using a ratio circuit for the final positional calculation. The positional signal for X is then expressed as

$$X = \frac{\sum_j \omega_j PMT_j(x, y)}{\sum_j PMT_j(x, y)}$$

where the denominator is the energy signal Z.

The energy signal Z is produced by summing all the unweighted PMT signals. This signal is then subjected to pulse height analysis as described earlier in this section and the XY signal is only allowed to pass to the processing system if the Z signal falls within the preselected energy window.

10.2.1.4 Correction Circuits

Image quality has been improved considerably in recent years by using microprocessor technology to minimise some of the defects that are inherent in a gamma camera. Exact methods vary from one manufacturer to another, the examples given below illustrate possible approaches.

Spatial distortion may be corrected by imaging a set of accurately parallel straight lines aligned with either the X- or Y-axis. The deviation of the measured position of each point on a line from its true position can be measured and stored as a correction matrix which may then be applied to any subsequent clinical image.

Similarly any variation in the energy signal with the position of the scintillation in the crystal can be determined by imaging a flood source and recording the counts in two narrow energy windows situated symmetrically on either side of the photopeak. If the measured photopeak coincides exactly with the true photopeak, the counts in each energy window will be the same. If the measured peak is shifted to one side, this will be reflected in a higher number of counts in the corresponding window. Once again variations in the measured photopeak from the true photopeak can be stored as a correction matrix for each part of the crystal.

Finally it is important to monitor and adjust the gains of the PMTs. One way to do this is to use light emitting diodes to flood the crystal with light.

10.2.1.5 Image Display

The output events generated by the pulse processing circuitry are recorded in a *digitised image* in which the image space is sub-divided into a matrix of pixels (usually 64 × 64 or 128 × 128 or 256 × 256 in nuclear medicine investigations). The total number of counts in each pixel is then recorded.

Early gamma cameras recorded events onto photographic film, but also incorporated an analogue to digital converter which permitted the final analogue image to be converted into a digital image when necessary. However, modern systems are fully digitised and are frequently referred to as *digital cameras*. Analogue to digital converters are fitted to each PMT thereby digitising much earlier in the imaging process. Block diagrams showing the components of three generations of gamma cameras equipped with reasonably comprehensive data processing facilities are shown in Figure 10.10.

The acquisition and processing terminals are generally based on Windows- or Linux-driven PCs with proprietary software running in the foreground. These enable the operator to control all aspects of the acquisition, such as single or multiple energy windows, or whether to terminate the image acquisition based on elapsed time or total count acquired.

10.2.2 Additional Features on the Modern Gamma Camera

10.2.2.1 Dual Headed Camera

One way to achieve greater sensitivity is to increase the area of crystal available for stopping gamma rays. This is one of the features of dual headed cameras which contain two high resolution rectangular detectors capable of acquiring full field of view anterior and posterior images simultaneously. The increased sensitivity may be used either to permit faster imaging times or to achieve the same counts in the image in the same imaging time with half the administered activity, and hence half the dose to the patient. In general, the imaging is reduced because a major limiting factor to image quality is patient movement and this is less likely to be significant if the imaging time is kept as short as possible.

The detector heads of these systems can be moved along a number of axes to allow a high degree of flexibility in terms of patient positioning. Typically, the heads can be positioned at 90° or opposed to one another for tomographic work. The detectors can often also be placed side by side for imaging patients on beds or trolleys and rotated perpendicular to the gantry for dynamic or static imaging of seated patients. Dedicated triple-headed camera systems are also available for brain work, although these do not have the flexibility of the dual headed systems.

10.2.2.2 Whole Body Scanning

This is a method to obtain a whole body scan, especially a bone scan, in a single image. Either the detector moves on rails along the length of the patient or, more typically, the

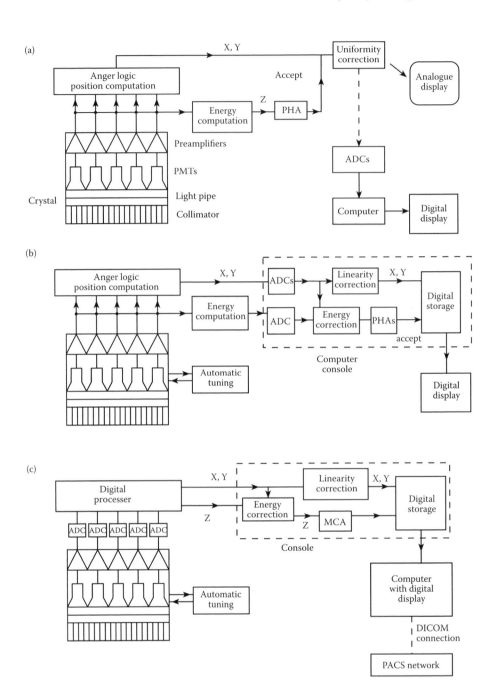

FIGURE 10.10
Block diagrams showing the development of data processing facilities with a gamma camera. (a) 1970s. Anger logic used to position events, as described in the text; pulse height analyser (PHA) discriminated against scatter; displays were mainly analogue with the occasional option of analogue to digital conversion (ADC). (b) 1980s to mid 1990s. PMTs were tuned individually; stand alone computer consoles were introduced with ADC as standard; two or three PHAs were provided to allow more than one gamma ray energy to be collected; linearity correction was introduced; images stored digitally. (c) Mid 1990s. ADCs fitted to the output from each PMT/ pre-amplifier; signals processed digitally throughout; multichannel analysers allow photons at many gamma ray energies to be collected simultaneously; networking becomes a possibility.

bed moves under the detector and the Y position signals have an offset direct current (DC) voltage signal applied to them which is a function of detector position. All the data are collected in a single pass of the camera over the patient thereby producing a non-overlapping image, hence facilitating interpretation, in the shortest possible scan time. Scan rates are typically 8–12 cm per min, so an average 1.7 m adult can be imaged in 15–20 min.

The large rectangular detectors of modern systems are big enough to cover the lateral field of view in a single pass, in all but the most extreme cases. Where the field of view is insufficient, the quality of the scan is likely to be poor in any case due to the degree of scatter and attenuation encountered in very large patients.

Most systems now incorporate automatic contouring to minimise patient-detector separation. This allows the resolution within the image to be kept approximately constant throughout by maintaining a constant distance between the patient and the detector. If the separation between detectors is constant, resolution will be degraded at the points where the detector is furthest from the patient. The contouring techniques used include infra-red beam, electrical impedence or 'learn mode' systems.

10.2.2.3 Tomographic Camera

As mentioned earlier the slip-ring design favoured for most current generation gamma cameras is ideally suited to tomographic imaging. By rotating the gamma camera around the patient and collecting data either continuously or at a fixed number of angles, a set of profiles may be collected and reconstructed to form sectional images. The technique of single photon emission computed tomography is discussed further in Section 10.5.

10.2.2.4 The Cardiac Camera

A recent development is the introduction of a number of compact dedicated cardiac camera systems. The marketing of these is addressed at reducing waiting time for cardiac imaging by allowing nuclear medicine departments to purchase a comparatively low cost system which will fit into a reduced space and can be used alongside the existing gamma camera systems.

Cardiac cameras typically have small field of view detectors (about 30 cm), and many have fixed detectors arranged in a 90° 'L-shaped' configuration. Since they are only used with radionuclides that emit low energy gamma rays, for example, Tc-99m and thallium-201 which emits gamma rays at 80 keV, the NaI (Tl) crystal can be thinner, typically 6–9 mm. Solid-state detector systems are now also becoming available with the inherent advantages that these bring in terms of energy resolution and sensitivity. Performance of most cardiac systems is comparable with that of a standard camera.

10.3 Factors Affecting the Quality of Radionuclide Images

As with all diagnostic images, the radiologist should always ask of radionuclide images 'Are the pictures of good quality—and if not, why not?' The numerous factors that affect image quality will now be considered.

10.3.1 Information in the Image and Signal to Noise Ratio

It is well known that the quality of an image depends on the number of photons it contains. Figure 10.11 shows three images of a simple phantom used in nuclear medicine with different numbers of counts in each image. Unfortunately, because injected radioactivity spreads to all parts of the body and is retained for several hours (in contrast to X-rays which can be confined to the region of interest and 'switched off' after the study), *in vivo* nuclear medicine investigations are always photon-limited by the requirement to minimise radiation dose to the patient. The number of useful gamma rays is further reduced by the heavy collimation that has to be employed. Hence a typical photon density in radionuclide imaging is of the order of 10 mm^{-2} compared with about 10^5 mm^{-2} in radiography and 10^{10} mm^{-2} in conventional photography.

Since the photon count is low, a minimum value for the noise in the image will be the Poisson error $n^{1/2}$ on the measured counts n (there will be other sources of noise). By considering a simple model of a small spherical region of activity in a uniform background of lower activity, it is possible to show (Sharp et al. 1985) that the signal to noise ratio depends on (i) the square root of the total counts, (ii) the lesion to background concentration ratio, (iii) the square root of the sensitivity of the imaging device.

Thus the information in the image can, in theory, always be increased by increasing the time of data collection but the signal to noise ratio only increases as the square root of time. Also this time will be limited by the length of time the patient can lie still, and the work load on the camera. In some situations physiological factors, for example, heart movement, may negate the potential gain in image quality from a long data collection time. Thus the primary objective is to obtain the maximum number of counts in the image in a given time with the maximum differential uptake into the organ or lesion of interest, subject to the limitation of an acceptable radiation dose to the patient. In achieving this objective, choice of radionuclide and choice of radiopharmaceutical are the two main factors to be considered.

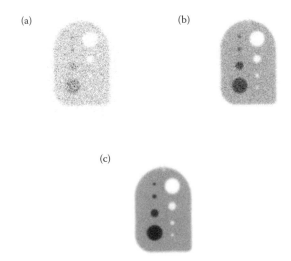

FIGURE 10.11
Images of a phantom at different count densities—(a) 20 kilocounts, (b) 200 kilocounts, (c) 2000 kilocounts. The cold areas on the left and hot areas on the right become sharper as the number of counts in the image increases.

10.3.2 Choice of Radionuclide

It is important to use a short half-life radionuclide so that, for a given injected activity, the radiation dose to the patient is as low as possible. Note, however, that the half-life should not be too short compared with the planned duration of the study, otherwise problems may arise with decay. Also, very short half-life materials may be subject to problems of availability.

For the same reason a radionuclide which decays to a non-radioactive or very long half-life daughter should be chosen. If both parent and daughter are radioactive, the ratio of their activities at equilibrium is the inverse ratio of their half-lives. Thus a long half-life daughter is excreted before any significant dose can arise from its decay.

The radionuclide selected should emit no β particles or, even worse, α particles. These would be stopped in the body due to their high linear energy transfer, adding to the radiation dose, but contributing nothing to the image.

The radionuclide should also have a high 'k' factor, the factor relating exposure rate and activity (see Section 13.7.3). This may seem paradoxical at first since a high 'k' factor implies a large absorbed dose. However, a high 'k' factor also means there are a large number of gamma rays being emitted and hence available to contribute to the image. Some radionuclides decay by more than one mechanism, so decay which produces useful gamma rays is preferable to decay which causes a dose but produces no gamma rays and hence no useful information.

Only gamma rays within a limited energy range are well suited to *in vivo* imaging. For example, they must be sufficiently energetic not to be absorbed in the patient—a lower practical limit is about 80 keV. Conversely, the gamma rays must be stopped in the detector or they will be wasted. The crystal used in a gamma camera becomes inefficient above about 300 keV (Table 10.1).

A range of radionuclides is used in diagnostic imaging (Table 10.2) but well over 90% of routine investigations are performed with Tc-99m. In addition to its short half-life and near monoenergetic gamma ray at 140 keV, Tc-99m emits no particulate radiations and decays to a long half-life daughter (Tc-99, $T_{1/2} = 2 \times 10^5$ years).

Availability of the 6 h half-life material is not a problem because it is possible to establish a generator system. As explained in Section 1.7, equilibrium activity in the decay series is governed by the activity of Mo-99 which has a half-life of 67 h.

$$\ce{^{99}_{42}Mo} \xrightarrow{\beta^- \, (67\,h)} \ce{^{99m}_{43}Tc} \xrightarrow{\gamma^- \, (6\,h)} \ce{^{99}_{43}Tc} \xrightarrow{\beta^- \, (2 \times 10^5 \, y)}$$

A Mo-Tc generator consists of Mo-99 adsorbed onto the upper part of a small chromatographic column filled with high grade alumina (Al_2O_3). When 0.9% saline solution is passed down the column, the Mo-99 remains firmly bound to the alumina but the Tc-99m, which is chemically different, is eluted. Essential features of a generator system are shown in Figure 10.12. Since the Tc-99m builds up fairly rapidly (see Section 1.7), it is possible to elute the column daily to obtain a ready supply of Tc-99m (Figure 10.13). The generator can be replaced weekly, by which time the Mo-99 activity will have decreased significantly.

The Mo-99 required for manufacture of the generator systems is reactor produced at a small number of sites worldwide. The vulnerability of this supply has become apparent in recent years as the reactors age and become more susceptible to unexpected failures. A number of unscheduled interruptions to generator manufacture have occurred with serious consequences for the nuclear medicine community. Obtaining a secure and reliable supply of molybdenum is now one of the main issues facing nuclear medicine in the coming years.

TABLE 10.2

Properties of Some Radionuclides Used for *In Vivo* Imaging

Nuclide	Half-Life	Type of Emission	Example of Use
Carbon-11	20 min[a]	β[+] giving 511 keV γ rays	[c]CO_2 for regional cerebral blood flow
Nitrogen-13	10 min[a]	β[+] giving 511 keV γ rays	[c]Amino acids for myocardial metabolism
Oxygen-15	2 min[a]	β[+] giving 511 keV γ rays	[c]Gaseous studies with labelled O_2, CO_2 and CO, labelled water
Fluorine-18	110 min[a]	β[+] giving 511 keV γ rays	Fluorodeoxyglucose for glucose metabolism
Gallium-67	72 h	92 keV, 182 keV, 300 keV γ rays	Soft tissue malignancy and infection
Technetium-99m	6 h[d]	140 keV γ rays	Numerous
Indium-111	2.8 day	173 keV, 247 keV γ rays	Labelling blood products
Iodine-123	13 h	160 keV γ rays	Thyroid and brain receptor imaging
Iodine-131	8.0 day	360 keV γ rays, β[-] particles	Metastases from carcinoma of thyroid
Xenon-133	5.3 day[e]	81 keV γ rays, β[-] particles	Lung perfusion studies
Thallium-201	73 h	Orbital electron capture[b] 80 keV X-rays and Auger electrons	Cardiac infarction and ischaemia

[a] Cyclotron produced positron emitter—see Chapter 11.
[b] T1–201 decays by orbital electron or K shell capture. This is an alternative to positron emission when the nucleus has too many protons and adjusts the balance by capturing an electron from the K shell. The initial capture process may not result in any emission of radiation but characteristic X-rays will be emitted as the vacancy in the K shell is filled. If the atomic number of the element is high enough (e.g. thallium Z = 81), this characteristic radiation may be of high enough energy to be useful for imaging.
[c] Not widely used at present.
[d] Generator produced. Note short half-life radionuclides that cannot be produced on site are of limited value for *in vivo* imaging.
[e] Since Xe-133 is used in gaseous form, the biological half-life is very short so the β[-] particle dose is small.

Insight

Potential Dose to the Patient from Tc-99

As mentioned earlier, the ratio of the activity of a mother-daughter radionuclide pair is the inverse ratio of their half-lives, if both are radioactive. Thus,

$$\frac{A_d}{A_p} \approx \frac{T_{1/2p}}{T_{1/2d}}$$

where A_d is the activity of the daughter, A_p is the activity of the parent, $T_{1/2p}$ is the half-life of the parent and $T_{1/2d}$ is the half-life of the daughter.

For the decay of Tc-99m to Tc-99 the half-lives are $T_{1/2p}$ = 6 hours and $T_{1/2d}$ = 2 × 10⁵ years. Therefore, the ratio becomes

$$\frac{A_{Tc\text{-}99}}{A_{Tc\text{-}99m}} \approx \frac{6}{2 \times 10^5 \times 365 \times 24} \approx \frac{6}{1.8 \times 10^9} \approx 3.4 \times 10^{-9}$$

This demonstrates that only a tiny fraction (approximately 3 parts in a billion) of the radiation dose to the patient is due to the contribution of the Tc-99. The biological half-lives of the radionuclides must also be considered. In this case, the biological half-lives are approximately equal, so the physical decay becomes the significant factor. However, if the daughter radionuclide had a significantly longer biological half-life, then the dose to the patient arising from it may become significant.

10.3.3 Choice of Radiopharmaceutical

For good counting statistics and a high signal to noise ratio, the radionuclide must be firmly bound to an appropriate pharmaceutical and the resulting radiopharmaceutical must achieve a high target:non-target ratio. In addition, it must satisfy criteria that are not generally relevant for non-radioactive drugs. It must be easy to produce, inexpensive,

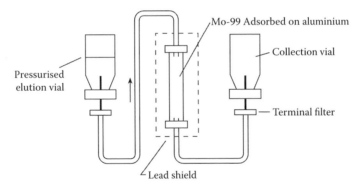

FIGURE 10.12
Simplified diagram of a generator system that operates under positive pressure.

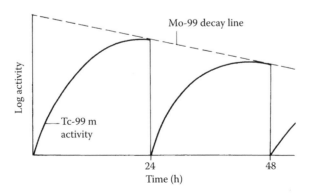

FIGURE 10.13
Curve showing the Tc-99m activity in a Mo-99/Tc-99m generator as a function of time, assuming the column is eluted every 24 h.

readily available for all interested users, have a short effective half-life and be of low toxicity. Very short half-life material may constitute a radiation hazard to the radiopharmacist if it is necessary to start the preparation with a high activity.

Radiopharmaceuticals concentrate in organs of interest by a variety of mechanisms, including capillary blockage, phagocytosis, cell sequestration, active transport, compartmental localisation, ion exchange and pharmacological localisation. The reader is referred to a more specialised text (e.g. Frier 1994) for further details. The exact mechanism of uptake into the organ of interest is often not vitally important, as long as sufficient is accumulated. However, if functional parameters are to be derived and quantified then a deeper understanding of the kinetic model underlying organ uptake is required (Peters 1998).

One disadvantage of Tc-99m is that, being a transition element, it is not easily bound to biologically relevant molecules and its chemistry is complex (Nowotnik 1994). Nevertheless in spite of the difficult chemistry, a wide range of pharmaceuticals has been labelled with Tc-99m (Britton 1995) and good target to background ratios are sometimes achieved. However, poor specificity of radiopharmaceuticals for their target organs remains a weak point in nuclear medicine imaging, with most commonly employed radiopharmaceuticals showing very poor selectivity, generally less than 20% in the organ of interest.

Note that the obvious elements to choose for synthesising specific physiological markers, hydrogen, carbon, nitrogen and oxygen, have no gamma emitting isotopes. Pharmaceuticals containing radioisotopes of some of these elements can be used for PET as discussed in Chapter 11.

10.3.4 Performance of the Imaging Device

Much has been written about the performance of the gamma camera and only the most salient features will be summarised here.

10.3.4.1 Collimator Design

As already explained, resolution and sensitivity are mutually exclusive. The inherently poor sensitivity, which may be as low as 100 cps per MBq for a high resolution, low energy collimator, is a major problem since, as explained in Section 10.3.1, the signal to noise ratio is proportional to the square root of the sensitivity of the imaging device.

10.3.4.2 Intrinsic Resolution

This is determined primarily by the performance of the scintillation crystal. Although a complex problem to treat rigorously, the following simplified explanation contains the essential physics. In principle, by arranging a large number of very small PMTs behind the crystal, one might expect to localise the position of a gamma ray event in the crystal to any required degree of accuracy. However, each 140 keV gamma ray only releases about 4000 light photons and if these are shared between 40 PMTs, the average number n reaching each tube is only 100. The process is random, so variations in the signal due to Poisson statistics of $\pm n^{1/2}$ or ± 10 will ensue.

Some PMTs will get more than 100 photons, some will get a lot less, but the 'error' on the signal from each PMT, which will contribute to the error in positioning the event, will increase rapidly if one attempts to subdivide the original signal too much. Typical modern gamma camera systems with digital heads have an intrinsic resolution of around 3.5 mm.

10.3.4.3 System Resolution

The resolution of the complete system, including the collimator, can be obtained by imaging a narrow line source of radioactivity—for example, nylon tubing of 1 mm internal diameter filled with an aqueous solution of Tc-99m pertechnetate. A result such as that shown in Figure 10.14 would be obtained and the spread of the image can be expressed in terms of the full width at half maximum height (FWHM) calculated as shown. For the arrangement in Figure 10.14 (c), which is perhaps the most realistic, and using a high resolution collimator the FWHM is 8.6 mm, which is substantially greater than the intrinsic resolution.

10.3.4.4 Spatial Linearity and Non-uniformity

The outputs from the PMT array must be converted into the X and Y signals that give the spatial co-ordinates of the scintillation. Any error in this process, caused perhaps by a change in the amplification factor in one PMT, will result in counts being misplaced in the ensuing image. This will result in distortion (or non-linearity) if a narrow line source of radioactivity is imaged, or in non-uniformity for a uniform extended source.

Correction circuits for non-linearity were discussed in Section 10.2.1. Non-uniformity in the image of a uniform flood source is a useful overall measure of the performance of the camera.

Several methods of expressing non-uniformity have been suggested. For example, integral non-uniformity is a measure of the difference between the maximum $U_{I(+)}$ and minimum $U_{I(-)}$ pixel counts in an image and the mean pixel count M

$$U_{I(+)} = 100 \times \frac{C_{max} - M}{M} \%$$

$$U_{I(-)} = 100 \times \frac{C_{min} - M}{M} \%$$

Measurements are usually made over the whole usable field of view and over the centre of the field of view which has the linear dimensions of the whole field of view scaled

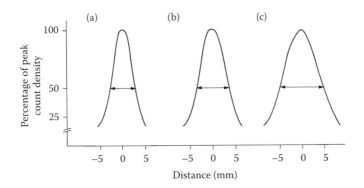

FIGURE 10.14

Derivation of system resolution from line spread function measurements and the effect of distance and scattering material on system resolution. The traces are typical images of a 1 mm line source of Tc-99m obtained under different conditions: (a) no scattering material, source on collimator face, FWHM = 5.7 mm, (b) 5 cm tissue equivalent scattering material, FWHM = 7.1 mm, (c) 10 cm tissue equivalent scattering material FWHM = 8.6 mm.

down by 0.75. The differential non-uniformity U_D is based on the maximum rate of change of count density,

$$U_D = \left(\frac{\Delta C}{C_{mean}} \right) \times 100\%$$

where ΔC is the maximum difference in counts between any two adjacent pixel elements.

The standard deviation (SD) of the pixel counts can also be calculated and the coefficient of variation is $100(SD/C_{mean})$ where C_{mean} is the mean of all pixels within the field of view. This is a useful measure for tracking change in non-uniformity over time. However, it may miss local defects in the image and should always be used in conjunction with visual inspection and another non-uniformity measure.

From the viewpoint of accurate diagnosis, camera non-uniformities must be minimised or they may be wrongly interpreted as real variations in the image count density. As for linear distortion, it is now routine to collect and store a uniformity correction matrix that can be applied to each image. For a well adjusted modern camera, integral non-uniformity over the centre of the field of view should be less than 2%. The introduction of the digitised detector heads has helped greatly in this respect, as they are far more stable than the previous analogue systems.

10.3.4.5 Effect of Scattered Radiation

Although pulse height analysis ought, in principle, to reject all scattered radiation, discrimination is far from perfect. One limitation of the scintillation crystal and PMT combination is that the number of electrons entering the PMT per primary X- or gamma ray photon interaction is rather small. There are two reasons:

1. About 30 eV of energy must be dissipated in the crystal for the production of each visible or ultraviolet photon.
2. Even assuming no loss of these photons, only about one photoelectron is produced for every 10 photons on the PMT photocathode.

Thus to generate one electron the photocathode requires about 300 eV and a 140 keV photon will produce only about 400 electrons at the photocathode. This number is subject to considerable statistical fluctuation ($N^{½} = 20$ or 5%). The result is that a monoenergetic beam of gamma rays will produce a range of pulses and will appear to contain a range of energies.

The result, as shown in Figure 10.15, is that, even in the absence of scatter, monoenergetic gamma rays produce light signals with a range of energies. This spread, expressed as the ratio of the FWHM of the photopeak spectrum to the photopeak energy, is a measure of the energy resolution of the system and is about 10% for a gamma camera at 140 keV. The spectrum is then further degraded by scatter in the patient (dotted curve).

Unscattered photons contribute information about the image so a wide energy window (typically about 20%) must be used. Unfortunately a wide energy window permits some gamma photons that have been Compton scattered through quite large angles, and may have lost as much as 20 keV, to be accepted by the pulse height analyser. The problem is greater for low energy gamma photons, for two reasons.

1. Fewer light photons are produced, so statistical variations are greater
2. The energy lost during a Compton interaction, for fixed scattering angle, is smaller

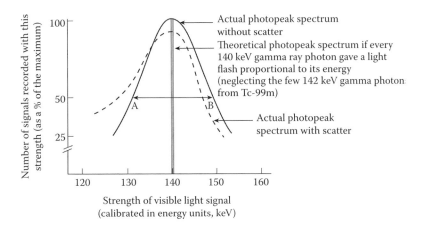

FIGURE 10.15
Graph demonstrating the energy resolution of the NaI(TI) crystal in a gamma camera. The FWHM (AB) is about 14 keV or 10% of the peak energy.

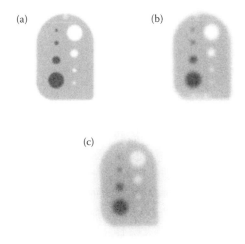

FIGURE 10.16
The effect of distance and scatter on image quality. (a) Test object in contact with collimator face. (b) 10 cm air separation. (c) 10 cm separation and perspex scatter material. In all images 2000 kilocounts were collected (cf Figure 10.11c).

Note that semiconductor detectors (see Sections 4.7 and 4.9) produce a narrower spread and much better energy discrimination. However, as was mentioned previously, these have only been produced commercially for small field of view gamma cameras designed for cardiac imaging.

Modern gamma cameras allow more than one energy window to be set, thus accepting several photopeaks. This can be useful when working with a radionuclide which emits gamma rays at more than one energy, for example, Ga-67 or In-111 or when attempting to image two radionuclides simultaneously.

As shown in Figure 10.16 scattered radiation causes deterioration in the image of a test object, especially when the scattering material also increases the distance to the collimator face. The patient is the major source of scattering material and there is an obvious difference in image quality for say a bone scan of a very thin person when compared with that of an obese person.

10.3.4.6 High Count Rates

The reasons for loss of counts at high count rates and implications for quantitative work will be discussed in Section 10.4.2. Some degradation of image quality also occurs. The main reason is thought to be that the system fails to separate in time two scattered photons whose summed energy falls within the pulse height analyser window. Thus an event is recorded at the weighted mean position of the two scattered photons but in a position unrelated to the activity in the patient.

10.3.5 Data Display

The quality of the final image is also influenced by the performance of the recording medium. Methods of displaying data will be discussed briefly.

10.3.5.1 Persistence Monitor

When setting up a study, an image of the distribution of activity may greatly assist patient positioning. This may be achieved using a video monitor on which each flash of light representing a collected gamma ray persists long enough for a transient image to be formed. This image is ideal for positioning but is unsuitable for diagnosis. Unfortunately, digitised detectors are less suitable for the generation of persistence images. Consequently, the images available for patient positioning have generally decreased in quality on modern systems.

10.3.5.2 Display and Hard Copy

Most nuclear medicine departments are now connected to *picture archiving and communication systems* (PACS). This has led to the rise of the 'filmless' department where reporting is undertaken directly from the screen and any hard copy required is generated from networked dry film printers. Wet film processing is now virtually redundant. See also Chapter 17.

The prevalence of digitised gamma camera systems has placed nuclear medicine in a good position to take advantage of the transition to PACS. However, the move to digital display and storage gives rise to a number of issues which must be fully considered before any change to purely digital archiving is undertaken. These include the storage capacity required, the resolution and performance of the display screens, the quality assurance of these screens and the interface between the PACS and the radiology information system (RIS) used for generation of the patient reports. The Royal College of Radiologists UK working groups have produced a number of useful reports on these issues which can be found on their website (BFCR(08) Reports 4, 6, 7 and 8, 2008). See also Chapter 17.

Compared to typical CT and magnetic resonance imaging (MRI) datasets, nuclear medicine images are generally quite small when transferred into a standard DICOM format. Most are a few tens of megabytes and, therefore, the entire image set can be readily transferred to the PACS. However, there are certain nuclear medicine image formats which do not readily transfer into the DICOM standard at present. Gated tomographic datasets and fused SPECT/CT datasets are some of the more common of these. This issue should be addressed in the next generation of the DICOM standard.

It might be assumed that due to the relatively low resolution of nuclear medicine data compared to CT and MRI, the performance of the display screen is not of such great importance. However, this would be erroneous as many nuclear medicine images are now reported by the radiologist with reference to previous imaging. Therefore, the display screen must be of sufficient performance to allow accurate assessment of images from other imaging modalities. A detailed report on the performance of display screens, the effect of the environment on their utility and their quality assurance has been produced by the American Association of Physicists in Medicine (Samei et al. 2005).

Where it is necessary to produce hard copy images, most departments now use dry carbon-based film which is readily utilised in networkable, high capacity printers. These are generally DICOM compatible which greatly simplifies the networking task and allows output in standard formats.

Digitised images permit graphical data to be produced and sophisticated forms of image processing are possible. The techniques are similar to those discussed in Section 6.11. If the images collected in a dynamic study are to be analysed quantitatively, they must of course be digitised. This aspect will be discussed in more detail in the next section.

10.3.5.3 Grey Scale versus Colour Images

In digitised images a range of count densities may be assigned to either a shade of grey or a spectral colour. Much has been written about the relative merits of grey scale and colour images and this controversial subject cannot be discussed fully here. The following simple philosophy suggests an approach to each type of display. The sharp visual transition from one colour to another may alert the eye to the possibility of an abnormal amount of uptake of radioactivity and colour images can be useful for this purpose. However, by the same token, this colour change may represent an increase or decrease of only one or two counts per pixel and may not be significant statistically. It thus follows that grey scale is generally preferable for unprocessed images such as bone scans but colour can be useful when looking at processed images, especially in functional imaging (see Section 10.4.1).

10.4 Dynamic Investigations

The potential for performing dynamic studies, in which changes in distribution of the radio-pharmaceutical are monitored throughout the investigation, was recognised at an early stage. However, two developments were essential before dynamic imaging became feasible on a routine basis. The first was an imaging device with a reasonably large field of view, sufficiently sensitive to give statistically reliable counts in short time intervals. The gamma camera satisfies these requirements although for dynamic studies it is not uncommon to choose a collimator design that increases sensitivity at the expense of some loss of resolution.

The second development was the availability of reasonably priced data handling hardware and software powerful enough to handle the large amount of data collected. Hence dynamic imaging has only been widely available in general hospitals since the late 1970s.

Important features of dynamic imaging will now be considered under two general headings, but with specific reference to some frequent dynamic investigations.

10.4.1 Data Analysis

Consider as an example the study of kidney function—nowadays usually performed by administering intravenously about 100 MBq of Tc-99m labelled MAG3 (mercapto acetyl triglycine) which is actively secreted by the renal tubules. Historically, fixed detectors of small area or 'probes' were used for such studies but the results were critically dependent on probe positioning. Using a gamma camera, a set of images can be obtained and regions of interest for study can be selected retrospectively. Thus methods of data display have been developed that will show both the spatial distribution of radiopharmaceutical and temporal changes in the distribution.

10.4.1.1 Cine Mode

A useful starting point is to examine individual image frames looking for aspects that require further detailed study, for example, in renograms to look for evidence of patient movement. This can be done by running a 'cine-film' of the frames using a continuous loop so that the display automatically returns to the start of the study and continues until interrupted. Data are usually collected in about sixty 20 s frames but it is sometimes useful to expand the initial time frame by collecting 1 s frames for the first minute.

10.4.1.2 Time-Activity Curves

The system will plot activity as a function of time for regions of interest selected by the operator. Examples of such regions of interest and the resulting curves, again taken from renography, are shown in Figure 10.17. Some other features of this apparently simple procedure must be mentioned. To achieve a better image (improved counting statistics) on which the regions of interest can be drawn, it may be necessary to add several sequential frames. Data smoothing will also help to keep noise to a minimum. Also it is important to

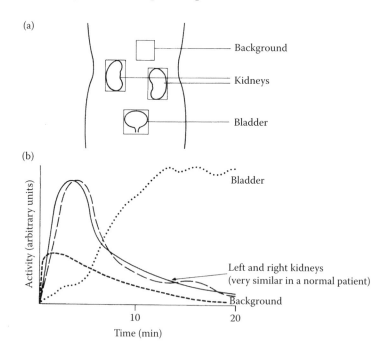

FIGURE 10.17
(a) Schematic drawings of regions of interest around kidneys and bladder for a renogram study. A background region is also shown. (b) Typical activity—time curves for such regions of interest.

subtract background counts arising from activity in overlying and underlying tissue. This is usually done by defining a region of interest representative of blood background.

10.4.1.3 Deconvolution

Although radioactivity is injected intravenously as a bolus, after mixing with blood and passing through the heart and lungs, it arrives at the kidneys over a period of time. Also some activity may recycle. Thus the measured activity-time curve is a combination (convolution) of a variable amount of activity and the rate of handling by the organ. It is possible to measure the mean transit time for the organ and deconvolution is a mathematical technique that offers the possibility for removing arrival time effects and presenting the result as for a single bolus of activity. However, in nuclear medicine the presence of noise limits the power of deconvolution methods. It has been shown that the handling of the noise by smoothing greatly increases the variability in measurements of this type (Houston et al. 2001) and they should, therefore, be treated with some caution.

10.4.1.4 Functional Imaging

Some dynamic studies now produce a large number of images, and although it may be important to examine these images in cine mode, methods of data compression will be required before a particular quantitative feature can be visualised, perhaps on a single image.

One approach to this problem is to choose a feature thought to be of physiological interest, for example, the time of maximum on a renogram curve, and then to examine the data sets, pixel by pixel mapping the time of maximum. The resulting image is a 'functional image' since it displays a feature of physiological rather than anatomical interest.

The most extensive use of functional imaging currently is probably in the analysis of multiple gated acquisition studies of the cardiac blood pool (MUGA). The patient's own red blood cells are labelled with Tc-99m and the gamma camera collects data in frames which are gated physiologically to the signal from an electrocardiograph attached to the patient. For a patient with a regular heart rate, 16–24 equally spaced frames between each R wave signal can be collected. The data may be collected in 'list' mode. That is to say the (x, y, t) co-ordinates of every scintillation accepted by the electronics are registered for subsequent sorting into the correct pixel and time frame. The number of counts collected during one cardiac cycle would be far too small for the statistics to be reliable so collection continues for several hundred cycles until there are about 200,000 counts per frame. Protocols vary between centres. One simple, standard procedure collects anterior and left anterior oblique (LAO) views only, with attention focused on the left ventricle, best seen in the LAO view.

One physiological feature of importance is the change in volume of the compartments of the heart through the cycle, especially of the ventricles. This property may be derived from a plot of the time-activity curve (TAC) which is an average cardiac cycle formed from the hundreds of cycles acquired. As shown in Figure 10.18 to a first approximation this may be assumed to be sinusoidal and the variation in counts $C(t)$ in a region of interest with time t is given by

$$C(t) = a + b \sin(2\pi ft + \phi)$$

where a is a baseline constant, b is the amplitude of the motion, f is the reciprocal of the number of frames per cardiac cycle and ϕ is the phase of the motion.

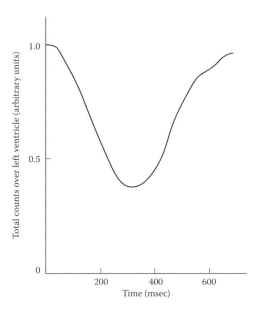

FIGURE 10.18
Time-activity curve over the left ventricle in a normal patient.

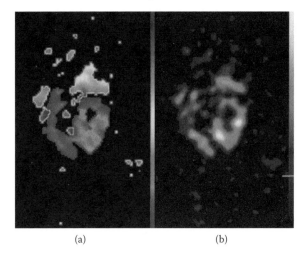

(a)   (b)

FIGURE 10.19
(See colour insert.) Functional image of a normal MUGA study. (a) The phase image represents the phase (or timing) of the contraction of the heart chambers. (b) The amplitude image represents the amplitude of the contraction. In this case it can be seen that the largest contraction occurs apically and along the lateral wall. Both of these parameters are derived from the Fourier fitted curve (Figure 10.18).

The most important quantity derived from the left ventricle TAC is the ejection fraction

$$LVEF = \frac{ED - ES}{ED - bgnd}\%$$

where ED = region of interest (roi) counts at end diastole, ES = roi counts at end systole. The choice of roi boundaries and the background (bgnd) roi are critical.

All the data may be analysed pixel by pixel to produce two functional images, one of which shows the phase of each part of the heart motion, the other showing the amplitude. Note that in practice the curves will not be truly sinusoidal but methods of Fourier analysis may be used to introduce further refinements if necessary. Such images are best displayed using a colour scale rather than monochrome, as the relative magnitude of the parameter in different parts of the image is more readily appreciated.

It can be seen in Figure 10.19 that for this normal patient, the phase is similar throughout each chamber and the atria are contracting out of phase with the ventricles. The amplitude (volume contraction) image demonstrates that the highest contraction is occurring apically and along the lateral wall.

A major difficulty with functional imaging is the choice of a reasonably simple mathematical index that is relevant to the physiological condition being studied.

10.4.2 Camera Performance at High Count Rates

When a dynamic study requires rapid, sequential imaging, the gamma camera may have to function at high count rates to achieve adequate statistics. However, the system imposes a number of constraints on maximum count rate. For example, the decay time of the scintillation in the crystal has a time constant of 0.2 μs and about 0.8 μs is required for maximum light collection. The electronic signal processing time is also a major limitation. The signal from the camera pre-amplifier has a sharp rise but a tail of about 50 μs so pulses have to be truncated or 'shaped' to last no longer than about 1 μs. The pulse height analyser and pulse arithmetic circuits have minimum processing times and it may be an advantage to by-pass the circuits that correct for spatial non-linearity, if any, to reduce processing time.

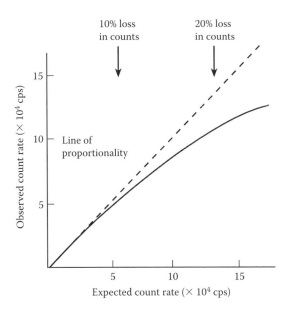

FIGURE 10.20

Curve showing how the various dead times in the camera system result in a count rate losses. In this example 10% and 20% losses in counts occur at about 6×10^4 and 13×10^4 cps, respectively. The maximum observed count rate would be about 30×10^4 cps after which the recorded count rate would actually fall with increasing activity.

For all these reasons, if sources of known, increasing activity are placed in front of a gamma camera under ideal conditions with no scattering material, a graph of observed against expected count rate for a modern camera might be as shown in Figure 10.20. In practice, performance would be inferior to this because the camera electronics has to handle a large number of scattered photons that are subsequently rejected by the pulse height analyser. Thus the exact shape of curve is very dependent on the thickness of scattering material and the width of the pulse height analyser window. Loss of counts can occur at count rates as low as 5×10^4 cps and this may be a problem when making quantitative measurements.

However, the likelihood of encountering count rate problems in routine clinical practice has greatly reduced in recent years. The developments in modern digital processing electronics have reduced the electronic processing time and, thus, increased the incident count rates at which significant losses occur. There is also a reduced clinical requirement for studies where high count rates may be encountered, such as first-pass cardiac studies. These have been replaced by non-radioactive alternatives including trans-oesophageal echocardiography.

10.5 Single Photon Emission Computed Tomography (SPECT)

The principles of SPECT are similar to those of CT which were covered in Section 8.3. As in CT, a number of projection views are obtained from many different angles around the object, then reconstructed to provide a representation of the detail within the object. However, in SPECT the detail obtained is not anatomical, but is a map of the concentration of the administered radionuclide which is varying continuously throughout the volume of interest.

If $C(x, y)$ is the number of counts per unit time recorded in a normal gamma camera image at an arbitrary point (x, y, z), this is related to the concentration of radioactivity (activity per unit volume $A(x', y', z)$) at some arbitrary point (x', y', z) in the same slice by the equation

$$C(x,y) = \int_{\substack{\text{all of the volume} \\ \text{occupied by activity}}} A(x',y',z)S(x-x',y-y',z)\, e^{-\mu t} \, dx'\, dy'\, dz$$

where $S(x-x', y-y', z)$ represents the response of the detector (at x, y, z) to a point source of activity (at x', y', z), μ is the linear attenuation coefficient of the medium and t is the thickness of the attenuating medium traversed by the gamma rays. The recovery of the function $A(x', y', z)$ for the whole slice from the available data $C(x, y)$ represents a complete solution to the problem.

Use of a single value of μ is of course an approximation. Ideally, it should be replaced by a matrix of values for the linear attenuation coefficient in different parts of the slice. To assist in obtaining these values, gamma camera manufacturers are now producing SPECT/CT systems where the tomographic gamma camera has a small CT system attached. Many of these are low-dose, non-diagnostic CT systems, but some are fully diagnostic systems. These have additional utility for imaging, similar to the PET/CT systems (see Chapter 11), but create a number of new problems for nuclear medicine departments in terms of radiation protection, room shielding and staff training.

Three fundamental limitations on emission tomography can be mentioned. The first is collection efficiency. Gamma rays are emitted in all directions but only those which enter the detector are used. Thus detection efficiency is severely limited unless the patient can be surrounded by detectors. As has been mentioned earlier, in Section 10.2.2, the development of dual headed camera systems has led to some improvement in this regard, although efficiency remains a fundamental limitation.

The second limitation of SPECT is attenuation of gamma rays within the patient (see Insight). The third limitation is common to all nuclear medicine studies, namely that the collection time is only a small fraction of the time for which gamma rays are emitted. Hence the images are seriously photon limited.

Insight

More on Attenuation Correction

Allowances for attenuation within the patient can be made and corrections simplified by adding counts registered in opposite detectors. As shown in Figure 10.21, for a uniformly distributed source a correction factor $\mu L/(1 - e^{-\mu L})$ where L is the patient thickness can be applied. However, experimental work indicates that the value of μ is neither that for narrow beam attenuation, nor that for broad beam attenuation, but somewhere between the two. Note—this analysis does not apply for a very non-uniform distribution. The reader can convince themselves of this by considering a point source that is not mid-way between A and B.

Although SPECT/CT systems have an advantage in determining the value of μ, there are still issues of accuracy regarding the narrow versus broad beam conundrum and the conversion from the values obtained at the kVp of the X-ray system to the monoenergetic photopeak energy of the radionuclide being used.

Note that accurate attenuation correction is only essential when SPECT is used quantitatively. Uncorrected images are generally acceptable for qualitative interpretation.

The projection data required are normally collected by rotating the gamma camera around the patient. Data is collected at fixed angular increments, typically 3° or 6°, for 15–20 min or during continuous rotation for the same time. The large field of view of modern detectors allows a large volume of the patient to be imaged in a single acquisition. The flexible positioning of the detector heads also allows acquisitions to be carried out with the heads at 90° to one another. This is typically used for cardiac SPECT where data are only acquired over a 180° arc, and allows the acquisition time to be reduced by a factor of two.

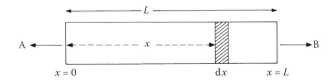

FIGURE 10.21

Correction factor to be applied for gamma ray attenuation in the patient when the source of radioactivity is uniformly distributed. If the total activity is I, then the activity per unit length is I/L and the activity in the strip dx is Idx/L. The signal recorded at A is $(I/L) \int_0^L e^{-\mu x} dx$ where μ is the linear attenuation coefficient of the medium. The signal recorded at B is $(I/L) \int_0^L e^{-\mu(L-x)} dx$. Both expressions work out to $(I/L)((1 - e^{-\mu L})/\mu)$ and since, in the absence of attenuation, the signal recorded at A and B should be I, the total activity in the strip, the required correction factor is $(\mu L)/(1 - e^{-\mu L})$.

Tomography places more stringent demands on the design and performance of gamma cameras than conventional imaging. For example, multiple views must be obtained at precisely known angles and the centre of rotation of the camera must not move, for example under its own weight, during data collection. The face(s) of the camera must remain accurately parallel to the long axis of the patient and the mechanical and electronic axes of the camera must be accurately aligned. Camera non-uniformities are more serious than in conventional imaging since they frequently reconstruct as 'ring' artefacts. If views are corrected with a non-uniformity correction matrix collected at a fixed angle, care must be taken to ensure that the pattern of non-uniformity does not change with camera angle. Such changes could occur, for example as a result of changes in PM tube gain due to stray magnetic fields.

For all these reasons, especially very poor counting statistics, resolution is inferior in SPECT to that in conventional gamma camera imaging and much inferior to CT. Resolution decreases with the radius of rotation of the camera. This is because the circumference is $2\pi r$ and for N profiles the sampling frequency at the edge is $N/2\pi r$. For body sections the resolution is about 8–10 mm so $3°$ sampling ($N = 60$) is quite adequate. For example, with objects 20 cm in diameter $N/2\pi r \approx 0.1$ mm^{-1}. If this is the minimum sampling frequency, then by the Nyquist theorem (see Section 8.4.2) the resolution limit set by sampling is $\approx 1/2\nu_{\mathrm{m}} \approx 5$ mm which is less than the limit set by other factors.

The clinical demand for SPECT has increased markedly in recent years as the development of slip-ring, dual headed gamma cameras has made data acquisition faster and simpler. Studies such as myocardial perfusion and regional cerebral blood flow imaging, which have to be carried out tomographically, have made up much of this demand. However, SPECT is often now carried out on studies such as bone scans, for which it previously would not have been considered feasible, due to the improvements in acquisition and processing systems. Table 10.3 contains a list of the ten most frequently performed examinations in a typical Nuclear Medicine Department of a University Teaching Hospital during 2008. This shows that a significant number of these examinations are performed tomographically.

The gamma camera manufacturers are now providing dedicated processing and analysis packages for many SPECT applications. Recent developments have included the production of resolution recovery software. This allows the production of reconstructed data of similar quality to the current standard using a reduced acquisition time. Times can typically be reduced by at least a half whilst maintaining image quality, allowing a significant increase in patient throughput.

10.6 Quality Standards, Quality Assurance and Quality Control

Quality standards are the standards that must be applied to the individual elements of a nuclear medicine service to ensure an agreed standard for the service overall. Quality assurance embraces all those planned and systematic actions necessary to provide confidence that a structure, system or component will perform satisfactorily in service. Quality control is the set of operations intended to maintain or improve quality. Although these are clearly separable aspects of quality, they are closely related and will be considered together in this brief overview of those aspects of the provision of a high quality nuclear medicine service where physical principles are important.

TABLE 10.3

Ten Most Frequently Performed Examinations in the Nuclear Medicine Department of a Teaching Hospital during 2008

Study	Radiopharmaceutical	Study Type	Number of Investigations
Bone scan	^{99m}Tc-MDP[a]	Wholebody	1769
		SPECT	28
Myocardial perfusion scan	^{99m}Tc-Tetrofosmin[b]	SPECT	1706
Ventilation and perfusion lung scan	^{99m}Tc-DTPA[c] aerosol ^{99m}Tc-MAA[d]	Static	1163
MUGA scan	^{99m}Tc-Pertechnetate[e]	Dynamic	641
Regional cerebral blood flow scan	^{99m}Tc-HMPAO[f]	SPECT	600
Renal function and drainage	^{99m}Tc-MAG3[g]	Dynamic and static	351
Sentinel node imaging and biopsy	^{99m}Tc-Nanocoll[h]	Static	277
Renal cortical anatomy	^{99m}Tc-DMSA[i]	Static	215
White cell scan for inflammatory processes	^{99m}Tc-HMPAO[f] labelled leucocytes	Wholebody	208
Dopamine transporter for Parkinsonian syndromes	^{123}I-Ioflupane[j]	SPECT	45

[a] Methylene diphosphonate
[b] 1,2-*bis*[di-(2-ethoxyethyl)phosphino]ethane
[c] Diethylene triamine pentaacetic acid
[d] Macro-aggregates of human albumin
[e] Oxoanion of ^{99m}Tc-TcO$_4^-$
[f] Hexamethyl propylene amine oxime
[g] Mercapto acetyl triglycine
[h] Human serum albumin colloid
[i] Dimercapto succinic acid
[j] Fluoropropyl carbomethoxy iodophenyl nortropane

The principal aim is to ensure that the requisite diagnostic information is obtained with the minimum dose to the patient. Prime areas of concern are the accurate measurement of the administered activity and the performance of the imaging device. However, consideration must also be given to the safety of staff, other patients and the public, especially since the patient themselves becomes a source of radiation, and to the release of radioactive waste to the environment.

10.6.1 Radionuclide Calibrators and Accuracy of Injected Doses

The main components of a radionuclide calibrator will be a well-type ionisation chamber (see Section 4.3), stabilised high voltage supply, electrometer for measuring the small ionisation currents, processing electronics and a display device. In the nuclear medicine department it will be used for a variety of purposes including the following:

(a) Determination of radiopharmaceutical activities after delivery by the manufacturer
(b) Dosage of solution for injection and oral application
(c) Checking eluate activities from generators (e.g. Tc-99m)
(d) Determination of attenuation of different materials, for example, glass and plastic containers
(e) Calibration of measuring equipment

A protocol for establishing and maintaining the calibration of medical radionuclide calibrators and their quality control has been prepared by the National Physical Laboratory (Gadd et al. 2006). Drawing on experience of traceability to national standards in radiotherapy, the report gives recommended methods for calibrating reference instruments at a large regional centre and for checking field instruments. It also gives guidelines on the frequency of quality control tests and acceptable calibration tolerances for both types of instrument. A 5% limit on overall accuracy is a reasonable practical figure for a field instrument.

A number of variables can cause significant errors in radionuclide calibrators. Two are particularly important.

(a) The container size and shape, and volume of fluid can be a problem with beta and low energy gamma emitters because of self-absorption. Even the thickness of the vial can affect calibration. These variables are less of a problem for energies above about 140 keV but it is important that the gamma ray energy being used for imaging is also the one being used for calibration by the calibrator. For example, the 160 keV gamma ray from I-123 is used for imaging but unless special precautions are taken to filter out low energy radiation, the calibrator will respond to the low energy characteristic X-rays at 35 keV.

(b) Contamination by other radionuclides can seriously affect calibrator accuracy if the calibrator is much more sensitive to the contaminant than to the principal product. Two examples are given in Table 10.4.

For Tl-201 1.5% contamination will overestimate the activity by 13%. For Sr-85, as little as 0.2% Sr-89 will overestimate the activity by 38%.

Great precision in respect of the isotope calibrator is of little value if there is uncertainty or variation in the amount of activity actually administered to the patient. Some possible causes for the wrong activity being given would be

(a) Variation in the volume injected

(b) Improper mixing of the radiopharmaceutical, for example, macroaggregates, colloid

(c) Retention of the radiopharmaceutical in the vial (stickiness) for example, methoxyisobutylisonitrite (MIBI), Tl-201

TABLE 10.4

Effect of Contaminants on Accuracy of Calibrator Measurement of Radionuclide Activity

Radionuclide	Typical Activity Present (%)	Calibrator Sensitivity pA MBq^{-1}
Tl-200	0.5	2.56
Tl-201	98.5	0.86
Tl-202	1.0	5.03
Sr-89	0.2	5.26
Sr-85	99.8	0.028

It is important to check that all operators can draw up and inject a specified volume, say 1 ml, accurately. With a syringe shield in place it is possible for an inexperienced operator to draw up almost no fluid at all.

10.6.2 Gamma Camera and Computer

The primary goal of quality control of the gamma camera and computer is to provide the physician and technologist with an assurance that the images produced during clinical studies accurately reflect the distribution of radiopharmaceutical in the patient.

Five elements can be identified in a good quality control programme—test to be performed, approximate frequency, accuracy and reproducibility of tests, record keeping and action thresholds.

Table 10.5 lists the more important tests and suggests an approximate frequency. There is some variation between centres. For further detail on these procedures see Bolster (2003) and Elliott (2005). Table 10.6 gives typical performance figures.

TABLE 10.5

Performance Measurements for a Gamma Camera

Test	Frequency	Comment
Physical inspection	Daily	
Photo peak position	Daily	
Visual uniformity	Daily	Failure of a number of functions will show as non-uniformities in a Tc-99m flood image
Quantitative uniformity	Daily	Both integral and differential uniformity should be checked over the whole field of view and over the centre of the field of view
Extrinsic uniformity	Monthly	Measured with the collimator in position
Centre of rotation	Monthly	Measured using an off-centre point source
Tomographic quality	Monthly	Assessed qualitatively using a suitable tomographic phantom
Spatial distortion	3 monthly	May be obtained from the same data set
Intrinsic resolution	3 monthly	
System spatial resolution	3 monthly	Measured with the collimator in position
Energy resolution	3 monthly	

TABLE 10.6

Typical Performance Figures for a Gamma Camera

Parameter	Value	Conditions
Intrinsic spatial resolution	3.7 mm	FWHM over the useful field of view
System spatial resolution	6.5 mm	FWHM with high resolution collimator, without scatter at 10 cm
Intrinsic energy resolution	10%	FWHM at 140 keV
	14%	FWHM at 140 keV with collimator and scatter
Integral uniformity	2.0%	Centre of field of view
Differential uniformity	1.5%	Centre of field of view
Count rate performance	130,000 cps	20% loss of counts without scatter
	75,000 cps	Deterioration of intrinsic spatial resolution to 4.2 mm
System sensitivity	160 cps MBq^{-1}	Tc-99m and a general purpose collimator

Note that quality assurance (QA) checks must not be unduly disruptive to the work of the department. Daily checks should take no more than a few minutes and monthly checks no more than 1–2 h.

There are many possible reasons for computer software failure, and complete software evaluation is extremely difficult. Since, however, the basic premise of software evaluation is that application of a programme to a known data set will produce known results; new software should be tested against

(a) Data collected from a physical phantom

(b) Data generated by computer simulation

(c) Validated clinical data

Any significant variation from the expected results can then be investigated.

10.7 Conclusions

The primary detector of gamma rays in nuclear medicine is invariably a sodium iodide crystal, doped with about 0.1% thallium. The advantages of this detector are as follows:

1. A high density and high atomic number ensure a good gamma ray stopping efficiency for a given crystal thickness.

2. The high atomic number favours a photoelectric interaction, thus a pulse is generated which represents the full energy of the gamma ray.

3. Thallium gives a high conversion efficiency of the order of 10%.

4. A short 'dead time' in the crystal generally permits acceptable counting rates except for very rapid dynamic studies, when dead times both in the crystal and elsewhere in the system can be important.

The instrument of choice for a wide range of static and dynamic examinations is the gamma camera. Its mode of operation may be considered in two parts.

1. A direct spatial relationship is established between the point of emission of a gamma ray in the patient and the point at which it strikes a large NaI (Tl) crystal by collimation.

2. An array of PMTs backed by appropriate electronic circuits is used to identify the position at which the gamma ray interacts with the crystal.

Over 90% of all nuclear medicine examinations are carried out with Tc-99m. The advantages of Tc-99m for radionuclide imaging are as follows:

1. A monoenergetic gamma ray is emitted—this facilitates pulse height analysis.

2. The gamma ray energy is high enough not to be heavily absorbed in the patient, hence minimising patient dose, but low enough to be stopped in a thin sodium iodide crystal.

3. No high LET radiations are emitted.

4. A decay product that delivers negligible dose.

5. A half-life that is long enough for most examinations but short enough to minimise dose to the patient.

6. Ready availability as an eluate from a Mo-99/Tc-99m generator.

The quality of radionuclide images is influenced by

1. The number of counts that can be collected for given limits on radiation dose to the patient, required resolution, and time of examination

2. The ability of the radiopharmaceutical to concentrate in the region of interest

3. The presence of, and ability to discriminate against, scattered radiation

4. Overall performance of the imaging device, including spatial and temporal linearity, uniformity and system resolution

Modern cameras collect all data digitally and all counts within one pixel are summed. For visual display it is then converted into an 'analogue type' image by interpolation and smoothing. Digitised images can also be used to extract, under computer control, functional data for specified regions of interest. The gamma camera is now used extensively for dynamic studies where important information is obtained by numerical analysis of digitised images on a frame-by-frame basis.

Tomographic imaging is becoming an ever more significant part of the workload in nuclear medicine. The large field of view digital detectors and improved computing power have removed many of the previous obstacles to SPECT, although the count limited nature of the process remains a fundamental obstacle.

References

BFCR, Reports www.rcr.ac.uk/publications.aspx?PageID=310, 2008.

Bolster, A., *Quality Assurance in Gamma Camera Systems. Report No. 86*, IPEM, UK, 2003.

Britton, K.E., Nuclear medicine, state of the art and science, *Radiography*, 1, 13, 1995.

Elliott, A.T., Quality assurance, in *Practical Nuclear Medicine,* 3rd ed., Sharp, P.F., Gemmell, H.G., and Murray, A.D., Eds., Springer, London, 2005, chap. 5.

Elliott, A.T. and Hilditch, T.E., Non-imaging radionuclide investigations, in *Practical Nuclear Medicine*, 3rd ed., Sharp, P.F., Gemmell, H.G., and Murray, A.D., Eds., Springer, London, 2005, chap. 4.

Frier, M., Mechanisms of localisation of pharmaceuticals, in *Text Book of Radiopharmacy Theory and Practice*, 2nd ed., Sampson, C.B., Ed., Gordon and Breach, London, 1994, 201.

Gadd, R. et al., *Protocol for establishing and maintaining the calibration of medical radionuclide calibrators and their quality control. Measurement Good Practice Guide No. 93*, NPL, Teddington, 2006.

Hart, D. and Wall, B.F., UK Nuclear Medicine Survey 2003–4, *Nucl. Med. Commun.*, 26, 937, 2005.

Houston, A.S. et al., UK audit and analysis of quantitative parameters obtained from gamma camera renography, *Nucl. Med. Commun.*, 22, 559, 2001.

Nowotnik, D.P., Physico-chemical concepts in the preparation of technetium radiopharmaceuticals, in *Text Book of Radiopharmacy Theory and Practice*, 2nd ed., Sampson, C.B., Ed., Gordon and Breach, London, 1994, 29.

Peters, A.M., Fundamentals of tracer kinetics for radiologists, *Brit. J. Radiol.*, 71, 1116, 1998.

Samei, E. et al., *Assessment of display performance for medical imaging systems, Report of the American Association of Physicists in Medicine (AAPM) Task Group 18*, Medical Physics Publishing, Madison, 2005.

Sharp, P.F., Dendy, P.P. and Keyes, W.I., *Radionuclide Imaging Techniques*, Academic Press, New York, 1985.

Exercises

1. What is a radionuclide generator?

2. A dose of Tc-99m macroaggregated human serum albumin for a lung scan had an activity of 180 MBq in a volume of 3.5 ml when it was prepared at 11.30 h. If you wished to inject 23 MBq from this dose into a patient at 16.30 h, what volume would you administer? (Half-life of Tc-99m = 6 h).

3. A radiopharmaceutical has a physical half-life of 6 h and a biological half-life of 13 h. How long will it take for the activity in the patient to drop to 15% of that injected?

4. List the main characteristics of an ideal radiopharmaceutical.

5. What are the possible disadvantages of preparing a radiopharmaceutical a long time before it is administered to the patient?

6. Why is the ideal energy for gamma rays used in clinical radionuclide imaging in the range 100–200 keV?

7. In nuclear medicine, why are interactions of Tc-99m gamma rays in the patient primarily by the Compton effect whilst those in the sodium iodide crystal are mainly photoelectric processes?

8. The sodium iodide crystal in a certain gamma camera is 9 mm thick. Calculate the fraction of the gamma rays it will absorb at (a) 140 keV and (b) 500 keV. Assume the gamma rays are incident normally on the crystal and that the linear absorption coefficient of sodium iodide is 0.4 mm^{-1} at 140 keV and 0.016 mm^{-1} at 500 keV.

9. Why is it necessary to use a collimator for imaging gamma rays but not for X-rays produced by a diagnostic set?

10. What are the differences between a collimator used to image low energy radionuclides and one used to image high energy radionuclides?

11. Why is pulse height analysis used to discriminate against scatter in nuclear medicine but not in radiology?

12. Compare and contrast the methods used to reduce the effect of scattered radiation on image quality in radiology with those used in clinical radionuclide imaging

13. How is the spatial resolution of a gamma camera measured and what is the clinical relevance of the measurement?

14. What factors affect
    i. The sharpness
    ii. The contrast of a clinical radionuclide image?

15. Explain, with a block diagram of the equipment, how a dynamic study is performed with a gamma camera

16. A radiopharmaceutical labelled with Tc-99m and a gamma camera system were used for a renogram. Curves were plotted of the counts over each kidney as a function of time and although the shapes of both curves were normal, the maximum count recorded over the right kidney was higher than over the left. Suggest reasons

17. What are functional images? Illustrate your answer by considering one application of functional images in nuclear medicine.

18. Explain how SPECT data is acquired and how the data are manipulated to produce a clinically useful image.

19. Describe the processes that must be undertaken to ensure that a tomographic gamma camera system is functioning correctly for the acquisition of SPECT data.

20. Describe the steps that must be taken to fully test a new gamma camera system after installation. Why may some of the steps previously required be unnecessary on modern systems?

11

Positron Emission Tomographic Imaging (PET)

P H Jarritt

SUMMARY

- Positron emission tomography is based upon positron emitting radiotracers which have a short half-life.
- The positron decay process inherently limits the resolution that can be achieved by an imaging system.
- The detector properties and configuration are important in determining the performance of an imaging system.
- Multiple data corrections are required to correct for limitations in the detection process and to provide quantitative images.
- PET system development has incorporated computed tomography (CT) scanners for the attenuation correction process.
- 'Fused' PET and CT imaging has enhanced the diagnostic accuracy and application of PET imaging, especially in radiotherapy planning.
- Radiation protection is an important issue in PET imaging for both staff and patients.

CONTENTS

11.1 Introduction

The development of PET as an imaging modality sprang from the recognition that emissions from radioactive decay processes could be used to measure metabolic processes *in vivo*. The radioactive tracer method was first applied by George de Hevesy in 1923 for which he was awarded the Nobel Prize in Chemistry in 1943. The process of positron decay was discovered in 1933 by Thibaud (1933) and Joliot (1933) and within 12 years positron emitting radionuclides were being used to undertake metabolic studies in animals using O-15 (Tobias et al. 1945) and within 20 years the first images from studies in man using coincidence detection were published (Anger and Gottschalk 1963). In these early years the potential and role of positron emitting radionuclides was formed. Radionuclides of carbon, C-11, nitrogen, N-13, oxygen, O-15 and fluorine, F-18 were produced and integrated in biologically active molecules to trace biochemical pathways and reactions. The desire for quantification was paramount and the application of coincidence detection of positron emissions provides the basis for the realisation of this goal today. Over the ensuing period a broad range of technological advances have occurred and this has led to the transition of PET from the research laboratory to a routine clinical imaging tool.

Progress in PET has been dependent on several factors:

(1) Development of the cyclotron and associated radiochemistry to routinely produce well-characterised radiopharmaceuticals targeted at particular clinical problems

(2) Development of the electronic and detector technologies needed to provide the high resolution and sensitivity for positron detection in the clinical environment

(3) Development of computer technology to provide the resource to process the very large data rates encountered in a timely manner

(4) Development of the image reconstruction and data analysis tools to provide reproducible quantitative measures of radionuclide distributions and kinetics *in vivo*

These developments alone were not sufficient to see the widespread adoption of PET imaging in a routine clinical setting. This came with the development in 2000 of two commercially available integrated PET and X-ray CT scanners to form the PET/CT scanner (Beyer et al. 2000). This added the advantages of a high resolution anatomical imaging modality (CT) to the highly sensitive functional imaging modality of PET, providing inherently registered and 'fused' images (Figure 11.1). This combination of function and structure has seen the widespread adoption of PET/CT in oncology as well as in neurological and cardiac applications.

Many of the challenges of the use of positron emitters remain. The application of the methodology is inextricably linked to the production and availability of well-characterised radiopharmaceuticals targeted to specific biochemical pathways and processes. Such tracers are important for diagnosis and disease staging but more importantly there is a

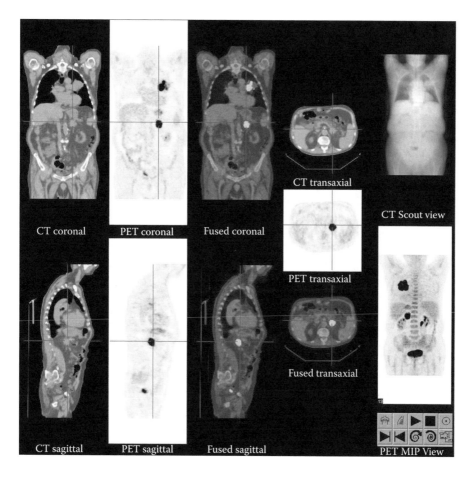

FIGURE 11.1
(See colour insert.) Typical PET/CT review screen showing CT, PET and fused image data sets. The bottom right hand image is a rotating maximum intensity reprojection image.

growing demand to be able to measure response to therapeutic interventions. This latter application is renewing the drive to discover and characterise 'biomarkers' which are uniquely sensitive to changes in disease state following treatment. Enhancements of the detector and image reconstruction technology still occur with continued developments in sensitivity, resolution and quantitative accuracy. These are combined with the exploration of other multimodality combinations such as PET and magnetic resonance (PET/MR) both in humans and high resolution animal systems.

This chapter provides an introduction to the principles behind PET imaging and the current developments and applications.

11.2 PET Radionuclide Production and Properties

Positron emitting radionuclides require an excess of positive charge in the nucleus and they are typically produced in a cyclotron by the bombardment of a target material with protons, deuterons or α particles which have been accelerated to high energies to initiate a nuclear transformation. The cyclotron consists of an ion source situated in the centre of two semi circular copper electrodes (dees) contained within a vacuum and mounted between the poles of a large electromagnet. The application of an alternating potential difference between the dees causes the charged ions to be accelerated within the dees to energies up to 20 MeV in a typical biomedical cyclotron. The high energy ions are extracted from the dees to produce a beam of positively charged ions.

The most common ion source is a negatively charged hydrogen atom (H⁻, a proton plus two electrons). The proton beam is generated by stripping the electrons from the H⁻ on extraction from the cyclotron to produce a beam of protons.

The short half-life of the resultant radionuclides typically requires direct transfer of the irradiated target material to radiopharmaceutical synthesis modules to produce tracers suitable for human use. An alternative source for some positron emitting radionuclides is a radionuclide generator where the required, short lived, radionuclide is continually being produced from a longer lived parent contained within the generator in a parent-daughter relationship (cf Mo-99/Tc-99m, see Section 10.3.2). The advantage of this approach is that the short half-life daughter product can be used remotely from the production site for the parent and used over a period of weeks. Examples of cyclotron and generator produced PET radionuclides are given in Table 11.1.

11.3 Principles of PET Imaging and Detector Technology

To understand the design choices of a modern PET tomograph it is necessary first to understand the process of positron decay and the interaction of the emitted positron and annihilation photons with surrounding material.

11.3.1 Positron Decay

Positron decay, sometimes called β⁺ *decay*, occurs where an atomic nucleus is unstable due to an excess of protons. The charge balance within the nucleus is restored by the

TABLE 11.1

Properties of Common PET Radionuclides Used for Clinical Imaging

Radio-Nuclide	Production Reaction	Decay Mode	Half Life	Positron Energy (MeV)	Approximate Effective Range in Water (mm)[a]
C-11	^{10}B (d,n) ^{11}C	β⁺, OEC	20.4 min	0.96	0.4
N-13	^{16}O (p,d) ^{13}N	β⁺	9.96 min	11.9	0.5
O-15	^{15}N (p,n) ^{15}O	β⁺	2 min	1.72	1.0
F-18	^{18}O (p,n) ^{18}F	β⁺, OEC	110 min	0.635	0.25
Ga-68	^{68}Ge-^{68}Ga	β⁺, OEC	68.3 min	1.9	1.2
Rb-82	^{82}Sr-^{82}Rb	β⁺, OEC	78 s	3.35	2.6

[a] see Derenzo 1986.

OEC = orbital electron capture, see Table 10.2.

Source: Some data from Derenzo S E. Mathematical removal of positron range blurring in high-resolution tomography. *IEEE Trans. Nucl. Sci.* 33, 565–569,1986.

transformation of a proton to a neutron with the emission of a positively charged electron and a neutrino.

$$p^+ \rightarrow n + e^+ + \upsilon + \text{energy}$$

$$^{18}_{9}\text{F} \rightarrow \, ^{18}_{8}\text{O} + \, ^{0}_{1}\beta + \, ^{0}_{0}\upsilon + 1.655 \text{ MeV}$$

The energy released in this transformation is shared between the neutrino, the kinetic energy of the positron, or positive electron and the annihilation photons. Whilst the energy released in the transition is characteristic of the transition, the kinetic energy of the positive electron ranges from 0 to E_{max} where E_{max} is the transition energy minus the energy of the annihilation photons (Figure 11.2).

The positron, after ejection from the nucleus, loses its kinetic energy through scattering interactions with the surrounding matter until it is effectively stationary. The range of travel of the positron will be dependent on its initial energy. Once at rest it will combine with a negative electron from the surrounding matter in an annihilation process in which the rest masses of the two electrons are converted to energy. The energy is released in the form of two annihilation photons each of energy 0.511 MeV. This characteristic of the positron decay process directly impacts upon the resolution that can ultimately be achieved in PET. The point of the annihilation event is remote from the nuclear decay event, and by inference the location of the radiotracer, by a distance determined by the energy of the positron. This separation is often described by an effective positron range which takes into account the distribution of positron energies and the perpendicular distance from the point of decay to the line of coincidence (Table 11.1).

The principle of conservation of momentum determines that, in an ideal situation, if the annihilation reaction occurs at rest the emitted photons will be emitted in exactly opposite directions in a single plane (co-linear) (Figure 11.3). In practice there is some residual momentum so the photons are emitted with a small angular distribution. The distribution of the angular deviation from co-linearity where the photons are 180° opposed is

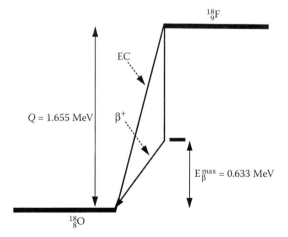

FIGURE 11.2
Energy level diagram for the decay of ^{18}F to ^{18}O via positron emission and orbital electron capture (OEC) routes (see Section 1.4 and Table 10.2). (Reproduced with permission from Cherry S R, Sorenson J A and Phelps M E. *Physics in Nuclear Medicine* 3rd ed. Saunders, 2003.)

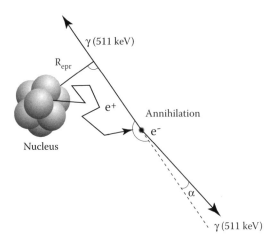

FIGURE 11.3
Illustration of the positron decay process showing positional errors in coincidence imaging caused by the 'effective positron range' and the non co-linearity of the annihilation photons.

TABLE 11.2

Example of System Resolution for F-18 with 90 cm Diameter Detector with a 3 mm Crystal Element

	R_{epr} (mm)	Diameter (cm)	$R_{co\text{-}lin}$ (mm)	R_{det} (mm)	R_{sys} (mm)
F-18	≈0.25	90	≈2.1	3	3.7

approximately Gaussian with a FWHM of approximately 0.5°. This non-co-linearity effect produces another constraint on the resolution that can be achieved with a coincidence detector system where co-linearity is assumed as part of the detection process. The degradation of resolution will be dependent on the separation of the detectors operating in coincidence. For a detector separation of 90 cm the effect is to produce an uncertainty of about 2.1 mm FWHM—see Table 11.2 and for further explanation Section 11.6.

The final property of the annihilation process that needs to be considered is the velocity of the annihilation photons. These have the characteristic velocity of light (3×10^8 m s^{-1}). For a pair of detectors the difference in the arrival time of the two photons will be dependent on the difference in the distance from the point of annihilation to each of the detectors. The accurate measurement of this difference would enable the exact specification of the position of the annihilation event. This is the principle that is adopted in PET detection systems with time of flight (TOF) capability. The accuracy with which position can be determined (Δd) is dependent on the timing resolution which can be achieved and is defined as follows:

$$\Delta d = \frac{\Delta t \times c}{2}$$

where c is the velocity of light, Δt is the timing resolution. For a timing resolution of 100 ps (1 ps = 10^{-12} s) Δd = 1.5 cm. Typical values for current TOF systems are 400–600 ps leading to positional uncertainties of 6–9 cm. It should be noted that this uncertainty is significantly less than the diameter of a PET tomograph and the implementation of TOF has the potential to improve quantitative accuracy for the detection process by reducing the uncertainty in the measurement.

11.3.2 Coincidence Detection

PET imaging is uniquely defined by its exploitation of the positron decay process and the construction of a detector system which can 'simultaneously' detect the two emitted photons to define a line along which the annihilation event occurred. This is known as coincidence detection and is illustrated in Figure 11.4.

For two opposed detectors a coincident or 'prompt' event is determined if a set of the following conditions are met:

(1) Two events are detected within a pre-defined electronic timing window known as the coincidence window.
(2) The energy deposited within the individual detectors is within a selected energy range.

The choice of electronic timing window is dependent on the detector characteristics (how quickly it can process an event) and on the uncertainties in signal transit time through the electronics. This is typically set to 3–9 nano seconds (1 nano second = 10^{-9} s). The longer the coincidence timing window the greater the number of random coincidences that will be detected. The use of energy discrimination enables coincidences between photons which have lost energy due to interactions in the object to be discarded. A true coincidence is defined as one where each detected photon has an energy of 511 keV. Due to the poor energy resolution of the detectors currently used, a wide energy window has to be selected leading to the inclusion of photons which have undergone a scattering event in the object. The choice of the energy discrimination window affects directly the apparent sensitivity of the device and the number of random coincidences which are detected. This leads to a loss of resolution and signal to noise ratio in the image.

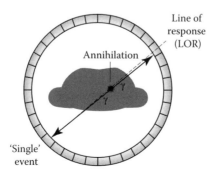

FIGURE 11.4
Illustration of the definition of line of response for a coincident event in a positron ring detector with discrete crystals. Each interaction within a crystal is termed a 'single' event.

11.4 Detector Geometry

PET detection systems have evolved from discrete opposed detector pairs, through large continuous detectors based upon opposed gamma camera pairs to full 360° static geometries using discrete detectors. Given that a decay event can project anywhere within a 4π geometry the most efficient detection geometry will approximate to a sphere. The practical implementation of this is a ring or multiple rings of detectors with a finite axial length as illustrated in Figure 11.5. This is the configuration found in all current commercial systems. Further discussion will be limited to the ring geometry but the same principles apply to all detector geometries.

11.5 Detector Construction

The detectors used in current clinical systems are based upon scintillation crystals arranged as block detectors connected to an array of photomultiplier tubes to enable the identification of the crystal element in which the interaction occurred and to measure the energy of the signal from the quantity of light emitted. The efficiency of the device is defined primarily by the efficiency of the detector in attenuating 511 keV photons. The count rate capability of the device is related to the rate at which the scintillator and associated electronics, can process an event. BGO (bismuth germanate), LSO (lutetium oxyorthosilicate) and GSO (gadolinium oxyorthosilicate) are all used in commercial systems.

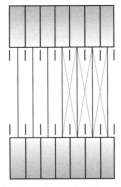
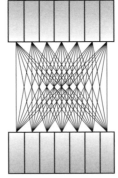
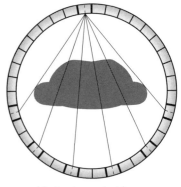

(a) Cross plane coincidences 2-D Mode

(b) Cross plane coincidences 3-D Mode

(c) In-plane coincidences (subset)

FIGURE 11.5

Illustrations of detector geometries; (a) Cross-section through the axial plane of a multiple ring PET detector showing limited coincident planes (crystal ring plane plus one adjacent ring only) in 2-D mode. (b) Cross-section through the axial plane of a multiple ring PET detector showing all potential coincident planes in 3-D mode. (c) Cross-section through the transverse plane of a PET ring detector showing angular limits for coincident events defining the field of view of the detector.

Insight

System Efficiency

It is important to note that the system efficiency is based upon the square of the individual detector efficiency as two events must be processed to provide a coincident event. The highest intrinsic efficiency is provided by BGO although the processing time for a scintillation event is relatively long. For high count rate applications LSO or LYSO (lutetium yttrium oxyorthosilicate) is used. It has a slightly lower efficiency but has a much faster processing time leading to improved energy and randoms discrimination. The same applies to GSO although it is the least efficient of the available detectors.

For TOF applications high speed scintillators are required to provide the timing resolution required to obtain the additional positional information. At present the scintillator still remains the rate limiting step. LSO, LYSO and GSO are all used in commercial TOF machines but faster crystals such as BaF_2 (barium fluoride) have been used in research systems with limited success due to their poor intrinsic efficiency.

The characteristics of a range of scintillators is given in Table 11.3.

11.6 Detector Resolution

For PET systems using discrete detectors, the size of the crystal is fundamental to the resolution that can be achieved in the imaging process. The positional accuracy that can be achieved is defined by the cross-section of the detector. At the detector face the resolution is defined by a box function of width d equal to the detector width. For two opposed discrete detectors the resolution at the mid point is defined by a triangular function with a FWHM of $d/2$. It should thus be noted that a PET detector will not have a uniform resolution characteristic throughout the detection volume. The limits of system resolution are determined by the physical geometry of the detectors (det), the effective positron range (epr) and the separation of the detectors (non-co-linearity effect [co-lin]). These can be combined to provide an overall estimate of the system resolution as follows:

$$R_{sys} \approx \sqrt{R^2_{(epr)} + R^2_{(co\text{-}lin)} + R^2_{(det)}}$$

It should be noted that for human imaging systems the non-co-linearity effect is the most significant and that its impact is significantly reduced for small diameter animal imaging systems (Table 11.2).

TABLE 11.3

Properties of Scintillators Used for PET Imaging

Scintillator	Relative Light Output (NaI ≡ 100%)	Decay Time (nsec)	Thickness for 90% Efficiency at 511 keV (cm)
NaI	100	230	6.75
BGO	15	300	2.42
LSO/LYSO	70	40	2.66
GSO	25	60	3.29
CsF	5	2.5	5.40
BaF_2	2	0.6	5.07

11.7 Detection Events

There are a number of events that can occur in coincidence detections as follows (Figure 11.6):

(1) *A single event.* This is the detection of a single event in a single detector. It is an important measure because these events will be directly proportional to the activity within the field of view of the detector.

(2) *A true coincidence.* This is the result of the detection of both photons from a single annihilation event with neither photon interacting with the surrounding material and which are detected within the coincidence timing and energy discrimination windows—the 'wanted' image forming events.

(3) *A random coincidence.* Such an event occurs when two or more photons from different decay events are recorded within the coincidence timing and energy discrimination windows with the remaining photons being lost to the detection system. The detection logic will assign these as valid events but they will not be related to the valid decay event. The number of random events will be a function of the activity contributing to events within the field of view of the detector.

The random event rate $R_E = 2\tau N_1 N_2$ where N is the single event rate for each detector and 2τ is the coincidence window width.

It is essential that a correction is applied to correct for random events which contribute to a loss of quantitative accuracy. A number of methods exist to correct for random events including estimates from the singles rate and an independent measure based on delayed coincidence windows for one or a pair of detectors. Random events which comprise of more than two simultaneous single events are discarded as these cannot be assigned to a pair of detectors unambiguously. Such events will also be related to the singles count rate.

(4) *Scattered events.* These events arise when one or both photons from a single annihilation event are detected within the coincidence timing and energy discrimination windows but have undergone a Compton scattering interaction (Figure 11.6). Whilst Compton scattering results in a loss of energy from the photon the poor

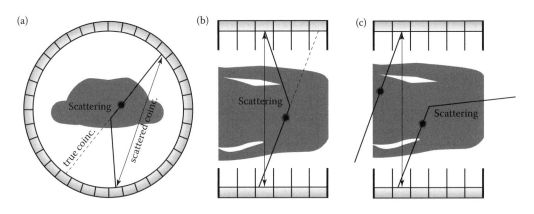

FIGURE 11.6
Illustration showing the positioning of a coincident event (line of response) due to the scattering of an annihilation photon; (a) in the transverse plane; (b) in the axial plane; (c) in the axial plane due to scattering of out-of-field activity.

energy resolution and the resulting wide energy discrimination window means that such events may not be rejected on the basis of their energy.

The event is thus assigned to a line of response (LOR) joining the two detectors in which the photons were detected which will not be correlated with the annihilation event. Scatter events reduce contrast within the image and result in a loss of quantitative accuracy in the reconstructed image. This definition is true for a radiotracer located within the field of view of the detector. However, a coincident event can also be generated by photons scattered from annihilation events outside the immediate field of view and by scattering from the gantry and bed environment. The number of scattered events is not a function of count rate but is constant for a particular object and radioactive distribution. The proportion of accepted coincidences that result from scattered events is known as the scatter fraction. The size of the scatter fraction is dependent on many factors including the geometry of the PET scanner, the energy discrimination window and the size and density of the scattering media and the out-of-field of view activity.

One method of reducing the scatter fraction is to operate the scanner in 2-D mode where each ring of detectors is separated from the adjacent rings by tungsten septa. Coincidences can then only be recorded between a restricted set of detectors resulting in a scatter fraction of approximately 15%–20%. All scanners can also be operated in 3-D mode where the septa are removed, yielding no physical restrictions on the possible lines of response. (It should be noted that the next generation of PET/CT systems will only operate in 3-D mode.) In this configuration the scatter fraction typically exceeds 35%.

A correction for scattered events is required to provide quantitative data. A variety of methods have been used to undertake scatter corrections. The methods can be divided into two distinct classes, first those based on estimates from multiple energy windows and second those based on simulations of the scatter component from an initial reconstruction of the data. This latter technique can be applied iteratively to refine the estimates. The reader is referred to Meikle et al. (2003) for a more detailed explanation of correction methods. Accurate scatter correction remains perhaps the most complex component of the PET imaging process.

11.8 Image Formation

The detection of a coincident event defines a 'line of response' in the detector geometry as the line joining the centres of the two involved crystal elements both within the plane of the ring and between rings. The possible lines of response (LORs) are restricted in the detection system depending on the scanner detection mode (2-D or 3-D) and on the permitted angle of acceptance for coincidences (Figure 11.5). For a ring detector the valid LORs are grouped according to angle around the ring. Events during the acquisition period are summed to provide a set of projection profiles covering 180°. There are typically 150–200 projection profiles depending on the detector construction (Figure 11.7).

The projection profiles are organised to form an image called a *sinogram* and this represents the raw data for each detection plane. In 2-D mode a set of projection profiles is created for each transaxial plane (ring plane) plus an additional set for each adjacent ring coincidences. This leads to $(2n - 1)$ projection sets where n is the number of detector rings. In 3-D imaging the cross-plane lines of response are re-assigned to match the 2-D transaxial planes for reconstruction purposes.

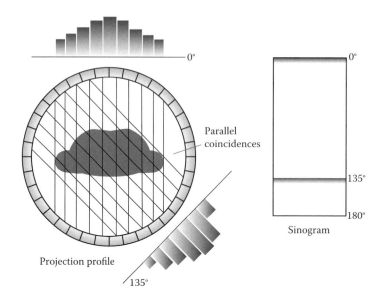

FIGURE 11.7
Illustration of the formation of projection profiles from parallel lines of response within a single plane of detectors (2-D imaging). For image storage and reconstruction purposes, this data is reorganised into sinograms where each projection profile forms a horizontal line in the image.

Each LOR has a particular probability of occurrence due to geometric and electronic considerations and a normalisation process must be implemented before reconstruction to correct for this effect (see Section 11.11). Once formed, the parallel projections are typically reformatted to yield sinograms (a single image formed as a stack of all projection profiles for a single plane). Image reconstruction can then proceed as for any projection image set (see Section 8.4.2).

Image reconstruction based on filtered back projection (FBP) and iterative methods have both been applied in PET imaging.

A sequence of corrections is usually applied to the data before or during reconstruction to yield quantitative results. These include the following:

(a) Normalisation factors

(b) Dead time corrections

(c) Scatter correction

(d) Attenuation correction

(e) Resolution recovery

(f) TOF corrections

The reader is referred to other texts (e.g. Defrise et al. 2003; Meikle et al. 2003) for a detailed explanation of methods used to implement these corrections.

The application of attenuation corrections using CT data is of particular interest. For a coincident event to be detected, both photons must have traversed the body unattenuated. The probability of attenuation for a particular LOR can be measured by comparing the count rates from a source when attenuated by the patient and when not attenuated by the patient (a blank scan calibration). Typically, rotating positron emitting rod sources were placed inside the gantry to provide these measurements. The introduction of CT provided an alternative method for attenuation determination with a better signal to noise ratio for the attenuation map and higher spatial resolution in a much shorter imaging time. The disadvantages of this method relate to the need to interpolate attenuation coefficients from CT energies (120–140 kVp) to 511 keV. This is not a linear transformation across the tissue attenuation range. Studies are often performed in the presence of intravenous or oral contrast and algorithms have been developed to ensure that these do not modify the attenuation coefficients used for PET data reconstruction.

The short acquisition times for CT data provide a further complication in that physiological motion due to respiration and cardiac function is frozen in space at the time of acquisition. The PET data is collected as a signal average typically over a period of 2–5 minutes. This will inherently lead to misalignments in the two datasets and is most noticeable for structures around the diaphragm. As system resolutions are improved this becomes an ever increasing problem. The development of motion correction methodologies is the most significant challenge in improving both the diagnostic and quantitative accuracy of the technique. Typically, PET and CT studies will be acquired as dynamic studies to permit the elimination of the motion component, thus ensuring optimal alignment of PET and CT datasets.

The response of a PET detector is not spatially invariant. This is typically manifest as variations in sensitivity and resolution within the field of view of the detector. The correction of these variations is particularly challenging especially in relation to variations in resolution.

11.9 Image Reconstruction

This is currently one of the key areas in the development of PET imaging. For 2-D imaging systems reconstruction with FBP algorithms was considered the gold standard for quantitative imaging. Whilst significant assumptions are made about the geometry of lines of response, the method is essentially linear across the activity range and reconstructions are not sensitive to changes in radioactivity distribution. By comparison the iterative methods are adaptive and are spatially variant in their response depending on radioactive distribution and signal to noise ratios. This can bring significant errors to quantitative studies especially dynamic studies where activity distributions and levels change throughout the study period.

However, the recent reliance on 3-D image detection has led to significant advances in iterative algorithms which permit many of the required corrections to be accurately modelled and applied to the data. In addition, the geometrical effects of the detector design and configuration can be accurately modelled leading to the potential to implement resolution recovery techniques and to effectively generate a spatially invariant response in the reconstructed data. The implementation of TOF reconstruction algorithms imposes another level of complexity into these algorithms. The methods require significant computing power which is now capable of introducing these methods into routine clinical practice although the reconstructions are still not available in 'real time'.

Insight

Related Technologies

Whilst this chapter has focused on PET and PET/CT technology it is necessary to place this in the context of other developments in imaging with radionuclides. Anger gamma cameras designed for tomographic (SPECT) imaging have been enhanced with the introduction of SPECT/CT systems in a similar manner to PET/CT. The broader range of tracers and radiopharmaceuticals would appear to offer many of the advantages of PET/CT systems. Whilst attenuation and scatter correction for single photon emitters is more complex than for positron emitters, the availability of an emission and transmission map will permit quantitative imaging using iterative reconstruction and system modelling. The limitations of a collimated system are that resolution and sensitivity are related by an inverse square law such that improvement in resolution by a factor of two will result in a four-fold decrease in sensitivity. In practical terms little progress is being made to overcome this physical limit. Sensitivity in relation to PET systems remains low.

Most SPECT systems comprise two opposed gamma camera detectors and a considerable investment was made to use these opposed detectors in coincidence mode to provide a dual SPECT/PET capable system.

The PET acquisition is achieved by the removal of the collimators, greatly enhancing the system sensitivity. The two gamma cameras are then operated in the equivalent of a 3-D mode in PET although a rotation of the detector through 180° is required to obtain full 360° sampling. The construction of such dual detection systems required the use of much thicker NaI crystals with typically 25 mm being used compared to 9 mm for standard SPECT systems. This was still insufficient to attenuate 511 keV photons effectively and compromised the resolution characteristics of the gamma camera for single photon use.

As with a 3-D PET system the data rates that are encountered are extremely high and the single crystal design of the gamma camera severely affected the data rates that could be processed compared to the discrete detector configurations in PET. This led to the need to reduce injected doses leading to increased study times.

Whilst the intrinsic resolution of the gamma camera is higher than for the PET detector the much poorer signal to noise ratios which could be achieved in clinical studies led to poorer image quality and effective resolution. It is no longer considered appropriate to undertake PET imaging using gamma camera based systems and the introduction of 511 keV photons into a routine nuclear medicine environment is not recommended due to the limited equipment and environmental shielding normally encountered.

It must, however, be noted that improvements in sensitivity for PET detector designs will only come with significant increases in detector area. This could be achieved by the introduction of additional detector rings or the introduction of PET-optimised panel detectors using scintillation or semi-conductor photomultiplier technologies.

As has been demonstrated, the use of 3-D imaging leads to higher proportions of random and scatter events. Increasing detector area increases this proportion further to a point where these events dominate the coincidences recorded, leading to a poorer detector performance as characterised by the noise equivalent count rate (NECR) that can be achieved at specific activity levels.

Such systems have been developed in a research and development context but have not yet been introduced as clinical systems. Whilst such systems would support rapid whole body imaging their development does not provide additional capabilities in relation to discrete organ imaging. Their development may be more useful in the development of simultaneous PET/MR systems where anatomical and functional data may be acquired in a truly simultaneous manner.

11.10 Multimodality Imaging

Multimodality imaging is often loosely used to define the combination of data from imaging techniques based on different physical principles. Images can be generated using a wide range of methods including X-rays, MR, ultrasound, thermography, radioactive

TABLE 11.4

Multimodality Imaging Developments

Combination	Method	Status
SPECT/CT	1 & 2	Routine clinical, animal work
PET/CT	1 & 2	Routine clinical, animal work
Optical/CT/SPECT/PET	1	Animal work
SPECT/PET/MR	1	Clinical research
PET/MR	2	Animal work/technical development

tracers and so on. Each will provide a unique view of the anatomy (physical characteristic) or physiological function contributing to the production of the image.

Method 1: Data can be acquired sequentially using separate acquisitions on different imaging systems. These can subsequently be combined using image alignment software to produce a 'fused' image from two or more imaging studies.

Method 2: The images can be acquired with a system capable of acquiring images using different imaging principles at the same time or closely related in time without needing to reposition the patient. Table 11.4 summarises the current status and combinations of technologies providing multimodality imaging.

The combination of PET and CT illustrate the principles involved. Integrated systems offer a number of advantages over the combination of images using software tools. First, and most importantly, the images are inherently aligned due to the aligned geometries of the acquisition systems. With adequate immobilisation no subsequent alignment of the images should be required. Second, the images are closely related in time, removing uncertainties due to changes in functional and anatomical configurations with time. Third, the use of alignment software to 'fuse' images can lead to artefacts and uncertainties due to the non-rigid transformations that are required.

For the clinician the superposition of functional and anatomic detail almost eliminates uncertainties in the location of structures with abnormal function. The interpretation of a functional image without correlated functional data can be difficult and lead to unnecessary further investigations.

For PET/CT and SPECT/CT the images are acquired sequentially without repositioning the patient. However, errors in alignment may still occur through patient motion and more importantly physiological motion due to respiratory and cardiac function. As has been discussed, the CT components in these combined modalities are used to provide attenuation correction maps. The misalignment of the emission and transmission maps due to physiological motion is currently an area of significant research interest with the introduction of 4-D imaging protocols where studies are dynamically acquired and reconstructed, including motion data.

11.11 Quality Control

Quality control of PET/CT systems can be broadly viewed as a series of tests which permit the performance of a scanner to be monitored on an ongoing basis. These tests are well specified and derive principally from the tests used to characterise a PET/CT scanner. These include for PET systems, spatial resolution, count rate performance, sensitivity and image quality with particular reference to the accuracy of attenuation and scatter corrections. For the CT system checks should normally include alignment of positioning lights, alignment of table and gantry, slice localisation, table increment accuracy, slice thickness, spatial resolution image quality, including noise and uniformity, and dosimetry verification (see Section 8.10).

For quality assurance purposes a series of routine assessments and calibrations must be performed. For the PET detector these include the following:

(i)   Detector function and electronics, checking coincidences, singles, dead time, timing and energy variables

(ii) Detector calibration to reset energy tuning, positional accuracy, coincidence timing and detector normalisation

(iii) 'Well counter' calibration for 2-D/3-D modes to relate coincidences to kBq/ml

For an integrated PET/CT scanner particular consideration should be given to the following:

(i)   The alignment of the PET and CT imaging planes and fields of view

(ii) Attenuation and correction values calculated from CT numbers and linearity, especially for applications in radiotherapy planning

(iii) Reconstructed image verification

Specific phantoms are used in each phase and can include an electron density phantom for CT number calibration, manufacturer alignment phantom and a solid Ge-68 cylindrical phantom for anatomical image quality assessment.

Some systems are equipped with an integral Ge-68 rod source which can be inserted within the PET field of view and rotated through 360° to permit the routine checking and re-calibration of detector performance. These tests are fully automated in routine operation and only require intervention when performance lies outside of pre-defined limits.

The 'Well counter' calibration typically requires a cylindrical phantom to be manually filled with a known concentration of F-18 and imaged using a preset protocol in each imaging mode. The concentration is usually calibrated by reference to an aliquot of the phantom solution measured in a fully calibrated, single sample well counter.

The frequency of testing will be dependent upon the stability of the system; however, in principle the calibration checks will be performed on a daily/weekly basis and in all cases ahead of quantitative studies.

11.12 Clinical Implementation—Radiation Safety Considerations for PET Imaging

11.12.1 Radiation Risks to Staff

The basic principles of radiation protection are of particular importance in the design and operation of a PET imaging facility. The dose rate constant for positron emitters is

TABLE 11.5

A Comparison of the Radiation Properties of F-18 and Tc-99m

	F-18	Tc-99m
Principal X-ray energy	511 keV	140 keV
Dose rate at 1 m from unshielded 200 MBq source in vial	32 μSv h⁻¹	4.4 μSv h⁻¹
Half value thickness for lead	≈6 mm	≈0.3 mm
Half-life	110 min	6 h

approximately 7–10 times greater than that for Tc-99m. It is thus essential that distance from sources is maximised, the time spent in the vicinity of radioactive sources is minimised and that shielding is sufficient to minimise the radiation exposure to staff. An effective design of a PET facility is key to the reduction of staff doses especially as the shielding required to attenuate 511 keV photons is very significantly larger than for Tc-99m. Relative half value thicknesses for lead are given in Table 11.5.

The conduct of a routine PET scan has a number of key components as follows:

(1) Receipt and preparation of a patient dose from a multi-dose stock vial

(2) Injection of the patient

(3) Uptake phase of the radiotracer in the patient

(4) Escorting and positioning the patient on the scanner

(5) Imaging the patient

Each of these phases of a PET study requires a different emphasis in radiation protection. The receipt and preparation of the patient dose requires high levels of shielding between the source and the operator and minimal operator interaction time. Typical shielding levels are 30–50 mm of lead around the stock vial or between the operator and the source. Whilst this process can be conducted manually with limited exposure of staff, semi- and fully automated dispensing systems are available to further reduce staff doses. The injection of the patient effectively returns the radiation source to an unshielded state. The minimisation of staff doses thus requires the minimum time to be spent undertaking the task at the maximum distance from the patient.

Commercial systems are now available which permit the automatic injection of the dose whilst the operator remains at 1–2 metres from the patient behind suitable shielding. A combination of dose preparation and injection in a single system is also available.

Once injected and during the uptake phase, the patient is usually accommodated within a well-shielded room with video surveillance and emergency call facilities. This precludes the need for regular staff access and also reduces the radiation levels within the environment. Typical dose rates outside the uptake area are designed at 1–2 μSv h⁻¹. Time and distance become the key factors to dose reduction in operation.

The acquisition of a study is usually undertaken from a shielded control room where doses are designed to be at safe public levels. Shielded viewing panels, intercom and video systems ensure patient contact is suitably maintained.

On completion of the study the patient is discharged from the unit as quickly as possible. The short half-life of the radionuclide means that dose rates have significantly decreased in this phase and except in rare circumstances no further precautions are required to protect the public, although staff should continue to minimise close contact with the patient to reduce their overall dose burden.

TABLE 11.6

Typical PET Operator Doses per Imaging Task (375 MBq F-18 FDG)

Technologist Task	Measured Whole Body Dose per Task (μSv)
Drawing up and measuring dose	0.29
Injecting patient	1.85
Observing patient from control room	0.12
Escorting patient to toilet	0.72
Positioning patient pre- and post-scan	1.8
Entering room during PET scan	0.29
Post-scan interaction	0.31
Total	**5.38**

Typical whole body effective doses to staff for studies using 375 MBq F-18 FDG (fluoro-deoxyglucose) are given in Table 11.6. Manual dispensing and injection techniques were used. For further information on effective doses see Section 13.5.

The annual dose l imit for unclassified workers in the United Kingdom is set at 6 mSv with typical investigation limits set at 4 mSv. From the data presented in Table 11.6 technologists would reach these limits after approximately 750 patient studies. This is much less than is typical of an efficient scanning unit and will lead to the need for staff rotation or the formal classification of staff as radiation workers.

Vigilance in radiation protection for staff is vital to maintain and further reduce staff exposure. The need to manipulate test sources should be minimised with adequate shielding available for storage. The short half-life of the radionuclide precludes any major difficulties in managing the environment as any contaminated materials or equipment will rapidly decay to safe levels.

The process of escorting the patient from the uptake area to the scanner room and positioning the patient again exposes the technologist to potentially high dose rates. This task remains one of the most intractable to effective solutions. Mobile shielding can be used but is ineffective where patients need significant assistance. The skill of the technologist in undertaking this task quickly is key to minimising radiation dose. The development of the use of PET/CT scanning for radiation therapy planning provides particular challenges in this phase of the study as radiation therapy planning studies may require the use of patient positioning aids and marking of the patient in relation to alignment lasers and the scanner plane. In practice, these processes take significantly longer than routine diagnostic studies and often require two technologists for verification purposes, increasing the overall radiation burden.

Insight

Risks for Research Studies

The conduct of some research studies may prove particularly challenging. The performance of dynamic studies in the scanner from the time of injection and with blood sample collection will impose a significantly higher radiation burden. Such studies, when combined with the use of short lived nuclides such as O-15, C-11 and N-13 may further increase staff doses. The routine monitoring of staff doses and practice is vital in such circumstances to modify practice and reduce staff doses effectively.

11.12.2 Radiation Dose to the Patient

On the basis of methodology developed by the Medical Internal Radiation Dose (MIRD) Committee of the Society of Nuclear Medicine (ICRP 2009), the effective dose from 100 MBq of F-18 for an FDG study is typically 1.9 mSv for an adult rising to 9.5 mSv for a 1 year old. Typically doses range from 150 to 400 MBq with modern 3-D scanners providing acceptable scan times and quality at the lower end of this range. Doses for studies using the shorter lived radionuclides are typically 2–3 times less. Whilst slightly higher than doses from routine nuclear medicine studies with Tc-99m the trend towards lower PET doses provides very similar effective doses to patients and reduces the radiation burden to staff.

The introduction of X-ray CT as part of the imaging procedure has significantly impacted on patient dose and the actual CT operating parameters are highly dependent on overall imaging strategies. The combination of diagnostic CT with a PET scan which precludes the need for a separate diagnostic CT may add no additional overall burden to the patient; however, the introduction of additional CT procedures will enhance the overall burden. At the very minimum a CT scan of sufficient quality to permit an accurate attenuation correction to be performed is required. For an irradiation of the patient from thigh to chin the effective dose, depending on the CT protocol used, is approximately 3–5 mSv. The use of diagnostic protocols can increase this to 15–20 mSv.

11.13 Current and Future Developments of PET and PET/CT

For recent reviews of clinical applications and potential applications of PET and PET/CT see Pan and Mawlawi, 2008, Mawlawi and Townsend, 2008, Ben-Haim and Ell 2009. The vast majority of routine applications are in oncology and account for more than 95% of studies performed. Applications include disease diagnosis, disease staging and re-staging, and therapy monitoring. PET and PET/CT provide just one tool in a wide range of imaging modalities, however, its effectiveness lies in the fact that functional changes as traced by radiolabelled compounds often precede significant anatomical changes and can thus contribute significantly to the management of patients. Applications are highly dependent on tracer availability which will reflect on whether a facility is restricted solely to commercially produced ^{18}F-FDG or has access to a local cyclotron and radiopharmaceutical production unit providing a much wider range of tracers. The vast majority of studies are still conducted using ^{18}F-FDG. Such studies have a well-defined role in the diagnosis and staging of lung, head and neck, oesophageal and colo-rectal cancers as well as lymphoma and melanoma. Whilst the presence of disease may be apparent from a range of diagnostic procedures, the location and extent of the disease is vital to determining the relevant treatment course and the prognosis for the patient. This is where the combination of PET and CT has proved especially effective. Identification of the correct treatment choice can impact on the use of health-care resources by preventing inappropriate surgery or radiotherapy treatments.

Of growing importance is the use of functional markers to monitor treatment. Lymphoma can in many cases be effectively treated with chemotherapy leading to remission. However, in some cases additional interventions are required. The selection of these patients can effectively be made using an ^{18}F-FDG study towards the end of the chemotherapy treatment. A failure to detect any uptake of the FDG is indicative of a successful treatment and no further interventions will be required immediately. For those with residual uptake of the FDG further treatments can be targeted appropriately.

In the wider context a clinical PET/CT facility must form part of a much bigger infrastructure including basic cell biology research, extensive pre-clinical assessment of drug efficiency using small animal imaging as well as *in vivo* methodology and the development and characterisation of PET radiotracers to support specific drug interactions. The outcome of these developments will require early phase investigation in man to validate and characterise the drugs before registration and wider testing in patients.

Another area of application which has developed rapidly with the introduction of integrated PET/CT systems is the use of PET/CT as the basis for radiotherapy planning. Most modern radiotherapy is planned from CT scans with the treatment volumes being determined on the basis of an interpretation of the anatomical information. As has already been described, functional changes and hence disease locations may precede anatomical changes and the introduction of a functional marker such as ^{18}F-FDG has the potential to modify the delineation of treatment volumes to modify outcomes. The characteristics of the functional marker are critically important to the success of this development. A functional marker with a high false positive rate for disease will result in larger treatment volumes with the irradiation of more normal tissue leading to higher normal tissue damage and potentially higher morbidity. Cure rates may, however, be higher. A high false negative rate will potentially lead to the non-treatment of tumours leading to poorer outcomes. Randomised controlled trials have not been extensively performed to verify this change in protocol, but, two effects have been validated. The introduction of PET results in changes to the delineated gross tumour volumes (GTV) with both increases and decreases when compared to CT planning only and intra-operator variability is significantly reduced. The potential to modify the final treatment volume after allowance for positioning errors and patient/organ motion is less clear and further work is still required to optimise the integration of PET/CT into radiotherapy planning (Benatar et al. 2000).

11.14 Conclusion

Significant clinical validation work remains to be undertaken to provide a firm evidence base for this technology. The introduction of motion detection and correction methods for imaging, planning and delivery systems for radiotherapy should further enhance the effectiveness of this application.

The applications for PET/CT will continue to grow although other technologies will begin to provide alternative methodologies to diagnosis, staging and therapy response monitoring. As has been the case with any radionuclide-based technique, the discovery and availability of specific and well-characterised radiopharmaceuticals and biomarkers are essential to the continued development of the methodology.

References

Anger H.O., Gottschalk A. Localisation of brain tumours with the positron scintillation camera. *J. Nuc. Med.*, 4, 326, 1963.

Benatar N.A., Cronin B.F., O'Doherty M.J. Radiation dose rates from patients undergoing PET: implications for technologists and waiting areas. *Eur. J. Nucl. Med.* 27(5), 583–89, 2000.

Ben-Haim S., Ell P. 18F-FDG PET and PET/CT in the evaluation of cancer treatment response. *J. Nucl. Med.* 50(1), 88–99, 2000.

Beyer T., Townsend D.W., Brun T., Kinghan P.E., Charron M., Roddy R. et al. A combined PET/CT scanner for clinical oncology. *J. Nuc. Med.* 41, 1369–79, 2000.

Defrise M., Kanchan P.E., Michel C. Image reconstruction algorithms in PET. In *Basic science and clinical practice in positron emission tomography*. Eds. Valk PE, Bailey DL, Townsend DW, Maisey MN. ISBN 978-1-85233-485-7, Springer, 2003.

Derenzo S.E. Mathematical removal of positron range blurring in high-resolution tomography. *IEEE Trans. Nucl. Sci.* 33, 565–69, 1986.

ICRP. Radiation dose to patients from radiopharmaceuticals, *International Commission on Radiation Protection no. 106*, Elsevier, 2009.

Joliot F. Preuve expérimentale de l'annihilation des électrons positifs. *C.R. Acad. Sci.* 197, 1622–5, 1933.

Ling C.C., Humm J., Larson S. et al. Towards multi-dimensional radiotherapy (MD-CRT): biological imaging and biological conformity. *Int. J. Radiat. Oncol. Biol. Phys.* 47(3), 551–60, 2000.

Mawlawi O., Townsend D.W. Multi-modality imaging—an update on PET/CT technology. *Eur. J. Nucl. Med. Mol. Img.* Suppl. 1, S15–29, Review 2009.

Meikle S.R., Badawi R.D., Valk P.E., Barley D.L., Townsend D.W., Massey M.N. Quantitative techniques in positron emission tomography. In *Positron emission tomography*. Eds. Bailey DL, Townsend DW, Valk PE, Maisey MN, Springer, 2003.

Pan T. Mawlawi O. PET/CT in Radiation Oncology. *Med. Phys.* 35(11), 4955–66, 2008.

Thibaud J. L'annihilation des positrons au contact de la matière et la radiation qu'en resulte. *C.R. Acad. Sci.* 197, 1629–32, 1933.

Tobias C., Lawrence J., Roughton F. et al. The elimination of carbon monoxide from the body with reference to the possible conversion of CO to CO_2. *Amer. J. Physiol.* 149, 253–63, 1945.

Further Reading

Bailey D.L., Townsend D.W., Valk P.E. Maisey M.N. (Eds). *Positron emission tomography*. Springer-Verlag, London, ch. 1–3, pp. 1–62, 2005.

Cherry S.R., Sorenson J.A., Phelps M.E. *Physics in nuclear medicine*. 3rd ed., Saunders, 2003.

Hamilton D. *Diagnostic nuclear medicine*. Springer-Verlag, Berlin, Heidelberg, pp. 163–204, 2004.

Kim E.E., Lee C.M., Inoue T., Wong W.H. *Clinical PET—principles and applications*. Springer, New York, chs 1–3, pp. 3–61, 2004.

Powsner R.A., Powsner E.R. *Essential nuclear medicine physics*, Blackwell Science, chs 8 & 9, pp. 114–135, 1998.

Exercises

1. What property of a nucleus makes it likely to emit positrons? Discuss the fate of a positron once it is produced in the body.

2. Explain how *TOF* of an annihilation pair of positrons can provide positional information. Show that the accuracy of positioning increases as the time of resolution of the detectors decreases.

3. What factors determine the system resolution (uncertainty of position of a positron decay) in a PET image? How should the uncertainties due to individual factors be combined, and why?

4. Explain the meaning of the terms *single events, scattered events* and *random coincidences* in PET.

5. If it is desired to obtain quantitative data from a PET scan, what corrections have to be applied to the raw data collected?

6. A clinical investigation requires images from the same patient to be collected using two different modalities. What are the two major methods? Discuss the strengths and weaknesses of each.

7. Explain how PET/CT can be used in radiotherapy (a) to improve delineation of the treatment volume, (b) to assess response to therapy.

8. What are the principal causes of radiation exposure to staff working in a PET facility?

12

Radiobiology and Generic Radiation Risks

P P Dendy and B Heaton

SUMMARY

- The reasons why ionising radiation is the most harmful form of radiation are presented.
- Radiation chemistry is explained briefly.
- Evidence that DNA is the primary target for radiation damage is presented.
- Ideas of radiobiological effectiveness (RBE) and radiation weighting factors based on survival curve theory are discussed.
- Evidence that radiation is a carcinogen and mutagen in humans and animals is presented and generic risk factors are established.
- The chapter concludes with a critical review of the evidence for and against risk at the very low doses characteristic of diagnostic radiology.

CONTENTS

12.1 Introduction

This chapter deals with a problem that is central to the theme of the book—namely that ionising radiation, even at very low doses, is potentially capable of causing serious and lasting biological damage. If this were not so, steps that are taken to reduce patient doses, for example, the use of rare earth screens, would be unnecessary and generally undesirable. Furthermore, the amount of physics a radiologist would need to know would be greatly reduced and this book might not be necessary!

Medical exposures are not the source of the highest radiation dose to the population. Some 200 million gamma rays pass through the average individual each hour from soil and building materials and about 15 million potassium-40 atoms disintegrate within us each hour emitting high energy beta particles and some gamma rays.

However, as shown in Table 12.1, medical exposures contribute a far greater proportion of the average annual dose to the UK population than any other form of man-made radiation and the majority of this can be attributed to diagnostic radiology. Since this radiation can cause deleterious effects, it is essential for the radiologist to know what these effects are and to be aware of the risks when a radiological examination is undertaken.

12.2 Radiation Sensitivity of Biological Materials

12.2.1 Molecular Basis of High Radiosensitivity

Evidence on the lethal dose to humans is rather limited. Some information comes from the effects of atomic bombs dropped on Japan, some from medical exposures and some from criticality accidents such as Los Alamos in 1958 and Chernobyl in 1986. However, there are often confounding factors. For example, many of the firemen at Chernobyl suffered severe but superficial beta 'burns' (see Section 12.2.4) so the precise cause of death is difficult to determine.

Extrapolation from experiments with animals, combined with the limited data available on humans, suggests that the dose of acute whole body radiation required to kill about 50% of a human population within the first 30 days (LD 50/30), without medical intervention would be about 4 Gy to the bone marrow. This would correspond to a much higher skin dose of 80 kVp X-rays owing to self-absorption in the body.

Several aspects of this statement need to be discussed, but first consider its meaning in terms of absorbed energy. When X-rays are absorbed in living material each ionisation event results in a cluster of ion pairs. Formation of one ion pair requires only 35 eV of energy (on average ~100 eV per cluster) so even the small exposure in a dental examination will generate millions of ion pairs in the exposed tissues. Furthermore, the ionisation process is such that this energy is not deposited uniformly. Relatively large volumes of

TABLE 12.1

Annual Contribution to the Dose to the UK
Population from Different Sources of Radiation

Source	Percentage
Natural	
Cosmic rays	12
Gamma rays from ground and buildings	13
Internal from body, food and drink	9.5
Radon and thoron	50
Man made	
Medical	15
Other (nuclear discharges, occupational, fall out)	0.5

Source: Watson S J, Jones A L, Outway W B et al., Ionising
radiation exposure of the U.K. population: 2005
review. Document HPA – RPD – 001, Health
Protection Agency, Chilton, Didcot, U.K. 2005.

each cell, and the macromolecules within it, receive no energy at all but at an ionisation site the energy deposited is much higher than that associated with cellular biochemical events. For example, the energy required to break a hydrogen bond in DNA is about 0.5 eV or 70 times less than the energy in an ion pair. The critical difference between ionising and non-ionising radiation is the size of the individual packets of energy not the total energy involved. The consequence is that even a few microgray of diagnostic X-rays in the energy range 20–100 keV will produce a large number of random sub-molecular events, any one of which may damage a sensitive macromolecule.

Because of this, in terms of absorbed energy ionising radiation is by far the most potent agent known to man. A lethal dose of 4 Gy corresponds to absorbed energy of 280 J in a 70 kg man and is about the same as the amount of energy in a small quantity of warm tea. The rise in body temperature may be calculated as energy absorbed/mass × specific heat. Using the specific heat capacity of water (4.2×10^3 J kg^{-1} K^{-1}), the temperature rise is $280/70 \times 4.2 \times 10^3 = 10^{-3}$ K. This would be virtually undetectable and is much less than diurnal variations in body temperature.

It is against this unique sensitivity of cells and tissues to ionising radiation that the use of X-rays in diagnosis must be assessed.

12.2.2 Cells Particularly at Risk

Cell populations may be sub-divided into several categories. A few of the more important ones are as follows:

(a) Closed static—composed entirely of fully differentiated cells with no mitotic activity

(b) Transit—a steady state is maintained by balanced in-flow and out-flow, some cell division may occur during transit

(c) Stem cells—a self-maintaining system with significant mitotic activity, producing cells for another population

There is a long-established 'law' in radiobiology, first proposed by Bergonié and Tribondeau (1906) which, in modern parlance, states that 'the more rapidly a cell is dividing, the greater its sensitivity'. From this it follows that the lower the degree of functional and morphological differentiation, the higher the radiosensitivity.

The law applies well to rapidly dividing cells such as spermatogonia, haematopoetic stem cells, intestinal crypt cells and lymphoma cells, which are all very radiosensitive. Differentiated cells are generally relatively radioresistant but three important exceptions should be noted. Small lymphocytes, primary oocytes, especially just before release, and neuroblasts are all radiosensitive.

At the lowest dose at which radiation-induced death is likely to occur, the primary effect will be severe depletion of the bone marrow stem cells. Hence dose to bone marrow is the most meaningful LD 50/30 to quote.

12.2.3 Time Course of Radiation-Induced Death

After a potentially lethal dose of about 4 Gy to the whole body a typical time sequence might be as follows:

(a) 0–48 h; loss of appetite, nausea and fatigue

(b) 48 h to 2–3 weeks; latent period of apparent well-being

(c) 2–3 weeks to 4–5 weeks; the manifest illness stage, which may include fever, loss of hair, extreme susceptibility to infection, haemorrhage and other symptoms

(d) 5 weeks onwards; the situation will have resolved itself one way or the other

This time scale reflects the changes taking place at the molecular and cellular level.

After exposure to ionising radiation, physical processes of absorption of photons of energy hf, ionisation and excitation will be complete within about 10^{-15} s. With any form of ionising radiation there is a possibility that this interaction will be directly with critical targets in the cell. Experimental irradiation with microbeams has shown that the cell nucleus, containing DNA and the chromosomes, is most sensitive to radiation injury so this is where the targets are likely to be located.

For diagnostic X-rays, however, it is more likely that the action will be indirect, the X-rays first interacting with other atoms or molecules in the cell to produce free radicals that are able to diffuse far enough to reach and damage the critical targets.

Since 80% of a cell is composed of water, indirect effects of radiation are most likely to involve water molecules. As a result of interaction with a photon of X-rays, the water molecule may become ionised

$$H_2O \rightarrow H_2O^+ + e^-$$

H_2O^+ is an ion radical with an extremely short lifetime, about 10^{-10} s. It decays to a free radical which is uncharged but still has an unpaired electron and hence is highly reactive. Further reactions may now occur, for example, if this free radical reacts with another water molecule, the highly reactive hydroxyl radical $OH^{\cdot}$ may be formed

$$H_2O^+ + H_2O \rightarrow H_3O + OH^{\cdot}$$

TABLE 12.2

Effects of Radiation Exposure on Nuclear DNA

Base modification/deletion—causing genetic defects and increased mutation rate in reproductive cells; or if not repaired or eliminated, increased risk of malignant cell transformation in somatic cells

Bond breakage—between complementary strands of DNA in the double helix, facilitating the loss of a base and changes in molecular shape and structure

Cross linkage—the additional covalent binding of two adjacent strands of DNA; this potentially inhibits semi-conservative replication of DNA

Single strand breaks—occur at random in either strand along the DNA double helix

Double strand breaks—may be formed either by a single event, e.g. when the track of a densely ionising particle passes through or close to the DNA helix, or more likely when X- or gamma rays, by random coincidence of two single events occurring at the same time on complementary DNA strands. This process becomes more probable the higher the X-ray dose and dose rate

Source: Reproduced with permission from Martin C J, Dendy P P and Corbett R H (eds.) *Medical Imaging and Radiation Protection for Medical Students and Clinical Staff,* British Institute of Radiology, London, 2003.

Hydrated electrons are also formed and are very reactive. Reactive radicals recombine rapidly, normally to re-form water, but they can diffuse over distances of a few nanometres, thus reaching and damaging DNA. Alternatively, they may react together to form toxic products such as hydrogen peroxide

$$OH^{\cdot} + OH^{\cdot} \rightarrow H_2O_2$$

Hence the body is flooded with toxins and this is why the general feeling of malaise results.

The steps leading from these initial physico-chemical changes to the observation of cell death (see Section 12.3) are still poorly understood. However, it is known that DNA is a major target for damage and a number of effects have been identified (see Table 12.2).

Further evidence for the direct involvement of DNA in radiation damage comes from the fact that many genes, for example, the tumour suppressor gene p 53, have been shown to be activated by even low doses of radiation. A summary of these effects is given in the 'insight'.

Insight

Genes Showing Early or Late Response to Radiation

Some effects of X-ray activation of early and late response genes are

- 'Housekeeping genes' repair point mutations and single strand DNA breaks using 'cut and patch' enzymes.
- 'Checkpoint genes' arrest cells in the cell cycle thus giving more time for repair before a single strand break can be converted into a more permanent double strand break during DNA synthesis.
- 'Checkpoint genes' may also prevent a cell proceeding to mitosis if double strand breaks cannot be repaired or may induce a cell to differentiate or be removed by apoptosis (see Section 12.7.2).
- Some patients, for example, those with *Xeroderma pigmentosum* or *Ataxia telangiectasia* have enhanced radiosensitivity which is probably linked to their reduced capacity for DNA repair.

In the longer term, according to the law of Bergonié and Tribondeau, the differentiated cells will have resisted the radiation well and will continue to fulfil their specialised functions, so a period of relative well-being should ensue. However, as these cells die, their replacement from the stem cell pool will have been severely depleted or will have stopped completely. The patient becomes manifestly ill when there is a marked loss of a wide range of differentiated, mature cells particularly in the circulating blood. Ultimately, the cause of death will usually be failure to control infection or failure to prevent haemorrhage.

The time scale of events for sub-lethal or just lethal radiation damage, together with detail of the techniques available to study the various stages, is shown in Figure 12.1.

12.2.4 Other Mechanisms of Radiation-Induced Death

Above 4 Gy, damage to the gastrointestinal stem cells becomes increasingly important and this effect dominates in the dose range 10–100 Gy with loss of body water and body salts into the gut being the major cause of death. The time scale is now much shorter and death will occur in about 3–4 days.

At doses of about 10 Gy to the skin, radiation 'burns' will occur as a result of damage to the single basal layer of rapidly dividing stem cells in the epidermis. Note that these are not 'burns' in the thermal sense but an acute radiation reaction.

At even higher doses, death may be caused by disturbance of the central nervous system (100–150 Gy), loss of lung function (150 Gy) and ultimately simultaneous disruption of the total body chemistry (above 200 Gy).

Survival time plotted as a function of dose might be as shown in Figure 12.2, but note that doses shown are only orders of magnitude and in any given situation death will probably be due to a combination of causes.

12.2.5 Transformation of Cells and Cancer Induction

Evidence is also accumulating on why cells are sometimes transformed, and may become malignant, rather than killed. Clinical cancer develops through a multi-stage process requiring initiation, progression and promotional events. Radiation may act by causing the over-mutation or over-expression of proto-oncogenes which are subsequently converted into full oncogenes. Some features of the increased risk resulting from radiation exposures are

- No particular genetic locus or chromosome is involved.
- Initiation is most likely caused by gross genome loss (see Sections 12.5.1 and 12.5.3).
- At least two events are required.
- Fifty or more cycles of the transformed oncogene cells are required before full malignancy is established.

For further comment on cancer induction at low doses and a model for oncogenesis based on these assumptions see Section 12.7.

12.3 Evidence on Radiobiological Damage from Cell Survival Curve Work

About 50 years ago, the first reports appeared of successful attempts to use a clonogenic assay of cell survival following irradiation. In the simplest version of this method a small

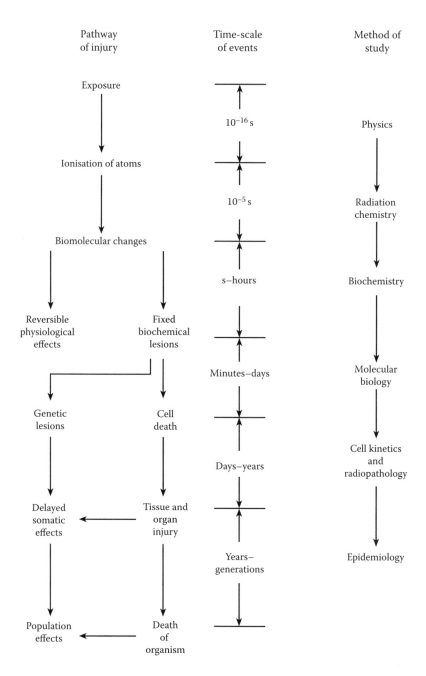

FIGURE 12.1
Time scale of events for radiation-induced damage.

number of single cells is placed in a Petri dish with growth medium and incubated for 10–14 days. The cells settle on the base of the dish and, if they are capable of cell division (reproductive integrity) they develop into sub-macroscopic colonies which may be counted. Not all cells are capable of growing into colonies, even if unirradiated. Suppose that 100 cells are seeded and 90 colonies grow. If now a second sample of the same cells is irradiated before

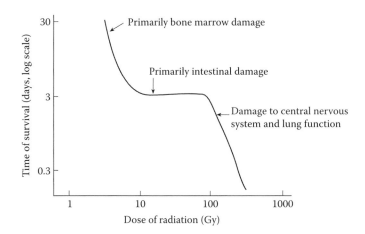

FIGURE 12.2
Approximate representation of survival time plotted as a function of dose for acute exposure to whole body radiation.

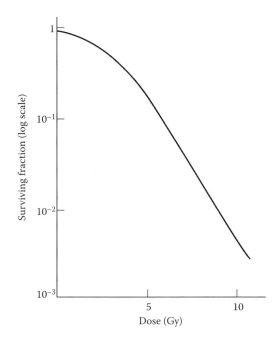

FIGURE 12.3
A typical clonogenic survival curve for mammalian cells irradiated *in vitro* with X-rays.

incubation, from 1000 cells only about 180 colonies might develop. The expected colonies from 1000 cells would be 900 and therefore 180/900 or 20% of the irradiated cells have survived. By repeating this experiment at different doses, a survival curve may be obtained and for mammalian cells exposed to X-rays it would resemble Figure 12.3.

For further details of the extensive literature on survival curves, including the evidence that qualitatively similar curves are obtained *in vivo*, the reader is referred to more specialised texts (e.g. Hall and Giaccia 2006; Mettler and Upton 2008). However, two aspects that are of great relevance to this chapter will be discussed.

12.3.1 Cellular Repair and Dose Rate Effects

After a dose of 1–2 Gy, some cells are killed and others damaged. It may be demonstrated convincingly that some damaged cells will recover by performing a 'split dose' experiment. In Figure 12.4 the dotted curve is reproduced from Figure 12.3. The solid curve would be obtained if a dose of, say 4 Gy were given but a time interval of 10 h elapsed before any further irradiation. The 'shoulder' to the curve has reappeared and the total dose to achieve a given surviving fraction is higher for the split dose than for the single dose. The recovery effect is detectable by 2 h and reaches a maximum by 24 h. Numerous other experiments have also shown that cells possess repair mechanisms.

One consequence of repair is that when radiation exposure is protracted, the effect may be dose rate dependent (see Figure 12.5). A simple explanation is that recovery is occurring during radiation exposure. Conversely, dose rate effects are evidence of recovery. They indicate that cell killing is occurring as a result of a sequence of events following interaction of more than one X-ray photon with the cell.

Cellular repair is observed *in vivo* but now another recovery mechanism is also observed. This has a much longer time scale and is primarily a result of homeostatic mechanisms involving stem cells in the whole animal. One consequence is shown in Figure 12.6. The LD 50/30 for acute exposure is about 4 Gy (4 Gy at 100 Gy per day corresponds to an exposure time of about 1 h), but the LD 50/30 increases when the radiation is protracted, first

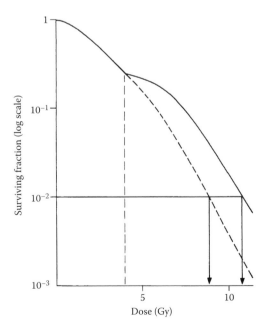

FIGURE 12.4
Typical results for a 'split dose' experiment to demonstrate recovery from radiation and cellular repair. If an initial dose of 4 Gy is delivered but there is a time delay before further irradiation, the survival curve follows the solid line rather than the dotted line. Note that the shoulder to the curve re-appears and the total dose to produce a given surviving fraction is now higher.

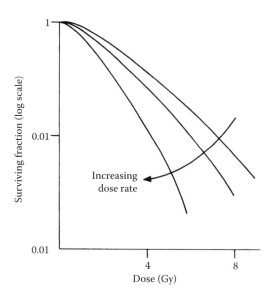

FIGURE 12.5
Graphs demonstrating that because of recovery the killing effect of X-rays may be dose rate dependent.

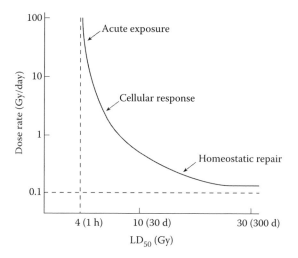

FIGURE 12.6
Curve to show that the LD_{50} *in vivo* is very dependent on the dose rate because of cellular and homeostatic mechanisms. The LD_{50} was measured in animals under conditions of continuous irradiation. The duration of irradiation is shown in brackets.

because of cellular repair and subsequently because of homeostatic repair. The results suggest that animals could tolerate 0.1 Gy per day for a long time.

12.3.2 Radiobiological Effectiveness

If the survival curve experiment is repeated with neutrons, the result shown in Figure 12.7 will be obtained. A smaller dose of radiation is required to produce a given killing effect and the curve has a smaller shoulder, indicating less capacity to repair sub-lethal damage.

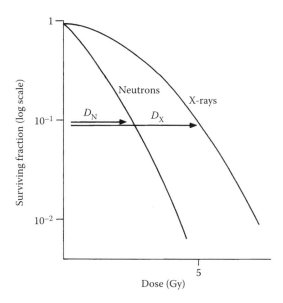

FIGURE 12.7

Comparative survival curves for X-rays and neutrons. If 220 kVp X-rays have been used as the reference radiation RBE = D_X/D_N for a given biological end point (10% survival in this example).

If 220 kVp X-rays have been used, then as shown on the curve, the RBE is defined as

$$RBE = \frac{\text{Dose of 220 kVp X-rays}}{\substack{\text{Dose of radiation under test to cause} \\ \text{the same biological end point}}}$$

The RBE of neutrons when determined in this way is frequently between 2 and 3.

Except at very high LET values (see Section 1.14), the RBE of a radiation increases steadily with LET. However, it is an incomplete answer simply to state that 'neutrons cause more damage than X-rays because they are a higher LET radiation and therefore produce a higher density of ionisation'. For equal absorbed doses measured in grays, the number of ion pairs created by each type of radiation in a macroscopic volume is the same. Therefore, for reasons not yet fully understood, and beyond the scope of this book, it is differences in the spatial distribution of ion pairs in the nucleus at the sub-microscopic level that cause the difference in biological effect. This is illustrated in Figure 12.8.

12.4 Radiation Weighting Factors, Equivalent Dose and the Sievert

Experiments on radiobiological effectiveness demonstrate clearly that, when attempting to predict the possible harmful effects of radiation, the purely physical concept of absorbed dose, as measured by the gray, is inadequate. However, in numerical terms RBE is a very difficult concept since it varies with dose, dose rate and fractionation, physiochemical conditions such as the presence or absence of oxygen, the biological end-point chosen, the biological species and the time after irradiation at which the measurement is made.

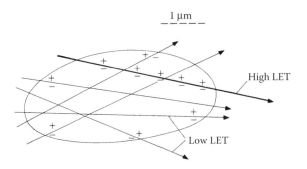

FIGURE 12.8
Illustration of the difference in spatial distribution of ion pairs across a cell nucleus for low LET and high LET radiations. In each case five ion pairs are formed (same absorbed dose) but whereas these are likely to result from the same high LET particle, and thus be quite close together, they are more likely to result from five different low LET photons and hence be much more widely separated.

Furthermore, since the shapes of the survival curves are different, inspection of Figure 12.7 shows that, at the very low doses that are important in radiological protection, the RBE may be somewhat higher than the value of 2 or 3 quoted for higher doses.

Largely for these reasons, the International Commission on Radiation Units introduced a new term 'Quality Factor' (see e.g. ICRU 1980). This is a dimensionless, invariant quantity for a given type of radiation, determined solely by the type of radiation. However, this quality factor was applied to absorbed dose at a point and in radiological protection it is the absorbed dose averaged over a tissue or organ and weighted for the radiation quality that is of interest.

Thus in 1990 the International Commission on Radiological Protection (ICRP 1991) introduced a new concept, the radiation weighting factor w_R. This should now be used to calculate the *equivalent dose* to a tissue T, H_T according to the equation

$$H_T = w_R \cdot D_T$$

where D_T is the dose averaged over tissue T from the radiation type R.

This is now considered to be an equivalent dose because for equal values of H_T the damage to tissue T will be the same for different types of radiation.

w_R is dimensionless so H_T still has the units J kg^{-1} but is now given the special name sievert (Sv).

Values of w_R are representative of the range of RBE values for that radiation in inducing stochastic effects (see next section) at low doses. A simplified version of the latest ICRP recommendations (ICRP 2007) is given in Table 12.3.

The concepts of radiation weighting factor and equivalent dose should only be applied in the context of radiological protection.

12.5 Radiation Effects on Humans

12.5.1 Tissue Reactions and Stochastic Effects

The biological effects of ionising radiation on humans can be divided into two general categories. For some effects there appears to be a definite threshold dose below which no

TABLE 12.3

Radiation Weighting Factors

Radiation Type and Energy Range	Radiation Weighting Factor w_R
Photons (all energies)	1[a]
Electrons	1[b]
Protons	2
Neutrons	See footnotes
Alpha particles, heavy ions	20

Notes: For neutrons w_R is a continuous function, with a maximum $w_R = 20$ at 1 MeV but falling to about 2.5 at lower and higher energies.

[a] There have been proposals, based on RBE data, that the w_R value for very low energy X-rays (e.g. ~10–25 keV used in mammography) should be higher. In the absence of unequivocal evidence for a higher value ICRP has opted for simplicity.

[b] An important exception to this table may be the radiation weighting factor for Auger electrons (Section 3.4.2), especially if they are emitted from a radionuclide that is closely bound to DNA. There is evidence that these electrons may be as harmful on an equi-absorbed dose basis as α-particles because the average distance between ionising events is about the same as the distance between DNA strands.

Source: ICRP, Recommendations of the International Commission on Radiological Protection, ICRP Publication 103, Ann. ICRP. 37, 2–4, 2007.

damage, in terms of measurable biological response, can be detected. Effects of ionising radiation such as skin erythema, epilation and opacification of the lens of the eye and death resulting from acute exposure are in this category.

Such effects are described as *tissue reactions* (or *deterministic effects*) and are characterised by a dose-response curve of the type shown in Figure 12.9a. They act at the tissue level, especially where the cells are mitotically active, and have the following features:

(a) 100 mGy or more are required to observe an effect.

(b) The dose threshold varies from one tissue to another.

(c) It is a somatic effect affecting the exposed individual.

(d) Repair and recovery can occur.

(e) The severity of effect depends on dose/dose rate/number of exposures.

(f) The effect usually occurs early, that is in days or weeks and may be repaired quickly afterwards.

(g) Mechanisms are relatively well understood, for example in radiotherapy.

Threshold doses for the most sensitive tissue reactions are summarised in Table 12.4.

These values are all well above the doses received by patients in conventional radiology. However, there have been isolated reports of radiation-induced skin injuries to patients resulting from prolonged fluoroscopically guided invasive procedures. Absorbed dose rates to the skin from the direct beam of a fluoroscopy X-ray system are typically in the range 0.01–0.05 Gy/min but may reach 0.2 Gy/min or even higher. A few minutes screening at this higher dose rate could cause early transient erythema and any additional fluorographic dose required for film or digital image recording must not be overlooked. For discussion of possible hazards from interventional procedures see Section 9.6.3.

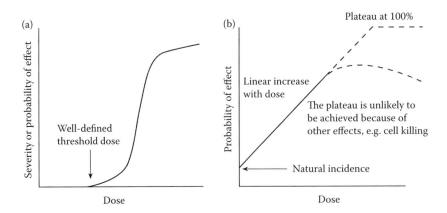

FIGURE 12.9
Idealised dose-response curves for (a) tissue reactions; (b) stochastic effects.

TABLE 12.4

Typical Threshold Doses for the Most Sensitive Tissue
Reactions

Tissue and Effect	Absorbed Dose for a Brief Exposure (Gy)
Testes	
Temporary sterility	0.15
Permanent sterility	>3.5
Ovaries	
Sterility	>2.5
Lens	
Detectable opacities	>0.5
Visual impairment (cataract)	>2.0
Bone marrow	>0.5
Depression of haematopoiesis	
Skin	
Early transient erythema	2
Temporary epilation	3

Stochastic effects, literally those governed by the laws of chance, are thought to occur primarily at the cellular level. Since a single ionising event may be capable of causing radiation injury, for example, to the DNA, and even the lowest dose diagnostic examinations cause millions of ionisations in each gram of irradiated tissue, it is normal to assume that there is no threshold dose for stochastic effects of ionising radiation. Thus the curve relating probability of effect to dose has the form shown in Figure 12.9b.

The two most important long-term effects of radiation, namely carcinogenesis and mutagenesis are thought to be stochastic in nature. The former is presumably the result of damage to a somatic cell, which either initiates or promotes a malignant change, the latter the result of damage to a germ cell. A further important feature of stochastic effects is that whereas their frequency will increase with increasing dose, their severity does not. Thus, for example, the degree of malignancy of a radiation-induced cancer is not related to the dose. A further reason for believing that radiation-induced carcinogenesis is a stochastic effect is that there is no evidence for a threshold dose for radiation-induced cancer in the Japanese survivors. There was little or no evidence for recovery effects in the multiple fluoroscopy work where patients received many small exposures over a period of time. Finally, a truly stochastic mechanism would exclude the possibility of recovery.

Note that carcinogenesis and mutagenesis may be contrasted in that the former is somatic, that is the effect is observed in the exposed individual, whereas the latter is hereditary, with the effect being detected in the descendants.

12.5.2 Carcinogenesis

There is ample evidence from a wide range of sources that ionising radiation can cause malignant disease in humans. For example, occupational exposure results in a greatly increased incidence of lung cancer among uranium miners, and in the period 1929–1949 American radiologists exhibited nine times as many leukaemias as other medical specialists. A frequently quoted example of industrial radiation-induced carcinogenesis is the 'radium dial painters'. They were mainly young women employed during and after the First World War to paint dials on clocks and watches with luminous paint. It was their custom to draw the brush into a fine point by licking or 'tipping' it. In so doing, the workers ingested appreciable quantities of radium-226 which passed via the blood stream to the skeleton. Years later a number of tumours, especially relatively rare osteogenic sarcomas were reported.

There is also evidence from approved medical procedures. For example, between 1939 and 1954, radiotherapy treatment was given to the whole of the spine for more than 14,000 patients suffering from ankylosing spondylitis. Statistically significant excess deaths due to malignant disease were subsequently observed, especially for leukaemia and carcinoma of the colon. Other data comes from the use of thorotrast, which contains the alpha emitter thorium-232 as a contrast agent in radiology; X-ray pelvimetry; multiple fluoroscopies and radiation treatment for enlargement of the thymus gland. Unfortunately, radiation exposure in medical procedures is associated with a particular clinical condition. Therefore it is difficult to establish suitable controls for the purpose of quantifying the effect.

Finally, there is information gathered from the survivors of the Japanese atomic bombs. This has been fully reported in a series of articles published over many years in Radiation Research (Preston et al 2007) and by UNSCEAR (2008).

Five key points, which all have a bearing on radiation risk, can be made from this follow-up:

1. The risk of cancer is not the same for all parts of the body, many parts are affected, but not all to the same degree.

2. There is a long latent period before the disease develops—extra cases of leukaemia peaked between about 5 and 14 years after exposure since when the relative risk (i.e. expressed relative to the natural incidence) has decreased. For solid tumours the relative risk has continued to rise (see Figure 12.10).

3. There is no evidence of a threshold dose.

4. In terms of relative risk, the lifetime risk of cancer is highest in those who were less than 10 years of age at the time of exposure and then falls steadily with age.

5. Evidence is in reasonably good agreement with that from other sources and permits a risk estimate.

Wall et al. (2006) listed 22 epidemiological studies of the effects of exposure to external low LET radiation. Eighteen of them were medical and most involved the relatively high doses

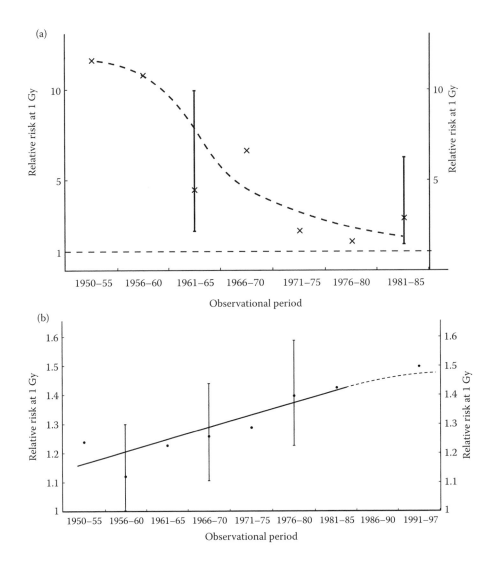

FIGURE 12.10
(a) Follow-up of relative risk of radiation-induced cancer (leukaemia) in the Japanese survivors. (Adapted from Shimizu Y, Kato H and Schall W J, Studies of the mortality of A bomb survivors. 9 Mortality 1950–1985, Part 2, Cancer mortality based on the recently revised doses DS 86, *Radiat. Res.* 121, 120–141, 1990.) (b) Follow-up of relative risk of radiation-induced cancer (all cancers except leukaemia) in the Japanese survivors Note that the straight line is drawn to indicate the trend—the incidence may be starting to plateau but is not falling. Some error bars are given to show the uncertainty of the data. (Adapted from Shimizu Y, Kato H and Schall W J, Studies of the mortality of A bomb survivors. 9 Mortality 1950–1985, Part 2, Cancer mortality based on the recently revised doses DS 86, *Radiat. Res.* 121, 120–141, 1990. With additional data derived from Preston et al., 2003.)

used in radiotherapy of malignant or benign conditions. The main sources of evidence that radiation causes cancer in humans are summarised in Table 12.5.

12.5.3 Mutagenesis

The circumstantial evidence that ionising radiation causes mutations in humans is overwhelming. For example, mutations have been observed in a wide variety of other species,

TABLE 12.5

Sources of Evidence That Ionising Radiation Causes Cancer in Humans

Occupational	Medical Diagnosis
Uranium miners	Prenatal irradiation
Radium ingestion (dial painters)	Thorotrast injections
American radiologists	Multiple fluoroscopies (breast)
Atomic Bombs	**Medical Therapy**
Japanese survivors	Cervix radiotherapy
Marshall islanders	Breast radiotherapy
	Radium treatment
	Ankylosing spondylitis and others

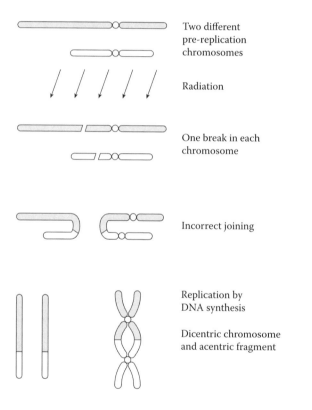

Two different pre-replication chromosomes

Radiation

One break in each chromosome

Incorrect joining

Replication by DNA synthesis

Dicentric chromosome and acentric fragment

FIGURE 12.11

Steps in the formation of a dicentric chromosome and acentric fragment as a result of radiation breaks and faulty rejoining.

including plants, bacteria, fruit flies and mice, and radiation is known to impair the learning ability of mice and rats.

Furthermore, radiation is known to cause extensive and long-lasting chromosomal aberrations. These may occur either pre-replication (chromosome aberrations) or post-DNA replication (chromatid aberrations). One mechanism by which these aberrations can arise from breaks and faulty rejoining is shown in Figure 12.11. Dicentric chromosomes, which occur when two different chromosomes break and then rejoin incorrectly (shown here) and ring chromosomes, when two breaks occur in the same chromosome, are particularly damaging since they are likely to prevent separation of the chromatids when the cell attempts mitosis. An important method of retrospective dosimetry is to score such aberrations in cells circulating in the peripheral blood (Edwards 1997). For a dose of 50 mSv whole body radiation, one or two dicentric chromosomes would be scored for every 1000 mitotic cells examined. The normal incidence is negligible.

Notwithstanding, researchers have so far failed to demonstrate convincing evidence of hereditary or genetic changes in humans as a result of radiation, even in the offspring of Japanese survivors. For the latter group four measures of genetic effects, ranging from untoward pregnancy outcomes to a mutation resulting in an electrophoretic variant in blood proteins have been studied. The difference between proximally and distally exposed survivors is in the direction expected if a genetic effect had resulted from the radiation, but none of the findings is statistically significant.

This failure to record a significant effect is presumably caused by the statistical difficulty of showing a significant increase in the presence of a high and variable natural incidence of both physical and mental genetically related anomalies. For severe disability the natural incidence is between 4% and 6%.

There are additional problems in assessing genetic risk arising from the diagnostic use of radiation. For example, only radiation exposure to the gonads is important and even this component can be discounted after child-bearing age. Second, many radiation-induced mutations will be recessive, so their chance of 'appearing' may depend on the overall level of radiation exposure in the population. Finally, the risk to subsequent generations will depend on the stability of the mutation once formed.

On the basis of the measured mutation rate per locus in the mouse, adjusted for the estimated comparable number of loci in the human, the effective dose required to double the mutation rate in humans is estimated to be at least 1 Sv. Hence 10 mSv per generation parental exposure might increase the spontaneous mutation incidence by about 1%. An alternative way to express the mutation risk will be discussed in the next section.

12.6 Generic Risk Factors and Collective Doses

Various international bodies have examined the data on cancer induction and genetic damage caused by radiation and used it to make an assessment of the risks of exposure to low doses of radiation. For example, as shown in Figure 12.12 the incidence of breast cancer in atomic bomb survivors, measured in women-years (WY), has been compared with three other sources of data (Boice et al. 1979; Upton 1987). All give reasonably good agreement that the probability of breast cancer in females is about 1% Sv^{-1}. (This notation will be used throughout to express risk factors. It means, literally, that if a large population of females were each exposed to 1 Sv, on average 1 in 100 would contract a radiation-induced breast cancer. Note, however, the remarks on collective doses later in this section.)

Not surprisingly, in view of the large statistical errors on available data and the extrapolation required from high doses, these estimates often vary considerably (see, for example, Table 9.1). It is important to keep this in mind in what follows. Some of the figures will be very approximate.

Since 1955 the ICRP has been making periodic reviews of all the available data on radiation risks to humans and making appropriate recommendations. The latest three reviews

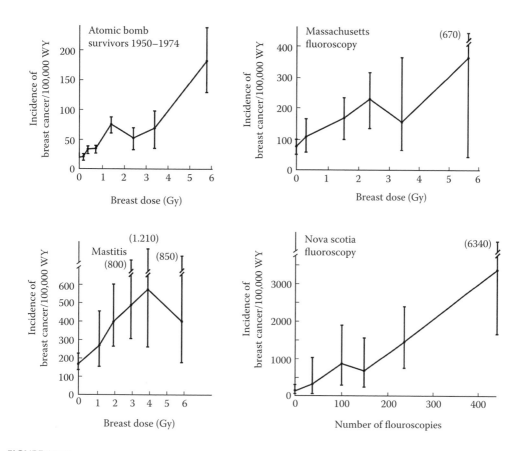

FIGURE 12.12

Incidence of cancer of the female breast as a function of dose in atomic bomb survivors, women treated with X-rays for acute *post partum mastitis* and women subjected to multiple fluoroscopic examinations of the chest during treatment of pulmonary tuberculosis with artificial pneumothorax. (Redrawn from Boice J R Jr, Land C E, Shore R E et al. Risk of breast cancer following low dose exposure, *Radiology* 131, 589–597, 1979.)

have been in 1977, 1990 and 2007 (ICRP 1977, 1991, 2007). The 1990 recommendations made some important changes to both risk factors and radiation protection terminology. In particular the risk estimate for radiation-induced cancer was increased by about a factor of four. Some of the reasons for this increase were as follows:

- Most of the information gathered between 1977 and 1990 had led to a higher estimate of risk. For example, dosimetry at Hiroshima and Nagasaki had been recalculated. This had shown that the previous estimates of neutron doses were too high, especially at Hiroshima. Because neutrons have a high radiation weighting factor, they make a relatively high contribution to the equivalent dose. Reducing the calculated neutron contribution increases the cancer risk per Sv.

- A second point was that prolonged follow-up had continued to show an excess of solid tumours in Japanese survivors exposed to radiation. Follow-up to 1985 gave better information on the increased risk to those who were under 10 years of age at the time of irradiation. In 1990, different risk factors were quoted for the work force (18–65 years) and the whole population.

- Third, the model for cancer induction was reviewed. Before 1990 it had been assumed to be additive. Now there is evidence that, for some cancers, for example breast, a relative risk model is more appropriate. This assumes there will be a proportional increase in cancer at all ages and as the population gets older and the natural incidence of cancer increases, the extra cancers will also increase on this model.
- Finally, in 1990 it was decided to make some allowance for the detrimental effect on quality of life of non-fatal cancers.

The 2007 recommendations contain relatively little data impacting significantly on risk factors and are, for the most part, a consolidation and clarification of earlier recommendations. The only substantial change in risk figures has been a reduction in the estimated heritable effects and most of this change is a result of a new method of presentation rather than new data—see Insight.

Insight

ICRP (2007) Revision of Heritable Effects

- Extended follow-up of offspring of Japanese survivors has provided little new data to alter conclusions of previous analyses (no effects have been demonstrated).
- Radiation-induced mutations are mostly large multi-gene deletions unlikely to result in live births.
- ICRP 60 (1991) assumed all genetic lesions should be treated as lethal. The lethality fraction has now been reduced to 80%.
- ICRP 103 (2007) has reverted to looking at the effect over two generations (as in ICRP 26 1977) because of severe scientific limitations on estimating an equilibrium value. This factor alone reduces the risk coefficient by about a factor of 2–3.

ICRP Recommendations are essential reading for radiation protection experts but can be quite heavy going. The reader may find a review of the 2007 recommendations by Wrixon (2008), which also contains a comprehensive list of references, is adequate. Nominal risk coefficients for 2007, 1990 and 1977 are compared in Table 12.6. Note that risks to the whole population are averaged over both sexes and a wide range of ages. Since cancer risk varies with age and sex, risks to individuals, for example, in diagnostic medical exposures, may

TABLE 12.6

Nominal Risk Coefficients in ICRP Recommendations

	2007 %Sv^{-1}		1990 %Sv^{-1}		1977 %Sv^{-1}
	Work Force	Whole Population	Work Force	Whole Population	
All cancers	4.1	5.5	4.8	6.0	1.25[a]
Heritable effects	0.1	0.2	0.8[b]	1.3[b]	0.4
Total (rounded)	4.2	5.7	5.6	7.3	2.0

Notes: In extrapolating from high doses ICRP considers that risks may be reduced by a factor of about 2 owing to dose and dose rate effects. This correction is known as a dose and dose rate effectiveness factor (DDREF) and has been made to the above figures which apply to low doses of radiation (see model A in Figure 12.13).

[a] Only fatal cancer was considered in 1977.

[b] The 1990 recommendations attempted to estimate an equilibrium value. Figures for 1977 and 2007 are for the first two generations only.

be somewhat different (but not necessarily outside the bounds of the considerable uncertainties attached to these figures).

There is a body of opinion that considers the potentially harmful effects of low doses of radiation are overstated. For example, the point is made that there is no direct evidence that the very low doses used in, say, dental radiology have caused any cancers at all. This is of course true. Observed excess cases have all been with much higher doses, for example the dose to breast tissue in early fluoroscopy work was estimated at 0.04–0.2 Gy and the number of examinations on one individual frequently exceeded 100.

Thus any extrapolation from observed data to lower doses is bound to be a model. Three possible models, which have all been widely discussed in the literature, are illustrated in Figure 12.13. However, we have no direct means of checking which of the three curves, or indeed any other shape of curve is correct. For further detail see the figure legend.

Resolution of this problem would require a much better understanding of the mechanism of carcinogenesis. However, if cancer arises as the final result of a series of successive changes, it is at least plausible that radiation-induced damage could be one key step in the chain of events. Since even a few micrograys of radiation will cause in excess of 10^8 ion pairs per gram of tissue and a single ion pair is capable of breaking a chromosome and hence causing a translocation, we must also accept that there is, effectively, no lower limit to the dose of radiation that will cause cancer. Clearly the vast majority of changes at the molecular level do not manifest themselves in disease!

This principle is extremely important in the diagnostic use of X-rays. If a *Linear response with No Threshold* is assumed, the LNT model—curve A in Figure 12.13, then doubling the dose of radiation will double the number of fatal cancers, however, low the dose. Also it is reasoned that 10 mSv given to 10,000 persons or 100 μSv given to a million persons are going to cause the same number of fatal cancers. A physical explanation for this reasoning would be that the same number of ion pairs has been created in each case. Therefore in

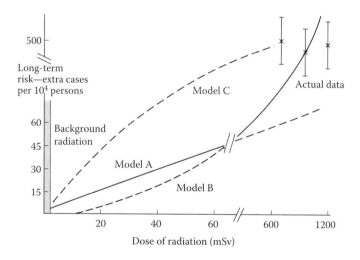

FIGURE 12.13

Three curves showing the possible variation in long-term risk with radiation dose. The solid line (model A) represents current thinking—a linear increase at low doses with no threshold (the LNT model). At higher doses (well above protection and diagnostic levels) the curve turns upwards. The dotted curves represent a non-linear increase, possibly with a threshold (model B) and a supralinear model in which low dose effects would be higher than predicted (model C). Typical doses from annual background radiation and simple radiological examinations are shown shaded.

situations where large numbers of persons are exposed to low doses, for example in diagnostic radiology it is the *collective dose* (100 personSv in this example) that will determine the detriment. Hence with a risk of fatal cancer of approximately 5% Sv⁻¹, a collective dose of 100 personSv to the whole population will cause five extra fatal cancers.

12.7 Very Low Dose Radiation Risk

Most of the quantitative evidence for radiation-induced cancer discussed in previous sections relates to the effects of relatively high doses—typically 1 Sv or more. Doses from radiological examinations, discussed in detail in Chapter 13, are equivalent to very low whole body doses of between 0.01 mSv and 20 mSv. This raises a key question 'When assessing radiation risk, to what extent is it valid to make a linear extrapolation from available data down to the much lower doses associated with diagnostic radiology?' To answer this question we must re-visit some topics already discussed. For a good review see Brooks et al. (2006).

12.7.1 Molecular Mechanisms

The typical clonogenic survival curve for mammalian cell irradiated *in vitro* (Figure 12.3) can be described by the equation

$$-\log S = k(\alpha D + \beta D^2)$$

where S is the surviving fraction, D the dose and k, α, β are constants.

However, there is increasing evidence that cell killing is not the only effect described by this equation. In different experimental conditions the yield of chromosome aberrations and the mutation frequency have both followed a similar equation. The similarity of response and reported correlations between effects, point to a common mechanism with both chromosome aberrations and mutations implicating DNA.

A relatively simple model of the way in which radiation may cause double strand breaks in DNA is that the double break may be caused by the passage of a single ionisation track (an effect which would be linear with dose) or two independent tracks (which would be quadratic with dose). Hence the number of breaks,

$$N = k'(\alpha D + \beta D^2).$$

Furthermore the model predicts that α/β will increase at low dose rates and for high LET radiation. Both effects have been observed in many systems. At very low doses the equation reduces to $N = k'\alpha D$, that is, the number of double strand breaks increases linearly with dose from zero dose upwards.

Double strand breaks also provide a starting point for a two-stage clonal expansion model, first proposed by Moolgavkar and Knudson (1981) and now commonly used for modelling radiation-induced and radiation-driven carcinogenesis. The model is illustrated in Figure 12.14. The organ of interest is assumed to contain N normal cells that have the potential to become malignant (M) in two rate-limiting stochastic steps (μ_1 and μ_2). In the intermediate stage (I) the growth advantage of cells on the pathway to malignancy is determined by the relative values of α (birth rate) and β (death rate) where $\alpha > \beta$. The growth time of

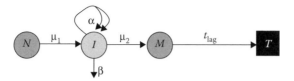

FIGURE 12.14
Schematic representation of the two-mutation model for cancer induction—for explanation see text. (With permission from Dendy P P and Brugmans M P J, Commentary: Low dose radiation risks, *Br. J. Radiol.* 76, 674–677, 2003.)

a malignant cell into a detectable tumour (T) depends on a deterministic lag time (t_{lag}). Spontaneous tumours result from spontaneous mutations. A single acute radiation exposure is assumed to increase the mutation rate in one of the steps but cannot influence both. At low dose, radiation is not thought to influence the clonal expansion stage. Variations in the relative values of α and β, and in t_{lag} can explain the well-documented variations between time of exposure and time of tumour appearance in the Japanese survivors.

In conclusion, experimental and theoretical work on double strand breaks in DNA strongly support the conclusion that cancer induction can be the ultimate outcome of a double strand break in DNA. The dose-effect curve is basically linear-quadratic but becomes linear at low doses extending back to zero dose.

12.7.2 Confounding Factors based on Radiobiological Data

There are an increasing number of reported effects which, whilst not destroying the theory of direct damage to DNA, suggest that the process is much more complex and that other effects could influence the response at very low doses.

12.7.2.1 Bystander Effects

It has been suspected for a long time, and fully confirmed in the past 10–15 years, that a cell does not have to be traversed by a charged particle to be influenced by it. Such effects are known as *bystander effects* (Mothersill and Seymour 2001).

Bystander effects have been demonstrated in two basically different ways:

1. Transferring culture media from irradiated cells to non-irradiated cells
2. Experiments with α-particle microbeams

One type of experiment with α-particles illustrates the principle.

- Use vital stains (i.e. ones which do not kill cells) to stain two groups of cells selectively—one group orange for the cytoplasm, the other blue for the nuclei
- Mix the cells and attach to a thin surface that can be penetrated by the α-particles
- Programme the microbeam so that only blue nuclei are hit—the microbeam is about 5 μm in diameter and the range of the α-particles so short that ionisation tracks will not extend outside the nucleus
- Fix 48 h later when micronuclei and chromosome bridges, implying significant chromosomal rearrangement, are observed in orange-stained cells which have not been directly irradiated

Cell-to-cell communication via gap junctions has been identified as an important contributing factor to the bystander effect. The effect is not observed when cells are sparse or gap junctions in close-packed cells are blocked.

There are no reports of similar effects *in vivo*, but if bystander effects occur *in vivo* the implications for extrapolation from high to low dose could be important. A dose of 1 mGy corresponds, on average, to about one ionisation track per cell so at low doses many nuclei are not hit. Two conflicting outcomes are possible, see Figure 12.15:

Either—the bystander effect causes harmful effects in unirradiated cells resulting in a supralinear effect (model D)

Or—the bystander effect amplifies adaptive processes (see next section) causing a sublinear effect (model E)

12.7.2.2 Adaptive Responses

Subsequent to the initial radiation damage, probably mostly to DNA, there are three possible biological outcomes, shown schematically in Figure 12.16.

1. The cell repairs the damage successfully.
2. The cell recognises serious damage and genetically activates the process of cell death (apoptosis).
3. The cell makes an erroneous repair. If the error is serious, cell death may still result (3a). If the cell remains viable and continues to divide, carcinogenesis, or in germ cell lines, mutagenesis, may result (3b).

The LNT hypothesis assumes that risk is only influenced by dose. However, Figure 12.16 shows that cancer is only one of four alternative outcomes. If the relative importance of

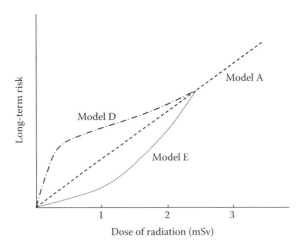

FIGURE 12.15
Expanded version of the very low dose region of Figure 12.13 showing possible consequences of the bystander effect. Model A is taken from Figure 12.13. In model D the bystander effect extends damage to unirradiated cells. In model E the bystander effect enhances adaptive processes. All curves are dotted to show there is no direct experimental evidence for any of them.

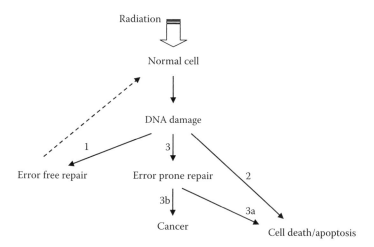

FIGURE 12.16
Possible outcomes of radiation damage at the cellular level—for explanation see text.

these alternative pathways varies with dose and/or time, the LNT hypothesis might not be valid.

Recent radiobiological research, both *in vitro* and in animals has shown a large number of 'adaptive responses' to low doses suggesting that alternative pathways can be altered. Further details are given in the 'insight' but the basic approach is to demonstrate that if a low priming dose of a few mSv is given, the system subsequently responds differently to both low and high doses. Overall the results show that

1. The adaptive effect of low doses reduces risk rather than increasing it.

2. Adaptive effects are not observed above 100 mGy.

3. Adaptive responses are not unique to mammals and may be part of an evolutionary response to background radiation.

Insight

Examples of Adaptive Responses

In Vitro
- The frequency of micronuclei in human skin cells exposed to a high dose of low LET radiation (4 Gy) is less if the cells have been exposed to a range of lower doses (1–100 mGy) 3 h previously.
- In non-dividing lymphocytes more cells die by apoptosis following a high dose (2 Gy) if the cells were exposed to 100 mGy 6 h earlier.
- In rodent cells a priming dose of 100 mGy delivered at low dose rate 24 h before a high dose (4 Gy) caused less malignant transformations than the high dose alone.
- Of particular interest to diagnostic radiology (because it involved no high doses) was the finding that in another strain of rodent cells single low doses of 1, 10 and 100 mGy all reduced the spontaneous malignant transformation frequency. Interestingly, the reduced frequency was the same at all doses, even though at 1 mGy not all cells would be traversed by an ionisation track. This may be evidence for the bystander effect.

In Vivo
  • A priming dose of 100 mGy at low dose rate the day before a carcinogenic dose of 1 Gy to
    genetically normal mice delayed the appearance of myeloid leukaemia.
  • The appearance of osteogenic sarcomas in genetically altered, cancer prone mice was
    delayed, relative to unirradiated controls, by doses of 10 mGy, but accelerated by 100 mGy.

12.7.3 Epidemiological Studies

Direct evidence on the risk of radiation-induced cancer in humans can only come from
direct observations on human populations, but there are many problems. One is that
because of the high natural incidence of cancer, large populations have to be studied to
demonstrate, statistically, an excess relative risk due to radiation. Another problem is that,
with one notable exception, large populations of humans exposed to doses of radiation
significantly higher than background do not exist.

The one exception is the survey of Japanese atomic bomb survivors, already referred to
earlier in the chapter, and this has become the major source of information for the following reasons:

  1. The study involves a large population of about 90,000 of all ages and both
     genders.

  2. About one-third of the survivors were exposed to doses in the range 5–100 mSv.
     100 mSv is about the highest dose of relevance in diagnostic radiology, correspond-
     ing to several CT scans.

  3. Cancer incidence and mortality data are available.

  4. Mortality follow-up is almost complete for adults, and about 50% complete for
     children.

  5. We hope it is unlikely that any comparable study can be performed in the future.

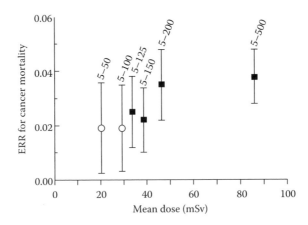

FIGURE 12.17
Excess relative risk (ERR) of cancer mortality for Japanese survivors at low doses. ERR is significantly different
from zero (■) for average doses of about 35 mSv (range 5–125 mSv) and above. For lower doses it is positive but
not significant (○). (With permission from Hall E J and Brenner D J, Cancer risks from diagnostic radiology,
Br. J. Radiol. 81, 362–378, 2008.)

Even for such a large population, cases had to be grouped into quite large dose bands. Thus the 5–50 mSv dose group was first compared with the controls (less than 5 mSv), then the 5–100 mSv group was compared with controls and so on. As shown in Figure 12.17 the excess relative risk became significantly different from zero for the 5–125 mSv group (mean dose 35 mSv). This is important because 35 mSv is comparable with some organ doses (35 mGy assuming $w_R = 1$ for X-rays) received in some high dose radiological procedures—for example, two CT scans or a long interventional procedure.

Some confirmatory evidence comes from a 15 nation study of 400,000 workers in the nuclear industry who received a mean dose of 20 mSv. Again the excess relative risk was above zero and almost identical to the figure for A-bomb survivors extrapolated to the same mean dose.

The best medical evidence comes from numerous studies of childhood cancer risk after foetal exposure to diagnostic X-rays. These have concluded that there is a significant increase in childhood cancer for foetal doses of 10–20 mGy.

12.8 Conclusions

In terms of absorbed energy ionising radiation is clearly the most harmful of all types of radiation. This is because of the unique way in which the ionisation process deposits packets of energy that are large compared to biochemical processes. Following the very early physico-chemical changes resulting from the radiation, double strand breaks in DNA are probably the most important early biological effect and manifest themselves in many different ways.

At high doses there is clear evidence that ionising radiation can cause cancer in humans and mutations in animals. Data from different sources are broadly consistent and permit risk estimates to be made. The absence of any major new ICRP recommendations between 1990 and 2007 gives confidence that radiation protection has matured into a sound, scientifically based discipline underpinning the systems of control of radiation exposure that have been introduced throughout the world.

The position at the very low doses associated with most diagnostic examinations is less clear cut. Epidemiological studies are no help below the upper end of the diagnostic range of doses and an increasing number of radiobiological experiments report adaptive processes at low doses. The ICRP (ICRP 2007) remains of the view that extrapolation of the LNT model (model A in Figure 12.13) below about 50 mSv is the most plausible scientifically. This is certainly the most robust approach for radiation protection planning purposes but care is required when applying it to estimate the risk from radiological exposures to patients. This subject is taken up again in Chapter 13.

References

Bergonié J and Tribondeau L, De quelques resultats de la radiotherapie et assai de fixation d'une technique rationelle, *C R Acad Sci* 143, 983, 1906 (Engl. Transl. Fletcher G H, Interpretation of some results of radiotherapy and an attempt at determining a logical treatment technique, *Radiat Res* 11, 587, 1959).

Boice J R Jr, Land C E, Shore R E et al., Risk of breast cancer following low dose exposure, *Radiology* 131, 589–97, 1979.

Brooks A L, Coleman M A, Douple E B et al., Biological effects of low doses of ionising radiation. In *Advances in medical physics*, eds. Wolbarst A B, Zamenhof R G, Hendee W R Medical Physics Publishing, Madison, 2006, pp. 256–286.

Dendy P P and Brugmans M P J, Commentary: Low dose radiation risks, *Br J Radiol* 76, 674–77, 2003.

Edwards A A, The use of chromosomal aberrations in human lymphocytes for biological dosimetry, *Radiat Res* 148 (Suppl.), 39–44, 1997.

Hall E J and Brenner D J, Cancer risks from diagnostic radiology, *Br J Radiol* 81, 362–378, 2008,

Hall E J and Giaccia A J, *Radiobiology for the radiologist* 6th ed., Lippincott, Williams and Wilkins, Philadelphia, 2006.

ICRP, Recommendations of the International Commission on Radiological Protection, ICRP Publication 26, *Ann ICRP* 1, 3, 1977.

ICRP, Recommendations of the International Commission on Radiological Protection, ICRP Publication 60, *Ann ICRP* 21, 1–3, 1991.

ICRP, Recommendations of the International Commission on Radiological Protection, ICRP Publication 103, *Ann ICRP* 37, 2–4, 2007.

ICRU, Radiation quantities and units, International Commission on Radiation Units and Measurements, *ICRU Publication* 33, 1980.

Martin C J, Dendy P P and Corbett R H (eds.), *Medical imaging and radiation protection for medical students and clinical staff*, British Institute of Radiology, London, 2003.

Mettler F A Jr and Upton A C, *Medical effects of ionising radiation* 3rd ed., Saunders Elsevier, Philadelphia, 2008.

Moolgavkar S H and Knudson A G, Mutation and cancer: A model for human carcinogenesis, *J Natl Acad Sci USA* 66, 1307–52, 1981.

Mothersill C and Seymour C, Radiation-induced bystander effects: Past history and future directions, *Radiat Res* 155, 759–767, 2001.

Preston D L, Shimizu Y, Pierce D A et al., Studies of mortality of atomic bomb survivors. Report 13: Solid cancer and non-cancer disease mortality 1950–1997, *Radiat Res* 160, 381–407, 2003.

Preston D L, Ron E, Tokuoka S et al., Solid cancer incidence in atomic bomb survivors 1958–1998, *Radiat Res* 168, 1–64, 2007.

Shimizu Y, Kato H and Schall W J, Studies of the mortality of A bomb survivors. 9 Mortality 1950–1985, Part 2, Cancer mortality based on the recently revised doses DS 86, *Radiat Res* 121, 120–41, 1990.

UNSCEAR (United Nations Scientific Committee on the Effects of Atomic Radiation), *Sources and effects of ionising radiation—2006 Report to the General Assembly with scientific annexes*, United Nations, New York, 2008.

Upton A C, Cancer induction and non-stochastic effects. In *Biological basis of radiological protection and its application to risk management, Br J Radiol* 60, 1–6, 1987.

Wall B F, Kendall G M, Edwards A A et al., Review article: What are the risks from medical X-rays and other low doses of radiation? *Br J Radiol* 79, 285–294, 2006.

Watson S J, Jones A L, Outway W B et al., Ionising radiation exposure of the U.K. population: 2005 review. Document HPA – RPD – 001, Health Protection Agency, Chilton, Didcot, U.K. 2005.

Wrixon A D, New recommendations from the International Commission on Radiological Protection—a review, *Phys Med Biol* 53, R41–R60, 2008.

Exercises

1. State the 'law' of Bergonié and Tribondeau on cellular radiosensitivity. Discuss the extent to which cell types follow the 'law'.

2. Explain the term *RBE*. List the factors that can affect the RBE value, giving a brief explanation of each.

3. What are the differences between tissue reactions (deterministic effects) and stochastic radiation effects? What is the evidence for believing that radiation-induced carcinogenesis is a stochastic effect?

4. Discuss the precautions that should be taken to minimise the risk of tissue reactions from radiation during interventional procedures and the actions that should be taken to ensure that they are identified if they occur.

5. Sketch the most common forms of chromosome defect detectable after whole body irradiation. Discuss the feasibility of chromosome aberration analysis as a method of retrospective whole body monitoring.

6. Review the evidence that ionising radiation can cause harmful genetic effects.

7. What are the four main sources of evidence that radiation can cause cancer in humans? Give *two* examples of each and discuss the strengths and weaknesses of each source.

8. With specific reference to diagnostic radiology, discuss the case for and against the linear-no-threshold (LNT) model of radiation risk.

9. What is the *bystander effect?* How may it be demonstrated and how might it affect low dose radiation risk?

10. Outline the evidence that adaptive processes are triggered in cells at very low doses.

13

Radiation Doses and Risks to Patients

K E Goldstone and P P Dendy

SUMMARY

- The reasons for assessing radiation doses to patients are discussed.
- Practical methods of measuring doses to patients are reviewed.
- Current data on patient doses from a wide range of radiological investigations is presented.
- The concept of diagnostic reference level (DRL) and its role in dose reduction is introduced together with practical methods of dose reduction.
- The concept of effective dose, its calculation and limitations in assessing risk is discussed.
- The special high-risk situation of irradiation of the foetus is reviewed.

CONTENTS

13.1 Introduction—Why Are Doses Measured?

As discussed in Section 12.5 there is ample evidence that ionising radiation is harmful to humans and it is important to remember that, almost always, the reason why doses are measured in radiology is to estimate the radiation risk. The reason for dose reduction is to minimise the risk, usually in respect of the possibility of stochastic effects, but occasionally the possibility of tissue reactions for interventional procedures carried out under X-ray control. As already discussed in several places in the book, especially Section 6.10 on quantum mottle, reducing the dose of radiation has an adverse effect on image quality. Therefore in the absence of risk to the patients much higher doses would frequently be used. The balancing of adequate image quality against dose and therefore risk is the process known as optimisation and we shall return to this idea in Section 13.6.

13.2 Principles of Patient Dose Measurement

13.2.1 Where Is the Dose Measured?

To estimate risk it is important to know the radiation dose (equivalent dose) to different body organs and tissues. This is because different organs have different radiosensitivites—a point we shall return to in Section 13.5. There is a very large amount of attenuation as a diagnostic X-ray beam passes through the body. Thus the exit dose will be typically between 0.1% and 1% of the entrance dose depending on the thickness of the body part being exposed and its composition. The doses to different organs within the beam will be very dependent on their depth and as shown in Figure 13.1 the dose to a critical organ may be substantially different for AP and PA projections.

In radiology there are two situations to consider, plain radiography, for example, a pelvis X-ray, where the field size and position are fixed so organs within and near the beam stay in the same position relative to the beam; and fluoroscopy (or complete examinations e.g. IVUs) where field size and position may change. Thus two kinds of measurement have developed and two practical dose quantities have been recommended (IPSM 1992)—entrance surface dose (ESD) for individual radiographs and dose area product (DAP) for complete examinations.

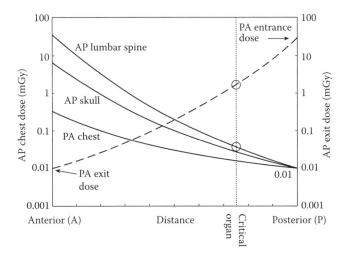

FIGURE 13.1
Attenuation of an X-ray passing through the body for three common X-ray projections (solid lines). The dotted line shows the effect on a critical organ dose, for example, the ovary if a PA projection were used instead for a lumbar spine view.

13.2.2 How Is the Dose Measured?

The ESD is defined as the absorbed dose in air at the point of intersection of the X-ray beam axis with the entrance surface of the patient, including back scattered radiation, and is measured in mGy. A well-defined protocol must be followed if results from different hospitals and different countries are to be compared. The Dosimetry Working Party of IPSM (IPSM 1992) recommended that direct measurements of ESD should be made using independently calibrated, thermoluminescent dose meters (TLDs) (see Section 14.7.1). They are small and can be stuck in a sealed plastic packet on the patient's skin. Since they are placed directly on the skin they measure the true ESD, including backscatter. The most suitable materials to use, lithium fluoride or lithium borate, are of low atomic number so they are unlikely to obscure diagnostic information and they cause no discomfort to the patient. TLDs generally provide a straightforward and accurate method of measurement and many major surveys in the past have used them. They are only suitable if a single view exposure is being made, for example, AP pelvis; they are not suitable for multiview procedures, for example, fluoroscopic procedures. Their disadvantages are (1) delay between exposure and read out; (2) possibly slowing patient throughput in a busy department, (3) expensive TLD read-out equipment is required.

Since about 2000 it has become common for all X-ray equipment, both radiographic and fluoroscopic, to be fitted with a DAP meter (see Section 4.6), or with a detection system which will calculate DAP. This approach can be used for both radiography and fluoroscopy.

Dose area product is defined as the air KERMA averaged over the area of the X-ray beam in a plane perpendicular to the beam axis multiplied by the beam area in the same plane, it therefore excludes backscatter. DAP may be thought of as a measure of the total energy fluence incident on the patient and may be related to the total energy absorbed in the patient. DAP is now the preferred measure of patient dose. It is usually expressed in Gy cm^2 but great care must be taken with units since they have not been standardised and results may be displayed in cGy cm^2 or μGy m^2 or a number of other combinations.

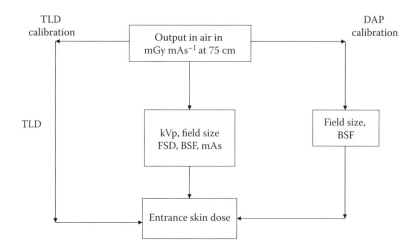

FIGURE 13.2
Direct and indirect approaches to assessment of entrance skin dose (FSD = focus-skin distance; BSF = backscatter factor).

A dose estimate may be required when a TLD or DAP measurement has not been made. Therefore indirect methods of dose measurement should not be overlooked. To do this the tube output is measured under specified conditions using an ionisation chamber in air, generally as part of routine quality assurance (QA) checks. A typical figure would be 100 μGy mAs^{-1} at 75 cm FSD for a modern tube operating at 80 kVp. The ESD can then be calculated from the knowledge of the exposure factors, applying any necessary correction for source-skin distance and a factor to allow for backscatter from the patient. Indirect methods also allow a large number of dose estimates to be made from a small number of measurements and may be useful at very low doses close to the detection limit of the TLDs or DAP meter. Indirect methods rely on knowing all the exposure factors and are difficult to apply to automatic control systems where the delivered mAs may not be recorded. The direct and indirect approaches for assessing entrance skin dose are summarised in Figure 13.2.

Note, (1) if the measurement is made with a DAP meter the result must be divided by the area of the beam (on the patient) to obtain the ESD and multiplied by the backscatter factor; and (2) the backscatter factor can be quite high and will typically be in the range 1.2–1.4. It is left as an exercise to the reader to explain why the backscatter factor will increase with increasing field size and increasing peak applied kVp/half value layer.

13.3 Review of Patient Doses

13.3.1 Entrance Doses in Radiography

Most dose surveys to date have shown a wide range of entrance doses for the same examination. Some of the spread is due to variation in patient thickness which has a significant effect because of the rapid exponential attenuation of the X-ray beam in the patient. However, the observed variation cannot be explained by this single factor alone.

In early surveys of individual patient doses the ESD for a lumbar spine (L5-S1) examination ranged from 2 mGy to 120 mGy. Equally surprising was that some hospitals had

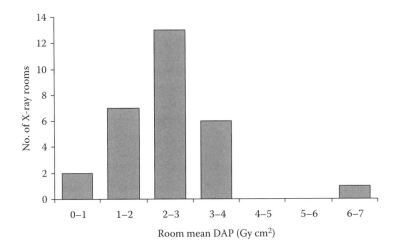

FIGURE 13.3
Comparison of room mean DAP distributions for lateral lumbar spine examinations from a survey carried out in East Anglia in 2007.

a very narrow range (less than 5 mGy), some a very wide range (up to 80 mGy). In dose surveys made in the 1980s, some of the dose variation was due to film-screen systems with different speeds being used, and departments changed to rare earth intensifying screens which were more sensitive and so less radiation dose was required to produce a diagnostic image. This pattern has been repeated, to a greater or lesser degree, for subsequent surveys with the overall trend being a reduction in dose and the range of doses. Results from a survey of lumbar spine DAP measurements undertaken in East Anglia, U.K., in 2007 are shown in Figure 13.3.

In an attempt to put downward pressure on patient doses a joint working party of the Royal College of Radiologists and National Radiological Protection Board (NRPB 1990) introduced the concept of reference values for entrance doses for standard radiological examinations. These levels are known as DRLs and are defined as 'dose levels for typical examinations for groups of standard sized patients or standard phantoms for broadly defined types of equipment' (IRMER 2000). UK National reference levels were set at the third quartile point from an NRPB survey conducted in the mid 1980s and were adopted by the European Union. In the UK employers are required to establish local DRLs (LDRLs) and to undertake local reviews and initiate appropriate action when they are consistently exceeded (see Section 13.6).

The NRPB (now the Health Protection Agency [HPA]—Radiation Protection Division) have continued to carry out 5 yearly reviews of doses to patients from X-ray imaging procedures in the UK, the most recent being published in 2007 covering the period January 2001–February 2006 (Hart et al. 2007). National reference doses recommended in 2007 in the review are shown in Table 13.1. In the majority of cases doses have reduced by a factor of 2 or more since the 1980s.

13.3.2 Entrance Doses in Fluoroscopy

As in radiography, the entrance dose to the patient will depend in part on the dose rate required by the image receptor (image intensifier or flat panel detector). Table 13.2 shows dose rates required at the image receptor for a range of systems.

TABLE 13.1

Recommended National Reference Doses for Individual
Radiographs on Adult Patients

Radiograph	Entrance Surface Dose (mGy)	DAP per Radiograph (Gy cm²)
Lumbar spine AP	5	1.6
Lumbar spine LAT	11	2.5
Lumbar spine LSJ	26	2.6
Chest PA	0.15	0.11
Chest LAT	0.6	0.3
Abdomen AP	4	2.6
Pelvis AP	4	2.1
Skull AP/PA	2	0.8
Skull LAT	1.3	0.5
Thoracic spine AP	4	0.9
Thoracic spine LAT	7	1.4

Source: Hart D, Hillier MC and Wall BF, *Doses to Patients from Radiographic and Fluoroscopic X-ray Imaging Procedures in the UK- 2005 Review.* HPA-RPD-029, Health Protection Agency, Radiation Protection Division, Chilton, UK, 2007.

TABLE 13.2

Range of Input Air KERMA Rates for Modern Image Intensifier
Systems under Automatic Brightness Control

System	Field Size (cm)	Dose Rate Settings (µGy s⁻¹)		
		Low	Normal	High
GE Advantx TC	30	0.2	0.35	0.78
Philips TeleDiagnost	38	0.15	0.3	–
GE legacy	32	0.23	0.5	1.02

Source: Adapted from Evans DS, Mackenzie A, Lawinski CP and Smith D, Threshold contrast detail detectability curves for fluoroscopy and digital acquisition using modern image intensifier systems. *Br. J. Radiol.* 77, 751–758, 2004.

One limitation of the definition of absorbed dose is that it changes very little with change in field size (there is a second order effect due to a change in the amount of scatter). However, the risk to the patient will clearly be greater if the dose is delivered over a bigger area, so the risk is more closely related to the total energy deposited as ionisation than to the absorbed dose. Therefore one of the variables that needs to be known if surface dose is measured with a TLD, is the field size. In fluoroscopy, field sizes are non-standard and may vary during a study so a dose measurement based on surface TLDs is not appropriate and a DAP measurement is used. In the most modern fluoroscopy equipment a measure of ESD is also displayed; this is a value calculated from the DAP assuming a specific focus skin distance. It is particularly useful in interventional procedures where the skin dose may be significant (see Section 13.3.3).

Recommended national reference doses for a number of more common fluoroscopic and complete examinations are shown in Table 13.3.

Note that the figure for coronary angiography considers all patients having this procedure, and includes a few patients who have already had a coronary bypass graft. For these patients doses may be of the order of twice the reference dose.

TABLE 13.3

Recommended National Reference Dose Area Product for a Selection of Fluoroscopic and Complete Examinations

Examination	DAP in Gy cm² per Examination	Fluoroscopy Time in Minutes per Examination
Barium (or water soluble) enema	24	2.8
Barium follow through	12	2.2
Barium meal	14	2.7
Barium meal and swallow	11	2.2
Barium(or water soluble) swallow	9	2.3
Coronary angiography	29	4.5
Femoral angiography	36	5.5
Fistulography	13	3.8
Hysterosalpingography	3	1
IVU	14	–
MCU	12	1.9
Nephrostography	12	4.8
Sialography	2	1.7
Sinography	9	2.1
Small bowel enema	40	9.2
T-tube cholangiography	8	1.9
Venography	7	2.2

Source: Hart D, Hillier MC and Wall BF, *Doses to Patients from Radiographic and Fluoroscopic X-ray Imaging Procedures in the UK- 2005 Review.* HPA-RPD-029, Health Protection Agency, Radiation Protection Division, Chilton, UK, 2007.

TABLE 13.4

Recommended National Reference Doses for Interventional Procedures on Adult Patients

Interventional Procedure	DAP in Gy cm² per Examination	Fluoroscopy Time in Minutes per Examination
Biliary drainage/intervention	50	15
Facet joint injection	5	1.8
Hickman line	3	1.4
Nephrostomy	14	5.1
Oesophageal dilation	11	2.8
Oesophageal stent	25	5.9
Pacemaker	11	8.2
PTCA (single stent)	50	13

Source: Hart D, Hillier MC and Wall BF, *Doses to Patients from Radiographic and Fluoroscopic X-ray Imaging Procedures in the UK- 2005 Review.* HPA-RPD-029, Health Protection Agency, Radiation Protection Division, Chilton, UK, 2007.

13.3.3 Doses in Interventional Procedures

In recent years, the use of radiological control while carrying out interventional therapeutic procedures has expanded greatly. Although, strictly speaking, DRLs are not applicable to interventional procedures since the patient is receiving treatment, DAP values for some of the more common procedures have been recommended (Hart et al. 2007) and are shown in Table 13.4.

Frequently the radiation field in such procedures is static for a considerable period of time so that not only is the total energy absorbed important but also the ESD since it may be high enough to cause skin damage (ICRP, 2000) see Sections 9.6.3 and 12.5.1. In some cases although the radiation fields may have different orientations, for example, in coronary angioplasty, there will be some areas of overlap. This can be illustrated and estimated using suitably calibrated film placed on the X-ray table under the patient (Morrell and Rogers 2006). Recently 'Gafchromic' film which requires no processing has been used for this purpose.

Morrell and Rogers and others have reported some ESDs in excess of 1 Gy for percutaneous transluminal angioplasty (PTCA) procedures.

13.4 Effect of Digital Receptors on Patient Dose

Computed radiography (CR) and digital radiography (DR) have come rapidly into common usage in radiology departments and there has been much debate over whether patient doses are or can be reduced using such systems.

Considering first CR systems, which have been around since about 1989, the overall consensus is that doses are comparable with film-screen systems of 300–400 speed. Since, immediately before changing to CR, radiology departments were commonly using 400 speed systems this suggests doses with CR may be slightly higher to produce a comparable image. Data submitted to the HPA Review (Hart et al. 2007) was in some cases obtained using digital systems rather than film-screen systems and it was possible to separate out results from CR and conventional systems but numbers were insufficient to recommend separate DRLs. Some results comparing mean values of doses collected from each room in the survey are shown in Table 13.5.

It is most important to optimise patient doses when using CR systems since high doses may go unnoticed giving excellent images but with excessive doses. Since CR systems are generally less sensitive than film systems at higher kVps, optimisation may involve modifying exposure factors (see Section 7.9.1).

Insufficient numbers of DR receptors were in use for radiography for any comparison to be included in the 2005 review. However, results from patient dose audits carried out in

TABLE 13.5

Comparison of Room Mean DAP per Radiograph for CR and Film

| Radiograph | DAP (Gy cm²) | |
	Computed Radiography	Film
AP abdomen	2.43	2.21
PA chest	0.13	0.1
AP lumbar spine	1.41	1.55
Lat lumbar spine	2.48	1.95
AP pelvis	2.08	1.95

Source: Hart D, Hillier MC and Wall BF, *Doses to Patients from Radiographic and Fluoroscopic X-ray Imaging Procedures in the UK- 2005 Review.* HPA-RPD-029, Health Protection Agency, Radiation Protection Division, Chilton, UK, 2007.

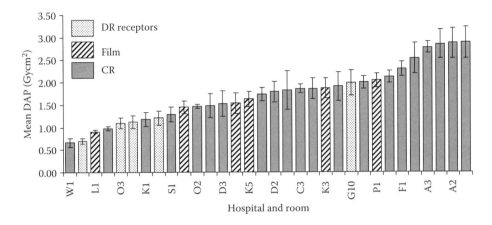

FIGURE 13.4
Comparison of room mean DAP for AP pelvis examinations carried out using DR, film or CR as the image receptor, from a survey carried out in East Anglia in 2007.

East Anglia between 2007 and 2008 for AP pelvis examinations suggest that a significant dose reduction can be achieved (Figure 13.4).

DR receptors have been used for fluoroscopy in place of conventional image intensifiers since about 2000. Although insufficient data were available in the HPA Review published in 2007 to recommend a DRL for such systems, some limited data were available from cardiac procedures. These tended to show an increase in DAP for digital detectors which may indicate inadequate image optimisation. The next 5-year review should provide a more solid indication of the dose saving or otherwise.

13.5 Effective Dose and Risks from Radiological Examinations

To be able to consider risk, a new dose term, *effective dose*, has to be introduced.

One of the main conclusions from the Japanese work was that many tissues are affected but not all to the same degree (see Section 12.5.2). Therefore, in situations where only part of the body is irradiated, some mechanism is required that will allow the risks from different patterns of exposure to be compared.

13.5.1 Tissue Weighting Factors

This is done by assigning to each tissue a *tissue weighting factor* w_T. Current recommended values for w_T are given in Table 13.6, together with the weighting factors that they superseded. The most significant change is to the weighting factor for the gonads where the risk for heritable disease is only considered up to the second generation and not beyond as was the case previously.

The effective dose, that is the whole body dose carrying the same risk, is then the weighted equivalent dose for all the organs and tissues in the body. Hence

$$\text{Effective dose} = \Sigma \, H_T \cdot w_T \text{ Sv summed over all organs and tissues}$$

where H_T is the equivalent dose in tissue T.

TABLE 13.6

Tissue Weighting Factors

Tissue or Organ	w_T (ICRP 2007a)	w_T (ICRP 1991)
Bone marrow (red)	0.12	0.12
Colon	0.12	0.12
Lung	0.12	0.12
Stomach	0.12	0.12
Breast	0.12	0.05
Remainder tissues	0.12	0.05
Gonads	0.08	0.2
Bladder	0.04	0.05
Oesophagus	0.04	0.05
Liver	0.04	0.05
Thyroid	0.04	0.05
Bone surface	0.01	0.01
Brain	0.01	Part of remainder
Salivary glands	0.01	Part of remainder
Skin	0.01	0.01
	Remainder = adrenals, extra-thoracic region, gall bladder, heart, kidneys, lymphatic nodes, muscle, oral mucosa, pancreas, prostate, small intestine, spleen, thymus, uterus/cervix	Remainder = adrenals, brain, small intestine, kidney, muscle, pancreas, spleen, thymus, uterus

13.5.2 Organ Doses

To estimate the effective dose and thus to estimate the radiation risk, the ESD or DAP must be used to calculate the dose to individual organs. This will depend on whether the organ is in the primary beam, the depth of the organ and the kV and filtration, that is, penetrating power, of the radiation. These days computer modelling is used to calculate organ doses and hence effective dose.

A general consideration of this problem is beyond the scope of the book but one specific situation that requires the calculation of an organ dose is when a patient has been exposed to X-rays during an unsuspected pregnancy. Although the exact position of the foetus depends on gestational age, available data on irradiation during pregnancy is generally not sufficiently refined to make calculations other than for the uterus.

Insight

Worked Example on the Dose to the Uterus

All practising radiologists are likely to be asked as some stage in their career to advise on a patient who has discovered she is pregnant after X-ray procedures involving the pelvis or lower abdomen have been performed. The worked example given below illustrates how the foetal dose may be estimated. It also serves as an important indicator of *all* the information (italicised) that must be on record in relation to the examination if the assessment is to be made reasonably accurately. The availability or otherwise of these data would be one measure of good radiographic practice.

- **Basic Data on Good Radiographic Set-Up**

  slim to average build lady estimated *7 weeks pregnant*
  2 × AP lumbar spine views at *70 kVp, 80 mAs, 100 cm FRD* (70 cm FSD estimated)

1 × lat lumbar spine view at 80 kVp, 140 mAs, 100 cm FRD (50 cm FSD estimated)
(L1-4)
1 × lat lumbar spine view at 90 kVp, 150 mAs, 100 cm FRD (50 cm FSD estimated)
(L5-S1)
These data are entered in columns 1–3 and 5 of Table 13.7
X-ray tube output data are entered in column 4

| kV | 70 | 80 | 90 |
| air KERMA rate (μGy mAs^{-1}) at 75 cm | 28.7 | 39.1 | 50.5 |

- **Calculation**

  1. Multiply the output factor (column 4) by the mAs and correct for the inverse square law to find the air KERMA at the skin (column 6)
  2. Use NRPB-R 186 (Jones and Wall 1985) to look up the backscatter factor (BSF) for this particular examination. This corrects for the fact that the output measurements in column 4 have been made in air whereas a significant amount of radiation will be scattered back to the skin from the patient
  3. The skin dose (column 8) is now equal to air KERMA at the skin × BSF × 1.06 (the ratio of mass absorption coefficients for soft tissue and air)
  4. Look up the uterus dose per unit skin dose for the appropriate examination in NRPB-R 186 (column 9)
  5. Calculate the dose to the uterus (column 10)

It is useful to check at this stage that the answer is consistent with information in HPA (2009) for the appropriate examination.

Notes
  (1) It would be unusual for a patient to have four views for a lumbar spine examination but they have been included here to illustrate different aspects of the calculation.
  (2) A period of screening would be dealt with similarly. A record of the screening time is essential. The kV and mA may have to be estimated if they are automatically adjusted by the equipment to maintain the correct image brightness. Ideally a DAP meter should be fitted.
  (3) Software has been produced, for example, Tapiovaara and Siiskonnen 2008, to expedite these calculations but there is some merit in working through them from first principles at least once!

TABLE 13.7

Foetal Dose Assessment from Recorded Exposure Factors

View	kVp	mAs	μGy per mAs	FSD cm	Air KERMA mGy	BSF[a]	Skin Dose mGy	Uterus Dose/Unit Skin Dose[a]	Uterus Dose mGy
AP × 2	70	80	28.7	70	2.64	1.36	3.8	0.179	0.684 × 2
lat	80	140	39.1	50	12.3	1.28	16.7	0.017	0.28
LSJ	90	150	50.5	50	17.0	1.27	22.9	0.014	0.32
Total									1.97

[a] from NRPB186.

Source: Jones, D.G. and Wall B.F., Organ Doses from Medical X-ray Examinations Calculated Using Monte Carlo techniques, NRPB-R186, National Radiological Protection Board, Chilton, Didcot, Oxfordshire OX11 0RQ, UK, 1985.

TABLE 13.8

Typical Effective Doses from Some Common Diagnostic
Examinations for Adults in the UK

Examination	Mean Effective Dose (mSv)
PA chest	0.02
PA skull	0.03
AP abdomen	0.7
AP pelvis	0.7
Lumbar spine (AP and lateral)	1.0
Barium enema	7

Source: Martin CJ, Dendy PP and Corbett RH, *Medical Imaging and
Radiation Protection for Medical Students and Clinical Staff*, British
Institute of Radiology 36: Portland Place, London, 2003.

13.5.3 Effective Dose

Unfortunately the effective dose is not a quantity that can be directly measured but conversion factors are available that allow directly measurable quantities, ESD or DAP, to be converted into effective doses (Hart et al. 1994). The conversion factors are based on the ICRP 1991 weighting factors but doses calculated using the programme of Tapiovaara and Siiskonnen take into account the 2007 factors.

Strictly speaking the concept of effective dose should not be applied to medical exposures because it is based on a reference person and a health detriment based over all ages and averaged over both sexes. This is obviously a very different situation compared with the individuals receiving medical exposures. We shall return to this topic when discussing risk in Section 13.8.

However, there are a couple of situations in which effective dose is useful. One is in the term *collective effective dose* which can be used for optimisation (see Sections 9.8 and 12.6). A second use of effective dose is as a parameter for comparing different examinations and procedures in the same or different hospitals and different technologies for the same procedure (as long as the patient populations are broadly matched for gender and age). Typical effective doses for some of the more common diagnostic medical procedures are shown in Table 13.8. Effective doses are given to 1–2 significant figures only since to do otherwise would give a false impression of the accuracy of data and parameters on which they are based.

13.6 Optimisation and Patient Dose Reduction

Reference doses have now become embedded in good radiological practice and setting local reference doses and investigating when they are exceeded is an important element of UK legislation (see Section 13.3.1).

Insight

Local DRLs

Legislation in the UK requires hospitals to set their own LDRLs to adapt local practices and ensure patient doses are being kept as low as reasonably practicable. This is not an entirely straightforward process and guidance has been published (IPEM 2004) on how it may be done.

The hospital, or hospitals if there is more than one in an organisation, first selects examinations on which it is appropriate for data to be collected. This may or may not be the same as the examinations for which there are national DRLs, although it should certainly include some of these examinations. The selection will depend on local activities, for example, a specialist chest hospital or a private hospital dealing largely with orthopaedic cases would select different procedures from a general hospital. Data is then collected from actual patient examinations, the minimum sample size is recommended as 10 patients per procedure per room with patient weights lying between 50 kg and 90 kg so that the mean weight of the sample is 70 ± 5 kg. The mean of the doses for each room is calculated, and then an overall mean calculated. If this exceeds the national DRL an investigation should be undertaken to determine the cause and if possible take corrective action. Some of the technical ways in which patient doses may be reduced are covered in this section but case mix and measurement methodology could be contributory factors as well as equipment and technique. For further detail the reader is advised to read the guidance.

What can be done when DRLs are consistently exceeded? Since this book is concerned with the physics of radiology, the emphasis will be on technical factors that affect the dose but other considerations must not be overlooked and will be mentioned briefly.

13.6.1 Technical Factors

- The choice of operating kilovoltage must be optimised and the appropriate amount of filtration must be used. The use of more filtration and higher kV will lower the ESD because of better patient penetration—but note that use of a grid may negate the dose reduction. The effective dose will not be reduced by as great a proportion as the ESD because the higher energy scattered radiation will travel more easily to more distant organs.

- For both analogue and digital systems the most sensitive receptor that will give the requisite image quality should be used. Note that the sensitivity of the detector varies with the kVp. Therefore, to achieve maximum dose reduction the kV should correspond to the maximum receptor sensitivity.

- It is important to measure the dose or dose rate entering the imaging system (i.e. the exit dose from the patient) since clearly the sensitivity of the receptor will have a major effect on the dose to the patient. Baseline and subsequent routine quality assurance measurements are essential to detect any deterioration in the sensitivity of the image receptor.

- In early dose surveys a significant cause of retakes with film, leading to unnecessary patient exposure, was poor image processing. This has been largely eliminated in digital systems, but care needs to be taken to ensure post-processing on images is optimal and images are not deleted before being transferred to the PACS system.

- Low attenuating components, for example carbon fibre, should be used where possible (e.g. for the couch, antiscatter grids, cassette fronts).

- Automatic exposure control devices should be used where practicable and must be checked regularly for reliability.

- If grids are essential, the lowest grid factor commensurate with adequate image quality should be used. The grid should be completely removed whenever possible—consider using an air gap as an alternative.

- The focus receptor distance should be optimal—see Insight.

- Automatic beam collimation should be used whenever possible.

- In fluoroscopy, select as low a detector input dose rate as possible to produce an acceptable image.

- Make full use of image storage to avoid continuous standing on the foot switch and to reduce fluoroscopy times.

- In the image acquisition phase in screening procedures, ESD rates may be of the order of 10 times those in fluoroscopy so acquisition should be used sparingly with as low a frame rate as possible for an acceptable image.

- The dose per pulse in image acquisition is variable, but is usually preset by the manufacturer. It is frequently possible to reduce the dose per pulse yet still produce acceptable images.

Periodic measurements of patient entrance skin dose should be made and the overall quality assurance programme for the department must be such that, when technical problems have been identified and corrected, patient doses are re-audited to demonstrate that the necessary dose reduction has been achieved.

Insight

Minimum Focus-Skin Distances

The reasons why minimum focus-skin distances are specified for certain examinations (e.g. 180 cm for chest films, 60 cm for mammography and 20 cm for dental films) are, first to reduce geometric unsharpness (see Section 6.9.1) and second to reduce patient dose. The reduction in patient dose is not always appreciated and is an important application of the inverse square law. Consider the situations shown in Figure 13.5 (in the right hand illustration the 'patient' has been placed extremely close to the X-ray focal spot for the purpose of illustration).

The dose at the image receptor, D say, must be the same in both arrangements to achieve the same receptor response. Attenuation through the patient will be the same in both cases and can be ignored. However, the effect of the inverse square law is to increase the dose at the front surface of the patient by $(120/100)^2$ that is, to 1.44 D on the left but by $(40/20)^2$, that is, to 4 D on the right.

There is unfortunately an adverse effect of the inverse square law. When longer focal-skin distances are used more X-rays (greater mAs) are required on the left to produce a dose (D) than on the right because the receptor is further from the focal spot.

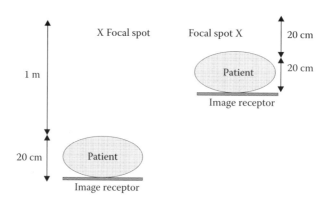

FIGURE 13.5
Illustrating how the inverse square law affects the entrance dose to the patient—for explanation see text.

13.6.2 Non-Technical Factors

- Probably the most frequent and certainly most avoidable unnecessary exposure to patients is the result of inappropriate radiological examinations. The guidelines in 'Making the best use of clinical radiology services' (RCR 2007) contains excellent advice on when X-rays are inappropriate or at least should be deferred for a few weeks.

- In most centres arrangements could be improved considerably to ensure investigations are not needlessly repeated. The advent of the digitised department enables searches to be made easily for previous images.

- Consideration must be given to using alternative imaging modalities such as ultrasound and magnetic resonance imaging (MRI) thereby avoiding the use of X-rays.

- Employment-related screening programmes should be used sparingly and clinically justified.

- All staff should be given appropriate training, covering both an awareness of the radiological examinations that carry the highest doses and of the procedures that are available to minimise those doses, for example, the use of gonad shielding where appropriate.

13.7 Procedures Requiring Special Dose Assessment/Measurement

13.7.1 Computed Tomography (CT)

CT is a relatively high dose examination. In the UK, by 2003, CT accounted for only 9% of the patient examinations but about 47% of medical contribution to the collective dose (Shrimpton et al. 2006). This represents a doubling over the previous 10 years and is a trend reflected internationally. It is therefore an important area in the radiation protection of patients.

Furthermore, the number of CT examinations per annum per million of the population is increasing sharply world wide. During the last 20 years or so there has been a 12-fold increase in the UK and a 20-fold increase in the US (Hall and Brenner 2008).

Since CT examinations involve the irradiation of thin transverse slices of the body by a rotating beam of X-rays, conditions of exposure are very different from those used in conventional radiology and different methods are required to measure radiation doses and calculate effective doses. Weighted CT dose index ($CTDI_w$), volume weighted dose index ($CTDI_{vol}$), and dose length product DLP are all practical dose quantities that can be used. Some or all of these quantities are displayed on the scanner console.

CT dose index is obtained from a measurement made in a 16 cm or 32 cm diameter polymethyl-methacrylate (PMMA) cylinder to represent a head and body, respectively. The measurement is made using a 100 mm active length pencil ionisation chamber. This measured quantity is denoted $CTDI_{100}$ and is the integration of the dose from a single rotation so it includes the spread of the radiation dose profile. Note that the X-ray beam will not have a dose profile that is an ideal rectangular shape but will have a more spread distribution (Figure 13.6). The $CTDI_{100}$ will vary across the field of view tending to be greater

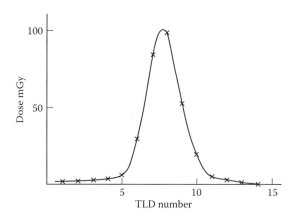

FIGURE 13.6
A CT dose profile (nominal slice width 2 mm) measured with a stack of TLD chips each 0.8 mm thick.

near the periphery of the phantom than at the centre. To account for this variation $CTDI_w$ is defined as one-third the $CTDI_{100}$ at the centre plus two-thirds the $CTDI_{100}$ at the periphery.

For a multislice scanner the CTDI is greater for narrow beam widths because of the effect of the beam penumbra (see Section 8.8).

CT dose index is a useful quantity when comparing two scanners, however, CT procedures rarely involve a single rotation and usually consist of a series of rotations which may have gaps or overlaps and these have to be allowed for; $CTDI_{vol}$ is therefore defined as

$$CTDI_{vol} = \frac{CTDI_w}{pitch}$$

where pitch is the total beam width/distance of travel per rotation.

A further practical quantity, dose length product (DLP), also has to be defined and is equal to the

$$CTDI_{vol} \times \text{irradiated length on the patient}$$

It is usually measured in mGy cm and is displayed on the scanner console.

To assess the effective dose and thus be able to estimate risk, it is necessary to know the dose to all the radiosensitive organs of the patient arising from the various highly localised patterns of exposure. Organ doses may be estimated from axial air doses using Monte Carlo techniques which have been developed into a computer programme by ImPACT 2004.

However, a simpler method has been developed whereby coefficients, depending on the region of the body irradiated and the age of the subject, have been derived using Monte Carlo techniques to convert DLP into effective doses. This method has been used in the 2003 review of doses from CT examinations in the UK (Shrimpton et al. 2006). Table 13.9 shows the coefficients used for adults and Table 13.10 reference levels for dose length products from data collected from actual patient examinations in the course of the review. As for conventional radiographic examinations the reference levels are based on the third quartile from the data collected.

Table 13.11 shows typical effective doses from a selection of examinations based on the mean dose values from the data collected in the 2003 review.

TABLE 13.9

Normalised Values of Effective Dose per Dose Length Product for Adults Over Various Regions

Region of Body	Effective Dose per DLP (mSv [mGycm] $^{-1}$)
Head and neck	0.0031
Head	0.0021
Neck	0.0059
Chest	0.014
Abdomen and pelvis	0.015
Trunk	0.015

Source: Shrimpton PC, Hillier MC, Lewis MA and Dunn M, National survey of doses from CT in the UK: 2003. *Br. J. Radiol.* 79, 968–980, 2006.

TABLE 13.10

National Reference Levels for CT Examinations on Adult Patients

Examination	DLP (mGy cm)	
	SSCT[a]	MSCT[b]
Routine head (acute stroke)	760	930
Abdomen (liver metastases)	460	470
Abdomen and pelvis (abscess)	510	560
Chest, abdomen and pelvis (lymphoma staging or follow up)	760	940
Chest (lung cancer: known, suspected or metastases)	430	580
Chest; Hi-resolution (diffuse lung disease)	80	170

[a] SSCT = single slice CT.
[b] MSCT = multislice CT.
Source: Shrimpton PC, Hillier MC, Lewis MA and Dunn M, National survey of doses from CT in the UK: 2003. *Br. J. Radiol.* 79, 968–980, 2006.

TABLE 13.11

Typical Effective Doses from CT Examinations for Adult Patients

Examination	Typical Effective Dose (mSv)
Routine head (acute stroke)	1.5
Abdomen (liver metastases)	5.3
Abdomen and pelvis (abscess)	7.1
Chest, abdomen and pelvis (lymphoma staging or follow-up)	9.9
Chest (lung cancer: known, suspected or metastases)	5.8
Chest; Hi-resolution(diffuse lung disease)	1.2

Source: Shrimpton PC, Hillier MC, Lewis MA and Dunn M, National survey of doses from CT in the UK: 2003. *Br. J. Radiol.* 79, 968–980, 2006.

As CT scanners developed, effective doses reduced. In part this was due to technical improvements, including high efficiency detectors, new beam shaping absorbers, rapid X-ray rise and fall times and the use of automatic exposure control systems with X-ray tube mA varied under computer control in response to changes in patient cross-section to give a pre-selected image quality. Also a greater awareness developed amongst radiologists and radiographers that CT is a high dose technique and by selecting suitable scan protocols operators can greatly assist in keeping doses as low as reasonably practicable. The introduction of multislice scanners has initially tended to increase dose (Yates et al. 2004) but with greater user awareness doses can potentially be reduced (ICRP 2007b).

Some protocols giving doses approaching or exceeding the reference levels generally involve a large number of slices. Thus CT examinations should be made with the minimum settings of the tube current and scan time to give adequate image quality and with the minimum irradiated volume, that is region of interest, required to give the necessary information. Particular care is required with high resolution modes because these are also high dose modes since, to maintain image quality for thin slices, a high dose is needed to maintain signal to noise ratio.

Note that scan projection radiography ('scout view') is a relatively low dose technique and is commonly used to determine the level at which CT slices are required. Modern scanners can traverse the length of the body in less than 10 s with a sub-millimeter aperture and generate good quality images.

In contrast, dynamic CT investigations (also called *CT fluoroscopy*) are usually high dose studies. Dynamic CT, which allows measurements of temporal changes in contrast density, for example, in blood vessels and soft tissues, after administration of contrast material, has much in common with dynamic studies in nuclear medicine (see Section 10.4). For example, a sequence of temporal images will be collected and analysed to map changes in CT number (analogous to changes in activity) as a function of time. Displayed images may be functional images, for example a map of time to maximum contrast but selection of suitable regions of interest to act as controls can be difficult.

The requirements for dynamic studies are a short scan time, short scan intervals, a high scan frequency and a large number of total scans. Furthermore resolution must be high and noise levels must be low so that small changes in density can be registered. These two conditions are only met if high mAs values are selected and, combined with the need for multiple images of the same slice, the net result may be a high effective dose. The effective dose can be reduced considerably by careful selection of the correct scan plane initially.

13.7.2 Mammography Doses

The breast is one of the more radiosensitive organs, and since mammography is used for screening large numbers of the healthy population, it is especially important to keep the radiation dose very low. The kVp used in mammography is also low so there is considerable attenuation in the breast tissue and the exit dose is very much less than the ESD. The radiation used is highly collimated to the breast and the concern is almost entirely confined to breast cancer. Therefore a quantity called *mean glandular dose* (MGD) is used to assess patient dose and effective dose is not relevant. For system comparison purposes and quality assurance measurements, before 2005 the MGD for the standard breast was calculated from measurements made using a 40 mm block of Perspex and multiplying by various conversion factors. In the UK the National Health Service Breast Screening Programme (NHSBSP) required the standard MGD to be less than 2 mGy. Young et al. (2005) reported

TABLE 13.12

Doses per Exposure Received in the NHSBSP
Mammography Screening Programme in 2001 and 2002

Projection	Mean MGD (mGy)	Average Breast Thickness (mm)
Oblique	2.23 ± 0.01	56.8 ± 0.2
Craniocaudal	1.96 ± 0.01	54.1 ± 0.2

Source: Young KC, Burch A and Oduko JM, Radiation doses received in the UK Breast Screening programme in 2001 and 2002. *Br. J. Radiol. 78*, 207–218, 2005.

the MGD to the standard breast (40 mm block) using a 28 kVp beam and a Mo target and filter was 1.42 ± 0.04 mGy per exposure for the 285 systems included in their survey.

To more adequately represent compressed breast thickness for the screened population, IPEM Report 89 (IPEM 2005) changed the standard measurement conditions to a block of 45 mm perspex and recommended the MGD should be less than 2.5 mGy.

Mean glandular dose to samples of women attending for mammography are regularly assessed and analysed by the NHSBSP (see Table 13.12). The actual doses received depend on the characteristics of the equipment, for example target and filter, film and screen, but also on individual factors such as breast size and composition and the degree of compression.

Recently digital imaging has been introduced into mammography. Results suggest that a reduction in MGD can be achieved using digital imaging although if the reduction is too great there will be significant effect on image quality especially in the detection of microcalcifications (Samei et al. 2007). For further comment on doses in mammography see Section 9.2.6.

13.7.3 Nuclear Medicine

When dealing with the dosimetry of radionuclides, we need a relationship between the activity of the source and the exposure rate at some point near the source. This depends on both the number and energy of gamma ray photons coming from the source in unit time.

The factor relating exposure rate and activity is variously known as the exposure rate constant, the specific gamma ray constant or 'k' factor. A suitable unit in which to measure 'k' would be coulomb per kg per h at 1 m from a 1 MBq point source. The corresponding absorbed dose rate in air or in tissue can then be calculated.

Note that some radionuclides have complex decay schemes. This is undesirable if some of the decays contribute no useful gamma rays to the image forming process but contribute locally to the dose to the patient. Thus, somewhat paradoxically, a high k factor is a desirable property for an imaging radionuclide.

Calculation of the effective dose due to radioactivity within the body follows the same principles as for external beams, that is the dose to each organ is computed and multiplied by the appropriate tissue weighting factor, but is rather more complicated. Some of the reasons for this are as follows:

(a) The total activity in the body will decrease as a result of a combination of physical and biological half-lives (see Section 1.8).

(b) The activity will redistribute throughout the different organs and tissues during its residence in the body.

TABLE 13.13

Effective Doses for Some Common Nuclear Medicine Examinations

Radiopharmaceutical	Study	Max Usual Activity (MBq)	Effective Dose (mSv)
51-Cr EDTA	GFR	3	0.006
99m-Tc MAG3	Renography	100	0.7
99m-Tc microspheres	Lung perfusion	100	1
99m-Tc microspheres	Lung perfusion	200(SPECT[a])	2
99m-Tc DTPA	Renography	300	2
99m-Tc phosphates	Bone	600	3
99m-Tc phosphates	Bone	800(SPECT[a])	5
99m-Tc Sestamibi	Myocardial imaging (MIBI)	300	3
99m-Tc Sestamibi	Myocardial imaging (MIBI)	400(SPECT[a])	4
99m-Tc iminodiacetates	Functional biliary system imaging (HIDA)	150	2
111-In white blood cells	Abscess	20	9
201-Tl ion	Myocardium	80	18.4
18-FDG	Tumour imaging	400	8

[a] SPECT = single photon emission computed tomography; for an explanation of other acronyms see Table 10.3.

(c) Activity in a particular organ will irradiate not only that organ, but also, if it is a gamma emitter, a number of other organs.

(d) Doses to other organs will depend on activity in the source organ, the inverse square law, attenuation in intervening tissues and the mass absorption coefficient of the target organ.

Methods of calculation are reviewed in the Administration of Radioactive Substances Advisory Committee (ARSAC) Guidance Notes (ARSAC 2006) and effective doses for adult patients for the majority of radionuclide investigations for the usual maximum administered activity (now called *Diagnostic Reference Level*, DRL) are tabulated. Values for some of the more common examinations are shown in Table 13.13.

Note that when administering radiopharmaceuticals it is important that, in addition to the effective dose being acceptably low, the doses to individual organs must also be checked to ensure that deterministic effects cannot occur. Biokinetic data and absorbed dose estimates to individual organs (both for adults and children) have been published for a wide range of radiopharmaceuticals by the ICRP. The most recent amendment (ICRP 2008) contains data on positron emitters, including 11-C, 15-O and 18-F compounds as well as newer compounds of gamma emitters. To obtain the effective dose values the weighting factors from ICRP 60 (ICRP 1991) have been used.

13.8 The Perception of Risk from Medical Radiation Exposures

Risks from medical exposures to radiation are justified in terms of the medical benefit. This is not readily quantifiable, certainly not by members of the public, therefore it is helpful to put radiation risks in context, that is in terms that the public can understand.

Both the Royal Society and the Health and Safety Executive in the UK have made studies of the public perception of risk. The Royal Society approach was to gauge public opinion on the risk associated with certain activities and then to work out the actual risk involved. Risks as low as 1 in 10^6 (1 in 1,000,000) a year are commonly regarded as trivial but an imposed risk at a level of 1 in 10^4 (1 in 10,000) is likely to be challenged. Typical lifetime risks from various causes not involving X-rays are shown in Table 13.14.

Taking the risk figure from the previous chapter of 5% per Sv for a reference person, an effective dose of 0.02 mSv is equivalent to a risk of 1 in 10^6 and a dose of 2 mSv equivalent to a risk of approximately 1 in 10^4. On the basis of rounded effective dose figures, risks from some common diagnostic procedures are shown in Table 13.15.

TABLE 13.14

Typical LifeTime Risk of Death from Various Causes for an Individual Aged 40 Years Assuming They Live to 75 Years

Activity	Risk Estimate
Killed by lightening	1 in 300,000
Anaesthesia (risk from single administration)	1 in 50,000
Work in service industry	1 in 6000
Accident on road	1 in 500
Pneumonia and influenza	1 in 30
Smoking 10 cigarettes a day	1 in 5

Source: Martin CJ, Dendy PP and Corbett RH, *Medical Imaging and Radiation Protection for Medical Students and Clinical Staff,* British Institute of Radiology 36: Portland Place, London, 2003.

TABLE 13.15

Lifetime Risks of Cancer from Some Common Diagnostic Procedures

Procedure	Effective Dose (mSv)	Risk of Fatal Cancer	Risk Classification
Limb and joint radiographs	<0.01	<1 in 2,000,000	Negligible
PA chest	0.02	1 in 1,000,000	Minimal
PA skull	0.03	1 in 700,000	Minimal
Hip	0.3	1 in 60,000	Very low
Tc-99m lung ventilation	0.4	1 in 50,000	Very low
Abdomen or pelvis	0.7	1 in 30,000	Very low
Tc-99m (MAA) lung perfusion	1.0	1 in 20,000	Very low
CT head	2	1 in 10,000	Low
Tc-99m Bone scan	3	1 in 7000	Low
CT Pelvis	10	1 in 2000	Low
F-18 tumour (PET)	10	1 in 2000	Low
Tl-201 myocardial perfusion	18	1 in 1000	Moderate

Source: Martin CJ, Dendy PP and Corbett RH, *Medical Imaging and Radiation Protection for Medical Students and Clinical Staff,* British Institute of Radiology 36: Portland Place, London, 2003.

However, these risk figures should be used with great care and some would dispute that they should be used at all when discussing risks of medical exposures since they rely on calculating the effective dose (Brenner 2008). Why should this be?

Effective dose was originally introduced as a means of describing the detrimental effect of radiation to a reference person from low doses of radiation. The detrimental effect is comprised of cancer induction (both fatal and non-fatal) and the hereditary effect. This has been averaged over all ages and both sexes. These factors have been incorporated into the weighting factors (w_T) which have been set out by ICRP and are based largely on data from the Japanese population irradiated as a result of the dropping of the atomic bombs in 1945. It is well recognized that radiation risks are age dependent with lifetime risks for children being significantly greater than for those exposed as adults. They are also gender dependent, although not to such an extent as age, with females overall being more radiosensitive than men.

Therefore, calculating effective dose for say an abdominal exposure for an individual patient involves significant errors. There are uncertainties involved in making the initial measurement, for example, DAP, on which dose is based. There are uncertainties in knowing exactly where a particular organ in the individual is situated so the computer calculation of organ dose may have significant errors and then there are the uncertainties introduced because the organ and tissue weighting factors are not based on a particular individual, say a 70-year old female weighing 50 kg, but on a generic, 'reference person'. So, although effective dose can be a useful concept in estimating risk, its shortcomings should be recognised and if it is considered necessary to assign a numerical value of risk to a particular individual patient, the individual organ doses combined with the relevant age-related and sex-related risk coefficients should be used. For this reason when advising patients about risk it is wise to use general terms as indicated in Table 13.15.

Note that where the radiation dose is delivered largely to a single organ, for example, the breast in mammography, the use of effective dose is inappropriate and equivalent dose should be used together with the organ-specific risk coefficient.

13.9 A Special High-Risk Situation—Irradiation *In Utero*

There have already been references to the enhanced radiation risk to children—see Section 12.5.

For irradiation *in utero* the position is more complex and there are several possible harmful effects to be considered depending on the stage of pregnancy at which the radiation exposure is received. A comprehensive review of the biological effects of radiation on the foetus was published by the ICRP in 2003 (ICRP 2003). This information has been incorporated into recent advice published by the HPA on the protection of pregnant patients (HPA 2009).

Initially there will be a small number of rapidly dividing, highly radiosensitive cells and damage is likely to result in pre-natal death. In the intermediate stages of pregnancy there may be sufficient cells overall for the embryo to survive but only small numbers are performing any one specialised function so radiation damage causes abnormalities and neonatal death. In the later stages, when the foetus has fully formed, the effects are likely to be qualitatively similar to those in the adult.

Each of the effects will be considered briefly.

13.9.1 Lethal Effects

At the pre-implantation stage this effect will be indistinguishable from a spontaneous abortion. At foetal doses of less than 100 mGy this effect is most unlikely. As the foetus develops it will increasingly become a deterministic effect and work with animals suggests that at the later stages of gestation the threshold dose could approach 1 Gy.

13.9.2 Malformations and Other Developmental Effects

In utero irradiation during the period of early organogenesis has been shown to cause a range of malformations, depending on the timing of exposure, in every animal model tested. Confirmatory evidence for malformations in the Japanese survivors is lacking but there is some evidence for a high incidence of miscarriages, still births, neonatal and infant deaths, as well as a decrease in the mean height and head circumference in offspring that had been irradiated *in utero* with fairly high doses.

ICRP (ICRP 2007a) have concluded that no significant deterministic effects will occur in humans below a dose of at least 100 mGy even for acute exposures in the most sensitive first month of pregnancy. In radiological examinations the risk must be negligible compared with the natural, spontaneously arising risk.

13.9.3 Radiation Effects on the Developing Brain

Most of the information on the effects of pre-natal irradiation on the nervous system comes from follow-up of the Japanese atomic bomb survivors. This is a very complex subject but basically the results demonstrate clearly that the period of maximum organogenesis is also the period of maximum radiation sensitivity.

The induction of severe mental retardation is greatest during the period from 8 to 15 weeks when there is a rapid increase in the number of neurones before they migrate to their ultimate development sites in the cerebral cortex and lose their capacity to divide. Data suggests a threshold of about 300 mGy and a loss in IQ points of about 25 points per Gy. Therefore, below about 100 mGy any effect on IQ is of no practical significance. Mental retardation is even less for radiation exposures before and after this critical period from 8 to 15 weeks post-conception.

13.9.4 Cancer Induction

The most detailed data on childhood cancer following radiation exposure come from the Oxford Survey of Childhood Cancers which was started in 1955 and remains one of the key studies on this topic (Wall et al. 2006). The recent advice (HPA 2009) compares the risk of cancer up to the age of 15 years, due to the radiation, with that of the natural cumulative risk of cancer in the UK up to that age (estimated to be 1 in 500). On this basis the increased excess absolute risk of childhood cancer is estimated as about 8% Sv^{-1}, (8 $\times 10^{-4}$ for 10 mSv). There is no distinction made between fatal and other cancers since survival rates in childhood cancers have significantly improved. The risk is considered directly proportional to the dose received by the foetus, and, after the first 3 or 4 weeks of pregnancy, is reckoned to be relatively independent of the stage of pregnancy at which irradiation occurred. The risk will be lower in the very earliest stages of conception (up to 3 weeks) but cannot be assumed to be zero. The lifetime risk will of course be higher and ICRP 60 (ICRP 1991) suggests 'at most a few times that for the population as a whole'.

If 'a few' were taken as 3, the lifetime risk of fatal cancer would be 15% Sv^{-1} or 15×10^{-4} for 10 mSv.

13.9.5 Hereditary Effects

There is no evidence to indicate that the foetal gonads are any more radiation sensitive than the adult gonads. Therefore the risk factor is assumed to be the same before and after birth and is estimated to be about 1 in 200,000 per mGy (0.5×10^{-4} for 10 mGy).

13.9.6 Summary of Effects of Radiation *In Utero*

The HPA in conjunction with the Royal College of Radiologists and the College of Radiographers have reviewed information on the effects of ionising radiation on the embryo and foetus and have published guidance on protection of pregnant patients (HPA 2009). They have advised that

- Radiation doses from diagnostic procedures present no risk of foetal death or other deterministic effects and negligible risk of hereditary disease in the descendants of the unborn child.
- For the majority of diagnostic procedures the risk of childhood cancer is very low.
- A few high dose examinations may result in doubling of the childhood cancer risk and therefore should be avoided if possible but if they do have to take place termination of the pregnancy on radiation dose grounds is not justified.
- Risks of childhood cancer following irradiation in the first 3–4 weeks post-conception are very low but, procedures where foetal doses may be in excess of about 10 mSv should be avoided if possible.

The guidance has also given some typical foetal doses from various examinations and put them into risk bands reflecting that risk factors cannot be calculated precisely. Some examples are shown in Table 13.16.

TABLE 13.16

Typical Foetal Doses and Childhood Cancer Risks from Some Common Procedures

Examination	Typical Foetal Dose Range (mGy)	Risk of Childhood Cancer
Chest X-ray	0.001–0.01	<1 in 1,000,000
^{81m}Kr Lung ventilation scan	0.001–0.01	<1 in 1,000,000
^{99m}Tc Lung ventilation scan (technegas)	0.01–0.1	1 in 1,000,000 to 1 in 100,000
Abdomen X-ray	0.1–1	1 in 100,000 to 1 in 10,000
Chest CT	0.1–1	1 in 100,000 to 1 in 10,000
Abdomen CT	1–10	1 in 10,000 to 1 in 1000
^{99m}Tc bone scan	1–10	1 in 10,000 to 1 in 1000
Pelvis CT	10–50	1 in 1000 to 1 in 200

Source: HPA, *Protection of Pregnant Patients during Diagnostic Medical Exposures to Ionising Radiation*, Advice from the Health Protection Agency, The Royal College of Radiologists and the College of Radiographers, Documents of the Health Protection agency RCE-9, 2009.

Insight

Balancing Risk for Foetus and Mother

Sometimes difficult decisions have to be made balancing the radiation dose and hence risk to the foetus against that of the mother. One example is in the diagnosis of pulmonary embolus which is a condition not uncommon in pregnancy. The choice is between a nuclear medicine V/Q scan which gives a maternal effective dose of 1.7 mSv although this can be reduced if a protocol is followed using a reduced activity for the perfusion scan and only proceeding to the ventilation study if defects are identified on the perfusion scan. The breast dose from a full V/Q scan is about 0.7 mSv. The foetal dose from a full V/Q scan is estimated to be of the order 0.7 mSv.

The alternative method of diagnosis is to use CT scanning with a suitably designed protocol. A survey of doses used in East Anglia gave an average maternal effective dose of 5 mSv. However, the equivalent dose to the breast from such an examination may be between 10 and 20 mSv. The foetal dose from such a scan is estimated to be up to 20 μSv in the first trimester, 30 μSv in the second trimester and 130 μSv in the third trimester (Winer-Muram et al. 2002).

So the choice is between a CT procedure that gives a much lower foetal dose than the nuclear medicine study but a higher effective dose and breast dose to the mother, and a nuclear medicine study where the foetal dose is higher but the maternal dose is lower than for CT. There may of course be other factors involved such as, in an emergency, which can be done sooner?

13.10 Conclusion

The potentially harmful effects of ionising radiation must be recognised and understood. Furthermore, it is important to appreciate that increasingly sophisticated experiments have failed to provide evidence of a safe level of radiation for the two most important long-term effects, namely carcinogenesis and mutagenesis. Therefore it is important that radiologists should have a good appreciation of the risks associated with the examinations they carry out. Indeed, the public is increasingly expecting to be kept informed of the risks involved.

Some situations carry a higher risk and the importance of minimising exposures to children and during pregnancy cannot be over-emphasised.

However, it is necessary to keep a sense of perspective and calculations show that when diagnostic X-rays are used correctly and doses are carefully controlled, risks are acceptable within the broader context of both the clinical value of diagnostic information gained and the risks associated with daily living.

References

Administration of Radioactive Substances Advisory Committee. Notes for Guidance on the Clinical Administration of Radiopharmaceuticals and Use of Sealed Radioactive Sources. ARSAC, Health Protection Agency, Chilton, UK, 2006 http://www.arsac.org.uk/ [accessed 18th January 2011].

Brenner, D.J., Effective dose: a flawed concept that could and should be replaced. *British Journal of Radiology* 81: 521–523, 2008.

Evans, D. S., Mackenzie, A., Lawinski, C. P. and Smith, D., Threshold Contrast detail detectability curves for fluoroscopy and digital acquisition using modern image intensifier systems. *British Journal of Radiology* 77: 751–758, 2004.

Hall, E.J. and Brenner D.J., Cancer risks from diagnostic radiology. *British Journal of Radiology* 81: 362–378, 2008.

Hart, D., Hillier, M.C. and Wall, B.F., Doses to Patients from Radiographic and Fluoroscopic X-ray Imaging Procedures in the UK- 2005 Review. HPA-RPD-029, Health Protection Agency, Radiation Protection Division, Chilton, UK, 2007.

Hart, D., Jones, D.G. and Wall, B.F., Estimation of Effective Dose in diagnostic radiology from Entrance Surface Dose and Dose-area product measurements, NRPB R262, National Radiological Protection Board, Chilton, Didcot, Oxfordshire OX11 0RQ, UK, 1994.

HPA, Protection of Pregnant Patients during Diagnostic Medical Exposures to Ionising Radiation, Advice from the Health Protection Agency, the Royal College of Radiologists and the College of Radiographers, Documents of the Health Protection agency RCE-9, 2009.

ICRP, 1990 Recommendations of the International Commission on Radiological Protection, ICRP Publication 60, *Annals of the ICRP* 21: 1–3, 1991.

ICRP, Avoidance of radiation injuries from Medical Interventional Procedures, International Commission on Radiological Protection, ICRP Publication 85, *Annals of the ICRP* 30: 2, 2000.

ICRP, Biological effects after Prenatal irradiation (Embryo and Fetus) International Commission on Radiological Protection (2003) ICRP Publication 90, *Annals of the ICRP* 33: 1–2, 2003.

ICRP, 2007 Recommendations of the International Commission on Radiological Protection, ICRP Publication 103. *Annals of the ICRP* 37: 2–4, 2007a.

ICRP, Managing Patient Dose in Multi-Detector Computed Tomography (MDCT), International Commission on Radiological Protection, ICRP Publication 102, *Annals of the ICRP* 37: 1, 2007b.

ICRP, Radiation dose to Patients from Radiopharmaceuticals, International Commission on Radiological Protection, ICRP Publication 106, *Annals of the ICRP* 38: 1–2, 2008.

ImPACT, CT Patient Dosimetry Calculator (version 0.99v). ImPACT, St George's Hospital, London. www.impactscan.org/ctdosimetry.htm, 2004 [accessed 18th January 2011].

IPEM, Guidance on the establishment and Use of Diagnostic Reference levels for Medical X-ray Examinations, Report 88, Institute of Physics and Engineering in Medicine, Fairmount House, York, 2004.

IPEM, The Commissioning and Routine Testing of Mammographic X-Ray Systems, report 89, Institute of Physics and Engineering in Medicine, Fairmount House, York, 2005.

IPSM, Dosimetry Working Party of the Institute of Physical Sciences in Medicine, National Protocol for Patient Dose Measurements in Diagnostic Radiology National Radiological Protection Board, Chilton, Didcot, Oxfordshire OX11 0RQ, UK, 1992.

IRMER, Ionising Radiation (Medical Exposure) Regulations 2000 (SI 2000 No 1059) London, HMSO, 2000.

Jones, D.G. and Wall B.F, Organ Doses from Medical X-ray Examinations Calculated Using Monte Carlo techniques, NRPB-R186, National Radiological Protection Board, Chilton, Didcot, Oxfordshire OX11 0RQ, UK, 1985.

Martin, C.J., Dendy, P.P. and Corbett, R.H., Medical Imaging and Radiation Protection for medical students and clinical staff, British Institute of Radiology 36: Portland Place, London, 2003.

Morrell, R.E. and Rogers A.T., Kodak EDR2 film for patient skin dose assessment in cardiac catheterization procedures. *British Journal of Radiology* 79: 603–607, 2006.

NRPB, Patient dose reduction in diagnostic radiology. Report by the Royal College of Radiologists and the National Radiological Protection Board. *Documents of the NRPB* Vol 1 No 3. 1990.

RCR *Making the best use of clinical radiology services. Sixth edition.* The Royal College of Radiologists, London, 2007.

Samei, E., Saunders, R.S., Baker, J.A. and Delong, D.M., Digital mammography: effects of reduced radiation dose on diagnostic performance, *Radiology* 243: 396–404, 2007.

Shrimpton, P.C., Hillier, M.C., Lewis, M.A. and Dunn, M. National survey of doses from CT in the UK: 2003. *British Journal of Radiology* 79: 968–980, 2006.

Tapiovaara, M. and Siiskonen,T., PCXMX-A Monte Carlo program for calculating patient doses in medical X-ray examinations (2nd edition) STUK – Radiation and Nuclear Safety Authority, Helsinki, Finland. 2008 http://www.stuk.fi/sateilyn_kaytto/ohjelmat/PCXMC/en_GB/summary/ [accessed 18th January 2011].

Wall, B.F., Kendall, G.M., Edwards, A.A. et al., What are the risks from medical X-rays and other low dose radiation? *British Journal of Radiology* 79: 285–294, 2006.

Winer-Muram, H.T., Boone, J.M., Brown, H.L., et al., Pulmonary embolism in pregnant patients: foetal radiation dose with helical CT. *Radiology* 224(2), 487–492, 2002.

Yates, S.J., Pike, L.C. and Goldstone, K.E., Effect of multislice scanners on patient dose from routine CT examinations in East Anglia. *British Journal of Radiology* 77: 472–478, 2004.

Young, K.C., Burch, A. and Oduko, J.M, Radiation doses received in the UK Breast Screening programme in 2001 and 2002, *British Journal of Radiology* 78: 207–218, 2005.

Exercises

1. Explain carefully why AP and PA chest examinations will result in different effective doses (refer to Figure 13.1).

2. Describe the main ways in which ESD can be measured or derived and discuss their relative merits.

3. What do you understand by a diagnostic reference level (DRL)? How are DRLs established and what is their purpose?

4. List the organs and tissues of the body identified in ICRP publication 103 (2007) as the most sensitive with regard to causing long-term detriment by ionising radiation. Give an approximate risk of fatal radiation-induced cancer for a whole body dose of 5 mSv.

5. Explain how the effective dose is calculated for a radiological examination. What are the disadvantages of using effective dose to estimate the risk to an individual patient?

6. What are the problems with estimating doses to patients in CT and how are they overcome?

7. Discuss the factors that influence the dose to the breast in mammography (refer also to Section 9.2.6).

8. What factors determine the effective dose to a patient when a radiopharmaceutical is administered? Suggest one way in which the effective dose can be reduced (without loss of image quality).

9. Review the potential sources of radiation injury to the developing foetus.

10. 'Risks from radiological examinations must be placed in the context of other environmental, sociological and occupational risks'. Discuss.

14

Practical Radiation Protection and Legislation

B Heaton and P P Dendy

SUMMARY

- The principles of radiation protection—justification, optimisation and limitation—are introduced.
- Application of these principles to patients, staff and the public is discussed.
- Relevant legislation, especially as it applies to medical exposures, is reviewed.
- The UK Ionising Radiations Regulations (IRR) (1999) are discussed in some detail.
- X-ray room design is discussed briefly.
- Special protection problems associated with unsealed radioactive materials (e.g. in nuclear medicine) are considered.
- Principles of personal dosimetry are explained and typical levels of staff doses are reviewed.
- Some variations in approaches to legislation in Europe and the US are noted.

CONTENTS

14.1 Introduction

Although the emphasis in this chapter is on radiation protection and safety it should be remembered that radiation safety is only a part of total safety and should not be over-emphasised to the exclusion of other aspects of safety.

Radiation exposures can be split into three broad categories: occupational, medical and public exposures. In radiology departments all three categories can be encountered. The current terminology used to describe any activity which increases overall exposure is a 'practice' and an activity which decreases an existing exposure is an 'intervention'. However, the latest set of International Commission on Radiological Protection (ICRP) recommendations (ICRP 2007) have moved away from practices and interventions and use a 'situation' based approach which characterises exposure as planned, emergency and existing exposure situations. ICRP does, however, continue to use the term practice to describe an activity which increases the risk of exposure or actual exposure to radiation. In future it recommends that the term intervention should only be used to describe protective actions that reduce exposure and uses the terms 'emergency' and 'existing exposure' to describe situations where such protective actions are required to reduce exposures. As in their previous recommendations (ICRP 1991) the aim is to keep the exposure of the human population as small as possible but these recommendations also start to address the consideration of doses to the natural environment. National legislation will start to incorporate these latest recommendations over the next 10 years.

The terms radiation or radiological safety and radiation or radiological protection are often used interchangeably. Strictly speaking radiation protection is concerned with the limitation of radiation dose whereas radiation safety is concerned with reducing the potential for accidents. Since the distinction is largely academic, both are described in this chapter without any attempt to distinguish between them.

14.2 Role of Radiation Protection in Diagnostic Radiology

14.2.1 Principles of Protection

The principles of radiation protection are common to all uses of ionising radiation. They are as follows:

14.2.1.1 Justification

Since there is no safe threshold dose for stochastic effects (see Section 12.5.1) ionising radiation should not be used in any 'practice' unless the net benefit from the exposure for an individual or society is greater than the radiation detriment; that is, the radiation should not do more harm than good. Although justification of a practice should be generic and broad in nature and should not necessarily be carried out each time a practice is undertaken, some aspects of it can require consideration each time the practice is undertaken. A patient must receive more benefit than detriment from an exposure. The benefit to the radiologist and radiographer from any dose they receive in the course of this investigation can only be quantified in terms of employment. The detriment to them must also be balanced against the benefits the patient and society gain from the exposure. Likewise the radiation doses members of the public receive, as small adventitious doses through being in the proximity of radiographic exposures or nuclear medicine patients, are justified by the benefit to society. The much larger doses which can be received by family or friends who are caring or comforting patients who are emitting ionising radiation must also be classed as a benefit to society.

Justification does not just apply to the detriment from the radiation dose, it also goes beyond radiation protection considerations.

14.2.1.2 Optimisation

This is applied to situations which have been identified as being justified. The magnitude of individual doses, the probability that exposures will occur and the number of persons that will be exposed should all be kept as low as reasonably achievable—the ALARA principle (sometimes called the ALARP principle—as low as reasonably practicable) taking into account economic and social factors. The process of optimisation is aided by setting 'dose or risk constraints'. Constraint doses are planning doses which should not be exceeded by a practice. They are not legal limits but are set by local or national bodies as levels of exposure which good working practice should achieve. They do not set levels which are considered to be unsafe if exceeded. If measurements show they are being exceeded an assessment should be made to ascertain why and remedial action, if deemed necessary, should be taken.

It is also recommended that under some circumstances a risk constraint should be set in the case of potential exposures to limit the risk to any one individual particularly in the event of an accident.

Justification is applied to the practice as whole whereas optimisation can be applied to individual components of practice.

14.2.1.3 Application of Dose Limits (Limitation)

After following through the processes of justification and optimisation there is, for occupational and public exposure, a legal limit on the radiation dose that an individual can receive (see Section 14.3.3). There are, currently, no limits on the doses that can be given for medical purposes. These are still at the discretion of the doctor involved who has to be prepared to clinically justify them if necessary (however, guideline doses in the form of constraints are recommended—see Section 14.2.2).

Insight

Constraint Doses

Constraint doses are referred to several times in this chapter. In all situations they are a planning dose which it is thought need not be exceeded when a planned operation or action involving ionising radiation is carried out. Although they may be based on national recommendations they are essentially a local limit and if exceeded need only be subject to a local investigation which need not be recorded.

The legal limits are set to restrict stochastic effects (Section 12.5.1) to acceptable levels and to be well below thresholds for deterministic effects.

The new ICRP recommendations (ICRP 2007) continue with the same dose limits as their previous recommendations but draw attention to the possibility that the eye may be more radiosensitive than previously was thought. They retain the option of recommending lower equivalent dose limits and although at the time of writing a formal statement has not been issued, it is suggested that a limit of between 20 and 50 mSv/y will be adopted in the near future.

14.2.2 Patient Protection

The most efficient way to reduce patient dose is not to carry out the radiographic examination!

Non-essential radiographs such as reassurance radiographs or radiographs for insurance purposes should not be taken. Past images should be utilised where possible and images should be moved with the patient between hospitals. National protocols such as those recommended by the Royal College of Radiologists (2007) should be used so that the minimum number of images required for a diagnosis to be made are taken. Where possible, techniques using non-ionising radiation such as ultrasound or magnetic resonance imaging (MRI) should be used.

Reduction of patient dose can be achieved by a wide variety of means but it must always be remembered that to obtain a radiograph a dose must be given to the patient. If the dose is reduced too far the radiograph does not contain the information required and in fact the dose given is wasted, as the benefit to the patient does not compensate for the risk from it.

i. *Constraints or 'investigation levels'.* Levels of dose have been set in terms of skin dose or dose area product dose (see Section 13.2.2) which should not be exceeded for common radiographic and screening examinations. These constraint doses are all some way above those required by adopting best practices and the circumstances under which they are not achieved should be quite rare. If circumstances are identified where they are being exceeded the techniques and equipment used should be investigated and remedial action taken as necessary.

ii. *Use of high kVp techniques.* As pointed out in Section 13.6.1 the use of higher kVp should reduce patient dose.

iii. *Collimation.* Reducing the volume of the patient irradiated not only improves contrast because of the reduction in scatter but can significantly reduce the effective dose. Ensuring that even a small border exists round the edge of a radiograph can significantly reduce the area exposed (coning the field to 1 cm inside an 18 cm × 24 cm cassette reduces the area irradiated by 18.5%). Whilst modern units which automatically adjust the field size to the cassette size ensure no unrecorded area of the patient is exposed, they do not necessarily produce the optimum field size for the anatomy being radiographed. The optimum size could sometimes be smaller.

iv. *Optimisation of imaging system.* All imaging systems can produce images using a greater radiation dose than is necessary by incorrect adjustment or use of the imaging parameters. In many of the digital systems this is often not apparent because of the inherent ability of the system to compensate. It is essential that care is taken when setting up these units to ensure that excessive doses cannot be used.

v. *Reduction of screening dose.* This can be achieved by keeping screening time to a minimum. For very old units this is essentially at the discretion of the radiologist but modern units allow a variety of methods to be used. The use of last frame hold can significantly reduce dose particularly during processes involving the insertion of catheters, orthopaedics or cardiology. Some units now allow the collimators to be repositioned using a last frame hold which further reduces doses. Pulsing the output of the tube during screening also reduces the dose. Obviously, the more pulses per second that are required to visualise the process involved the larger the patient dose per unit time. To see movements of the heart requires many more pulses per second than to see movement in the large bowel.

The method of automatic control of kVp and mA to maintain the dose rate of an image intensifier input should also be optimised (see Section 5.14.3). Initially during most investigations the kVp should be raised to the maximum value that still provides acceptable contrast with further output, if necessary obtained by raising the mA. This will ensure that the dose to the patient is minimised whilst diagnostic information is maintained.

vi. *Filtration.* The use of filtration above 2.5 mm of aluminium for normal diagnostic work is mandatory in X-ray units. As described in Section 3.9 this removes the lower photon energies from the X-ray beam. Special K-edge filters can be used in paediatric radiography (Section 9.7) and in mammography (Section 9.2) to remove unwanted high and low energy photons. Some sophisticated screening units now automatically insert between 0.1 mm and 0.2 mm of additional copper filtration to further harden the beam. These units effectively calculate the attenuation of the patient and choose the thickness, if any, of copper to insert. Depending on kVp, the first 0.1 mm of copper reduces the patient skin dose by 30%–40% and the next 0.1 mm by approximately 10%. (The effective dose reduction will vary but will be no more than approximately half of these figures.)

vii. *Output wave form.* The more closely the output of the generator approaches a steady (DC) voltage supply the higher the quality of the beam (Section 2.3.4) and the fewer low energy photons are present. Even quite simple X-ray units are now using medium or high frequency generators, which produce essentially a DC supply. It should be noted that high frequency generators do not require a three phase electrical supply.

viii. *Use of low attenuation materials.* After the X-ray beam leaves the patient it may travel through several structures before reaching the imaging medium. Each of these can absorb some of the beam. It can be estimated that by replacing the covers on cassettes, table tops and the spacing material in grids by carbon fibre based materials a dose reduction of up to nearly 60% could be achieved, depending on tube kVp.

ix. *Choice of grid.* As described in Section 6.8.1 the higher the grid ratio the greater the dose required to produce the required film density.

x. *CT slit width.* The dose from CT scans is high, so the number of slices, or the beam width when using a multi-detector array, should be kept to a minimum and repeating the scan with contrast medium should only be undertaken if definitely required. Note that using several narrow beams is generally less dose efficient than using a wide beam (see Figure 8.20).

   xi. *Gonad shields*. These must be used whenever possible.

14.2.3 Staff Protection

Whilst all staff working with ionising radiation should be trying to keep their radiation dose as low as possible, radiography and radiological staff also have a vested interest in keeping the dose to the patient as low as possible because in principle their own doses are proportional to the patient dose. As demonstrated in Table 14.3 (Section 14.7.3) despite the very high dose rates produced by diagnostic X-ray equipment, by taking relatively simple precautions, the doses received by staff are kept very low. The dose received is equal to the dose rate times the period of exposure so a reduction in either dose rate or period of exposure reduces the dose received.

The design of modern units ensures that it is virtually impossible to receive an inadvertent whole body exposure during normal operations. Some radiologists still occasionally image their own fingers because they move them from just outside the main beam to just inside during the investigation. (Some orthopaedic pinning operations are very difficult to do without occasionally 'seeing' the surgeon's fingers.)

Scattered radiation is present whenever an X-ray set is operated so protection against scatter is important. The amount of scattered radiation produced and its distribution varies with the kVp (see Section 3.4.3). The scattered radiation dose in the vicinity of the patient when taking a large film could be reduced by 50%–65% by increasing the kVp from 50 kVp to 90 kVp.

14.2.3.1 Reduction of Dose Rate

Times—It is self-evident that the shorter the time of exposure, the lower the dose.

Distance—As described in Section 1.12 the dose rate in a radiation field from a point source falls as the square of the distance. For X-ray units this rule only strictly applies to the main beam where the focal spot can be assumed to be a point source. The scatter from the patient and other equipment is not from a point source and the rate of fall off with distance is more complicated. However, the general rule still applies that the further away from the X-ray tube, patient and table, the lower the dose rate in the radiation field. As a useful principle, during an exposure personnel should stand at least 2 m from the source of the scattered radiation whenever possible.

Shielding—Section 3.2 showed how materials could reduce the intensity of a radiation field. At diagnostic energies lead is one of the most efficient materials because it has a large linear attenuation coefficient. As described in Section 14.5 it is incorporated into some fixed shielding in cubicles and doors and it is also used in personal protective clothing. Lead/rubber aprons are used for body protection. These can have a lead equivalence of up to 0.5 mm. However, these aprons are heavy and uncomfortable to wear and most X-ray rooms would have aprons of no more than 0.35 mm lead equivalence. A thickness of 0.35 mm is laid down in the Medical and Dental Guidance Notes (2002) as the minimum allowed when X-ray photons are generated above 100 kVp. (Below 100 kVp only 0.25 mm lead equivalence is required.) Lead rubber gloves should have a lead equivalence of at least 0.25 mm and thyroid collars a similar lead equivalence to aprons. If using an over-couch screening unit lead glass spectacles or a lead glass screen must be used if eye measurements show high doses are being received. The ICRP (2007) statement (Section 14.2.1) would imply that the routine use of eye protection for all screening examinations might be prudent.

Protective clothing is not designed to shield against the main beam. It will only protect against scattered radiation or the beam after it has passed through a patient. Protective clothing must be properly stored so that it does not become cracked and must be periodically checked to ensure that it is in good condition. Suffice to say it must also be worn when provided. If neither protective clothing nor a permanent screen is available, the person should leave the room. 'Hiding' behind someone wearing a lead/rubber apron is not acceptable.

14.2.4 Public Protection

Members of the public are distinguished from patients if they receive an exposure but gain no clinical benefit. This can happen in waiting rooms, changing cubicles or as a consequence of being in the corridor adjacent to the X-ray department. These doses are kept to a minimum by permanent shielding incorporated into the walls of the X-ray rooms and doors. When mobile units are being used in wards the situation is rather different (and more difficult). Here the radiographer must ensure that the primary beam is not directed towards a patient in an adjacent bed. Any patient in a nearby bed who cannot be moved should be protected from scattered radiation by laying a lead apron over them. The Approved Code of Practice and Guidance Notes to the IRR 1999 (IRR 99) lay down a limit to any member of the public of 5 mSv over any 5 consecutive years, from adventitious radiation of this nature.

Members of the public who hold or support patients (such as parents holding children) knowingly and willingly after the risks are pointed out to them are not subject to the 'member of the public' dose limit set out in Section 14.3.3 or the limit identified above. These persons come into the category of 'comforters and carers'; they must, however, be subject to a dose constraint. Compliance with the dose constraint must be demonstrated by suitable monitoring.

Members of the public who become involved in medical research are also not subject to the normal dose limits. If the radiation exposure received is part of the normal treatment protocol then it is of direct benefit to the patient and is considered as a normal medical exposure. If it is additional to the normal treatment protocol or is to be delivered to a healthy volunteer, then the Ethical Committee considering the research project must take any risk into account.

The Euratom Directive 97/43 (1997) recommends that member states set dose constraints for research projects. A useful document is ICRP 62 (1992). This has a section dealing with the use of ionising radiation in medical research and identifies, in broad terms, the benefits to science and society that must be gained for the exposure of volunteers to different radiation doses.

A prerequisite of research work is that the informed consent of the volunteer must be obtained. Presentation of the possible risks of an exposure, extrapolated from the information given in Chapters 12 and 13, in a form that is informative but not frightening, is difficult. It is, however, essential.

14.3 European Legislation

14.3.1 Introduction—the ICRP

Increasing recognition of the harmful effects of ionising radiation led to the establishment, in 1928, of the ICRP. This commission is a non-governmental body and through an expert committee structure uses the knowledge of independent advisers, from any country in the World, to give guidance on all aspects of radiation protection. This guidance is issued in the form of recommendations and can range from broad concepts to detailed review of scientific research. Because of this structure ICRP recommendations are not mandatory. However, they are very influential internationally and few countries adopt regulations dealing with radiation protection which differ to any extent from them. On a worldwide basis the legislation controlling the use of ionising radiation is generally very similar. The major differences are found in the methods of enforcement.

The legislation in all countries currently uses the recommendations laid down in ICRP 60 (1991) which reviewed the epidemiological data available at the time, introduced new maximum permissible doses and developed the conceptual framework of justification, optimisation and limitation. This review process has now been repeated and the new publication, ICRP 103 (2007), takes account of the latest biological and physical information available and sets the framework for legislation for the next decade. The new recommendations are not fundamentally different to the ones they replace and consolidate and develop rather than change. The changes to any national legislation to reflect them are therefore not likely to be large. A useful review of the new recommendations has been presented by Wrixon (2008).

An ICRP report (2010), currently in draft form, which will be of direct relevance to readers of this text and which will influence future legislation is entitled Radiation Protection Education and Training for Healthcare Staff and Students. (It should be noted that at the time of writing a reference is not available for this.)

14.3.2 European Legislation

ICRP recommendations are promulgated within the European Community through a series of Directives. Member states must respond to a Directive, within a laid down time scale, by introducing national legislation. This legislation must, as a minimum, comply with the Directive but can be more restrictive if nationally deemed necessary.

14.3.3 Basic Safety Standards Directive 96/29/Euratom (1996)

This applies to all practices which involve a risk from ionising radiation arising from natural or manmade sources. The basic requirements are summarised as follows:

(a) All practices above the minimum levels below which the Directive does not apply must be reported and some must be specifically authorised.

(b) Occupational and public exposures are distinguished from medical exposures and those of carers and comforters (other than those carrying out the tasks as part of their work). Directive 96/29/Euratom states that if the dose is willingly and knowingly (i.e. the risks have been explained) received then the 'member of the public' dose limits do not apply. Constraint doses for carers and comforters should be set, however.

(c) The dose limits for exposed workers and students or apprentices above the age of 18 are shown in Table 14.1.

TABLE 14.1

Maximum Permissible Occupational Dose Limits as Specified in
96/29/Euratom

Effective dose in any consecutive 5-year period	100 mSv
Effective dose in any year (the option for a yearly effective dose of 20 mSv is maintained)	50 mSv
Equivalent dose to the lens of the eye in any year	150 mSv
Equivalent dose to any 1 cm² of the skin in any year	500 mSv
Equivalent dose to extremities in any year	500 mSv
Foetal dose once a pregnancy has been declared	1 mSv

Note: In addition there is a requirement that during breast feeding there is no risk of bodily contamination.

(d) For students and apprentices below the age of 18 years who are obliged to use ionising radiation, the effective dose limit is 6 mSv/y and the equivalent dose limits are 50 mSv/y for the lens of the eye and 150 mSv/y for any 1 cm² of the skin.

(e) The dose limits for members of the public are an effective dose of 1 mSv/y (in special circumstances this can be exceeded but the dose must not exceed an average of 1 mSv/y over any 5-year period), an equivalent dose to the lens of the eye of 15 mSv/y and an equivalent dose to any 1cm² of skin of 50 mSv/y.

(f) Work places where there is a possibility that the exposure to ionising radiation could exceed the limits set for members of the public must be classified as *controlled* or *supervised*. Controlled areas must be suitably demarcated and delineated with controlled access. Appropriate work instructions (*local rules*) must be issued for operations in these areas. Radiological monitoring is required.

For a supervised area there is a requirement to monitor the work force, but the other requirements of a controlled area are discretionary.

(g) Exposed workers are divided into two groups.

Category A—those who are liable to receive an effective dose greater than 6 mSv/y or an equivalent dose greater than 3/10 of the limit for the lens of the eye, skin or extremities.

Category B—exposed workers who are not Category A workers.

(h) Category A workers must be monitored. Sufficient personal or area monitoring must be carried out to demonstrate that category B workers do not need to be category A workers.

(i) All exposed workers, apprentices and students who, in the course of their work are dealing with sources of ionising radiation, must be given adequate training. Women must be informed of the need to declare if pregnant or breast feeding so as to avoid any hazard to the foetus or neonate.

(j) A suitably qualified expert must be appointed to advise undertakings on radiation risks and assess radiation protection arrangements (this person is frequently called the *radiation protection adviser*, RPA).

(k) All work places using ionising radiation must have suitable monitoring equipment available.

(l) Records of monitoring must be kept and made available to individuals. (For Category A workers this is for at least 30 years following termination of work involving exposure or until the age of 70 years).

(m) Category A workers must receive suitable medical surveillance. This must include a medical before starting work and a periodic review of health. Medical records must also be kept until the person is 75 years of age or for 30 years after work involving exposure to ionising radiation ceased.

(n) A system of inspection to enforce the requirements of the Directive must be established.

14.3.4 Medical Exposures Directive 97/43/Euratom (1997)

This Directive acknowledges the benefits that the use of ionising radiation bring to medicine but identifies that it is the major source of exposure to artificial sources of ionising radiation. Therefore medical practices need to be undertaken in optimised radiation protection conditions.

The Directive applies to all medical exposures including those for medico/legal reasons and the exposure of carers and comforters. It places considerable emphasis on the justification of medical exposures, requiring that they show a net benefit for the individual or society, identifying in particular

(a) Consideration of new types of practice

(b) Consideration of the specific objectives of the exposure and the characteristics of the individual involved

(c) A requirement that exposures for research purposes must be examined by an Ethics Committee

(d) A requirement that exposures for medico-legal reasons or other reasons with no direct health benefits to the individual should not be undertaken

The Directive requires all doses to be optimised. To this end it recommends that diagnostic reference levels for radiodiagnostic examinations should be established and used for members of the public. Whilst establishing that carers and comforters are not subject to dose limits (see Section 14.3.3) it requires that national constraint doses be established for these persons. It also requires that nuclear medicine patients or their guardians are issued with written instructions with regard to the steps to be taken to minimise the exposure of anyone coming in contact with that patient.

The responsibilities of both the person prescribing and clinically responsible for the exposure and the person carrying it out are identified. Both groups must have suitable training. This applies to qualification under the regulations and to on-going education after qualification. Some emphasis is placed on having written protocols for standard radiological procedures and recommendations for referral criteria. Medical physics experts must be available to give advice on all aspects of exposure from ionising radiation in medicine.

The equipment used for medical exposures must be included in an inventory kept by the competent authority, must be subject to acceptance testing before it is used for medical exposures, must be kept under radiation protection surveillance and must be subject to regular quality control. The Directive prohibits the use of direct fluoroscopy without an image intensifier and recommends that all fluoroscopic dose rates be controlled. It also recommends that, where practicable, all new radiodiagnostic equipment shall have in-built dose measuring devices.

The Directive draws particular attention to the requirements to keep doses to a minimum for paediatric exposures, health surveillance programmes, potential high dose techniques such as CT scanning and interventional radiography, pregnant women and, with regard to nuclear medicine investigations, breast feeding females.

14.3.5 Outside Workers Directive 90/641/Euratom (1990)

This Directive was designed to avoid the situation where itinerant workers could receive above the yearly maximum dose by working on the premises of several different employers in several different countries. It requires that a Category A employee of one employer working in a controlled area of another employer must carry a passbook in which his/her previous doses are recorded. On leaving that controlled area the controller of that area must enter into the passbook an estimate of the dose received. It might apply in a situation where a medical worker, for example, an interventional radiologist, works in two hospitals controlled by different employers. It also applies to X-ray engineers (if classified) working in an X-ray room.

14.3.6 New Euratom Basic Safety Standards

This document is currently being revised and is only in draft form at the time of writing. It has consolidated all of the above Euratom directives and includes the latest recommendations from ICRP 103. There is very little in its current form that is operationally or materially different to the individual directives outlined above.

14.4 UK Legislation

UK legislation complies with the Euratom Directives in various ways.

14.4.1 Radioactive Substances Act (1993)

This act essentially controls the holding and disposal of radioactive materials and has been formulated primarily to protect the environment. It essentially lays down three conditions:

(1) Before an undertaking can hold radioactive sources, the premises must be registered under section 7 of the Act with the Environmental Protection Agency acting in the country concerned. This registration certificate is very specific. It specifies the radionuclides, the activity of the nuclides and, for sealed sources, the number of sources that can be held. Occasionally it may allow a general beta/gamma emitting radionuclide holding for a small amount of activity to accommodate new projects. Records must be kept to show that the registration is not exceeded and inspectors visit the premises periodically.

Occasionally, a sealed source is used in more than one location. An example would be an americium-241 source used to check the lead equivalence of the walls of X-ray rooms. In these situations a mobile registration under section 10 of the Act must be obtained. One normal condition of a section 10 registration is that a clear record must be kept of the location of the source at any time.

(2) Radioactive waste can only be accumulated if the user is authorised under section 15 of the Act and only disposed of if the user is authorised under section 13 of the Act. Both authorisations are generally placed in the same certificate. The authorisation is very specific. It will state where the waste can be accumulated, the route of disposal, the radionuclides that can be disposed of and the activity that can be disposed. Again records must be kept to show compliance.

Currently there are various exemptions to the Act. The two exemptions probably most relevant to the Health Service are as follows:

(a) The Radioactive Substances (Testing Instruments) Exemption Order (1985): This order allows small sealed sources used in measuring instruments or for testing instruments to be exempt from registration. Examples are the external standard source in a liquid scintillation counter and the 3 MBq Co-57 sources used as markers in nuclear medicine.

(b) The Radioactive Substances (Hospitals) Exemption Order (1990): This order allows the occasional patient from a peripheral hospital to be investigated in a central main nuclear medicine department without having to transfer patients to the main hospital and without having to obtain a registration and authorisation for the peripheral hospital. It lays down conditions about the storage of any sources, procedures for emergencies and the maximum amounts that can be held on the premises. The maximum amount of most relevance is the 1 GBq limit for Tc-99m. There are similar conditions with regard to any waste produced, the limit of most relevance is 1 GBq for Tc-99m Technetium-99m discharged to the sewage system of the premises in a month. If the amounts exceed these limits then a full registration or authorisation must be obtained.

The Act was amended in 2005 by the High Activity Sealed Sources and Orphan Sources Regulations which put additional requirements on the conditions of registration for high activity sources (and other sources of a similar level of potential hazard). These essentially require a greater level of security which has to be agreed by the police before a registration is granted and the provision of a financial guarantee that funds will be available to dispose of the source when necessary.

At the time of writing this Act is under review and will be totally replaced in October 2011.

14.4.2 The Medicines (Administration of Radioactive Substances) Regulations (1978)

These regulations are operated through a committee called the Administration of Radioactive Substances Advisory Committee (ARSAC) which advises Health Ministers on the issue of certificates under the regulations. Because of this Committee these are commonly called the ARSAC regulations and the certificates issued under the regulations ARSAC certificates. These regulations are primarily concerned with the protection of patients or volunteers during clinical or research use of radioactive substances. They cover all administrations of radioactive materials no matter how small the amount being administered. Only a doctor or dentist can hold a certificate to administer radioactive materials but it is acceptable that other persons acting on their behalf and under their guidance may actually carry out the administration. This transfer of authorisation to administer should be a formal written transfer even though the responsibility still rests with the certificate holder. Certificates are normally only issued to senior staff of consultant status and are time limited.

Notes for Guidance (2006) on these regulations and application forms are available from the ARSAC. These guidance notes are particularly useful for routine tests as they give descriptions of classes of radioactive materials that are considered acceptable for given purposes. As long as the maximum recommended activity is not to be exceeded no justification need be given for the request to use the radioactive material for the purpose described in the Notes for Guidance. Application for an ARSAC certificate must include a description of the applicant's experience and training, a description of the equipment available to carry out the tests, the agreement and description of the scientist(s) associated with the work and the agreement of the RPA. If a research proposal is being considered or a routine test not listed in the Guidance Notes then, in addition to the above, the ARSAC also require a full justification and description of the use, including effective dose calculations.

These certificates are specific to the appointment held at the time of application and the hospital in which the investigation or treatment will be done. If either of these changes permanently the ARSAC should be informed, as the certificate will require amendment. If, in a specific case, for clinical reasons, a change in hospital premises must be made to administer the radiopharmaceutical this is permitted with the agreement of the local RPA.

A certificate from the ARSAC to carry out a research project that involves the administration of a radioactive substance does not remove the requirement for Ethical Committee approval. Most Ethical Committees would require an ARSAC certificate to accompany the application or assurance that an application has been made to the ARSAC.

14.4.3 The Ionising Radiation (Medical Exposure) Regulations (2000)

These regulations are generally referred to as the IRMER regulations. They were introduced in response to the Medical Exposures Directive described in Section 14.3.4.

They apply to all diagnostic and therapeutic exposures, including research exposures. The implementation requires the production of protocols and procedures which clearly identify the responsibilities of everyone involved in the examination or treatment of patients or research volunteers using ionising radiations. Although the Directive applies to two types of justification, that of the practice and that of the individual patient exposure the Regulations only address the latter. Key persons are defined in the Regulations and must be referenced in the protocols and procedures referred to above. They are as follows:

> The *employer* (normally the NHS Trust). This is essentially the body responsible for the patient as opposed to the normal definition under employment law. (The Trust retains the responsibility under the Regulations even when contract staff are involved.)
>
> The *medical physics expert*. This is a suitably trained person who can advise on such matters as patient dosimetry, unintended exposure evaluations and the development of techniques. This person will often be the RPA but the post can be carried by someone who is not an RPA.
>
> The *operator*. This is anyone who carries out a practical aspect of an exposure. Any exposure will probably have several operators involved because it includes not just the person who controls the actual exposure but other persons such as physicists, technicians and radio-pharmacists who may influence the patient dose. Their roles and responsibilities must be clearly identified in the procedures.
>
> The *referrer*. The employer must identify in the procedures the persons who can act as the referrer requesting a particular medical exposure. Some (such as doctors) may be

allowed to request a very wide range of examinations or treatments, others (such as speciality nurses) will be restricted in what they can request.

The *practitioner*. The practitioner is the person who reviews the request from the referrer and agrees (by signature) that the request is justified for the patient concerned in the light of the clinical information provided by the referrer.

The requirement above regarding the clinical information required is emphasised in the regulations. Currently much emphasis is placed by the IRMER inspectors on the production of written operating procedures. These establish the framework under which all the health-care professionals involved in the imaging process can practice.

(a) Referral criteria and exposure protocols must be established. A common level of detail is required across all departments no matter how large or small the department.

(b) The employer must establish diagnostic dose reference levels using local measurements where possible but taking into account published national figures (see Section 13.3.1).

(c) Reference level doses apply equally to research investigations.

(d) Training and training records are an important part of the regulations and the employer has the responsibility for ensuring that all their employees and any outside contractors working with their patients are adequately trained.

(e) Any incidents where a dose 'greater than intended' is given must be investigated and reported. The definition of what is greater than intended is given in the Health and Safety Executive (HSE) Guidance Note PM 77 (2006). It should be used for all incidents such as mistaken identity, wrong view and lost images. There is a clear direction that when a dose is 'greater than intended' the patient must be told.

(f) All research programmes involving the use of ionising radiations must be submitted to a Local Research Ethics Committee. A dose constraint must be set as part of the programme to limit the dose that can be received by any patient participating. This dose must not be exceeded. The risks of participating in such a programme must be explained to all persons asked to participate in a manner that allows them to understand them. This can make research involving children or mentally incapable adults very difficult to carry out.

(g) All medical diagnostic exposures must have a clinical evaluation carried out on them. If it is known that this will not happen then the exposure is not justified. Even if the employer is also the practitioner/operator (as in say a dentist) a record must still be created of the outcome. This record should also include sufficient exposure details, for example, kVp, mA or DAP reading to allow an estimate of the effective dose to the patient to be made.

(h) An inventory of all the X-ray units in a department must be held along with the staff training records referred to above for all staff.

14.4.4 Transport Regulations for Radioactive Material

These will only concern a few readers working in nuclear medicine departments. They differ from all other regulations in that they must comply with international standards as laid down in International Atomic Energy Agency Regulations (IAEA 1996) and the

European Agreement (The International Carriage of Dangerous Goods by Road [ADR 2007]). They are implemented in the UK by the Carriage of Dangerous Goods and Use of Transportable Pressure Regulations (2009). They do not apply if a radionuclide is contained in the body of a person undergoing medical treatment or is a luminous device worn by a person. Because of the exemption applied to patients, many hospitals will never be directly concerned with these regulations because they do not send radionuclides to other hospitals. A hospital that sends out radionuclides will often only send sufficient activity for a single patient. The drivers of vehicles carrying these packages require training and should understand the hazard of the material they are carrying. Part of the quality assurance system for the transport operation would refresh this training annually. Of particular importance is that the driver knows what actions to take in the event of an emergency.

Anyone considering transporting radionuclides away from a hospital site should consult their Medical Physics Expert or RPA well before the event as it can take some time to make the necessary arrangements.

14.4.5 Ionising Radiations Regulations (1999)

The system of regulation adopted for worker protection in the UK for all aspects of safety is based on three tiers of documentation. The first tier is legislative where a statutory instrument is passed by parliament. The regulations in such an instrument are mandatory and can only be changed by a new statutory instrument in parliament. The second tier is in the form of Approved Codes of Practice (ACOP). These are prepared by the HSE and approved by the Health and Safety Commission and support the relevant statutory instrument by expanding each regulation with an interpretation as to what is required to comply. In legal terms an ACOP is evidentiary and one does have the option of non-compliance with it. However, if this option of non-compliance is chosen it could be necessary to prove in a court that the action taken was at least as good as the one recommended in the ACOP. The advantage of an ACOP is that it can be changed by the Commission without recourse to parliament. The third tier is called 'Notes for Guidance'. For the IRR 1999 the HSE have issued a combined Approved Code of Practice and Guidance Notes which between them explain in more detail how the legislation should be interpreted and implemented. A second level of 'Notes for Guidance' exists which have been prepared by a committee of experts drawn from operators, advisers and legislators. These notes for guidance, for example, the Medical and Dental Guidance Notes (2002) written by the Institute of Physics and Engineering in Medicine give a practical guide on how the legislation and the HSE ACOP and Guidance can be implemented and complied with in that area of use. The notes for guidance do not have a direct legal standing.

The IRR apply to all users of ionising radiation and medical users must comply fully. The implementation of the regulations is the responsibility of the employer who will be prosecuted in the event of a breach. There are some 41 Regulations and 9 Schedules in the regulations. The following are identified as the ones of most interest in radiological practice.

14.4.5.1 Regulation 1

This defines the persons covered by the regulations. The regulations apply to workers who receive a radiation dose in the course of their work and to members of the public who are exposed due to a work activity. Limited sections apply to patients, comforters and carers.

14.4.5.2 Regulation 6

Before starting work with ionising radiations the HSE must be told at least 28 days before the work starts. This applies to any new X-ray unit which is situated in a building complex that has not previously contained an X-ray unit, to the knowledge of the HSE.

14.4.5.3 Regulation 7 Risk Assessment

It is the duty of the employer to have in place a risk assessment for every activity, both new and old, which is covered by the regulations. Paragraph 44 of the ACOP identifies the matters that need to be considered. An integral part of any risk assessment is the identification of foreseeable accidents and the production of contingency plans to deal with them. These must be included in the local rules (see below). Risk assessments must be periodically reviewed to ensure that they are still valid.

Insight

Regulation 8 with Schedule 4 Facility Planning

These identify dose limits and the need for employers to keep doses as low as reasonably practicable by engineering controls and suitable systems of work. The ACOP identifies the importance of planning new facilities properly in conjunction with the RPA. The ACOP also forbids the use of direct vision fluoroscopy under any circumstances and requires the use of warning devices for X-ray units. In Regulation 8 the requirement not to exceed the dose limits specified in Schedule 4 of the Act is laid down. This regulation places emphasis on the use of proper engineering controls to restrict radiation doses. It identifies the need for warning devices such as lights which are, whenever possible, automatic. These engineered safety features should, as far as possible, be supported by systems of work to be followed by persons working or present in the area. These systems of work are incorporated into the Local Rules (see Regulation 17). Under this regulation personal protective equipment (PPE) must be provided. The HSE expect PPE to be used unless wearing it renders the wearer susceptible to another, greater risk.

The expansion of this regulation in the ACOP regulation covers the exposure of patients in a hospital to radiation from another patient as may happen from a nuclear medicine scan. The ACOP recommends that a maximum of 5 mSv can be received under these circumstances from a course of treatment. This is five times the normal member of the public limit.

Once a female employee declares that she is pregnant the foetus must not receive a dose that is greater than 1 mSv during the rest of the pregnancy. For workers in a typical X-ray department this is broadly equivalent to a dose of 2 mSv to the surface of the abdomen. An employer must also take into account the potential risks to the newly born child of a female working with unsealed radionuclides.

If any employee receives a radiation dose in excess of 15 mSv in a year the employer must carry out an investigation.

14.4.5.4 Regulation 10 Engineering Controls

Any engineering controls of radiation dose and PPE (such as lead rubber aprons) must be checked at least once a year. It is good practice to record these checks and to keep the records for at least 2 years.

14.4.5.5 Regulation 11 and Schedule 4 of ACOP Dose Limits

This regulation lays down the dose limits for workers and members of the public—see Section 14.3.3. It is the responsibility of the employer to ensure that these limits are not exceeded. The ACOP lays down the exemption from dose limits of comforters and carers (as defined in Regulation 1).

14.4.5.6 *Regulation 13 Radiation Protection Adviser*

Since all X-ray departments contain controlled areas (see below) an RPA must be appointed, under regulation 13. The ACOP outlines the specific matters on which the RPA should be consulted.

14.4.5.7 *Regulation 14 Training*

Suitable training must be given to all persons using or exposed to ionising radiation in the course of their work. The level of training will be advised by the RPA. Records of this training must be kept. Particular training must be given to Radiation Protection Supervisors (see Regulation 17).

14.4.5.8 *Regulations 16 and 18 with Schedule 4 Designation of Controlled and Supervised Areas*

Section 14.3.3 describes the criteria adopted for designation of controlled and supervised areas. In practice a controlled area is any area where the dose rate is in excess of 7.5 μSv/h or potential contamination levels are such that an effective dose of 6 mSv may be received in a year. This means that all X-ray rooms are classified as controlled areas. A controlled area must have an engineered barrier and access must be under the control of the employer. This is generally by the walls of the X-ray room. All entrances must have clear warning signs. The ACOP, however, acknowledges that where it is not reasonably practical to demarcate a controlled area, (such an area may be when using a mobile X-ray unit on a ward) then if access to the area can be visually controlled and verbal warnings given then this will be satisfactory.

Access is limited to classified workers, persons working to a written arrangement and patients. The written arrangement controls working practices such that it ensures that the 3/10 dose limit cannot be exceeded. In medical circumstances virtually all employees work to written arrangements and are not classified. To demonstrate that the arrangements are successful everyone regularly exposed to ionising radiation is generally issued with a personal dosimeter.

Supervised areas can be identified by warning signs but there is no restriction on access for employees. Supervised and controlled areas must be identified in the Local Rules (see below).

14.4.5.9 *Regulation 17 Local Rules and Radiation Protection Supervisors (RPSs)*

Every employer who undertakes work with ionising radiations must have, in writing, a set of *Local Rules*. Local Rules ensure both compliance with the requirements of the regulations in practice and that exposures are kept to a minimum. The Local Rules are specific to an area, not general to an establishment. A newcomer to a department should be able to identify quickly the procedures that must be followed. Local Rules should be clear and concise and the essential contents are as follows:

(a) A clear description of the area to which they apply, identifying which areas are controlled and which are supervised.

(b) The name of the RPS of the area (see below), the name of the RPA for the area and the method of contacting them.

(c) The responsibilities of the RPS.

(d) The working instructions for the area, including a clear system of work if non-classified persons are entering controlled areas.

(e) The procedures to deal with any emergencies which may be anticipated.

In addition, it may be useful to add to the rules the local methods of ensuring compliance with other requirements of the regulations. For possible inclusion would be a description of the management and supervision structure for the area, the method and the frequency of testing safety controls and warning devices, the method of measuring dose rate and, particularly, contamination, the testing of instruments and the arrangements for personal dosimetry.

The RPS has local responsibility on behalf of the employer for ensuring compliance with the working procedures identified in the Local Rules. The RPS must be appointed in writing and must have sufficient experience, knowledge and status in the management structure to be able to ensure the implementation of the Local Rules and to action the emergency procedures should an emergency arise.

14.4.5.10 Regulations 19 and 21 Dose Assessment and Monitoring

An employer must supply adequate area monitoring equipment, which must be calibrated at least every 12 months by a 'qualified person'. Records of any monitoring undertaken must be kept for at least 2 years.

Suitable personal dosimeters should be worn (see Section 14.7). Employees must be told of the doses recorded on personal dosimeters.

14.4.5.11 Regulation 27 Sealed Sources

Sealed sources must be securely and properly stored. All sealed sources must be leak (wipe) tested at least once every 2 years and records of the test kept for at least 2 years. Some sources used in nuclear medicine require testing.

14.4.5.12 Regulations 31 and 32 Duties of Manufacturers and Equipment Requirements

A manufacturer has a responsibility to ensure that any equipment they supply is both constructed and designed to keep all exposures to staff as low as reasonably practicable. To ensure that this is the case, a 'critical examination' must be undertaken by an RPA (employed by either the manufacturer or the employer buying the equipment). This examination will check the correct operation of the safety features and warning devices, ensure that there is sufficient protection for staff and that adequate operational information about use, testing and maintenance has been provided.

Employers must ensure that any equipment used for medical exposure is designed installed and maintained such that the objectives of use, for example, diagnosis, treatment or research, are achieved with the minimum practical exposure. The word 'equipment' is interpreted quite liberally and includes not only the ionising radiation generating device but also ancillary equipment such as image receptors, calibrators and image processing units.

It is incumbent on persons buying such equipment to take into account potential patient dose when deciding which equipment to purchase. It is a requirement that all X-ray equipment has installed on it a device for recording the dose received by a patient. For many, but not all, X-ray sets this is by having a DAP meter installed.

A quality assurance program must also be in place to ensure that the equipment is working as originally intended. Action levels must be established so that any unacceptable equipment can be adjusted or replaced.

If a dose that is greater than intended is given (see Guidance Note PM 77) due to equipment failure then an investigation must be carried out and a report submitted to the HSE.

14.4.5.13 Regulations 33 and 34 Employee Responsibilities

These regulations impose on an employee a requirement to comply with the regulations and not knowingly to expose themselves to a radiation dose that is more than necessary to carry out their work.

- No one must recklessly or intentionally misuse radioactive materials or X-ray equipment.
- The recommended personal protective equipment (PPE) must always be used unless a justifiable reason can be given not to use it.
- Personal dosimeters must be worn and replaced when requested.
- Any doses to patients that are 'greater than intended' must be reported through the hospital incident reporting system and the RPS.
- All equipment defects (including PPE) must be reported to the RPS.

14.5 X-ray Rooms

14.5.1 Introduction

X-ray rooms must be designed to accommodate the equipment being used in them. This design requires the co-operation of the radiologist, the radiographer and the physicist. In general, the walls and the door constitute the barrier of the controlled area produced by the X-ray unit and access is controlled through the doors of the room. The shielding required in each wall, the ceiling and floor depends on the fraction of the total work load for which the X-ray beam is pointing in each direction. Only the total workload needs to be considered when calculating the shielding required from the scattered and leakage radiation as these are considered isotropic. This workload can be averaged over a suitable period (8 h in the UK) but many countries do not allow any adjustment for occupation of the area being shielded and always assume 100% occupation. This is a very conservative assumption in all but a very few cases.

The maximum permissible dose rates outside the walls after shielding depend on whether a member of the public or a hospital employee could be present. They would normally be 2.5 μSv/h and 7.5 μSv/h, respectively using the averaging technique of workload and use identified above.

14.5.2 Points of Note on Room Design

(i) The shielding can be achieved by a suitable thickness of concrete blocks, barium plaster on bricks or plaster lath or lead sheet. Applied lead sheet is not common because of cost and construction difficulties. It is, however, being increasingly used pre-bonded to plaster or plywood board.

(ii) Doors must have an equivalent shielding to the walls. All cracks in the door frame must be covered by lead sheet. Sliding doors are attractive for saving space but can be difficult and tiring to move. This can result in them not being closed properly in a busy room.

(iii) The radiographers' cubicle in the room should have sufficient shielding to ensure that the dose rate does not exceed 1 μSv/h. It should be positioned and have sufficient shielded windows to ensure that the radiographer can see the patient at all times. It should also be large enough to accommodate every person who will require to stand behind a shield unless supplementary mobile lead shields are also present in the room.

(iv) A warning light should be positioned at the entrance to each room. Ideally this would be a two stage device with a 'Controlled Area' warning illuminating when the electricity is switched on at the unit and a 'Do Not Enter' warning illuminating immediately before exposure on 'preping' the X-ray tube. Rooms relying on a single warning light must also have a notice indicating the action to be taken when the light is illuminated.

(v) If there are two X-ray units in the room, warning lights must be placed on the head of each unit to indicate which one has been selected.

(vi) Hanging facilities for lead/rubber aprons must be provided and storage for lead/rubber gloves and gonad shields should be provided.

(vii) Emergency 'cut off' and aid buttons must be positioned in suitable locations.

(viii) On completion of building or modification of a room, an RPA should carry out a critical examination to ensure that all safety devices are working properly and that the shielding is as designed. The leakage dose rate from the unit in the room must not exceed 1 mGy/h at 1 m from the X-ray head.

The methodology of room design is covered in several publications but that of Sutton and Williams (2000) is recommended for further reading.

14.6 Nuclear Medicine

14.6.1 Introduction

The radiation hazards in a nuclear medicine department are the external hazard from the radiation field around the radiopharmaceutical container (bottle or patient) and the internal hazard that would result from ingestion of some of the radiopharmaceutical. Ingestion of the radiopharmaceutical could take place in one of three ways; through the mouth, through the nose or through the skin if a cut or abrasion is present. Some radioactive elements that can be found in a nuclear medicine department, in particular the iodine isotopes, can actually diffuse through unbroken skin. Once in the body the radio-nuclide is accumulated in the organ or tissue which would normally accumulate the stable element or, for some radionuclides, a close chemical analogue. Once incorporated, it is virtually impossible to remove the radionuclide medically. It will only leave by natural biological processes. As the radionuclide decays, all of the energy from any β particles emitted will be absorbed in the body together with a proportion of the gamma photon energy. Small ingested amounts of a radionuclide can result in significant radiation doses.

14.6.2 Potential Internal Doses

The dose received following ingestion of a radioactive material cannot be measured in the same way as external radiation doses. ICRP recommends that the committed effective dose to a tissue T over an assumed working lifetime of 50 years ($H_{50,T}$) following intake of a radionuclide should be calculated. By summing over all tissues, allocating all this dose to the year of intake and ensuring the dose limit for that year is not exceeded, even for radionuclides with a very long effective half-life, the limit for any year is not exceeded

The ICRP introduced the concept of an annual limit on intake (ALI) for radionuclides. This intake can be taken every year and the effective dose limit for the year will not be exceeded.

In some countries, because of its usefulness, the ALI has continued to be used but in others it has now been superseded by the methodology described in Section 14.6.3 below.

Insight

ALI and Derived Limits of Airborne Contamination

For the purposes of this example, let us make the following assumptions:

1. All exposure arises from internal sources.
2. The committed *effective dose to tissue* T over 50 years for unit intake (say 1 MBq) is $H_{50,T}$.
3. The intake is *I* (measured in MBq).

Then the total committed effective dose is $I \cdot w_T H_{50,T}$ summed over all tissues. Hence the ALI is that value of *I* which satisfies the equation

$$I \sum w_T \cdot h_{50,T} = 20 \text{ mSv (for 20 mSv/y effective dose limit)}$$

To find the value of *I*, $H_{50,T}$ must be calculated. To do this, the factors to consider are as follows:

1. The total energy radiated per unit time by the radionuclide, perhaps separated into that from charged particles and that from gamma rays
2. Absorbed doses from gamma rays, originating both in tissue T and elsewhere, and taking into consideration geometrical factors, the inverse square law and the mass absorption coefficient
3. The distribution of radionuclide within the body
4. The time scale of exposure. If the radionuclide localises quickly, this may be expressed in terms of the effective half-life, but note that after leaving an organ the activity may localise elsewhere—for example, the kidney excretes via the bladder

When the ALI is known, the permissible derived air concentration (DAC) can be calculated if simplifying assumptions are made. Assuming a working year of 2000 h (50 weeks at 40 h), a rate of breathing of 0.02 m³ per min and that inhalation is the only route of intake

$$DAC = \frac{ALI}{2000 \times 60 \times 0.02} = \frac{ALI}{2.4 \times 10^3} Bq\,m^{-3}$$

For Tc-99m the ALI is 1×10^9 Bq, hence the maximum permissible DAC, rounded to the nearest whole number, is 4×10^5 Bq m⁻³. For I-131 the ALI is 2×10^6 Bq and the DAC 7×10^2 Bq m⁻³ (ICRP 1994).

Note that the concept should be used with care since it only applies to the ICRP reference man working under conditions of light activity and makes a number of other assumptions about the metabolic breathing pattern.

The derived working limit for surface contamination that will ensure the maximum permissible DAC is not exceeded varies from one radionuclide to another. For radionuclides used in nuclear medicine, it may be as low as 370 Bq spread over an area of 10⁻² m². This causes two problems:

1. The activity 'seen' by a small detector may be no more than 20–30 Bq. A Geiger-Müller counter is insufficiently sensitive to detect this level of activity and a scintillation crystal monitor must be used.
2. It may be impossible to decontaminate to maximum permissible levels by washing and cleaning after even quite a small spill (say 1 MBq). If the radionuclide has a long half-life, contaminated equipment will then have to be removed from service. Fortunately for Tc-99m (6 h half-life) the contamination may decay to an acceptable level overnight.

14.6.3 Calculation of Ingestion Dose

The ICRP has published a series of reports, the last one being ICRP 72 (1996), in which it has compiled the dose coefficients for several age groups of the public and for workers for ingested and inhaled radionuclides. These are based on the latest ICRP models for the alimentary tract and respiratory tract and are calculated for standard persons. The coefficients give the effective dose in Sv per Bq of the radionuclide in question being inhaled or ingested.

When calculating ingestion dose one has often to assume an activity inhaled or ingested as direct measurement is not possible. The errors of this assumption are generally far greater than those introduced by using dose estimates for the standard man and no further adjustment to the received dose need be made. If, however, a direct measurement can be made then the dose commitment can sometimes be refined by using the biological half-life for the radionuclide for that particular person rather than that of the standard man.

14.6.4 Special Precautions in Nuclear Medicine

With unsealed radionuclides the two hazards identified in the introduction to this section must be controlled. Although small doses from the external hazard are most frequent, the internal hazard is potentially far more serious.

The external hazard can be greatly reduced by following relatively simple precautions:

1. Always keep bottles containing radionuclides in lead pots in a shielded area—never leave the vial unshielded

2. Never handle bottles containing radionuclides directly but always use tongs. (This also helps avoid contaminating gloves because of a contaminated stock bottle.)

3. If possible stay at least 0.5 m away from a patient who has received radioactivity, and check that nurses do not remain unduly close to the patient for unnecessarily long periods. The dose rate 25 cm away from a patient who has received 500 MBq of Tc-99m for, say, a bone scan is about 33 µSv/h. At a metre the dose rate is only 9 µSv/h. (Note that the inverse square law is not obeyed for an extended source.)

4. Use shielded syringes to give injections.

5. If a shield cannot be used, extra care should be taken when handling the syringe not to hold the end containing the radionuclide.

The internal hazard can be avoided by simple good house-keeping practices. Protective clothing, especially gloves, should always be worn. Syringes do backfire and contaminated skin is difficult to clean. All manipulations of the radionuclide from bottle to syringe must be performed over a tray so that any drops can be contained. Syringes must always be vented into a swab, never squirted generally over the room. All contaminated and potentially contaminated materials must be disposed of in a container that has been clearly labelled as suitable for the purpose. Hands must always be washed before leaving a radioactive area.

14.6.5 PET Facilities

An increasing number of hospitals now have facilities for PET imaging (see Chapter 11). These often also involve the installation of a cyclotron in a specially shielded area to produce the short-lived positron emitters that are to be used. The design of a PET facility is beyond the scope of this book but suffice to say it must be carried out by a suitably experienced RPA. The potential causes of radiation doses are similar to those in general nuclear medicine but the risks are higher. The activities manipulated and administered to patients tend to be higher because of the short half-lives involved. The potential radiation doses when handling the materials in a PET facility are very high. The gamma photons produced by the annihilation of the emitted positrons are much higher energy than in a general nuclear medicine department (0.51 MeV) and the consequent radiation fields in the vicinity of patients during and after the injection of the positron emitter are high. Much greater care is therefore required when working in a PET facility to keep doses to the staff involved, particularly those of the technicians involved in the injection process, to an acceptably low level. A more detailed description of the problems is given in the chapter on PET facilities by Heaton et al. (2011).

14.7 Personal Dosimetry

There are now several types of personal dosimeters available but the two most common ones used in hospitals are still the film badge and the thermoluminescent dosimeter. However, this will change in the near future because the main supplier of dose monitoring film has decided to cease production. The physical principles of these two techniques of radiation measurement were discussed in Chapters 4 and 5, so only those aspects of their performance that are relevant to personal dosimetry are referred to here. This will be done by considering the requirements of an ideal personal dosimeter and the extent to which each method satisfies this ideal after briefly reviewing their characteristics. The two techniques, described in 14.7.2, should also be considered relative to these requirements.

14.7.1 Thermoluminescent Dosimeters (TLDs) and Film Badges

14.7.1.1 Thermoluminescent Dosimeters

In TLD materials the electron traps in the forbidden region (see Section 4.7.1) are normally empty. When the material is exposed to ionising radiation electrons are excited from the valence band into the conduction band from where they fall into one of these electron traps.

Here many are stored almost indefinitely and only released when the TLD is heated, generally to about 300°C–400°C. When heated they escape to the conduction band from where they fall back to a luminescence centre in the valence band giving off light. The amount of light emitted is proportional to the exposure of the TLD material. Only the traps relatively close to the conduction band can be emptied at room temperatures and to avoid this affecting the magnitude of the light output, the signals from these traps are ignored during the read out cycle. TLDs have wide application for personal dosimetry. They are also used extensively for measuring the patient skin doses received during X-ray examinations as discussed in Chapter 13. The method of read out is the same in all applications. Here use of TLDs as general-purpose dosimeters is considered. In this capacity they are generally used in pairs in a holder. One of the dosimeters is open to the air and effectively measures skin dose, the other is under a thickness of material to simulate a dose measured at 10 mm depth in tissue.

14.7.1.2 Film Badge Dosimeters

The properties of photographic film as a recording medium for X-rays were discussed in Section 5.5. For use as a general-purpose dosimeter, film has a number of disadvantages, notably, variation in sensitivity with photon energy (see below). However, it does have important applications as a personal dosimeter and it is instructive to compare the extent to which films and TLDs satisfy the requirements of an ideal personal dosimeter.

14.7.1.3 Range of Response

The monitor must be sensitive to very small exposures since they will be the most likely to occur, but it must also be capable of recording accurately a high exposure should this arise in an accident. Hence a wide range is required.

Because of the shape of its characteristic curve, photographic film is useful over only a limited range of doses. However, the range can be extended if a fast film is backed by a much slower film. If the fast film is over-blackened, it is carefully removed and readings are obtained from the slower film.

There are no such problems with TLD which has a wide range of response from 0.1 mGy upwards.

14.7.1.4 Linearity of Response

If the response is linear with dose, measurement at just two known doses will allow a calibration curve to be drawn. This is possible with TLDs although not recommended, but for film, because of the shape of the characteristic curve, calibration is necessary at a large number of doses so that the exact shape of the curve can be established. Furthermore, calibration is necessary each time a batch of film is developed because of potential variations in film blackening with development conditions.

14.7.1.5 Calibration against Radiation Standards

A measure in terms of a fundamental physical property—for example, temperature rise—would be desirable. However, neither film blackening nor thermoluminescence is in the category. Therefore calibration against a standard radiation source is necessary on each occasion.

14.7.1.6 Variation of Sensitivity with Radiation Energy

As explained in Section 4.10, the sensitivity of a monitor will vary with photon energy unless it is 'tissue, or air equivalent'. Because of the presence of silver and bromine in film, blackening per unit dose is much higher at low photon energies where the photoelectric effect dominates than at higher photon energies. Similar differences in sensitivity exist for electrons. Therefore the film badge holder contains a number of filters (Figure 14.1a), which not only extend the range of radiation energies over which the blackening per unit dose is approximately constant (Figure 14.1b), but also provide data which may be used to calculate the dose for low energy photons and electrons. The cadmium filter will capture neutrons with subsequent emission of gamma rays so additional blackening under the filter is evidence of neutrons.

The sensitivity of a lithium fluoride TLD is independent of radiation energy down to about 100 keV, then increases slightly due to the small difference in atomic number between LiF and soft tissue. At even lower energies this effect is counter-balanced by self-absorption in the lithium fluoride/ceramic chip and the overall change in sensitivity with energy is unlikely to exceed 20%.

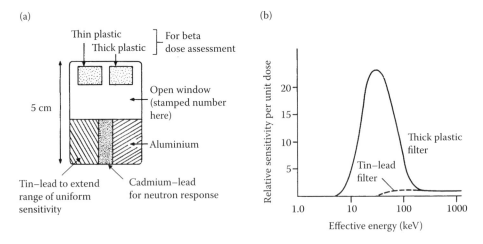

FIGURE 14.1
(a) Details of a film badge holder. (b) Curves showing the variation of sensitivity (film blackening per unit dose) with radiation energy both without and with filter. Note that in the diagnostic range, correction for variation in sensitivity has to be made.

14.7.1.7 Sensitivity to Temperature and Humidity

The fog level of film can increase markedly under conditions of elevated temperature or humidity and this may be misinterpreted as spurious radiation. TLDs show no such effects.

14.7.1.8 Uniformity of Response within Batches

Provided care is taken over storage and development, photographic films are nowadays uniformly sensitive within a batch. The sensitivity of a TLD can vary within a batch so individual calibration may be necessary. Also careful annealing is required after use or the TLD tends to 'remember' its previous radiation history.

14.7.1.9 Maximum Time of Use

This is primarily governed by the risk of latent image fading when radiation is accumulated in small amounts over a long period of time. For both systems the effect is negligible provided that the time scale for calibration is comparable with the time scale for use. The effect of temperature on film, however, limits the use to a period of 2 months. TLD can easily be used for 4 months and in some circumstances they are used for longer.

14.7.1.10 Compactness

As shown in Figure 14.1 a film badge holder is quite small. TLDs can be extremely small (Figure 14.2) and are especially useful for monitoring exposure to the hands and fingers. This also allows a dosimeter to be placed onto a patient to record their skin dose without significantly affecting the radiographic image.

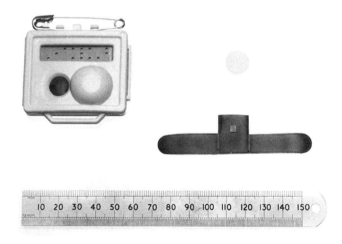

FIGURE 14.2
Examples of TLD monitors. On the left is a badge holder for whole body monitoring. On the upper right is the LiF ceramic disc that would be useful for measuring extremity doses. A very small LiF chip is shown on the lower right.

TABLE 14.2

Relative Merits of Film Badges and Thermoluminescent Dosimeters as
Alternative Methods of Personal Monitoring

	Film Badge	TLD	
1	Range of usefulness	0.2 mGy–6 Gy	0.1 mGy–10^4 Gy
2	Linearity of response	No	Yes
3	Calibration against radiation standards?	Yes	Yes
4	Response independent of radiation energy?	No	Yes (except at low kV)
5	Sensitive to temperature and humidity	Yes	No
6	Uniformity of response within batches	Yes	Yes (with care)
7	Maximum time of use	2 months	12 months
8	Compactness	Small	Very small
9	Permanent visual record?	Yes	No
10	Indication of type of radiation?	Yes	No
11	Indication of pattern of radiation exposure?	Sometimes	No

14.7.1.11 Permanent Visual Record

When investigating the possible cause of a high reading (e.g. is it real or has it arisen because the dosimeter but not the individual has been exposed?), access to a permanent visual record can be useful. This is readily available with photographic film but with TLDs all the raw data are essentially lost at read out.

14.7.1.12 Indication of Type of Radiation

Similarly, the pattern of film blackening can give useful information on the type of radiation as discussed above. Unless a number of TLD dosimeters are used under several different filters, each one read individually, this is not possible for a TLD dosimeter. Such a dosimeter would be very expensive.

14.7.1.13 Indication of Pattern of Radiation

Again, this information is occasionally useful and can be obtained from a film but not from TLD. For example, if a filter casts a sharp shadow, this suggests a single exposure from one direction but a diffuse shadow suggests several smaller exposures from different directions. The presence of small, intense black spots on the film suggests contamination with unsealed radioactive material.

All these factors are summarised in Table 14.2. There is no 'best buy'—each method is well suited to certain applications and both are used widely in the UK at the present time.

14.7.2 Optical Luminescence and Electronic Dosimeters

14.7.2.1 Optically Stimulated Luminescence

The principle of operation of these devices is similar to thermoluminescence (see Section 5.3) in that an ionising event causes a change in the material and on reading out the dosimeter a light signal is produced which is proportional to the dose received. These devices do, however, have several other useful properties. The most commonly used optical luminescence device uses a cold luminescent technique and is based on aluminium oxide. When read out by using selected frequencies of laser light all the information on the dosimeter is not removed as in a TLD. This allows several read out operations to be carried out on the same dosimeter, all with the same degree of accuracy. The dosimeter also does not require annealing before use. It is used with a variety of filters (open window, copper and plastic) to allow for the measurement of beta and photon doses. It has a very high dose range (from 10 μGy to 100 Gy). The pattern of the filters allows the exposure conditions of the dosimeter to be assessed under some circumstances. The tolerance of these dosimeters to harsh environmental conditions is also very high.

14.7.2.2 Electronic Personal Dosimeters (EPDs)

For some time there have been relatively small, direct reading EPDs available based on small Geiger-Müller tubes which can be clipped to a coat or carried in a pocket. Although reasonably accurate at high photon energies, because of the energy response of the Geiger-Müller tube, they very much underestimate the dose at low photon energies (below approximately 50 keV). They cannot therefore be used to show compliance with regulations.

An EPD using simple photodiodes and microelectronics, which has an energy response as good as that of TLD and film, has, however, been developed. The photodetectors produce pulses of charge when exposed to ionising radiation. These are integrated and the dose calculated by the device using an algorithm in the processor contained in each device. Each dosimeter has to be individually calibrated and adjusted. They can also record dose rate and activate in-built alarms if required to do so. Once recorded by the detector, no fading of the record takes place so the dosimeters can be worn for an indefinite period (subject only to battery life) and still give the accurate accumulated dose for the period. Although they can give no information as to how the dose was received, they are capable of measuring doses in the microsievert range which is lower than either film badges or many TLD dosimeters can record. A special read out instrument communicates with the detector using an infra-red beam. It is possible, by using linked computers through a modem, to read out the dosimeter remotely thus removing the necessity to return the dosimeter to a central dosimetry service for read out other than the periodic dose data check, battery change and function check.

Currently EPDs are expensive to use in areas where there are no problems associated with operating a system using film badges or TLD dosimeters. However, they may find application in the clinical environment where there is a need for immediate knowledge of a received dose or a very short-term measurement such as a parent holding a child during an X-ray examination.

14.7.3 Staff Doses

In many countries the number of classified workers exceeding 20 mSv/y is very small (less than 10 in total for all occupations in the UK in the latest records available). However, some hospitals will have a few persons who could potentially exceed 6 mSv/y and need to be classified (although in practice this does not actually happen, see Table 14.3). Likely candidates would be workers in PET facilities, interventional radiologists, orthopaedic surgeons and possibly cardiologists.

Table 14.3 shows the distribution of doses for different staff groups, collated from a large number of the major diagnostic radiology departments in the United Kingdom. Very few radiographers exceed 1 mSv/y, just less than 1% of the total reported. For general radiologists

TABLE 14.3

Occupational Exposure in Several Diagnostic Radiology Departments in the UK during 2001

	Number of Workers in Dose Range (mSv)					
Occupational group	0–1	1–5	5–10	10–15	15–20	>20
Radiographers	4581	30	1	0	0	0
Radiologists (General)	456	11	0	0	0	0
Radiologists (Interventional)	63	4	0	0	0	0
Cardiologists	544	29	0	0	0	0
Other clinicians	1178	19	0	0	0	0
Department nurses	2120	21	0	0	0	0
Science/technical staff	590	2	0	0	0	0
Others	804	5	0	0	0	0

Source: Watson S J, Jones A L, Oatway W B and Hughes J S. *Ionising Radiation Exposure of the UK Population: 2005 Review.* HPA-RPD-001 Health Protection Agency Didcot, 2005.

the number exceeding 1 mSv/y is about 2.5% but none exceed 5 mSv/y. Interventional radiologists are a little higher with just over 6% exceeding 1 mSv but again none exceeding 5 mSv. Compared to previous years the doses to cardiologists have now fallen and only 5% exceed 1 mSv/y and none exceed 5 mSv/y. Nurses now have the highest percentage exceeding 1 mSv/y at 10%. This may reflect the difficulty in reducing their doses because of the role they carry out whereas the doses to the other staff involved have been reduced over recent years.

As can be seen from Table 14.3, it should be relatively easy for radiographers who follow good working practices to continue working when pregnant and to stay well within a dose of 2 mSv to the abdomen once the pregnancy has been declared. Radiologists and nurses may have to be a little more careful and avoid interventional investigations especially in a screening room which uses an overcouch X-ray tube because of the risk of being exposed to scattered radiation (see Figure 3.7). For all females working in radiology departments it is prudent to tell the RPS or Superintendent as soon as a pregnancy is confirmed so that any changes to work activity can be made as soon as possible. However, recall from Section 13.9 that the risk to the foetus of a dose of 1 mSv is extremely small compared to the natural risk of a physical malformation or mental defect occurring.

Appendix

A Perspective on Current Regulatory Arrangements Concerning Medical Uses of Ionising Radiation in the United States[*]

Philosophically, radiation practices in medicine use a 'radiation management' approach rather than the 'radiation safety' approach required for US industry; it is understood that

---

[*] Kindly contributed by Ian S Hamilton PhD and Douglas A Johnson MS, Foxfire Scientific Inc, www.foxfirescientific.com

dose optimisation rather than minimisation is necessary for proper clinical efficacy. This attitude towards dealing with the unique health and safety issues concerning ionising radiation has long been dominant for therapeutic applications. As implementation and use of more radiation-based diagnostic modalities has flourished, radiation management has become a more important controller for this aspect of the use of radiation in medicine. Much of the current US regulatory framework for medical radiation reflects this approach.

The practice of medicine, including areas that utilise radiation, is fairly homogenous across the United States. This consistency is primarily due to

- An umbrella of federal rules and guidelines to which the states adhere
- A commonality of organisations and associations to which individual physicians and radiation professionals belong
- A commonality of requirements for hospital accreditation

Concerning the first two items above, written regulations tend to lag behind professional best practices and voluntary controls, such as those implemented by the larger community of medical radiation users. Thus, improved management of radiation practice is often first achieved through dissemination by professional organisations and national and international guidance bodies, which constantly strive to add relevant recommendations for the use of ever-newer radiation devices and procedures.

The American constitution recognises the 50 states as having final regulatory authority over all but a few activities. As such, individual state governments improve, update and revise rules and guidelines based largely on the experience and value demonstrated by the aforementioned professional organisations. States also share membership with one another in organisations to form working groups that generate 'state draft rules', that is, templates used for individual legislatures to adopt and impose.

One area where requirements are commonly specified in detail is new quality assurance standards for devices. Another is increased training for radiation users. Increasingly, states are requiring additional certifications, particularly for non-physician medical radiation practitioners. Once certification is in place, regulatory requirements are indirectly increased each time certifying organisations increase the rigor and requirements to earn the certification.

Along side the evolution of regulations derived from voluntary professional standards, regulations continually arise from more politically pure sources. These are primarily from legislative actions that result when members of the public or public interest groups or professional lobbyists act through one or a group of politicians to enact new laws. They differ from the evolved professional good practice regulations in that they occur relatively more quickly, and are more likely to have "unintended consequences" than the former. The genesis of a community-generated regulation is typically that one, or a few, high profile incidents occur that have some mix of public outcry, attention from the media and politicians willing to get involved. The science behind this type of regulation may not stand the test of time. Conversely, regulations that occur as a result of professional lobbying tend to protect or advance an industry or organisation but are also more likely to be better crafted, and have greater longevity and applicability.

No matter the source of new rules, adoption remains slow because of the complex and bureaucratic nature of the regulatory process. Hearings, draft regulations and public comment periods all dampen the system and effectively prevent many, if not most, of the worst rules from attaining ultimate passage. This same inertia also slows the removal of unwanted or obsolete rules from regulatory codes. Frequently, very old rules or allowances remain valid for operators who may be in violation of current rules, but because the licensee or registrant was/is compliant with a previous code they are conditionally absolved from compliance. This process is known as 'grandfathering'.

Constant advances in radiation-related technologies pose special challenges. As new modalities become available, existing regulations may be inadequate to address new concerns or may impose restrictions that are no longer relevant. This is a difficult problem because some improvements are made that simplify and automate functions on equipment, while regulations may still require actions based on operations or technologies that no longer exist. In general, operators and regulators in the US adapt to these changes; however, until new regulations are in place, inspectors may be placed in a compromising position wherein they will have to "pass" a facility (or device) that does not conform to current (non-updated) rules simply because the regulations do not apply to the newer type of equipment and no regulatory exception exists.

One solution to broad and inclusive regulations is to create separate regulations based on the specific use of the medical radiation source. Independent sections of licenses apply only to the use of certain devices like bone densitometers, nuclear medicine imaging devices, computed tomography (CT) and mammography.

Mammography is an interesting exception to the way most regulations evolve in the US, in that the federal government legislated new rules for states to adhere to several years ago and these have had significant impact across the country. New regulations demanding much higher standards were added faster than normal and have resulted in

> States creating separate mammography programmes that comply with federal requirements
> Increasing the training required by inspectors and often creating a separate inspector category for mammography
> Denying 'grandfathering' and increasing license cost, which rapidly eliminated older devices in medical practice and, at least initially, significantly reduced the number of facilities nationwide that perform mammography
> Creating incentives for manufacturers to push their technology to increase image quality while reducing patient dose

The final unifier noted in the second paragraph, hospital accreditation, is driven by government requirements that certain inspections are a prerequisite to be eligible for reimbursement with public funds. Similarly, insurance companies may also require compliance with 'volunteer' accreditation programmes as a *de facto* standard of quality before permitting third party reimbursements.

In general, the pace of regulatory change in the US appears to be accelerating. While this has largely mirrored technological changes, any future socio-political climate will have an impact in ways that are yet to be known. What can be predicted is that, as in Europe, US regulations concerned with the safe use of ionising radiation in medical diagnosis and therapy will continue to be guided by recommendations from organisations such as the International Commission on Radiological Protection (ICRP), while focusing on patient dose-justification and optimisation.

References

Administration of Radioactive Substances Advisory Committee (2006) *Notes for Guidance on Radioactive Substances to Persons for Purposes of Diagnosis, Treatment or Research* (Department of Health UK).

Carriage of dangerous Goods and Use of Transportable Pressure Equipment Regulations (2009) *Statutory Instrument 2009 No. 1348.* HMSO, London.

Directive 90/641 Euratom (1990) *Outside Workers.* Official Journal of the European Communities No. L349.

Directive 96/29 Euratom (1996) *Basic Safety Standards Directive.* Official Journal of the European Communities No. L159.

Directive 97/43 Euratom (1997) *Medical Exposures Directive* Official Journal of the European Communities No. L180/22.

Guidance Note PM77 (2006) *Fitness of Equipment Used for Medical Exposure to Ionising Radiation,* 3rd edition. HSE, London.

Heaton B, McCallum S and Michael W (2011) Cyclotron and PET Facilities. *IPEM Report 63 Ch 11* Institute of Physics and Engineering in Medicine York.

IAEA Safety Series No. TS-R-1 - 1996 edition (revised) *Regulations for the Safe Transport of Radioactive Materials.* IAEA, Vienna.

International Commission on Radiological Protection (1991) *Recommendations of the International Committee on Radiological Protection Publication ICRP 60.* Annals of the ICRP **21** (2/3) Pergamon Press, Oxford.

International Commission on Radiological Protection (1992) *Radiological Protection in Biomedical Research ICRP 62.* Annals of the ICRP **23** (2) Pergamon Press, Oxford.

International Commission on Radiological Protection (1994) *Dose Coefficients for Intakes of Radionuclides by Workers.* Annals of the ICRP **24** (4) Pergamon Press, Oxford.

International Commission on Radiological Protection (1996) *Age-dependent Doses to Members of the Public from Intake of Radionuclides: Part 5 Compilation of Ingestion and Inhalation Dose Coefficients.* Annals of the ICRP **26** (1) Pergamon Press, Oxford.

International Commission on Radiological Protection (2007) *The 2007 Recommendations of the International Commission on Radiological Protection Publication ICRP 103.* Annals of the ICRP **37** (2–4) Elsevier, Amsterdam.

Ionising Radiations Regulations (1999) *Statutory Instrument 1999 No. 3232.* HMSO, London.

IR(ME)R Ionising Radiation (Medical Exposure) Regulations (2000) *Statutory Instrument 2000 No. 1059.* HMSO, London.

Medical and Dental Guidance Notes (2002) IPEM, York.

Radioactive Substances Act (1993) HMSO, London.

RCR (2007) *Making the Best Use of Clinical Radiology Services,* 6th edition. The Royal College of Radiologists, London.

The European Agreement concerning the International Carriage of Dangerous Goods by Road (ADR) 2007 United Nations, Economic Commission for Europe.

The Medicines (Administration of Radioactive Substances) Regulations (1978) *Statutory Instrument No. 1006.* HMSO, London.

The Radioactive Substances (Testing Instruments) Exemption Order (1985) *Statutory Instruments No. 1049.* HMSO, London.

The Radioactive Substances (Hospitals) Exemption Order (1990) *Statutory Instruments No. 2512.* HMSO, London.

Watson S J, Jones A L, Oatway W B and Hughes J S (2005) *Ionising Radiation Exposure of the UK Population: 2005 Review.* HPA-RPD-001 Health Protection Agency Didcot.

Wrixon A D (2008) New ICRP recommendations. *Journal of Radiological Protection* **28**, 161–168.

Further Reading

Farr R F and Allisy-Roberts P J (1997) *Physics for Medical Imaging.* W B Saunders & Co.
Institute of Physics and Engineering in Medicine (1997) *Recommended Standards for the Routine Performance Testing of Diagnostic X-ray Imaging Systems. IPEM Report No. 77.* Institute of Physics and Engineering in Medicine York.
Sutton D G and Williams J R (2000) *Radiation Shielding for Diagnostic X-rays.* British Institute of Radiology, UK, The Charlesworth Group.

Exercises

1. What do you understand by the ALARA principle? Give examples of its application in an X-ray department.

2. What is the purpose of local rules? What would you expect to find in the local rules for an X-ray department?

3. Discuss the responsibilities placed on the employee by the Ionising Radiations Regulations 1999.

4. Summarize the radiological techniques that can reduce the dose to the patient.

5. Comment on the validity of the statement 'Any dose can be justified in diagnostic radiology'.

6. What precautions should be taken regarding radiation protection in a paediatric X-ray clinic?

7. As the consultant in charge of a department planning a new suite of rooms for neuroradiological investigations, what considerations would you have with regard to radiation protection when deciding on the structure and lay out, and on the equipment installed?

8. List the precautions that must be taken when using a mobile X-ray image intensifier in an orthopaedic theatre.

9. What are the requirements of an ideal personal dosimeter?

10. What are the advantages and disadvantages of a film badge personal dosimetry system when compared with a thermoluminescent system?

15

Diagnostic Ultrasound

A C Fairhead and T A Whittingham

SUMMARY

- The physics of ultrasound propagation in the body is presented.
- The properties of tissues which cause image formation with ultrasound are discussed.
- The mechanisms of action of ultrasound probes are discussed.
- Technical aspects of B-mode ultrasound are presented and B-mode artefacts are explained.
- More advanced techniques—tissue harmonic imaging, compound imaging, coded imaging—are explained briefly.
- Ways in which information can be obtained from the Doppler effect are considered.
- The chapter concludes with considerations of the safety of ultrasound.

CONTENTS

15.1 Introduction

Energy can exist in many forms, and in some forms has the ability to move from one place to another. When energy moves as a wave through a medium, this is known as radiation. Since sound is a mechanical vibration within an elastic medium, and since such vibration of any part of that medium will be passed on to neighbouring parts, it is valid to refer to sound (and ultrasound) as radiation, though not of course *ionising* radiation.

Many of the differences between X-ray and ultrasound imaging stem from the nature of the energy being used, and how it propagates. Whereas X-rays can be considered either as discrete photons or as waves, ultrasound is solely a wave phenomenon. It behaves like light waves in some ways, although it carries a different form of energy. For example, it travels in straight lines unless it is caused by some material to change its direction; it undergoes partial reflection at the boundary between two media, and is subject to refraction, diffraction, scattering and absorption, all of which are relevant to diagnostic imaging. However, sound waves are fundamentally different from light, X-rays and all other electromagnetic waves, in that they cannot travel through a vacuum. They exist only as vibrations of the particles of a medium and their propagation depends on the density and stiffness of the medium.

Narrow beams of sound tend to be used in medicine, though not necessarily remaining narrow for long distances, as we imagine laser light does. The partial reflection and scattering, which are crucial to imaging, occur whenever the travelling ultrasonic wave meets a change in the density or stiffness of tissue. The result in many instances is that an *echo* is sent back to the transducer, and diagnostic ultrasound relies on these echoes to form its image. The pulsed ultrasound waves emitted from the transducer travel at a fairly constant speed in their given direction, so that the range and direction of the features that produce the echoes can be measured and plotted. The image is therefore an 'echo-map'.

The reflected ultrasound waves offer another diagnostic possibility: if scattered from moving blood cells within the body, they will undergo *Doppler shift* (a change in frequency). Measurement of the shift allows the blood flow in specific parts of the body to be examined, thus providing functional information.

One of the principal attractions of ultrasonic imaging is its freedom from the harmful effects associated with ionising radiations. Its growth to its present dominant role (in terms of number of examinations performed) is also due to it being relatively cheap and mobile, to its 'real-time' nature, and to an ever-increasing appreciation that it can provide unique diagnostic information about the structure, properties and mechanical behaviour of tissue and blood flow. The major limitations to ultrasonic investigations are the barriers presented by gas and bone, and the dependence on a relatively high degree of operator skill and experience. In addition, mapping distortions and artefacts occur, so some understanding of the basic physics of sound propagation and of the techniques used in scanning equipment is necessary if high quality scans are to be produced and their limitations appreciated. This chapter will provide an initial understanding of the principles and the technology, so that the reader is better prepared for further study of more advanced texts, for example, Hoskins et al. (2003).

15.2 The Ultrasound Wave and the Principles of Echo Mapping

Sound consists of vibrations propagating within a medium. The source of sound (the transducer—Section 15.7.1) is itself a piece of material that is deliberately made to vibrate. It must be directly connected to a medium for its sound to propagate, and that medium must be elastic (compressible). Soft tissues in the human body are clearly compressible, but even bone and metals are compressible by the tiny amounts required by a sound wave.

It is useful to consider the medium as a set of connected particles which can move slightly relative to each other, and because of their elasticity can also be squeezed and stretched. The momentary forwards push of the transducer—a tiny distance is adequate for diagnostic ultrasound—increases the pressure in the medium at the point of contact. As a result the particle at that point is pushed in the same direction, and because it is elastic, is compressed as well. Subsequently, the vibrating transducer will move backwards, causing the pressure at the point of contact to fall below normal; so the particle that had moved forward is pulled back and stretched. The resulting vibration of that particle affects its neighbour in the same way, and so on with each particle in turn. Thus energy is propagated through the medium. At any instant, there will be regions (*compressions*) where the particles are compressed and therefore close together; and other regions (*rarefactions*) where they are stretched and further apart.

If viewed along a set of particles, the movement (and deformation) travels as a *longitudinal wave*, so called because the particles move forwards and backwards, and the energy propagates forwards, along the same line (Figure 15.1). Note that although the wave travels, no part of the medium moves permanently from its rest position. The wave is simply travelling energy, consisting of both kinetic energy (manifested in local

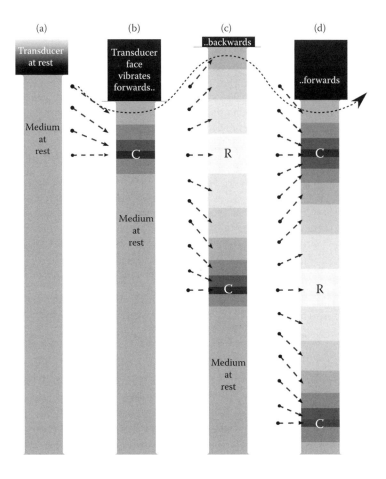

FIGURE 15.1

Schematic representation of a propagating longitudinal sound wave produced by a vibrating transducer. (a) Before the vibration begins, the medium is undisturbed. (b) The transducer face has moved forwards, pushing forwards 'particles' of the medium and compressing them. (c) The face has moved backwards, allowing the particles to relax backwards and stretch. Meanwhile, the compression C has continued to propagate. (d) A further compression is produced by the next forward push of the transducer face, and so on until the transducer stops vibrating. The movement, compressions C, rarefaction(s) R, and therefore energy, are passed from one layer of particles to the next at the speed of sound c. Note that the layer next to the transducer (and each subsequent layer) is displaced sinusoidally with time about its rest position (dotted curve). (The movement and compression/stretching of particles is exaggerated.)

movement of the tissue) and potential energy (in the local compression or expansion of the tissue).

As noted above, ultrasound imaging depends on the emission of a burst, or 'pulse', of sound into the region of interest, and the return of echoes of that sound from tissue boundaries and inhomogeneities. It is a *pulse-echo* technique. Because the speed of sound in the medium is known (or assumed) to be a particular value (1540 m s^{-1}—see Insight), the distance (depth) of an echoing feature can be calculated from the time that the echo takes to return. This time will be 1.3 μs for each mm of distance between the feature and the transducer. Such times are easily measureable with electronic circuitry.

Insight

The Speed of Sound, and What Determines It

In soft tissue, the speed of sound is approximately 1540 m s^{-1}. In any medium, it is determined primarily by two properties of the medium—the density and compressibility; and to a lesser degree by the conditions of the medium such as ambient temperature and pressure. Density ρ describes the mass of the medium per unit volume, and stiffness K (also known as the bulk modulus) is the pressure required to produce a given fractional change in volume. K is high for relatively incompressible media such as solids, but low for compressible media such as gases. The speed of sound c may be expressed in terms of K and ρ thus $c = \sqrt{K/\rho}$.

Note that the speed should be slower in a dense medium but we often observe sound travelling faster in dense tissues. This is because they also tend to be stiff. In fact, tissues differ more in stiffness than in density, so although bone, for example, is denser than muscle, it has a higher speed of sound because it is much stiffer than muscle. The speeds in individual tissues (Table 15.1) vary from about 5% above the mean value of 1540 m s^{-1} (for some samples of muscle) to about 10% below it (for some samples of fat). The speed of sound in water is also in this range, being 1482 m s^{-1} at 20°C, but increasing with temperature at the rate of about 3 m s^{-1} per °C.

To be able to separate the echoes from different boundaries that are close together, the echo should be a short sound, so the emission should be a short pulse of only a very few vibrations. A related benefit of the short echo is that its time of arrival can be measured more precisely.

However, an ultrasound image cannot be produced by measuring only the depth of echoing features—it has to show the direction in which each feature is located. It is therefore an equally important principle that the sound pulses must be sent along precisely known but different paths in turn, so that, after a particular emission, the only features sending back echoes will be those along a known path. The physics of diffraction tells us that only high frequency waves can be confined to narrow paths. The ultrasound used for medical imaging normally has a frequency in the range 2–15 MHz. Frequencies outside these limits might be used for special imaging purposes.

This basic ultrasound pulse-echo method is used in various ways to obtain diagnostic information. The production of a real-time, two-dimensional (2D) map of the boundaries and other features of tissue within a section of the body is known as 2D or B-mode imaging. The sections up to 15.8 describe the nature of the diagnostic ultrasound pulse,

TABLE 15.1

Ultrasound Properties (Approximate) of Some Human Tissues at 37°C, and Other Media

Medium	Speed of Sound c (m s^{-1})	Characteristic Acoustic Impedance Z (kg m^{-2} s^{-1} or Rayls)	Attenuation Coefficient D (dB cm^{-1} MHz^{-1})	Notes
Air	331	400	1.6 (see note)	At 20°C, 10% humidity D varies as MHz^{-2}
Water	1480	1.5×10^6	0.002 (see note)	At 20°C, 1 atmos D varies as MHz^{-2}
Blood (whole)	1580	1.7×10^6	0.15	
Liver	1580	1.6×10^6	0.4	
Muscle (skeletal)	1580	1.6×10^6	0.6	
Fat	1460	1.3×10^6	1.0	Values are variable
Bone	3300	6×10^6	23	Values are variable

Source: Adapted from Duck FA *Physical Properties of Tissue*, Academic Press Ltd., London, 1990.

its propagation in tissue, its reflection from boundaries and other features, and the probes that produce it and scan it to cover the region of interest. Sections 15.8 onwards describe the different diagnostic modes, such as B-mode and Doppler.

A further principle of B-mode echo mapping is that the brightness of any echo displayed on the map indicates the strength of that echo. This is how B-mode gets its name.

15.3 Quantities That Describe an Ultrasound Wave

15.3.1 Describing the Vibration of the Medium

Since ultrasound is a cyclical, longitudinal displacement of the particles of a medium which we shall assume for now propagates faithfully through a medium, an adequate description of it is a graph of the displacement of any particular particle with time (Figure 15.2a). The amplitude of the displacement is tiny—typically just a few micrometres.

In the axial direction (the direction of the wave's travel), particles further from the transducer take up the vibration described in Figure 15.2a later than those close to it. By considering a given instant when the further particles might be just beginning the vibration and the closer ones might be just finishing it, we see that the vibration history becomes spread

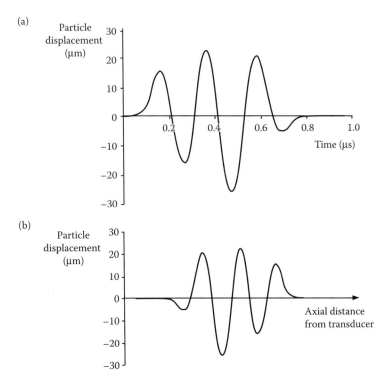

FIGURE 15.2
(a) Particle displacement versus time waveform of a typical ultrasonic pulse. (b) Showing how, due to energy propagation in the medium, adjacent particles progressively exhibit the same displacement-time waveform. At any instant, as shown here, the graph of their displacement versus axial distance is the reverse of (a).

out in space. Another way of picturing the wave is therefore to draw a graph of the different displacements of particles versus axial distance into the medium at a given instant (Figure 15.2b). This appears as the reverse of the time-graph. The pattern of displacement shown by the spatial graph moves to the right with speed *c*, representing the progress of the wave.

Alternatively, ultrasound can be described in terms of the velocity *u* with which the particles move as they vibrate. Velocity is the derivative with respect to time of displacement, so *u* also varies cyclically with time, from positive (forwards) values to negative (backwards). This alternating velocity *u* of the particles during their vibration must not be confused with the steady velocity *c* with which the vibration travels through the medium. The velocity that the particles reach (the velocity amplitude) is related to their displacement amplitude and to the frequency of vibration. It is typically a few m s^{-1}.

A graph of the velocity of a particle with time would be similar to the graph of displacement versus time (though the time of maximum velocity does not coincide with the time of maximum displacement). A plot of the velocity, at some instant, of all particles along the axial direction of the wave would appear as the reverse of the temporal velocity graph.

Acceleration of the particle, and other higher derivatives of movement, also varies cyclically with time. The variation of any one of the above quantities (along with knowledge of the propagation speed *c*) adequately describes the wave. However, the quantity most commonly used to describe the wave is its varying excess pressure.

15.3.2 Excess Pressure

When discussing sound waves, the term 'pressure' normally means the excess pressure or acoustic pressure *p* which is the departure from the ambient pressure at any point in the medium as the sound wave passes. This alternates between positive and negative values. It is the cause of the wave motion discussed above, but is also a consequence of that wave motion. This inter-relationship between local pressure in the medium and particle movement is both how and why the wave propagates.

As for other wave variables, we can plot the excess pressure as it varies either with time at a particular point in the medium (Figure 15.3a) or with distance into the medium at some instant (Figure 15.3b). Positive excess pressure signifies that the medium at that point and at that instant is compressed, that is, the particles at that point in the medium have temporarily moved towards each other. Negative pressure signifies a rarefaction, where the particles have moved further apart at that point and time. Typically, peak positive and negative pressures are up to about a megapascal (MPa)—about 10 times atmospheric pressure.

The excess pressure of an ultrasound wave is of interest because it is related to physical effects the wave might have on the medium, and it allows quantities such as the energy and intensity of the wave (see Insight) to be calculated. It is also measurable in a more straightforward way than the other wave variables, using a hydrophone.

Insight

Energy, Power and Intensity

The transducer does work, giving energy to the first layer of particles of the medium as it moves and deforms them. The energy is passed from particle to particle (though some of it is absorbed by each particle, so that the wave becomes weaker as it propagates—see Section 15.6.2). A single pulse from a diagnostic ultrasound probe might typically carry with it a few microjoules (µJ) of energy. As the pulse is repeated, the probe transmits a certain amount of energy within any specified time period. The amount of energy being transferred per second by the transducer or

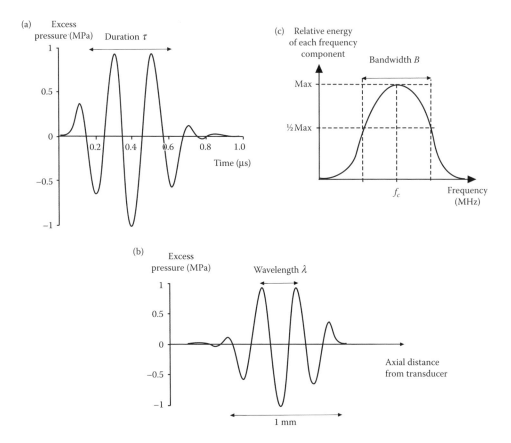

FIGURE 15.3
Graphs of excess pressure of an ultrasound pulse versus (a) time, and (b) axial distance from the transducer. Note that the peaks of pressure do not coincide in time with those of particle displacement for the same wave, shown in Figure 15.2. The wave has a nominal frequency of 5 MHz, and a wavelength of about 0.3 mm. (c) Energy spectrum of a typical pulse. The centre frequency f_c and bandwidth B are indicated.

wave is the power (P). The unit of power is the watt (W), where 1 W = 1 J s^{-1}. A medical scanner might typically transmit a few thousand pulses per second, so its output power may be up to a few hundred mW.

Because ultrasound is directional, a quantity that is often of more interest than power is its intensity, defined as for X-rays (see Section 1.10). Strictly the SI unit of intensity is the watt-per-square-metre (W m^{-2}), but ultrasound intensities are usually quoted in W cm^{-2} or mW cm^{-2}. Although defined in terms of an area, intensity describes the situation at a point. It equals the total power that would be measured within a square cm if the power at the centre of it extended throughout that area.

The intensity varies in different parts of the medium and also with time, because the emission is pulsed. It can even be regarded as varying with the cyclical wave motion, though this is not useful in practice. It is better to average out these rapid variations over a chosen timescale such as the period of the wave which still allows the variation during the pulse to be described. This 'smoothed' intensity can be shown to be related to the pressure-amplitude p_{max}:

$$I(t) = \frac{p_{max}^2}{2\rho c} = \frac{p_{max}^2}{2Z}$$

where p_{max} itself is a function of time. The product ρc is given a single symbol Z, and is known as the *characteristic acoustic impedance* of the medium (see Section 15.5.1).

From a graph or measurement of $I(t)$ the temporal-peak intensity (I_{TP}) as the disturbance passes a particular point can be identified, or with further averaging, the temporal-average intensity (I_{TA}) that occurs over a period of time at that point. I_{TA} is important for determining the temperature rise that might be produced in a sound absorbing medium. The largest value of I_{TA} to be found at any point in the scanned area is known as the spatial peak temporal-average intensity (I_{SPTA}). The value of I_{SPTA} that an ultrasound scanner might produce depends on the settings of the machine controls and the choice of scanning mode. If a machine is set such that its I_{SPTA} is maximum, this is typically 30 mW cm^{-2} for B-mode, 100 mW cm^{-2} for M-mode, and 1000 mW cm^{-2} for pulsed Doppler.

15.4 The Scale of the Diagnostic Ultrasound Pulse in Time and Space, and Why This Is Important

The varying quantities considered above testify to the disturbance that the wave causes as it propagates through the medium. It is useful to establish the scale of the disturbance in time and distance, because this has an important bearing on the resolution of the echo map that will result.

The previous section showed how the spatial distribution of vibration along the direction of propagation (the axial direction) within the medium is related to the temporal sequence of vibration. Since compressions (or any other chosen part of the wave cycle) repeat with a very short period—just $1/f$ µs, where f is the frequency of the wave in MHz—they will become spaced in the axial direction of the medium by c/f m (the distance that they move forward in one period). This distance is the wavelength λ, and is shown in Figure 15.3b. It is between about 0.1 mm (for 15 MHz waves) and 0.5 mm (for 3 MHz waves).

This explains why a short pulse of just a few ultrasonic vibrations is emitted into the body for diagnostic imaging. Its duration is just a few periods, perhaps 1 µs in total, and its axial length is just a few wavelengths, probably less than 1 mm in total. If the pulse duration is τ, its axial length is τc, and the axial resolution $d_{a\,min}$ is approximately

$$d_{a\,min} \approx \frac{\tau c}{2}$$

The axial resolution is half the pulse length because if two features are separated in the axial direction by $\tau c/2$ m, the distance travelled by sound to each of them and back in the pulse-echo process differs by τc m (see Figure 15.4a). This means that the echoes returning to the transducer from each of them are separate in time, not overlapping, and the scanner will plot them as separate features on the image. The axial resolution is thus approximately 0.75 mm for each microsecond of pulse duration. By increasing the ultrasound frequency, the pulse can be made even shorter in time and axial length, giving better axial resolution.

The short emissions necessary for diagnostic ultrasound contain a wide spectrum of frequencies (see Insight). The corollary is that a transducer cannot provide good axial resolution unless it can emit a broad band of frequencies. From the insight, axial resolution can also be expressed as

$$d_{a\,min} \approx \frac{c}{2B}$$

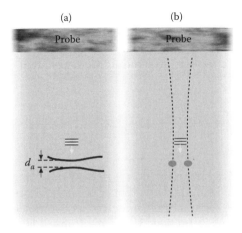

FIGURE 15.4
(a) Axial resolution depends primarily on pulse length τc (see text). In the illustration, $d_a > \tau c/2$, and so the two interfaces are resolvable. (b) Lateral resolution depends primarily on the lateral extent of the travelling pulse. In the illustration, the two features both send back echoes at the same time, and are not resolvable.

where B is the pulse bandwidth. This is a more fundamental relationship than the one given above (see Section 15.13).

Insight

Pulse Waves, Energy Spectra and Bandwidth

Section 15.4 explains why the pulsed waves used in medical ultrasound generally have a length of only about two cycles. This pulse must contain a spread of frequencies, because it is a direct consequence of Fourier analysis (see Section 6.11) that only a continuous wave (CW) has a unique frequency. Figure 15.3c shows the energy spectrum of a pulse. Two useful characteristics of it are the centre frequency f_c, at which the spectrum has its maximum height, and which is the frequency we readily observe from the waveform; and the bandwidth which is defined as the width of the energy spectrum at half its maximum height. An important rule is that, as the pulse gets shorter in time, its bandwidth increases. In fact

$$\text{Pulse bandwidth} \approx \frac{1}{\text{Pulse duration}}$$

For a typical two cycle imaging pulse, the bandwidth is about 50% of f_c. Thus, a '5 MHz' imaging pulse really means a pulse with a centre frequency of 5 MHz, but containing substantial energy at frequencies between about 3.8 MHz and 6.2 MHz.

However, good axial resolution on its own cannot ensure a detailed B-mode image: good resolution in the lateral direction is also required. If the emitted disturbance intercepts two features that lie side-by-side in the lateral direction (Figure 15.4b), they will both send back echoes to the transducer that will arrive simultaneously, and will not be resolved. Therefore, *the lateral extent of the disturbance in the medium must be narrower than the distance between the features we want to resolve.* Section 15.7.2 will describe how the ultrasound probe is designed to propagate the sound along a narrow path or 'beam'. The lateral beamwidth is related to, among other things, the wavelength of the ultrasound, suggesting that to improve lateral resolution, a higher ultrasound frequency should be used.

15.5 Production of Echoes

The disturbance generated by the transducer is narrowly confined in both axial and lateral directions so that features within the body that send back echoes of the disturbance can be accurately located, and also separated from each other. No reflection occurs while the disturbance is travelling through a homogeneous medium—it is only when the wave meets a different medium that some of its energy is reflected. Specifically, a reflection occurs at the boundary between two media if they have different values of characteristic acoustic impedance.

15.5.1 Characteristic Acoustic Impedance of a Medium

The varying excess pressure p of the wave has an intimate relationship with the varying quantities that describe the motion of the particles in the medium (Section 15.3.1). Its relationship with particle velocity u is simple and important. The two variables are in phase with each other so that, at any instant, the particle velocity is directly proportional to the excess pressure, that is,

$$\frac{p}{u} = \text{constant } Z \text{ at every instant}$$

The constant of proportionality Z depends on the particular tissue, and is known as its characteristic acoustic impedance.

Z depends on two properties of the medium: density ρ and stiffness K, that is,

$$Z = \sqrt{\rho K}$$

Z can also be shown to be equal to density multiplied by the speed of sound in the medium concerned, that is,

$$Z = \rho c$$

It is independent of any other property of the wave, including frequency.

Despite the name impedance, Z must not be interpreted as having anything to do with attenuation of the wave's energy as it travels through the medium. It simply allows particle velocity to be calculated from excess pressure and vice versa using the above relationship, and also the intensity of a wave to be calculated from measurements of its pressure amplitude (see Insight in Section 15.3.2). Table 15.1 shows the values of Z for various materials, where it can be seen that soft tissues and water have similar Z, because of their similar stiffness, density, and speed of sound. (The rather unwieldy formal SI unit for Z, that is, $kg\ m^{-2}\ s^{-1}$, is often referred to as a rayl, in honour of Lord Rayleigh who did much early work on the theory of sound.)

However, the absolute value of Z is less important than the fact that the strength of the echo produced at a reflecting interface is determined by the *difference* in the values of Z in the two media on each side of the interface (see next section).

15.5.2 Reflection

An interface between two tissues in the body which is smooth (i.e. has flat areas considerably larger than the wavelength of the sound involved), is called a *specular reflector*, because it reflects sound waves in the same way that a mirror (or partial mirror) reflects light waves.

The reflected wave has a certain fraction of the intensity of the incident wave. If I_r and I_i are the intensities of the reflected and incident waves, an *intensity reflection coefficient R* can be defined as the ratio I_r/I_i. If the incident wave is perpendicular to the specular reflector (Figure 15.5a), then

$$R = \frac{I_r}{I_i} = \left[\frac{Z_2 - Z_1}{Z_2 + Z_1} \right]^2$$

The numerator shows that the reflection coefficient depends on the difference between Z_1 and Z_2, the characteristic acoustic impedances of the media on either side of the interface. Because the numerator is squared, it doesn't matter whether Z_2 or Z_1 is the larger. (The sign of the difference does affect the phase of the reflected pressure wave, however, which undergoes a 180° shift if $Z_2 < Z_1$, but not if $Z_2 > Z_1$.)

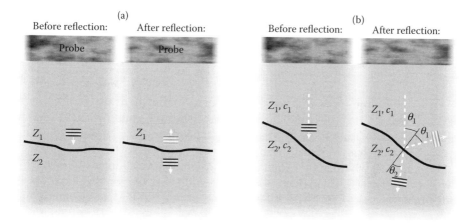

FIGURE 15.5
Partial reflection occurs when an ultrasound beam meets the boundary between two media of different characteristic impedances. (a) Perpendicular incidence is assumed in the definition of reflection coefficient. (b) With non-perpendicular incidence, the echo will not return to the receiving transducer. If the speed of sound is different in the two media, as assumed here, the transmitted beam is refracted ($\theta_2 \neq \theta_1$—see Section 15.6.1).

Since the energy that is not reflected must be transmitted through the interface, the transmitted intensity I_t is equal to $I_i - I_r$. An *intensity transmission coefficient* (T) can therefore also be defined:

$$T = \frac{I_t}{I_i} = 1 - R$$

Clearly, big differences in impedance must be avoided to transmit appreciable intensities of ultrasound across an interface. For example, the very high reflection coefficient at an interface between gas and tissue (0.999 or greater) is the reason why ultrasound will not penetrate beyond the first gas-filled bowel or air-filled lung surface encountered, and why even the thinnest layer of air between the probe and the patient's skin must be excluded by use of a coupling gel.

On the other hand, since the acoustic impedances of all soft tissues are rather similar, only a small fraction of the energy is reflected at a boundary between soft tissues. Even if two rather different soft tissues are considered, such as fat and muscle (Table 15.1), a boundary between them would have a reflection coefficient of around only 0.01. In other words the reflected wave would be only 1% as strong as the incident wave. Some soft tissue pairings, such as blood and blood clot, can produce reflection coefficients as low as 1×10^{-6}. Given the problems that strong reflecting boundaries cause, a low reflection coefficient can be regarded as an advantage in medical imaging—as long as the transducer is sensitive enough to pick up the weak echo—because it means that most of the wave's power travels on towards further boundaries.

If the sound hits a specular reflector at other than normal incidence, the expression for the reflection coefficient is more complicated than that given above, but in fact the echo would not return to the transducer because, as at a mirror, the angle of reflection equals the angle of incidence (Figure 15.5b). A practical consequence is that a curved tissue boundary may be incompletely visualised (Section 15.10.6).

15.5.3 Scattering

For many tissue boundaries, incident ultrasound is scattered over a wide range of angles by small surface irregularities. Such boundaries are described as *diffuse reflectors* by analogy with the way that a matt surface or ground glass plate produces diffuse reflection of light. Fortunately for pulse-echo imaging, some of the scattering will return to the transducer, even if the boundary is not perpendicular to the incident wave direction. These 'backscattered' echoes are weaker than those from a specular reflector, but allow the scattering boundaries to be imaged even when non-perpendicular to the beam (Figure 15.6a).

Ultrasound is also scattered by the innumerable tiny microstructures within tissue (Figure 15.6b), that is, wherever a wave meets a small feature, about a wavelength or less in size, with a different Z. An analogy is the scattering of ripples on the surface of a pond by a reed, where the scattered waves take the form of circular ripples spreading out from the reed.

Early scanners were not sensitive enough to detect most backscattered echoes—they tended to image only organ boundaries. B-mode scanners are now sensitive to a much wider dynamic range of echoes (see Sections 5.5.4 and 15.9.1), and translate that range to a 'grey-scale image', giving the vital possibility of differentiating between different tissue types and pathologies.

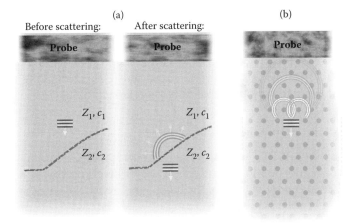

FIGURE 15.6
(a) A rough boundary scatters sound over a range of angles, so that backscattered echoes may return to the transducer even if the sound is not incident perpendicularly to the boundary. (b) Scattering is also produced by small-scale inhomogeneities with the tissue parenchyma. Backscattered echoes are received that depend on the structure of the tissue.

When the scattering structure is much smaller than a wavelength, the process is called *Rayleigh scattering* in which, in theory, the power of the scattered wave is proportional to f^4 and to a^6, where a is the scatterer size. However, this relationship may not be observed in tissue which consists of many different sizes of scatterer. Nevertheless, different tissues and pathologies do lead to different strengths and textures in the complex speckle patterns produced by scattering from within them, and this offers a way of differentiating between them. An unwelcome aspect of scattering is that these speckle patterns are largely artefactual (Section 15.10.1), and do not give a true image of tissue structure, so quantitative tissue characterisation using ultrasound is difficult to achieve. (Note that although the terms 'specular' and 'speckle' sound similar, they have nothing in common, and must not be confused.)

15.6 Other Aspects of Propagation

15.6.1 Refraction at a Boundary

Ultrasound can be refracted if it meets, at an angle other than 90°, a boundary between materials where its speed of propagation changes. As in optics, Snell's law applies:

$$\frac{\text{The sine of the angle of incidence}}{\text{The sine of the angle of transmission}} = \frac{c_1}{c_2}$$

where c_1 and c_2 are the speeds of sound in the first and second media respectively, and angles are measured from a line perpendicular to the boundary. The law shows that the transmitted beam is deflected further away from the normal when $c_2 > c_1$, or towards the normal (see Figure 15.5b) when $c_2 < c_1$. If $c_2 = c_1$, or if the beam strikes the boundary at right angles (regardless of the values of c_2 and c_1), then no refraction takes place.

Refraction produces effects in ultrasound imaging that are not necessarily obvious to the user. There might be an acoustic lens within the probe, playing an essential part in the production of a narrow beam for the sound pulse to travel along (see Section 15.7.2). Refraction can also cause various artefacts (see Section 15.10.10).

15.6.2 Attenuation of Ultrasound

The reduction of intensity of a wave as it travels away from its source has huge implications for all modes of diagnostic ultrasound. For a wave travelling through the body, attenuation might be caused by divergence of the beam, and by partial reflection and refraction at tissue interfaces. However, these effects are not easy to quantify. Two causes that may be quantified are absorption and scattering. As for X-rays (see Section 3.3.2) both depend on the particular tissue the wave is travelling through.

Scattering in tissue (Section 15.5.3) is a cause of attenuation because it diverts energy away from the original wave. Since scattering increases with frequency, so does the attenuation due to scattering. However, it is usually less significant than absorption as a cause of attenuation.

Absorption does not cause ionisation (cf. X-rays) but converts the wave energy to heat in a tissue. It depends on the nature of the tissue. If a medium has a relatively simple molecular structure, like water, it reacts to the pressure changes of a sound wave in an elastic fashion, passing on the pressure stimuli without loss. However, the large and complex molecules in tissue respond to changes in pressure in more complex ways, for example, by changing their structure or vibrational behaviour. Such changes are reversible and would be of no consequence to the wave if they took negligible time to occur. However, the high frequency of the ultrasound vibration means that the time occupied by each cycle is similar to that needed for these changes to occur, so the wave energy 'borrowed' to implement them during the time of increasing pressure might be given back as an increase in pressure just as the pressure due to the original wave is falling again. This partially cancels out the amplitude of the ordered pressure swings, resulting in attenuation of the wave, and an increase in random motion, that is, heat energy.

In many tissues, absorption increases in proportion to frequency. Furthermore, the amount of energy that the wave loses through absorption at any point depends on the amount of energy it has at that point.

15.6.3 Calculating the Effect of Attenuation

Calculations of the decrease in intensity of a wave as it travels through tissue usually take account of absorption and scattering within the tissue, but not the other three possible causes of energy loss, for the reasons given in the previous section.

For absorption and scatter, it is a matter of experimental observation that when ultrasound with intensity I passes through a small thickness Δl of tissue, the loss of intensity ΔI is proportional to $I \, \Delta l$. When Δl is small, this may be written $dI = -\mu_s I \, dl$, where μ_s is a constant for the particular tissue. This is the differential form of the exponential equation

$$I = I_0 \exp(-\mu_s l) \tag{15.1}$$

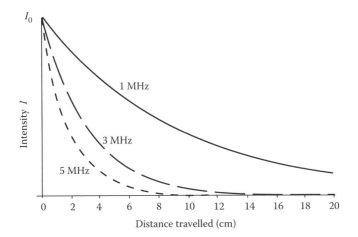

FIGURE 15.7
Attenuation of an ultrasound wave in tissue causes intensity to decay exponentially with the distance travelled. The solid curve illustrates the situation for a 1 MHz wave propagating through tissue that attenuates 10% of its intensity per cm. Higher frequency waves are attenuated more.

that has already been discussed in Section 1.5 (radioactive decay) and Section 3.2 (attenuation of X-rays). All the mathematics presented in Section 1.5 can be applied equally well to the intensity of ultrasound, including the simple graphical method for solving problems involving exponential changes. For ultrasound, the y-axis will be the ultrasound intensity, and the x-axis is axial distance into the medium (see Figure 15.7). Table 3.1 could be extended to attenuation of ultrasound using the symbols μ_S and $H_{S\frac{1}{2}}$ to represent the linear attenuation coefficient and half-value thickness, respectively. However, half-value thickness is not generally used in ultrasound—attenuation in a particular tissue is usually described as the proportional loss of intensity per cm of that tissue.

As explained in the previous section, for soft tissues attenuation is generally proportional to frequency and Equation 15.1 can then be written:

$$I = I_0 e^{-afl} \tag{15.2}$$

where a is independent of frequency. a is scaled so that values of f in this equation are conveniently in MHz and values of l in cm. Thus a is the attenuation coefficient at 1 MHz. (Unfortunately the notation is not standardised—μ_S in Equation 15.1, the attenuation coefficient at f MHz, is sometimes given the symbol α.)

To avoid continual use of the exponential in attenuation calculations it is routine to use the special unit *decibel* (see Insight). Typical decibel values of the attenuation coefficient D are shown in Table 15.1. At ultrasonic frequencies attenuation is a considerable problem for the pulse-echo technique.

Insight

Decibels

The decibel (more useful in practice than the *bel*, from which it is derived) allows the calculation and expression of the result of a series of proportional processes more simply than the serial calculation or expression of each individual process. In Section 15.6.3, the attenuation of intensity of an f-MHz wave travelling through l cm of tissue is shown to take the form

$$\frac{I}{I_0} = e^{-afl}$$

where I_0 is the initial intensity, I is the resulting intensity, and a is the attenuation coefficient of the tissue at 1 MHz.

Converting the proportion I/I_0 to decibels involves taking the value of $\log_{10}(I/I_0)$ and multiplying by 10, thus:

$$10 \log_{10}\left(\frac{I}{I_0}\right) = 10 \log_{10}(e^{-a}) \times f \times l.$$

Since e^{-a} is by definition the ratio I/I_0 of the intensities of a 1 MHz wave after and before travelling 1 cm, we can write

$$10 \log_{10}\left(\frac{I}{I_0}\right)_{\substack{\text{after } l \text{ cm} \\ \text{at } f \text{ MHz}}} = 10 \log_{10}\left(\frac{I}{I_0}\right)_{\substack{\text{after 1 cm} \\ \text{at 1 MHz}}} \times f \times l$$

$$= -D \times f \times l,$$

where D is the decibel (dB) value of the tissue's attenuation coefficient. Its units are dB cm^{-1} MHz^{-1}. It is a positive number because it describes an attenuator (the tissue). It can be shown to be equal to $4.34 \times a$. Its effect on ultrasound intensity would of course be $-D$ dB cm^{-1} MHz^{-1}. The benefit of this expression is that, to calculate the decibel proportion of the intensity remaining after a wave of any frequency f MHz travels any distance l cm, we simply take the negative value of the three factors on the right-hand side, multiplied together.

There are two further benefits of using decibels for calculations. First, the fact that they are logarithmic means that very large ratios can be expressed compactly. In scanning situations where one echo might have an intensity that is 10^{10} times that of another, it may simply be said that its intensity is 100 dB greater (since $10 \log_{10} 10^{10} = 100$). The second benefit derives from the fact that, where numbers are multiplied or divided, their logarithms are simply added or subtracted. Thus, suppose the energy of a pulse is reduced by a factor of 1/2 due to one process (which might be travelling through the half-power thickness of a tissue) and then by a further factor of 1/10 by a second process. Overall, the energy has been reduced by a factor of $1/2 \times 1/10 = 1/20$. The alternative approach is to express the two factors in dB and simply add them together. Thus the first factor (1/2) is a reduction of –3 dB, and the second factor (1/10) is a reduction of –10 dB, giving a total reduction of –13 dB. This result could be converted back into the form of a ratio (using antilog$_{10}$ 1.3 = 20), but would often be left in dB form.

Note that, since intensity varies with the square of the amplitude, doubling the amplitude causes a 6 dB increase in intensity.

Therefore, although higher ultrasonic frequencies provide better image resolution (see Section 15.4), their higher attenuation means that the choice of ultrasound frequency in practice is always a compromise. It must be low enough for the sound to penetrate to (and return from) the deepest tissue of interest, but high enough to provide the best possible resolution.

Finally, note from Table 15.1 that not all materials exhibit an attenuation coefficient that is proportional to frequency. For bone, not only is the value of D enormous (perhaps 70 dB cm^{-1}

at 3 MHz), it increases more rapidly with frequency than for soft tissues. The attenuation coefficient of water, although very low at ultrasonic frequencies, increases according to f^2.

15.6.4 Non-Linear Propagation

It is usual, as above, to develop the concepts of ultrasound propagation assuming that the waveform is propagated faithfully by the medium. For this to be true, the relationship between acoustic pressure and the variables that describe particle motion must be linear. Although the linear model adequately describes most of the major effects relevant to diagnostic ultrasound, in reality it is only accurate for waves of small amplitude, and not for diagnostic ultrasound pulses with pressure amplitude of the order of 1 MPa. The effect of the non-linear relationships between the wave disturbance and the response of the medium is that the wave becomes distorted as it propagates.

This effect has been ingeniously exploited for diagnostic ultrasound in a process called *tissue-harmonic imaging* (Section 15.11).

15.6.5 Dispersion

Wave dispersion occurs if the wave contains different frequency components, and the frequencies travel at different speeds through the medium. An ultrasound pulse in a dispersive medium would lose its narrow confinement in space and time—the dispersion of its frequency components in time would mean that they no longer synthesized a high amplitude, short-duration pulse. Thus the ultrasound system would lose sensitivity and axial resolution. Fortunately, the media involved in medical ultrasound are non-dispersive, that is, c is independent of frequency.

15.7 Ultrasound Probes, and How They Work

A diagnostic ultrasound probe that the operator holds or inserts into the patient must fulfil the following technical requirements: emit a short pulse, confined as much as possible to a single direction; receive echoes of the pulse that arrive from the same direction; and repeat this process over a sequence of directions to cover a 2D tomographic (i.e. sectional) 'field of view' within the patient. These requirements are met through three separate aspects of probe design.

15.7.1 The Transducer Element

An ultrasound transducer is a device that converts electrical signals into ultrasound waves, and vice versa. Strictly, the word applies to a specific component of the probe, but sometimes the complete probe is referred to as the transducer.

Up to the time of writing, transducers used in diagnostic ultrasound have been based on the piezoelectric effect (although other techniques are being pursued, particularly capacitive micro-machined ultrasound transducers—CMUTs). Piezoelectric materials change in size when a voltage is applied across them and, conversely, generate a voltage in response to an applied pressure. Modern medical ultrasound probes usually contain the material lead zirconate titanate (PZT), an efficient piezoelectric ceramic that can be cast or machined into discs, rectangular slabs, bowls or any other desired shape. A single disc-shaped transducer was originally used, but now most diagnostic probes contain arrays of thin, rectangular PZT slabs, known as elements (Figure 15.8).

The front and back faces of each transducer element are coated with a conducting layer of silver to form electrodes, to which electrical leads are attached. Across these leads an alternating voltage is applied at a particular frequency. The piezoelectric slab deforms in sympathy, expanding and contracting in thickness (by only a few µm). Both faces of the slab therefore act as piston sources of ultrasound waves that have the same frequency as the applied voltage. In reception, the pressure variations produced by a returning echo cause the element to expand and contract accordingly, and thus generate proportional ('analogue') voltage variations between the two electrodes. These form the electronic echo signal that is further amplified and processed by the receiver.

A disadvantage of PZT for diagnostic transducers is its high characteristic acoustic impedance compared with the materials that might be adjacent to it, such as air or tissue. The large reflection coefficient (Section 15.5.2) this produces at its faces means that a wave generated inside the transducer is significantly internally reflected at each face. The process is known as reverberation. If the transit time of the wave across the transducer and back were equal to the period of the applied electrical excitation, then subsequent vibrations would be reinforced and the reverberation would build up to a significant resonance. This particular resonance of a thin slab is called *half-wave resonance*, because the transducer thickness must be half a wavelength. Thus a PZT slab with a thickness of 0.67 mm would exhibit half-wave resonance at 3 MHz (taking c in PZT as 4000 m s^{-1}).

In some applications of ultrasound, the transducer is designed to be resonant at the required frequency, because this results in highly efficient energy conversion. However, resonance must be deliberately suppressed in a pulse-echo transducer, because short acoustic pulses are necessary for good axial resolution (Section 15.4). Even if a short pulse of alternating voltage were applied, a resonant transducer would continue to vibrate ('ring') at its resonant frequency for some time.

The ringing time is reduced by placing an absorbing backing layer immediately next to the back face of the transducer. The backing material is chosen to have an impedance close to that of the transducer, so that the wave energy trapped in the transducer passes easily into it, and is absorbed. Thus the vibration dies away quickly, and the pulse emitted from

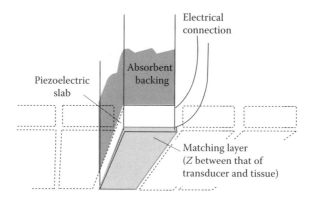

FIGURE 15.8
Construction detail of a typical ultrasound probe. Rectangular slabs of piezoelectric material are grouped together to form an emitting and receiving *aperture*. The electrical connection to the front of each element, the matching layer, and the absorbent backing would all be continuous across the elements.

the front face is short. By deliberately introducing such energy loss, the transducer's efficiency at resonance is lost and it is hardly any more responsive at its resonant frequency than it is for a wide band of frequencies on either side. However, this means that it is well suited to producing and receiving short pulses which necessarily must contain a wide band of frequencies (see Insight in Section 15.4).

If the front face of the PZT transducer were coupled directly to tissue, its large reflection coefficient would further impair the transducer's efficiency as a transmitter and receiver. The problem is reduced by having a thin impedance-matching layer (also known as an anti-reflection coating) immediately next to the front face of the PZT. The simplest 'quarter-wave matching layer' is made from a slab of low attenuation material a quarter wavelength thick. The wave reverberates within this layer, undergoing a two-way delay that amounts to half its period. At the back face, the reverberating wave comes up against the high acoustic impedance of the transducer, and is internally reflected with a phase inversion (180° phase shift—Section 15.5.2). The combined result of delay and phase inversion is that the reverberating wave reinforces waves entering the matching layer from the transducer, and so improves the transmission efficiency of the transducer. In reception, the passage of echoes back into the PZT is similarly enhanced by the matching layer.

It can be shown that if the material of the matching layer is given a characteristic impedance of $\sqrt{Z_{PZT} \times Z_{tissue}}$, that is, the geometric mean of the impedance of PZT and tissue, then all the energy entering it from the transducer is eventually transmitted into the tissue. Thus the layer entirely eliminates reflection losses between the transducer and tissue. However, this is only true at the single frequency for which it is designed. For short pulses, with their broad frequency spectrum, it is necessary to use multiple matching layers, consisting of several layers with various thicknesses and impedances.

Thus each transducer element is designed to produce short pulses of ultrasound. However, it cannot by itself direct that pulse along a defined narrow path, as required for pulse-echo methods.

15.7.2 Directing Ultrasound along a Narrow Beam

Although a diagnostic ultrasound beam is often shown diagrammatically like a beam of light, it does not exist in its entirety at any time, because the transducer is not emitting continuously. Such diagrams should be regarded as outlining the path that the emitted pulse will follow, and also the region from which the transducer, or group of transducers, will be sensitive to returning echoes. Confining the emitted pulse to a narrow path in a specific direction, and making the probe sensitive only to echoes returning along that narrow path, is known as beamforming.

15.7.2.1 Principles of Beamforming

The spatially varying distribution of wave amplitude or intensity in a medium coupled to an emitting ultrasound transducer, or group of transducers, is known as the ultrasound *field*. Although the front face of the transducer may be vibrating as a simple piston, the resulting field is shaped by diffraction. This occurs because medical ultrasound waves are *coherent* which means that the various waves contributing to the field are so closely related in frequency that the resulting field strength is significantly influenced by interference between them. At the moment, coherence distinguishes ultrasound from most other forms of medical radiation, apart from a laser beam.

Diffraction was originally explained by Huygens for coherent light. Applied to ultrasound, *Huygens' principle* is that all points on the face of the transducer (or group of transducers) should be considered to be radiating separate waves, and that the instantaneous pressure at a given point in the field is equal to the sum of the pressure contributions at that point from every such wave. The radiating face is often called the *aperture*. Although all points on the aperture may be vibrating in phase with each other (i.e. at the same part of the pressure cycle at the same instant), geometry shows that the separate waves they produce travel different distances to arrive at any particular point in the field. They therefore arrive at that point with different phases, because the aperture is generally larger than the wavelength of the ultrasound being emitted. The field at that point will be strong due to constructive interference if most of the contributing waves arrive in phase, but weak due to destructive interference if many waves arrive producing positive pressures at the same time as many producing negative pressures.

The shape of the field is therefore determined by the size and shape of the aperture, by the wavelength, and by the fact that in practice, accidentally or deliberately, the amplitude and phase of the waves across the aperture are not constant (i.e. the transducer face is not a simple piston). A pulsed field which contains waves of many frequencies is further complicated by being the sum of the fields of the individual frequencies. This summation fortunately has an averaging effect, producing a simplified field pattern which can be summarised as follows.

Consider the plane beyond the transducer containing the tissue of interest, the 'scan plane'. (The same theory applies to any plane containing the axis normal to the transducer aperture, but the scan plane is the most relevant.) Let the distance across the aperture in that plane be $2a$. If pulsed ultrasound of nominal wavelength λ is emitted from the transducer, its path in that plane will be outlined by a slightly converging beam for a distance a^2/λ (Figure 15.9a), and then by a diverging beam outlined by the angles $\pm\theta$ to the axis, where $\sin\theta = 0.6\,\lambda/a$. Unless a is significantly larger than λ, the converging path will be short, and the divergence will be severe. The aperture must be several wavelengths wide to prevent divergence until the pulse is near the back of the section being imaged, because any beam divergence will produce a loss of location accuracy in a pulse-echo system. A given size of aperture is larger in terms of wavelength if it is emitting a higher frequency

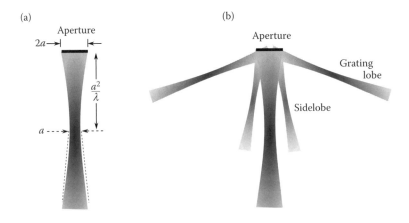

FIGURE 15.9
(a) Beam shape produced by an aperture of half width a. (b) Sidelobes are also produced, and if the aperture is formed by a group of transducers, then grating lobes as well.

(smaller λ). High frequency beams are therefore better collimated than low-frequency ones, the converging part of the beam is longer and the subsequent divergence is less severe.

The distance a^2/λ at which divergence begins is called the *Rayleigh distance*, and marks the boundary between the *near field* or *Fresnel zone* and the *far field* or *Fraunhofer zone*. Here, the width of the beam achieves its narrowest value of about a, as measured between points where the intensity is half that on the axis (the '-3 dB width'). Since the total power in any beam cross-section should be the same at all ranges (assuming no absorption or scatter), this is therefore the range at which the on-axis intensity reaches its maximum. This region is sometimes called the *last axial maximum*, although the term is more relevant in the field of a continuous wave.

As noted in Section 15.7.1, the aperture is formed by a single disc-shaped transducer, or more usually by a group of transducers. Both arrangements are described in more detail in Section 15.7.3. There may be between 20 and 128 transducers in a group, and each has its own pulse generator which delivers the oscillating voltage to excite the element and produce the ultrasonic pulse, and its own pre-amplifier for giving initial amplification to the echoes.

15.7.2.2 The Receive Beam, and the Principle of Reciprocity

The discussion so far has considered a transmission beam. Transducers acting as receivers have receive beams, defined as the region in which a point source of sound would produce a detectable signal. Fortunately, the principle of reciprocity applies—if a transducer both transmits and receives (at different times, of course), the shape and size of the receive beam will be identical to that of the transmit beam. (Remember though that a contour plot of the receive beam would show the variation of receive sensitivity to a point source at different positions, whereas a contour plot of the transmit beam shows the peak pressure or intensity that would be measured at each point.) Realise also that the principle of reciprocity described here is not the same as that relating to the latent image-fading of film (see Section 5.7). In practice, the transmit and receive beams have different requirements for sidelobe reduction and focussing (described next). Array probes (Section 15.7.3) will have different sizes of aperture for transmission and reception, with different patterns of amplitude and phase across them.

15.7.2.3 Sidelobes and Grating Lobes

The above simplified description of a beam refers only to its *main lobe*; but in practice this is flanked by weaker *sidelobes* (Figure 15.9b) which arise because it is theoretically impossible with a finite-sized aperture to confine all the energy to a single main beam. Sidelobes effectively widen the beam, possibly bringing spurious off-axis echoes back to the transducer, and thus reducing the spatial resolution and contrast of the image. Their strength can be reduced by a technique known as apodisation, in which the transducer face is designed to vibrate more strongly at its centre than at its edges. If the aperture is formed from a group of transducers, the independent electrical access to each element provides an effective way of applying apodisation. In transmission the pulses are fed to individual elements with different amplitudes, and in reception each of the individual pre-amplifiers is given a different gain (amplification factor).

A group aperture also suffers from *grating lobes*—beams which emerge at greater angles to the axis than sidelobes. Pulse-echo mapping can erroneously occur along these, adding more artefacts to the image. They are so-called because they are a consequence of the aperture being formed from separate slabs rather than a single continuous piece—it is effectively a diffraction grating. The angle of these lobes relative to the main lobe is inversely proportional to the spacing of the element centres. If the elements can be made narrow enough that their spacing can be <0.5 λ, the grating lobes are at such a wide angle that they do not intrude significantly into the field of view. They can also be suppressed by varying slightly the spacing of the elements, or restricting the elements' response at wide angles.

15.7.2.4 The Need for Focussing

The Rayleigh distance (Figure 15.9a) may be regarded as the intrinsic focal point of a plane aperture, but since the aperture size $2a$ must be several wavelengths so that the Rayleigh distance is sufficiently deep within the patient, then the minimum beamwidth a will not be narrow enough for accurate echo mapping. Extra focussing is required to produce a narrow beam, otherwise the system will have poor spatial resolution.

A single disc-shaped transducer can be focussed by making it concave, or by putting an acoustic lens in front of it. Thus the relatively weak individual pulses from all parts of it arrive at roughly the same time at a 'focal zone' (see Figure 15.10a), where they reinforce each other and produce a large resultant pulse. The focal zone can extend along the axis for a considerable distance, but at any point that is only a small distance away from the axis, the individual pulses may arrive at different times, and interfere destructively. Thus the narrower beamwidth and increased intensity in the focal zone provide better lateral resolution and sensitivity there. Beyond the focus, the beam diverges, losing its intensity more rapidly than an unfocussed beam, and becoming unsuitable for pulse-echo methods.

Note that the focal length cannot be made greater than the Rayleigh distance. If it is made less than half the Rayleigh distance, such as in Figure 15.10a, then the beam is said to

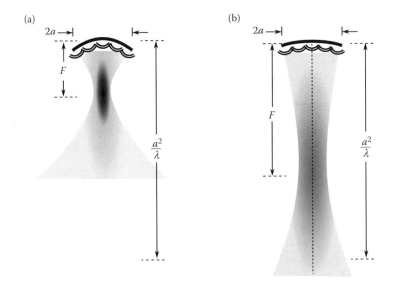

FIGURE 15.10
(a) Strong and (b) weak focussing of a single-element transducer. Pulsed waves are shown leaving each part of the transducer, as in Huygens' principle, and will reinforce each other at the focus. The focal length F is measured from the transducer to the focus, at the narrowest part of the beam.

be 'strongly focussed', with a narrow but relatively short focal zone. At the focus, the width (w_F) of a strongly focussed beam from a transducer of radius a and focal length F is

$$w_F = \frac{F\lambda}{a}$$

Thus, as for unfocussed beams, high frequencies (small λ) produce narrower beams. This equation also shows that a wider aperture would be required to achieve the same focal beam width for a deeper focus.

If the focal length is more than $0.5\,a^2/\lambda$ (Figure 15.10b), then the beam is said to be 'weakly focussed', with a wider but longer focal zone. Where the focal length cannot be varied, weak focussing is preferred to strong focussing. This is because the region of greatest clinical interest might be anywhere from just beneath the probe to near the limit of penetration, and hence a long, moderately narrow focal zone is of more value than a very narrow but short one.

An aperture formed from a group of transducers allows the focal depth to be varied. In transmission, the operator can position the focus at the same depth as any region of interest, thereby improving the lateral resolution and sensitivity there. The trigger signal undergoes different time delays on the way to the pulse generator for each element, the delays being chosen to compensate for differences in path length (and hence travel time) between that element and the selected focal point (Figure 15.11a). This is equivalent to the mechanical methods described above that equalise the path lengths from a single-element transducer to its focus, but has the advantage that it can be adjusted. The delays are pre-programmed for each possible setting of the 'transmission focus' control.

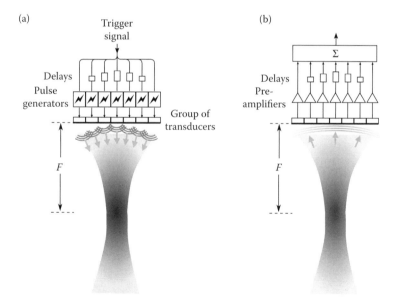

FIGURE 15.11

(a) Focussing of a group aperture in transmission. The transmission of each element is delayed appropriately so that the individual ultrasound pulses form a curved wavefront. Note how the delays form a pattern which is similar in shape to the required wavefront. (b) The reception beamformer uses the same pattern of delays to re-synchronise the echo signal which is shown arriving from distance F, reaching the transducer elements at different times.

In reception, a variable receive focus is achieved in a similar way (Figure 15.11b) by introducing time delays into the electrical signal paths of the returning echoes. These delays are pre-calculated to compensate for the small differences in travel time between the required receive focus and each element. When the delayed echo signals are summed together they produce a strong response from any reflector at or near the focus. However, in contrast to the transmission situation, the operator is not required to select a receive focus. Instead, the beamformer automatically changes the pattern of delays every few µs over the period during which echoes return, advancing the range of the receive focus at an average rate of 1 mm every 1.3 µs. This ensures (see Section 15.2) that the receive focus always coincides with the depth of origin of the echoes. This *dynamic focussing* technique produces an effective receive beam that is focussed along its length, maintaining good lateral resolution and sensitivity from near the surface to the maximum depth of penetration.

In dynamic focussing, the number of elements in a group is also increased as the focus advances in depth. Recalling the equation $w_F = F\lambda/a$, this allows the beamwidth w_F at each focal distance F to be kept constant, by increasing the aperture $(2a)$ in proportion to F. The maximum number of elements in the group is limited by the machine's number of processing channels, which is typically between 20 and 128. A large number of channels implies a complex and expensive beamformer, but produces better lateral resolution and sensitivity for deep targets.

In some modes of ultrasound it might be useful to have many focal positions in the transmit beam, effectively forming a beam which is narrow along its length. However, this cannot be achieved in the same dynamic way as it can for the receive beam. Instead, if the operator requests *multiple-zone focussing in transmission*, several transmit-receive sequences must be carried out along each scan line. In each sequence, the transmitted pulse is focussed at a different range, and only those echoes arriving at times corresponding to ranges close to that focus are used to form the image. Clearly, the more focal zones that are chosen, the greater the range of depths that will have improved lateral resolution. However, the 'real-time' aspect of the mode will suffer, because each additional transmission focus increases the time spent interrogating each beam direction.

15.7.2.5 *Reducing the Slice Width of the Beam*

So far, beamforming has only been considered in the scan plane, with focussing techniques improving the lateral resolution of the pulse-echo system. However, we must also consider the plane containing the axis of the beam, but perpendicular to the scan plane, the *elevation plane*. The width of the beam in the elevation plane should be as narrow as possible, otherwise echoes will be detected not just from targets lying in the plane of interest, but also from targets outside that plane. The beam shape in the elevation plane is again governed by diffraction. Because the aperture's size, say $2b$, in this direction must be sufficient to ensure the Rayleigh distance $(b^2/2\lambda)$ in the elevation plane is deep within the patient, the beam in the elevation plane will be too wide (Figure 15.12a) without extra focussing.

The single-element disc-shaped transducer has focussing applied to reduce the lateral beamwidth (see Figure 15.10), and this will reduce the elevation beamwidth in the same way. An aperture formed from a group of transducers has versatile electronic methods applied to focus the beam laterally (see Figure 15.11), but these do not work in the elevation plane, because the group is not orientated in that direction. It is necessary to put a weak focussing lens in front of the group (Figure 15.12b), with curvature only in the elevation direction so that it does not influence the lateral focussing. The lens, usually a rubber material, is chosen to have an acoustic impedance near to that of soft tissue, so that the matching layer behind it (Section 15.7.1) is still effective.

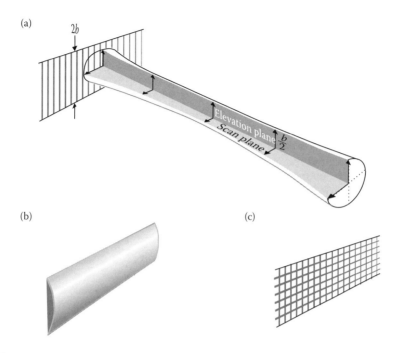

FIGURE 15.12

(a) Group aperture and its unfocussed elevation beam. Note that the beam is focussed more effectively in the lateral (scan) plane than in the elevation plane. In the latter, its total width is no narrower than b, the half width of the aperture in that plane. (b) Use of a lens for elevation beam focussing. (c) A 1½-D linear array has several rows of transducer elements, allowing dynamic beam forming in the elevation plane, and hence producing an effective slice thickness that is narrow over a large depth range.

Some modern probes provide adjustable elevation focussing and apodisation by having the aperture divided into elements in the elevation direction as well as in the lateral direction (see Figure 15.12c). However, since every sub-division of the aperture in the elevation direction represents an additional row of transducer elements with its associated connections, trigger generators, delays and pre-amplifiers, it is not technologically feasible to provide more than about 10 such rows of elements. The resulting matrix of elements, typically 10×100, is therefore called a *1½-dimensional array*.

15.7.3 Scanning Probes

Most ultrasound probes are required not just to send a pulse of sound and receive its echoes along a narrow, focussed path in one direction, but also to repeat that process in a sequence of directions.

15.7.3.1 Mechanically Scanned Probes

This earliest form of scanning-beam probe is rarely used now. A disc-shaped transducer, with a weak fixed focus set at approximately half the intended maximum penetration depth, is turned through a sector of angles (Figure 15.13) by an electric motor and suitable coupling mechanism. Hence the beam oscillates through an angle of typically 45° on each side of the probe's principal axis. Another part of the mechanism monitors the direction the transducer is pointing, so that the echo-map image can be accurately constructed. Each

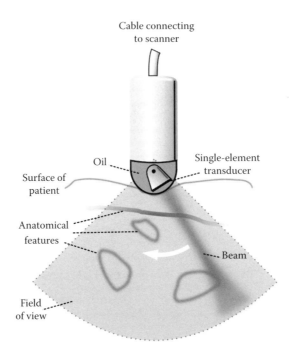

FIGURE 15.13
Simplified representation of a mechanically scanned probe. The single-element transducer has electrical connections, a matching layer, and absorbent backing just as the elements in a group aperture do (see Figure 15.8). The assembly is mounted on a pivot in an oil-filled container. The beam scans over a sector-shaped field of view.

complete sweep of the transmit-receive beam across the sector provides a new frame of the real-time image, showing all echo-producing structures in their current positions.

To couple the sound between the moving transducer and patient, the transducer must be contained in a chamber of oil or water with an acoustically transparent shell which is held against the patient. The difficulties of eliminating air bubbles from the chamber, and the gradual wear of the mechanism for moving the transducer, are the reasons that this type of probe has been largely superseded by the various array probes that scan the beam electronically with no moving parts.

15.7.3.2 Linear and Curvilinear Array Probes

A simplified diagram of a linear array probe is shown in Figure 15.14. It might have up to 200 narrow rectangular transducer elements, each less than two wavelengths wide. The transducer array is usually constructed by taking a long slab of polarised PZT with electrodes on both sides and cutting channels through it to form separate elements. There is a separate electrical lead to the rear electrode of each element, but all elements share a common front electrode and lead. An active group of say 20–128 adjacent elements (not the whole array) is electronically selected to act as the beamforming aperture (Section 15.7.2). The active group, and hence the beam, is stepped across the array by progressively switching off an element at one end of the group and bringing in the next one at the other end.

The rectangular field of view means that linear arrays are not suitable for imaging wide sections in the body, but they are popular for scanning superficial structures such as the neck, breast, scrotum and limbs. However, curvilinear (or convex) array probes which are basically

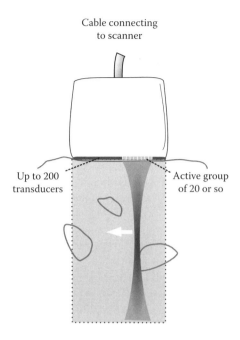

FIGURE 15.14

A linear array probe, showing an active group of elements which is about the same size as the single element in Figure 15.13. The beam, scanned along the array, covers a rectangular field of view.

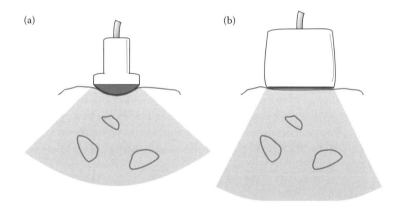

FIGURE 15.15

(a) A curvilinear array, and (b) a linear array with peripheral beam steering, offer a wide field of view superficially, becoming even wider at depth.

linear array probes constructed with some degree of curvature (Figure 15.15a), provide not only a usefully wide field of view close to the probe but an even wider one at depth. They are popular for obstetric and some abdominal applications, but they cannot be used where the need to push the convex front face into full contact with the patient would cause unacceptable distortion of superficial structures. This problem has been overcome in the trapezoidal scan format now being offered by some linear array probes (Figure 15.15b). This involves a combination of linear array beam stepping with the phased array technique described below.

An important advantage of electronically scanned arrays over mechanical probes is their ability to combine different scanning modes. (These modes are described later.)

There is no mechanical inertia associated with the scanning, so the beam can be made to jump virtually instantaneously from scan line to scan line in any sequence. The independent sequences of beam directions needed for each mode are interleaved, so that the two (or more) modes are seen to proceed together in real time.

15.7.3.3 Annular Array Probes

These are mechanically scanned probes in which the single disc-shaped element is replaced by an array of concentric ring-shaped transducer elements (Figure 15.16). Using the same electronic focussing technique as the linear array, the annular array achieves focussing of the beam in the elevation direction as well as the lateral, because of its circular symmetry. However, the problems of mechanical scanning outlined above have prevented this idea from being widely adopted.

15.7.3.4 Phased Array Probes

Like the linear array, a phased array or *electronic sector* probe uses electronic means to form and scan the beam, but the beam is steered rather than stepped, producing a sector-shaped field of view (Figure 15.17a). It may contain up to 128 elements, but these are less than half a wavelength wide, and therefore the probe face is shorter than that of the linear array. For these reasons, the probe is useful for imaging the heart, either trans-thoracically (through intercostal spaces) or trans-oesophageally as described in the next section.

The beamforming principles described above apply—variable aperture, apodisation, and focussing are possible in both transmission and reception. The difference is that the transmission and reception foci are made to lie along a sequence of angled scan lines, and all the elements are used to generate all beams in the sweep. Beam direction is controlled by delaying the pulses to and from each element to compensate for the extra path length between that element and the desired focus (Figure 15.17b). The technique originated with radio antennae (another form of wave aperture), generally operating at a single frequency, for which the delays could be regarded as phase-shifts; hence the name. For broadband apertures such as an ultrasound probe, the name 'delay-steered array' would be more appropriate.

As with the other array probes (apart from the annular array), such electronic beamforming only affects the beam shape in the scan plane. A cylindrical lens provides fixed, weak focussing

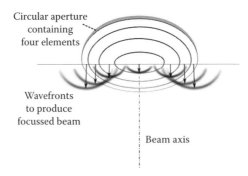

FIGURE 15.16
Basic configuration of an annular array of transducers. Wavefronts, delayed by different amounts in an electronic beamforming network similar to that in Figure 11a, are shown leaving the array, and will come to a focus in both the lateral and elevation planes. The annular array probe can only scan its beam mechanically in an oil-filled probe similar to that in Figure 15.13.

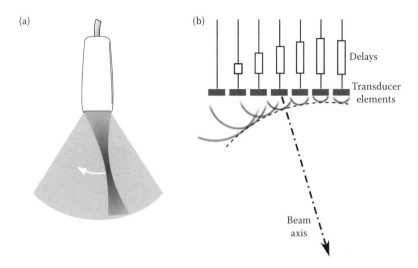

FIGURE 15.17
(a) A phased array probe scans its beam across a sector-shaped field of view. (b) The principle of beam steering is similar to electronic focussing (see Section 15.7.2), but the delays are chosen to produce wavefronts that focus on an angled beam axis. Dynamic focussing in reception involves generating a sequence of receive foci, closely spaced along the same scan line. (The 7-element aperture shown here for clarity could in practice consist of at least 50 elements.)

in the elevation plane. Alternatively, some modern phased arrays are '1½-dimensional' (see Section 15.7.2), having the transducers subdivided across the array as well as along it.

A disadvantage of the phased array is that the transmit and receive beamwidth increase, and hence the lateral resolution becomes poorer, towards the lateral edges of the sector scan. This is because the effective width of the aperture is smaller when 'seen' from the direction of a scan line at an angle to the probe axis, and the focussed beamwidth is inversely proportional to this width (see Section 15.7.2). Suppression of grating lobes is also crucial, otherwise as the main lobe is scanned to one side, a grating lobe might enter the field of view.

15.7.3.5 Intra-Corporeal Probes (Endoprobes)

Intra-corporeal probes may allow the transducer to be positioned nearer to a region of interest in the body. The shorter propagation through overlying tissue allows a higher frequency to be used, so lateral and axial resolution are improved (Section 15.4). Image quality is further improved by the reduction of beam aberration (Section 15.10.10) in the intervening tissue. An oesophageal probe for scanning the adult heart might have a centre frequency of 6 MHz, compared to the 3 MHz that would be more typical for a probe scanning through the chest wall. Endoscopic and intra-catheter probes are also available.

Any of the beamforming and scanning methods discussed above may be incorporated into an intra-corporeal probe. There might be a suitably compact mechanically scanned transducer (single disc or annular array), a linear or curvilinear array or a phased array. A further possibility is *radial scanning*, in which the beam is scanned through 360° around the axis of the probe, in the manner of a circling lighthouse beam (e.g. 30 MHz intra-arterial probes). Older radial scanning probes used a single-element transducer of fixed focal length, mechanically rotated within the probe, but modern ones have an array of transducers round their circumference, allowing electronic scanning and electronic control of focal length. These probes may be thought of as an extreme form of curvilinear array, in which the curvature extends through an entire 360°.

The particular scanning method chosen, the frequency, and the overall shape and size of the endoprobe depend on the anatomical constraints of the particular application. Thus a forward-looking curvilinear array mounted at the end of a trans-vaginal probe is suited to imaging the ovaries or an early pregnancy, but other orientations of the aperture are required for imaging the prostate gland trans-rectally. Sometimes, as in trans-oesophageal imaging of the heart, two compact phased arrays are mounted at right angles to each other, providing orthogonal sections of the target area. Endoscopic probes may have either radial or longitudinal fields of view, the latter being useful for guiding biopsy.

15.8 Overview of Diagnostic Ultrasound Modes

15.8.1 Review of Principles

All modes are based on the principles of pulse-echo mapping (Section 15.2). Real-time capability is achieved if the rate of transmission of pulses, called the *pulse repetition frequency* (PRF), is sufficient to allow the scanner to keep up with any changes in the anatomical region of interest, or with movement of the probe. However, an upper limit to the PRF is set by the requirement that there should be time for the echoes from each pulse to return from the deepest reflecting or scattering structures before the transmission of the following pulse. Otherwise late echoes from one transmitted pulse will be confused with early echoes from the following pulse. Most systems assume the sound velocity in tissue is 1540 m s^{-1} which means that an echo from a structure x mm deep will return at a time 1.3x μs after transmission. Thus, for diagnostic information down to say 15 cm deep, the system must wait approximately 200 μs between emissions, and the highest PRF that would avoid range ambiguity is approximately 5000 pulses per second. Higher PRFs are clearly possible for scanning more superficial regions.

15.8.2 A-Mode

In an A-mode scan, the amplitudes of echoes along a single scan line are represented as positive vertical deflections of a real-time graph on a display screen (Figure 15.18). A-mode scans provide the most accurate way of measuring the amplitudes of echoes and the distance between two targets. Dedicated A-mode equipment is rarely manufactured today, but A-mode is sometimes provided as an additional feature on some B-scan equipment, for example, in eye scanners, for measuring various axial dimensions of the eye.

15.8.3 B-Mode

In a B scan, a probe held in the operator's hand produces a sequence of beams which automatically sweep across a chosen scan plane in the patient's body, defining a sectional (tomographic) field of view. The varying angle and position of the scan lines are electronically controlled so that, as echoes are received, the targets' directions and ranges can be established. Echoes obtained from all scan lines are displayed in their correct respective positions, to form a 2D image of the scan plane (Figure 15.19). The amplitude of each

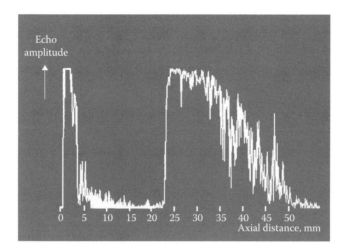

FIGURE 15.18
A-mode display from ultrasound beamed into the eye. Echo amplitude is plotted against axial distance of the echoing feature, calculated from the elapsed time since transmission.

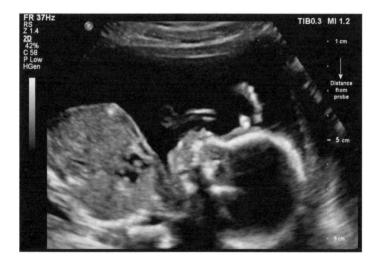

FIGURE 15.19
A B-mode ultrasound image is a tomographic section through the tissue beneath the hand-held probe. In this case it shows the head and chest of a foetus *in utero*. Note the safety indices in the top right-hand corner of the display (see Section 15.18).

echo determines the brightness (grey level) of the corresponding point on the display. Hence the name B-mode. This is the most widely used type of ultrasound imaging, and is described more fully in Section 15.9. It is often used in conjunction with other modes, such as M-mode and Doppler.

15.8.4 M-Mode

M stands for motion, and M-mode shows how the positions of reflecting surfaces along a *single* scan line vary with 'physiological time', that is, on a timescale of seconds rather than

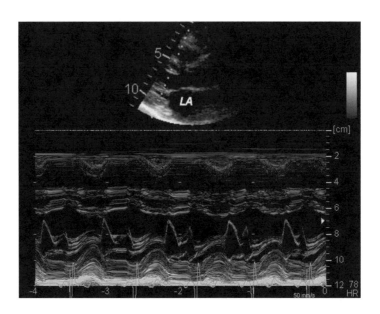

FIGURE 15.20
M-mode display, with associated B-mode image above it. On the B-mode image is the left atrium (*LA*) of the heart, and to the left is its outlet to the left ventricle, across which the mitral valve opens and closes. The dotted line shows the direction of the M-mode beam, chosen to intercept the mitral valve, so that the M-mode display allows dynamic study of its functioning. The vertical coordinate of the M-mode display represents range (depth) while the horizontal coordinate represents time (the most recent 4-s period). Brightness (or grey level) indicates echo amplitude. The leaflets of the valve can be seen opening and closing repetitively with time. An ECG waveform at the bottom of the trace allows reference to the cardiac cycle.

the microseconds associated with ultrasound travel time (Figure 15.20). It is used primarily in cardiac work, in conjunction with B-mode, via which the direction of the M-line can be monitored and adjusted to run through the moving interfaces of interest (e.g. valve leaflets, chamber walls). On activating M-mode, the B-mode scanning action stops and all further transmitted pulses are sent down the selected scan line. The echoes resulting from each transmission pulse are presented as brightness modulations along a corresponding vertical line of the M-mode display, with the brightness (grey level) indicating echo amplitude. Each transmission-reception sequence results in a new M-mode line, displayed alongside the previous one, the data being continuously scrolled across the display. Most machines save the most recent say 10 s-worth of the data that has been scrolled off the display, allowing it to be replayed if the acquisition of new information is frozen. There is often provision to show other physiological waveforms, such as an echocardiogram (ECG) or intra-cardiac pressure waveform, on the same timescale as the M-mode display.

15.8.5 Doppler Modes

Spectral, or pulsed-wave, Doppler allows the operator to choose on a B-mode display a particular point in the heart or in a blood vessel, and displays how the blood velocity at that point is varying with physiological time. It provides functional information about the heart or vascular system. Colour flow mapping and power Doppler imaging show blood flow within a chosen region of the scan plane, generally as a colour overlay on the grey-scale B-mode image. These modes are discussed further in Section 15.17.

15.9 Technical Aspects of B-Mode Ultrasound

15.9.1 Some Factors that Affect the Quality of a B-Mode Image

The width of the imaged region (the field of view) is determined by the type of probe, linear probes offering the greatest width for superficial targets, and others giving greater widths for deep targets (see Figures 15.13 through 15.17). The maximum depth from which diagnostically useful echo information can be obtained is called the *penetration*. Use of a higher frequency means an increase in attenuation and hence a decrease in penetration, but the penetration to be expected for a particular frequency has increased with better impedance-matching of transducers (see Section 15.7.1) and with the new technique of coded excitation (see Section 15.13). Penetration depth is determined by the sensitivity of the scanner, a measure of the weakest scatterer or reflector that can be distinguished from noise. This depends on the setting of the output power (which controls the amplitude of the emitted pulses) and overall gain control (see Section 15.9.3). It is itself a function of depth, generally being greatest close to the transmit focus. The *dynamic range* of a scanner is the difference in decibels between the intensity of the largest ultrasonic echo that can be handled by the receiver electronics and that of the weakest echo that can be distinguished from electronic noise. Modern systems have an impressively high dynamic range—perhaps approaching 100 dB. It can be deliberately reduced below its maximum by the dynamic range control (see Section 15.9.7).

The degree of spatial detail that can be seen within an image is described by the *spatial resolution*. Section 15.4 explains the two different forms of this for a B-mode image—axial and lateral—and that both are better at higher frequencies. The ability to focus at depth depends on the number of channels in the scanner receiver which is effectively the maximum number of transducer elements in the probe that can contribute to the beamforming aperture.

Since lateral resolution describes the minimum separation for two reflecting point features at the same depth to be separable on the image, it also describes the lateral extent of the image of a single point feature. The *point-spread function* of B-mode ultrasound is always wider in the lateral direction than the axial, by a factor of at least two even at the focus of the beam. Lateral spreading occurs because the angular separation of the emitted beams is less than the angle subtended by individual beams. A point target may send back echoes in response to several beamed emissions (Figure 15.21), provided it lies within the sensitivity pattern of the corresponding receive beams, and each echo from the target is displayed on the axis of the respective beam because the receiver has no better information about where the target is actually located. The spreading depends on the echoing strength of the feature concerned, being worse for strongly reflecting targets. It will occur for the image of any strongly reflecting features, not just points, and is known as the beamwidth artefact (Section 15.10.4).

Beamwidth and therefore lateral resolution is also affected by *acoustic aberration* in superficial tissue (Section 15.10.10). This is a limiting factor in B-mode performance because of the distorting effect it has on beam patterns, and therefore on spatial resolution and sensitivity.

Contrast resolution is the ability of a scanning system to differentiate between two echoes, or two regions of a scan, on the basis of their brightness on the image. This depends on the machine's dynamic range, on the *grey-scale transfer characteristic* selected by the operator, and on the *slice thickness* (Section 15.7.2).

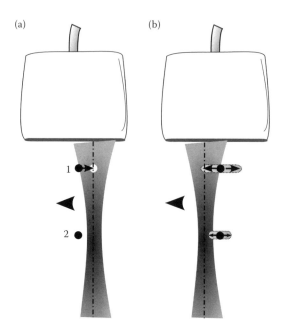

FIGURE 15.21
Lateral resolution depends on the width of the ultrasound beam. The beam is scanning from right to left. There are several beam positions between those shown in (a) and (b), all of which intercept the point target '1' which will therefore be imaged on the axis of each of them. The resultant image is a horizontal 'streak' with a length equal to the beam width, as shown in (b). The point target '2' is at the narrowest (focal) point of the beam, and will be imaged with a narrower streak.

Temporal resolution describes the ability to follow changes with time in the imaged tissue. It depends on the frame rate (which itself depends on the depth and width of the field of view and on the scan line density) and on the degree of frame averaging (see Section 15.9.9) and spatial compounding (see Section 15.12.1) that is selected.

The sections that follow describe the various parts of the scanner, and explain how the above factors can be controlled individually. It is important to realise, however, that image quality results from the combination of these separate factors. It is difficult and of questionable value to consider the separate effects that they have on image quality. Because they interact in a complicated way with each other and with the operator's perception, it is even more difficult to determine what value each factor should have for a particular clinical application. Imagine trying to decide whether contrast resolution is more important than spatial resolution (or vice versa) for a particular application, and what the optimum combination of these should be.

15.9.2 The Beamformer

This is a crucial part of all B-mode scanning systems that use array probes. It contains the delays, switches, transmission pulse generators, and pre-amplifiers concerned with generating, focussing and scanning the transmitted beam and with receiving similarly scanned and focussed beams. It also contains the software algorithms and circuitry to control these components. It is not required for mechanically scanned single-element probes, because their beam is defined (and fixed) by the shape of the transducer, and its direction by the scanning mechanism.

As explained in Sections 15.7.2 and 15.7.3, the selection of the transducers forming a transmitting or receiving group, and the relative amplitudes and delays of their signals, is pre-programmed for each possible beam direction and focal point. This beamforming was traditionally implemented with analogue electronic circuitry (so-called because the oscillating voltage it generates for a transmitting transducer or processes from a receiving transducer is, in effect, an analogue of the ultrasonic pressure variation at the transducer). However, beamforming's requirement for rapid but accurate changes is much better met by digital electronic technology, and this has become possible as digital microcircuits have been developed by the computer and defence industries. In modern scanners, beamforming is more likely to be digital than analogue, or at least a combination of the two technologies. A necessary part of this digital revolution has been the development of circuits that can accurately, and at very high rates, convert the analogue electronic signals produced by echoes reaching the transducers, to digital format. However, these 'analogue-to-digital convertors' first require the very weak and noisy signals from each transducer to be conditioned in various ways, as described next.

15.9.3 Radio Frequency Amplification and Time-Gain Control

The weak electronic echo signals produced by each receiving transducer are always accompanied by random voltage variations (electronic noise). Contributions to this noise are generated at all frequencies, so the wider the range of frequencies that an electronic circuit can amplify, the more noisy the output of that circuit will be. Thus, to keep the signal-to-noise ratio as high as possible, the amplifier is designed to handle only the range of frequencies that is in the spectrum of the echo pulses. This type of frequency-selective amplifier is known as a radio frequency (RF) amplifier, since it was first developed to amplify signals generated by radio waves. The signals in ultrasound systems have much in common with radio signals, having an underlying oscillation at a MHz frequency, and conveying information by changes in amplitude (B-mode systems) or frequency (Doppler systems).

The substantial and frequency-dependent attenuation suffered by ultrasound in tissue has implications for the design of the RF amplifier. Echo signals from targets deep within the body are much weaker than those from identical targets closer to the surface. Thus, if pulsed ultrasound is used to image two identical targets in liver (attenuation coefficient of 0.6 dB cm^{-1} MHz^{-1}), one being 15 cm deeper than the other, the 3 MHz component (for example) of the two echoes will differ in intensity by 54 dB (=0.6 dB cm^{-1} MHz^{-1} × 15 cm × 3 MHz × 2, allowing for a two-way trip); 54 dB represents a huge difference in intensity, 250,000 times (see Insight in Section 15.6.3). Because the strength of the reflection or scatter, after further processing, will be used to indicate the nature of a target, this reduction in echo intensity with target depth would make interpretation very difficult. It is an *artefact* of the imaging process. To overcome this problem, *swept gain*, or *time-gain compensation (TGC)* is always used. The gain of an amplifier is the ratio of the amplitude of the output signal to that of the input signal. Swept gain means that the gain of the RF amplifier is increased with time so that the strong, early, echoes arising from targets close to the surface are amplified by only a small gain factor, but the later echoes from more distant targets are amplified by a large gain factor (Figure 15.22). The gain versus time function is preset for each clinical application, but can also be adjusted by the operator, so that scattered echoes from a particular type of tissue show at approximately the same brightness at all depths.

The dependence of echo strength on depth is even more marked at higher frequencies because attenuation in tissue increases with frequency. For the same two targets in liver as

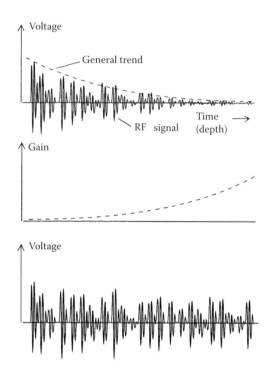

FIGURE 15.22
TGC attempts to compensate for the effect of attenuation by progressively increasing the amplification (gain) received by echoes as they return from greater ranges.

above, a 6 MHz frequency component from the deeper target would be 108 dB (6.3 × 10¹⁰) weaker than that from the nearer target. A consequence, therefore, is that the spectrum of a deep echo will have relatively less high frequency energy than that of an echo from a closer target. The centre frequency and bandwidth of echoes arriving at the transducer therefore become progressively lower as they return from increasingly greater depths. This means the beamwidth and pulse length become greater, and hence the lateral and axial resolutions become poorer (Section 15.4), for deeper targets.

This variation of echo spectrum with target depth is taken into account in high performance ultrasound scanners, where the centre frequency and bandwidth of the RF amplifier is continuously changed with time to match the expected changes in the echo spectra. The progressive reduction in amplifier bandwidth reduces electronic noise, leading to worthwhile improvements in sensitivity, contrast resolution and penetration for deeper targets.

An overall gain control is also provided to set an underlying amplification for all echoes, irrespective of depth, before the TGC is applied. It controls all receiver channels, and partially determines the sensitivity of the scanner.

15.9.4 Digitisation

The RF signals, which are the electrical analogues of the ultrasound echoes received by each transducer, must now be converted into digital form for beamforming and further processing. This process is similar to that for X-rays described in Section 5.2. Two main factors describe the performance of an analogue-to-digital convertor (ADC): the rate at which it can repeatedly sample the analogue signal and produce a corresponding digital number; and the accuracy or resolution of that number for representing different voltages.

Nyquist's theorem (see Section 8.4.2) states that an analogue signal is not adequately represented by the digital sequence unless its maximum frequency component is sampled at least twice per cycle. As the RF signals in an ultrasound receiver might well contain frequencies higher than 10 MHz, the ADC must perform its conversions extremely quickly.

As in digital radiography (see Section 5.9), a binary scale is normally used. Recalling that the dynamic range of a scanner is the ratio of the largest to the smallest of the signals that can be processed, a 12-bit ADC will have a dynamic range of 4095:1, but an 8-bit ADC only 255:1. For ultrasound, the dynamic range limit imposed by the ADC is often expressed in decibels. Every additional bit is equivalent to an increase of almost exactly 6 dB (see Insight in Section 15.6.3), because it allows a doubling of the largest number representing echo amplitude. Thus the dynamic range following an 8-bit ADC is 48 dB, while that following a 12-bit ADC is 72 dB.

Collecting ultrasound echoes from the body is demanding of electronic hardware and the amount of data it has to process. Each receiving channel, of which there might be up to 128, connecting the group of receiving transducers via the probe cable to the beamformer, requires RF amplification and a high-speed ADC. Each channel is processing a similar amount of information to that processed by a television receiver. Ultrasound operators wonder whether they will ever have the pleasure of using a wireless probe, that is, one that does not need the cumbersome cable connecting it to the scanner. Unfortunately, it seems inconceivable that technology will ever be developed to achieve this. Not only would it require the probe to emit radio waves equivalent to 128 TV stations, it would also require the transducers to be excited and controlled wirelessly, and the probe to be supplied with electrical power to allow all this. The bulk of the electronic receiving hardware must therefore be within the scanner itself.

15.9.5 Write Zoom

This causes the beamformer to restrict the scanning action to a smaller region of interest which is either the central region of the image, or is defined by the operator positioning a box on the un-zoomed image. The screen is then filled with a magnified real-time image of the selected area. The time saved by not scanning outside the zoomed region leads to an improvement in frame rate. In some applications, part of the potential increase in frame rate is traded for an increase in lateral resolution. For example, the time consuming technique of multiple-zone focussing in transmission (Section 15.7.2) may be used to produce a narrower effective transmission beam.

15.9.6 Amplitude Demodulation

Apart from the target position, the only echo information that is required for the B-mode display (or A- or M-mode display) is the echo amplitude. It is therefore necessary to remove the RF oscillations within the sampled digital signal from the beamformer, to leave a train of simple pulses, each with the same amplitude as their RF counterparts (Figure 15.23). The process that achieves this is called *amplitude demodulation*, and the simplified train of echo pulses it produces is called the *demodulated signal*. Since the demodulated pulses have the same duration as the RF pulses, their bandwidth is similarly a few MHz, but it is centred on 0 Hz instead of the nominal emitted frequency. Demodulated signals are called *baseband signals*, to distinguish them from their RF counterparts, or alternatively called *video signals* because their frequency range is similar to that of the signals in video systems.

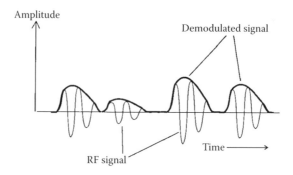

FIGURE 15.23
Amplitude demodulation produces a simplified echo waveform with an outline corresponding to the peak amplitude of the RF signal.

15.9.7 Dynamic Range Compression

The demodulated echoes from within the patient have a large dynamic range. Even after the effect of attenuation has been removed by the TGC, the echoes reflected from gas or bone interfaces could be up to 100 dB stronger than echoes scattered from microstructures within tissue or from groups of blood cells. B-mode ultrasound involves displaying echo strength as brightness. However, no monitor could display an image with such a large dynamic range, and the eye cannot appreciate a dynamic range of brightness of more than 30 dB at any one time. It is therefore necessary to 'pre-process' the demodulated signal, that is, to compress its dynamic range down to 30 dB, to be compatible with the display monitor and the eye.

The 'pre-processing', 'compression' or 'dynamic-range-control' algorithm takes the digitised value of each signal sample and converts it to a different number, such that the total range of output numbers is less than the total range of input values. Its rule for converting numbers—its transfer characteristic or look-up table (see Section 6.3.3)—is usually preset according to the clinical application, but can also be changed by the operator. Since this must convert a large range of numbers into a smaller range, it is inevitably a logarithmic function. It is usually controlled by a single number loosely called the 'dynamic range' which can be adjusted between say 40 and 90 dB. This refers to the dynamic range of input values that will be compressed into the 30 dB output range (Figure 15.24). Selecting a dynamic range of say 50 dB means that only echoes with amplitudes within 50 dB of the maximum would be displayed. All echoes weaker than this would be given the value zero by the pre-processor.

A low 'dynamic range' (≤50 dB) represents a considerable loss of information, but it may be appropriate if weak echoes are not of interest, for example, in cardiac imaging (echocardiography). Other applications require such echoes to be visible on the image, and therefore a wide dynamic range.

The compressed echoes can be adequately represented by 8-bit digital numbers. Since they all had different places of origin in the patient (given by their time of arrival and the direction of the beam axis), the next task is to assign locations to them which indicate where they should be positioned on the image. Thus the process of building the image begins.

15.9.8 Image Memory, Frame Store, and Scan Conversion

The digital numbers representing the echoes are stored in locations of an 'image memory' or 'frame store' that correspond to their point of origin (Figure 15.25). The content of the

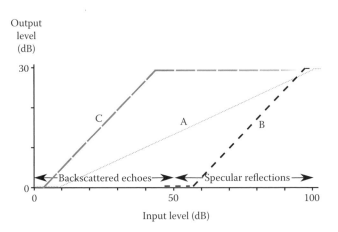

FIGURE 15.24
Simplified examples of pre-processing transfer curves, determined by the 'dynamic range' (DR) control. These allow the large dynamic range of echoes (horizontal axis) to be reduced to a range compatible with a viewed image (vertical axis). Curve A shows a large DR setting (~90 dB), allowing most tissue echoes to be displayed on the image, although not necessarily with high contrast. Curve B represents a low DR setting (~40 dB) which might be used for cardiac imaging, where significant boundaries are of more interest than tissue backscattering. Curve C shows a similar dB range centred on a lower value. This process is analogous to choosing a window width and window level in CT (Figure 8.4). (Note that because the axes are in logarithmic decibel units, the sloping lines represent non-linear curves.)

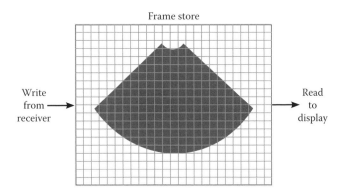

FIGURE 15.25
The frame store is a block of computer memory in which the image is assembled, ready for display. The shaded area (the shape of which depends on the probe being used) gets filled with echo information arriving via the receiver, one scan line at a time. Text and graphics information about the machine settings and the patient are added around the scan area. The scanner's display reads its information from the frame store row by row. In practice, the store has many more locations than shown here—perhaps 1024 × 1024.

frame store is therefore a preliminary version of the eventual image which will be presented as a matrix of small picture elements (pixels), each having a grey level determined by the amplitudes of the echoes from the corresponding part of the scan plane.

The process of storing echo amplitude information in the correct memory locations is known as writing into the store. The sequence the locations are written to is defined by the beam-scanning format. At the same time, the process of reading takes place, in which the memory locations are interrogated row by row, for compatibility with the monitor screen's line-by-line display of the image. Because the read sequence is different from the write sequence, the process achieves 'scan conversion'.

15.9.9 Facilities Based on the Image Memory

During the process of writing the echo samples from each beam into the image memory, *edge enhancement* (see Section 6.11) can be achieved in the axial direction by adding to each number a proportion of the difference between itself and its neighbour. The amount of edge enhancement is usually preset for the application, but can be changed by the operator.

Another choice available to the operator is how much the number stored for each pixel should change from frame to frame. This number is not simply replaced on a new for old basis in every frame because electronic noise accompanying the genuine echo information may cause a pixel to be unduly bright in one frame and unduly dark in the next frame. This noise may be reduced by *frame averaging* (see Section 6.11) which leaves unchanging genuine echo information unaffected, but averages out the time-varying noise component. The value stored for each pixel is the average of the previously stored value and that generated in the new frame. Averaging a large number of frames reduces noise effectively, but leads to a longer persistence (poorer temporal resolution). Averaging fewer frames gives a noisier image but a faster response to tissue motion, or to movements of the probe from one site to another.

By selecting *'freeze'*, the operator can prevent any further echo information being written into the image memory. The reading action continues, allowing the image monitor to receive and display the now unchanging stored image data. Freezing the image also stops the probe from transmitting and thus prevents the patient receiving unnecessary ultrasound exposure.

A facility known as *'read zoom'* is often provided. This allows the operator to select an area of interest within the frozen image and to magnify it to fill the display. The reading of the image memory is restricted to the reduced set of memory locations corresponding to the defined area. The data from this smaller area is thus displayed across the full extent of the screen. Interpolation between data from adjacent memory locations is necessary, otherwise the pixels of the zoomed image would be large and conspicuous. Read zoom does not have the benefits of improved frame rate and lateral resolution that are provided by write zoom.

The extremely valuable *measurement* possibilities on an ultrasound image (e.g. the measurement of anatomical distances, areas and volumes from B-mode images), are also based on the frame store. They involve the operator placing graphical cursors on the image to identify both ends of a distance measurement, or outlining an area to be measured. These cursors are generated within the machine by software that relates their position to particular pixels in the frame store. Since the machine can also relate these pixels to actual positions within the body (relative to the probe), it is able to compute the desired measurements. Automatic analysis of the image held in the frame store is being developed, to allow particular measurements to be made with the minimum of operator intervention.

15.9.10 Post-Processing, or Grey-Map Selection

Between the image memory and the image display and recording devices is a look-up table for modifying the relationship that determines how the numbers stored in the image memory will appear as brightness or grey level on the display. A selection of 'grey maps' is usually available. A 'reject' control is also available to suppress (display as black) echoes below a certain threshold which might be judged not to contribute useful information to the image, and might even be electronic noise.

As with most scanner adjustments, the grey map and reject are usually preset, but can be adjusted by the operator if required. Post-processing, unlike pre-processing, can be changed even when the image is frozen. However, it further complicates the overall transfer characteristic of the scanner which already depends on many other factors such as gain, dynamic range setting and the transfer characteristic of the display itself (see next section). Maps may even be available that provide tints of colours other than grey, although coloured B-mode imaging would not normally be selected unless preferred by the operator.

15.9.11 The Display Monitor

Some time ago, analogue TV-type display monitors gave way to video graphics array (VGA) monitors and their variants. These display a digital image developed within the monitor from an analogue video signal provided by the scanner. More recently, digital video interface (DVI) monitors have appeared, no longer using a bulky cathode ray tube, but a liquid-crystal display (LCD). The benefit of the digital interface is that the image pixels can be transferred from frame store to display (via the post-processing algorithm) without conversion to analogue video and back again.

The monitor brightness and contrast settings are crucial to the quality of the ultrasound image, and yet are often poorly adjusted. The brightness control determines how brightly the lower end of the monitor's dynamic range will be displayed. This is the background level, at which anechoic parts and the surround of a B-mode image will appear. This should be dark, but not so dark that weak but possibly important echoes are not displayed on the image. The contrast control determines how bright the white parts of the image will be displayed against this background. It must obviously provide sufficient contrast for a distinctive display of the compressed dynamic range of echo strengths, but dazzling white areas will be uncomfortable and tiring for the operator. The optimum settings of both controls depend on the ambient light in the scanning room which should be as dim as possible to allow full use to be made of the monitor's dynamic range.

15.9.12 Storage of Images, Patient Details and Examination Reports

Modern health care is changing the requirements for archiving and reviewing patient details, their images, and reports of examinations. 'Hard copy' paper or film used to be the medium for this, but digital storage is now usually available in the scanner, or in a computer workstation connected to the scanner. For more permanent storage, the patient archive could be 'backed up' onto magneto-optical discs, for example. However, not only is there concern about the security of patient data on such portable media, but even these have proved inadequate in busy departments, particularly those performing echocardiography which requires cine-loop sequences ('video clips') of real-time images to be recorded. There is also more need to have the archive available to other departments or hospitals. These requirements are increasingly met by picture archiving and communication systems (PACS) on hospital-wide computer networks, to which the scanners are connected (see Section 17.4). Echocardiography cine-loops pose a problem even for PACS systems, because of the amount of storage they require.

15.10 B-Mode Artefacts

Image artefacts are misrepresentations of tissue structure in the ultrasonic image. An ultrasound image is prone to artefacts because of the artificial way in which it is produced. It is not an image in the sense of a photograph, but a map, obtained by plotting features according to simple and fallible rules. Whenever the scanner receives an echo, it assumes there is a feature in the direction in which the most recent pulse was emitted, at a distance determined by the time of arrival of the echo. Clearly, if either the direction of arrival or the time of arrival is not what it should be for a particular echoing feature, an artefact will result. The displayed appearance of tissue may also be artefactually altered. Experience and a good knowledge of the relevant anatomy allow most artefacts to be recognised. Some common artefacts and their causes are discussed below.

15.10.1 Speckle Pattern

Scattering tissue parenchyma show on a B-mode image as a complex mottled (speckled) pattern of bright and dark regions (Figure 15.26). Although, for stationary tissue, this pattern repeats itself exactly from frame to frame, it is spatially random and there is no one-to-one correspondence between an individual bright speckle and an individual scatterer in the tissue.

This familiar speckled image of tissue is actually an artefact. Individual scatterers are too close together (and their individual echoes too weak) to be resolved on an image. At any instant, echoes will be arriving from targets distributed over a volume defined by the time since transmission, the beam cross-section and the pulse duration. These dimensions define the *resolution cell* of the scanner. Targets in different resolution cells will be resolved

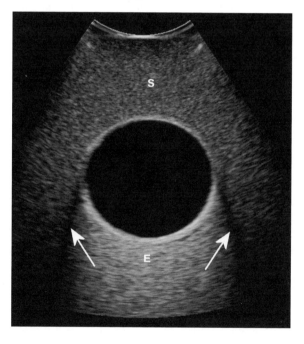

FIGURE 15.26
Example of a B-mode image showing the following artefacts: speckle pattern **S**, post-cystic enhancement **E**, and edge-effect shadows (arrows).

from each other, but those within the same cell will not. Unless very-high-frequency ultrasound can be used, the resolution cell will be larger than individual parts of tissue structure, and thus tissue structure will not be resolved on the image.

The individual pressure waves from scatterers within the same resolution cell will interfere at the transducer to form a resultant signal that is either weak or strong depending on the precise spatial distribution of the scatterers that caused them. The resultant signal will therefore vary randomly between high amplitude (a bright display) and small amplitude (a dark display) along each scan line. Bright intervals on several adjacent scan lines make up a bright speckle.

The average axial and lateral speckle dimensions are related to the corresponding dimensions of the resolution cell. Thus the speckle pattern is predominantly a characteristic of the scanner and probe, not the tissue being imaged. The structure of the tissue has only an indirect effect on speckle size, by governing the average brightness of the speckle according to the strength of backscattering it produces.

Speckle adversely affects both the detection of small fluid-filled cysts, and the contrast resolution of different tissues.

15.10.2 Reverberation (Multiple Reflections)

When a transmitted pulse encounters a pair of relatively strongly reflecting parallel interfaces which are normal to the beam, the echo from the deeper interface can be partially reflected at the nearer interface to form a new, albeit weaker, 'transmitted pulse' travelling back into the body again. This will be partially reflected again at the deeper interface, and so on (Figure 15.27), resulting in each target producing not one echo on the image, but a series of reverberation echoes at regular spacings, gradually getting weaker with range.

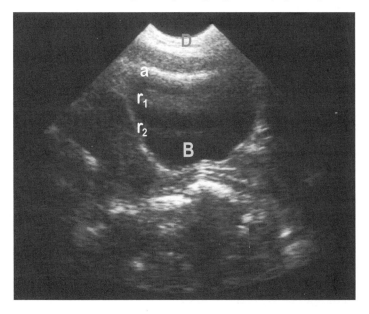

FIGURE 15.27
Reverberations between strongly reflecting interfaces produce strings of uniformly spaced false echoes. In this case, the emitted sound is reverberating between the anterior wall **a** of the bladder **B** and the probe itself, producing reverberations r_1 and r_2 within what should be a dark fluid region. Another reverberation artefact **D**, called *dead zone*, can be seen at the top of the image.

The regular spacing of the reverberation echoes helps to identify them as such. It is equal to the spacing of the pair of reflectors responsible for the reverberation. A common reverberation involves the probe face itself acting as the nearer reflector, and the anterior wall of a liquid-filled vessel acting as the further reflector. As in Figure 15.27, reverberation may produce acoustic noise within what should be an echo-free image of a liquid-filled structure.

Reverberation of a different appearance is the *comet-tail artefact*, due to reverberation between the proximal and distal edges of a small structure such as a small bone, a metal implant, or a foreign body, with different characteristic acoustic impedance from its surroundings. It may be helpful in pin-pointing an implant or foreign body.

Yet another form of reverberation occurs within the layered structure of the transducer itself (Section 15.7.1). The transmission produces an immediate sequence of closely spaced echoes at the start of each scan line, thereby obscuring extremely close echoes right across the image. This *dead zone* is visible at the top of Figure 15.27, but is much reduced (see Figure 15.26) in modern transducers which use composite materials (periodic structures of piezoelectric ceramics and non-piezoelectric polymers) with impedance closer to that of tissue, and multiple matching layers.

15.10.3 Mirror Image

A strongly reflecting boundary can act as a mirror, deflecting both the transmitted and reflected pulses. Echo-producing targets on the near side of the boundary then appear not only in their proper position, but also as virtual images on the far side of the boundary. Thus the air backed diaphragm acts as a mirror to make structures in the liver appear also in the lung.

15.10.4 Beamwidth Artefacts

Because the image of strongly reflecting features tends to be spread in the lateral direction (Section 15.9.1), two close targets at the same range may appear as one elongated target. This can affect the common measurement of femur length in the foetus, for example, which is used to monitor the growth of the foetus. Alternatively, a hole in a lateral tissue boundary may be obscured due to the merging of the 'streaks' extending from the boundary tissue on each side.

15.10.5 Slice Thickness Artefact

The scanning beam does not interrogate an infinitely thin plane, but a finite-thickness slice (see Section 15.7.2), and the elevation beamwidth at any depth defines the slice thickness at that depth. The spurious echoes from targets within the slice, but not actually on the scan plane, are a form of acoustic noise, reducing the contrast resolution of the image. The effect is mainly noticeable when the scan plane passes through liquid, but when there is solid (scattering) tissue within half a beam width of it (Figure 15.28). Thus the image of a small spherical cyst, or a longitudinal section through a blood vessel that is narrower than the slice thickness, is likely to show acoustic noise within it.

15.10.6 Incomplete Boundaries

As mentioned in Section 15.5.2, smooth reflecting boundaries may not be visualised unless the beam is almost perpendicular to the interface. For example, the walls of arteries are often incompletely visualised, where the ultrasound scan lines do not meet them at normal incidence.

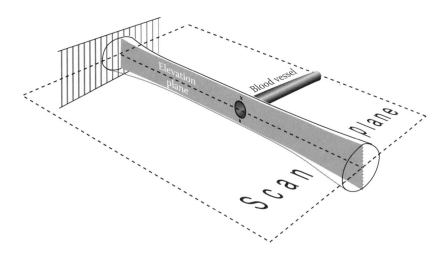

FIGURE 15.28
Slice thickness artefact. The blood vessel lies in the scan plane, and should appear dark on the image, in contrast to the surrounding tissue. However, echoes from tissue × within the scanned slice, but not in the scan plane itself, appear on the image as speckle and degrade contrast resolution.

15.10.7 Acoustic Shadows

A structure such as a bone or a calcified gall stone can reflect and/or absorb virtually all the ultrasound power incident upon it, creating a region of little or no echo strength in the area on the image beyond it. The problem can usually be solved by moving the transducer to another position from where the sound wave can avoid the obstruction. However, it can sometimes be of diagnostic value since it identifies the mass as being of a highly absorbing nature, like some tumours, or strongly reflecting, like a calcified gall stone (Figure 15.29).

15.10.8 Post-Cystic Enhancement

This is the opposite of shadowing in that the area behind a weakly attenuating structure is shown with enhanced brightness. It occurs where the TGC has been set up to compensate for absorption in solid tissue, but part of the field of view consists of a liquid region or a tissue with low absorption. Echoes returning from beyond the liquid are given the benefit of extra gain, but have not actually suffered the attenuation that the extra gain was intended to compensate for. The artefact is useful in that bright echoes in a region beyond a relatively echo-free area are a good indication that the echo-free area represents a liquid (see Figure 15.26). Enhancement can also be seen in Figure 15.29.

15.10.9 Axial Registration Error

In a B-mode scan the speed of sound is assumed to be the same in different tissues. In reality there can be an error in range, due to an echo returning at a time that is different from that predicted by the assumed speed (axial registration error). The eye in particular contains tissues with different speeds of sound, and for accurate measurements of its dimensions the speed of sound assumed by the scanner might have to be altered accordingly.

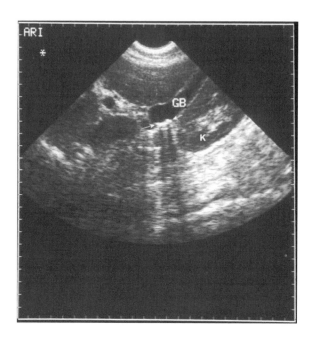

FIGURE 15.29
Shadow and enhancement artefacts. Strongly reflecting and/or absorbing features such as small stones (arrowed) in the gallbladder (GB) produce a shadow directly behind them, where the tissue cannot be imaged by conventional pulse-echo methods. Post-cystic enhancement is produced behind the kidney (K), because the depth-gain control over-compensates for reduced attenuation in the gall bladder and kidney.

15.10.10 Refraction Artefacts

15.10.10.1 Lateral Registration Error, and Double Image

The deviation of an emitted pulse from its intended path due to refraction (Section 15.6.1) is generally only slight, because variations in the speed of sound in soft tissues are small. However, it might cause features to be incorrectly located on the image. With certain anatomical structures, the beam could be bent onto a feature from two different angles, causing that feature to appear twice on the image.

15.10.10.2 Edge-Effect Shadowing

This is noticeable around cysts or other fluid-filled structures, at the point where the boundary runs in line with the beam direction (see Figure 15.26). There may be several reasons for it, such as the beam being refracted into the anechoic fluid, or being reflected off the curved boundary divergently, and therefore becoming weak. It is unlikely that the sound has continued in the expected direction.

15.10.10.3 Beam Distortion, or Aberration

The name aberration is borrowed from optics for this particular effect of refraction. It is caused by small-scale inhomogeneities such as the numerous fatty inclusions that are encountered in the subcutaneous tissue of some patients. This is a crucial region of propagation, close to the probe, where the sound disturbance should be gathering into a tightly confined packet of energy, to travel narrow and straight into the body as far as required. The precisely calculated beamforming process (Section 15.7.2) relies on several component waves travelling from different parts of the transducer with the uniform speed of sound expected in soft tissue; whereas the fatty inclusions have a significantly lower speed of propagation (Table 15.1). The result, worrying for the operator, is a de-focussed image. It is one of the reasons why some patients make poor scanning subjects. Sophisticated machines provide a degree of aberration correction, by allowing the operator to specify the nature of the superficial tissue (e.g. fatty or dense), and adjusting the beamformer accordingly.

15.11 Tissue-Harmonic Imaging (THI)

Many of the artefacts mentioned in Section 15.10 occur because the emitted sound has to be strong. This is to make the echo from weakly backscattering features detectable in spite of attenuation along the pulse-echo path. A possible consequence is that, as well as travelling along its expected path, the sound might have sufficient strength to travel an unexpected path and still return as a detectable echo. In some cases, an unusually strong reflector will help it. (It is left as an exercise for the reader to determine which artefacts arise in this way.) These particular artefacts would be less likely to occur if the sound could somehow be generated within the patient's tissue, rather than having to be delivered trans-cutaneously; and this is effectively achieved in THI.

In THI, as in normal pulse-echo methods, sound consisting of a *fundamental band* of frequencies (see Insight in Section 15.4), is emitted into the patient. THI exploits an effect known as non-linear propagation (Section 15.6.4) which causes the pulsed waveform gradually to become distorted as it propagates through the tissue. A distorted waveform can be shown by Fourier analysis to contain not only its original fundamental frequencies, but bands of harmonic frequencies as well. On the frequency scale, these are at integer multiples of the fundamental band (Figure 15.30). Harmonic frequencies do not exist in the wave emitted from the transducer, but gradually develop in the waveform as it travels through the tissue.

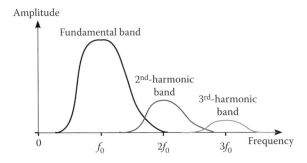

FIGURE 15.30
Spectrum of an emitted pulse after having undergone non-linear propagation. Frequency bands have appeared which are harmonics of the fundamental band. The desire for wide-bandwidth pulses means that the second-harmonic band may overlap with the fundamental, requiring signal-processing methods to isolate the harmonic echoes.

The second harmonic band of frequencies (at double the fundamental frequencies) is of greatest interest: higher harmonics are generally too weak to be useful. Echoes returning to the transducer, depending on where they originated, may or may not contain harmonics. Many of the artefactual echoes do not, as will be explained below. If the receiver is therefore tuned to the second-harmonic frequencies rather than the fundamental frequencies, it will not detect these artefactual echoes. In many cases, THI provides a 'cleaner' image than does conventional B-mode.

The potential of each tissue to distort a sinusoidal waveform is described by its *non-linearity coefficient* β. However, as soon the second harmonic begins to propagate in the tissue, it suffers from attenuation. The actual build-up of its amplitude is therefore determined more by the tissue's attenuation at the second-harmonic frequency than by its non-linearity coefficient (Duck 2002).

Another vital factor is the amplitude of the fundamental wave. As its name suggests, non-linear propagation occurs if the particles of the medium do not move with a velocity exactly proportional to the varying excess pressure they experience and do not exert an exactly proportional version of that pressure waveform on their neighbour. This does not happen with small amplitude waveforms, but is an increasingly strong effect as the amplitude increases. Essentially, the distortion occurs at the amplitude peaks of the waveform. Consider that the stiffness of the tissue increases when it is under high pressure, but decreases during rarefaction. Since propagation velocity is proportional to the square root of stiffness (see Insight in Section 15.2), the high-pressure parts of the waveform travel faster than the low-pressure parts, and so the shape of the wave becomes distorted (compare Figure 15.31 with Figure 15.3).

These three factors—the amplitude of the wave, and the non-linearity and harmonic attenuation coefficients of the tissue—interact in a complex but fortuitous way to determine

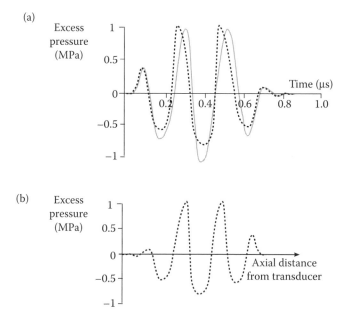

FIGURE 15.31
(a) and (b) Distorted versions of Figure 15.3a and 15.3b, due to non-linear propagation. Note the characteristic sharp peaks and blunt troughs.

which echoes returning to the transducer will contain harmonics, and which will not. Genuine echoes from reflecting and backscattering features may contain detectable harmonics. However, reverberation echoes may have travelled far enough for their harmonics to have been attenuated. The same may be true for mirror-image echoes. Sidelobes of the beam will not be strong enough to provoke harmonics. A similar argument applies to the edges of the main lobe, but THI does not necessarily improve lateral resolution or slice width because it might use a lower emission frequency than conventional imaging to reduce the overall attenuation.

It is not obvious whether THI reduces the problems of aberration in superficial fatty tissues (Section 15.10.10). Aberration tends to divert energy away from the main beam into sidelobes. Although THI would be expected to reduce artefactual echoes from the sidelobes (Tranquart et al. 1999), damage to the main beam would still occur.

Theoretical aspects of harmonic ultrasound fields (Duck 2002) are beyond the scope of this chapter. The practical outcome, however, is that a second-harmonic field can provide a vehicle for pulse-echo imaging, and THI is now available on all but the cheapest scanners.

The separation of the harmonic component of echoes from the fundamental component, so that the latter can be rejected, can be achieved in either of two ways. The earliest method, and still the only one available on cheaper machines, uses *frequency-selective filters*. The bandwidth of the emitted pulse is restricted so that the fundamental and second-harmonic bands do not overlap in the frequency domain (see Figure 15.30). It is then relatively simple with an analogue or digital electronic filter to admit the harmonic band into the echo receiver, but not the fundamental band. The disadvantage of this method is that, by definition, bandwidth-restriction requires an emitted pulse of longer duration, and thus sacrifices some axial resolution. Paradoxically, though, because the transducer must be capable of both emitting the fundamental band and receiving the second-harmonic band, the requirement for a wide overall frequency response pushes transducer design techniques to the limit.

Pulse-inversion or *phase inversion* is a more modern method of producing a response to the second-harmonic component of the echo while rejecting the fundamental component. It involves the emission of two pulses in succession along each scan line, the second of which is an inverted (anti-phase) version of the first. Upon reception, the echoes of each pulse are simply added, and cancel by destructive interference. However, any second-harmonic components of the two echoes will be in phase, because they have two cycles for every one of the fundamental (Figure 15.32). Their addition therefore results in constructive interference of the second-harmonic component which can then be processed further to produce the image. This method allows more of the transducer's bandwidth to contribute to axial resolution, because it doesn't matter if the fundamental and harmonic bands overlap.

In further developments of pulse-inversion THI, each emitted pulse consists of two frequency bands, one centred at a nominal frequency f_0, and one at $2f_0$. Non-linear propagation not only produces harmonics of these, but also produces other frequency bands due to *intermodulation* of the two emitted bands (Figure 15.33). These intermodulation bands are centred at the sum of the centres of the two emitted bands (i.e. $3f_0$) and at the difference between the two centres (i.e. f_0). Intermodulation is an inevitable consequence of the non-linear response of the tissue, but only occurs when there are at least two frequency components in the wave. In this mode the receiver is tuned to the difference-frequency band, centred at f_0. Any relics in the echoes of the undistorted emitted wave, centred at f_0 and $2f_0$, are eliminated by the pulse-inversion process described above, and thus artefacts

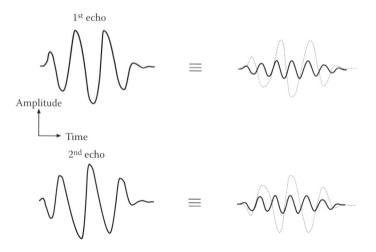

FIGURE 15.32

Pulse-inversion harmonic processing. Two anti-phase echo pulses return from each feature. They are shown broken down (by Fourier analysis) into fundamental (grey) and second-harmonic (black) components. The fundamental components are in anti-phase, so when added together they cancel; but the second-harmonic components are in phase, and will reinforce each other when added.

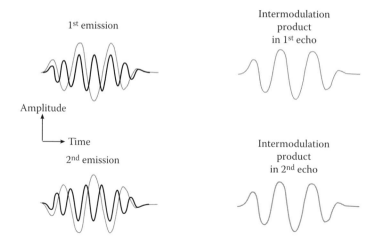

FIGURE 15.33

Advanced pulse-inversion harmonic methods exploit intermodulation products. Two emissions both consist of a fundamental (grey) and second-harmonic (black) component. The second emission is in anti-phase to the first. The two echoes returning from each feature contain harmonics of both components (not shown) which, being in anti-phase, are cancelled by the summation process. However, they also contain an intermodulation product, from non-linear interaction between the two emission components. Its instantaneous phase is equal to the instantaneous phase *difference* between the two emission components. It has the same phase in both echoes, so addition provides a reinforced echo signal for image formation.

are suppressed. However, the difference-frequency components in the pair of echoes from each range turn out to be in phase (see Figure 15.33), and therefore progress reinforced through the receiver. The advantage of imaging echoes via their difference-frequency component rather than their second-harmonic component is that it is attenuated less in tissue, and therefore allows better penetration.

15.12 Compound Imaging (Compounding)

This new technique reduces two types of image artefact—speckle and the incomplete imaging of tissue boundaries. The name implies that the image is not formed by simple B-mode principles, with its known deficiencies, but from a combination of pulse-echo processes. As in the compound eye of certain insects, the deficiencies of each individual process would be expected to be slightly different, and so the combined image should have more genuine content and less 'missing information'. There are two ways of compounding pulse-echo processes.

15.12.1 Spatial Compounding

This uses sets of beams from linear or curvilinear probes that are swept at different angles across the 2D plane of interest (Figure 15.34), rather than one set of beams normal to the probe face as in the standard B-mode method. Each displayed frame of the image is the result of averaging the frames produced by the different sets of beams. The beams are angled from the normal by using the phased array steering technique (Section 15.7.3) within the transmit and receive beamformers.

Thus, spatial compounding provides a more complete image of tissue boundaries that have different orientations in different parts of the image, as it is more likely that some beams will strike them at normal incidence, and therefore produce strong reflections. This is particularly useful in musculo-skeletal imaging, where incomplete boundaries could be misinterpreted as tissue damage.

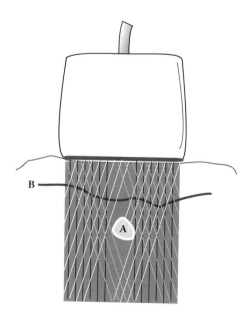

FIGURE 15.34

B-mode imaging with spatial compounding. This example uses three sets of beams, shown in black, grey and white, but more might be used in practice. Notice how the shadow behind a strongly attenuating or reflecting feature **A** will be reduced by the beams that reach behind it, and more of a surface such as **B** is insonated by beams at normal incidence.

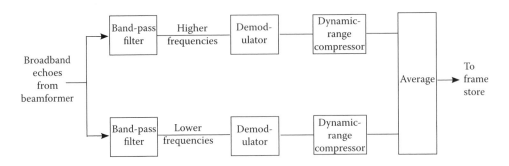

FIGURE 15.35
Frequency compounding in a B-mode echo receiver. Part of the receiver is split into two channels that process different sub-bands of the echo spectrum. The result is a smoothing of speckle and a reduction of noise.

Statistical considerations show that the method also reduces the brightness variation of the speckle artefact from tissue, producing a smoother image. Each set of angled beams will result in a different spatial distribution of bright and dark regions which, when averaged together, produce a pattern with less variance in brightness, that is, a smoother image of tissue parenchyma. The effect of electronic noise is also reduced. Like speckle, it will have a different pattern in the component images, and therefore reduced variance in the compound image.

According to Section 15.10.1, a smoother speckle pattern should provide better detection of small cysts, and better contrast resolution of different tissues. On the other hand, spatial compounding has poorer temporal resolution, since each frame requires several beam-sweeps from the probe. If too many angled sets of beams are used, operators describe the resulting image as 'swimmy'.

15.12.2 Frequency Compounding

This provides speckle- and noise-reduction without compromising frame rate. After the receive beamformer, bandpass filters split the wideband echo signals into narrower sub-bands (Figure 15.35). The sub-band signals then pass through separate demodulators and dynamic range compressors before being averaged together again and entered into the frame store to form the compound image. The processed echoes that would have produced speckle on the conventional B-mode image do not emerge identically from the separate processors, because the different frequency components of the emitted ultrasound will have fared differently as they propagated through the tissue. Therefore, the sub-band processors provide different speckle patterns, and the averaged pattern for the resultant image has reduced variance, that is, a smoother texture. The separate electronic-noise outputs of the sub-band processors are similarly smoothed in the final image.

15.13 Coded Excitation

Here, 'excitation' refers to the emitted pulse, and 'coded' means that the waveform is given a definite pattern, longer and more complicated than a simple pulse such as that in Figure 15.3. The aim is to lessen the conflict between desired resolution and required penetration (Section 15.6.3), by allowing higher frequencies to penetrate further.

Greater penetration of a given frequency in a given tissue can only be achieved by putting more energy into the emitted pulse. This can be achieved either by making the pulse longer, or by increasing its amplitude. The latter is the only sensible method for a simple pulse emission, because to increase its duration would degrade axial resolution. However, the theory of *matched filtering* or *pulse compression* shows that a longer pulse need not degrade axial resolution, as long as its waveform has a more complicated pattern than a simple burst of a single frequency, and as long as that pattern (the 'code') is precisely known. The waveform within the pulse must be *modulated* in some way. This is a technique borrowed from the field of communications and radar. Three common forms are amplitude, phase, and frequency modulation, and all of these have been applied to diagnostic ultrasound emissions. For example, a frequency-modulated emission might have a duration of 10 µs or so, and a gradually increasing frequency throughout, starting at f_1 MHz say, and finishing at f_2 MHz (Figure 15.36a). Such a waveform is known as a *chirp*, from the way it would sound if it were audible.

A simple, single-frequency emission of duration 10 µs would provide axial resolution of 7 mm at best—inadequate for B-mode imaging. However, its fundamental limitation is not its long duration, but its narrow bandwidth of 0.1 MHz (see Insight in Section 15.4). A coded waveform can be given a bandwidth B that is unconnected with its duration T. The chirp's bandwidth, for example, is approximately the frequency-sweep $f_2 - f_1$. If this is 2 MHz, for example, a chirp of long duration is capable of providing the same axial resolution as a 0.5 µs simple pulse.

If echoes of the long pulse are produced by features that are closer together than $Tc/2$ in the axial direction, they are not separate when they reach the transducer. A 'matched filter' within the receiver takes the frequency components of the echo pulses and adjusts their phase relationship so that they form a short pulse with duration $\tau = 1/B$ (Figure 15.37). The result is equivalent to a conventional pulse-echo system having emitted a short pulse into the body. The axial resolution is $\tau c/2$. However, the coded excitation with duration T has T/τ times the amount of energy, and this energy is conserved during compression by the

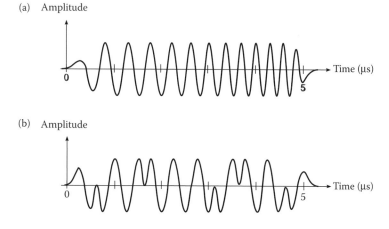

FIGURE 15.36
Examples of high-energy coded-excitation waveforms (a) is a chirp and (b) is encoded via phase-modulation, whereby the phase is reversed periodically in a pre-determined sequence. Although both have a duration T of 5 µs, their bandwidth is not 1/T = 200 kHz, but 2 MHz. They can therefore provide the same axial resolution as a simple pulse of duration 0.5 µs, but have ten times the amount of energy.

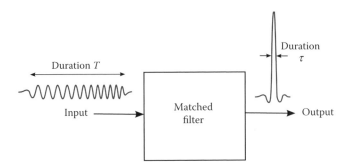

FIGURE 15.37
Coded-excitation processing system. Echoes of the chirp emission in Figure 15.36a are fed into a matched filter which compresses their duration down to approximately 0.5 μs, so that even if they overlap in time on entry to the matched filter, they can be separated at the output to an axial resolution equivalent to 0.5 μs. Note also the increase in amplitude which increases their detectability.

matched filter, so the short pulse is boosted in power by the same factor T/τ. This means that even echoes which return to the transducer with amplitude too weak to be detected by a conventional receiver may be detected after pulse compression, thus penetration is improved.

The 'power-boosting' factor T/τ of the coded-excitation or pulse compression system is an important figure of merit. Since $\tau = 1/B$, this factor can be expressed as TB, the *time-bandwidth product* of the coded emission waveform. In the example given above, the time-bandwidth product is 20 (13 dB), meaning that, in theory, the system would have that much extra sensitivity to weak echoes. Commercial coded-excitation systems might use pulses with 5 MHz centre frequency for abdominal imaging, where conventional systems would be restricted to 3.5 MHz.

Coded waveforms are also suitable for other modes of diagnostic ultrasound, and are compatible with the harmonic and compounding techniques described above. Both chirps (Behar 2004) and phase-modulated waveforms (Nowicki 2003—see Figure 15.36b) are used. There are challenges posed by coded excitation, however. One is the complicated interplay between the coded waveform and beamforming, both of which rely on precise phase relationships within and between signals, and their different requirements might not be completely reconcilable in the near field of the transducer. Another issue is the potential of the coded waveform to produce physical effects in tissue. Although its amplitude in the body is no higher than that of a conventional pulse, its duration is longer, and therefore the energy deposition is greater. This must be controlled, as for all diagnostic ultrasound (see Section 15.18).

15.14 Contrast Media—Imaging and Therapy

There are particular cases in ultrasound imaging, as in other modalities, when it is desirable to make particular tissues more visible than their surroundings in the image. When this is achieved by the introduction into the body of substances that enhance the appearance of those tissues, the general method is known as contrast imaging. In ultrasound imaging, it can be applied to blood in the body, or certain other fluids that do not otherwise send back appreciable echoes. In many applications of B-mode imaging, this lack of echogenicity is far from a disadvantage, because the relative darkness of vessels containing these fluids contrasts inherently with the surrounding echogenic tissue. However, that intrinsic contrast does not provide visibility of fluid regions that are similar to or smaller in cross-section than the speckle pattern or resolution cell (Section 15.10.1). There are, of course, many such small channels in the body. The small blood vessels that perfuse organs, and the ducts in the female reproductive system are particular examples where visibility would provide useful diagnostic evidence of patency. This would be realised if the fluids within them could be made *hyper-* rather than *hypo-*echoic.

The established way of making fluids backscatter strongly is to infuse them with microscopic bubbles containing gas or air. The gas has such a different characteristic acoustic impedance from the fluid that the bubbles cause strong backscattered echoes to be returned to the transducer. The backscattering is even stronger if the bubbles are designed to resonate in the undulating pressure field of the ultrasound pulse, which is achieved by making their diameter about $3.3/f$ microns, where f is the nominal ultrasound frequency in MHz. Microbubbles of this size are safe in the bloodstream, and can enter small blood vessels. They are injected into a peripheral vein, usually in the hand, wrist or arm. They must be stabilised with a shell, usually of lipid or albumen, otherwise they would be absorbed into the blood, particularly on passing through the lungs. Contrast agents (Quaia 2005; Claudon 2008) have found particular use in studies of the myocardium, and in helping to visualise focal lesions in organs such as the liver and kidney. It is the increased vascularity of such lesions that leads to their contrast from surrounding tissue when perfused by the microbubbles.

A particular behaviour of the resonant bubbles in the ultrasound field allows a further improvement in the contrast of tissues containing them. The radius of the bubbles changes non-linearly with excess pressure, so the backscattered wave is a distorted version of the incident waveform. The echoes returning from the contrast medium may therefore contain harmonics, even for a low-amplitude emission, and contrast imaging is often done in harmonic mode.

The pulses for contrast imaging are generally emitted at low amplitude for another reason. Bubbles become unstable under high amplitude pressure variations, and might collapse, which of course would end their role as a contrast agent. Occasionally, in fact, this collapse is deliberately provoked by a brief sequence of high amplitude emissions known as 'bubble-burst mode'. The collapsing bubbles emit shockwaves, the ultimate form of non-linear wave, whose broad bandwidth and high amplitude cause the locations of the collapsing bubbles to appear brightly on the B-mode image. As more contrast bubbles are perfused into the region with time, further 'bubble-burst' sequences will provide further snapshots of the perfused region. Stringing these images together gives the equivalent of time-lapse photography, showing more quickly a process (in this case perfusion) that evolves over a period of time.

The ability of ultrasound to burst bubbles when they have reached a particular location in the body offers the possibility of *targeted therapy* (Lanza and Wickline 2001). The contrast-agent bubbles are manufactured to contain cancer-controlling drugs or other therapeutic agents with a tendency to bind chemically to particular cells in the body. When they have reached their target, as evidenced by the enhanced image, they are burst to release the drug.

15.15 3D and 4D Ultrasound

15.15.1 Introduction

Despite the great utility of B-mode and the other 2D imaging modes of ultrasound, there is the obvious fact that the body and its anatomy are three-dimensional (3D). 3D imaging is not necessarily better than 2D, because at the moment, display technology is 2D, and poses challenges for presenting 3D information. Nevertheless, there are strong arguments for trying to extend a 2D imaging modality into the third dimension, and even the fourth. (Four-dimensional [4D] ultrasound aims to provide 3D information in real time.)

It is appropriate to present different information in a 3D image from that in a 2D image. It is not possible to present, at any one time, a 3D version of the information on tissue structure that B-mode ultrasound gives; such information would have to be presented as a sequence of sectional images. A 3D image, however, can give an impression of gross anatomy (surfaces, volumes and spatial relationships) more effectively than a 2D image can. In this way, 3D/4D ultrasound not only complements 2D, but has special applications.

15.15.2 3D and 4D Probes and Modes

The first requirement is to perform pulse-echo mapping throughout a 3D volume. The principle of emitting a sequence of narrow beams in known directions must still apply. Various methods have been developed to extend the scanning of a standard array probe (Section 15.7.3) into three dimensions. For example, the probe may be moved manually to cover a set of 2D sections which are then assembled into a 3D volume of data. The machine has to know the plane of each section, and movement and position sensors have been used for this. The most practicable method, however, is to mount the transducer array on a pivot in an oil-filled case, so that its scan plane can be automatically swept through a sequence of known directions. The mechanical movement and the oil bath have disadvantages described in Section 15.7.3, but such systems are the most common 3D implementation at the moment. Both extra- and intra-corporeal probes are available, and there are versions of the 360° intra-luminal probes which are mechanically or manually moved along the lumen.

The alternative to the mechanically scanned 3D probe is one with full electronic scanning. Since a standard 'one-dimensional' array of transducers can form and scan beams throughout a 2D section, to do the same throughout a 3D volume requires a 2D transducer array. (The 1½D arrays described in Section 15.7.2 are only capable of scanning in one plane.) Despite the increase in receiver channels and beamformer processing power they demand, such probes are available, particularly in phased array format for echocardiography applications. (These are not always used for full 3D imaging, but often to scan two 2D image planes at right angles, providing simultaneous coronal and sagittal sections of cardiac valves and chambers.) Beamforming theory shows that the 2D transducer arrays do not have to contain a full complement of transducers which might be of the order of 100^2. With careful statistical design, a *sparse array* with perhaps only a few hundred transducers can steer a narrow beam in any direction within a 3D volume.

The other technological challenge in 3D/4D ultrasound is the computing involved in rendering the image, and this has only become possible in recent years. Various *volume*

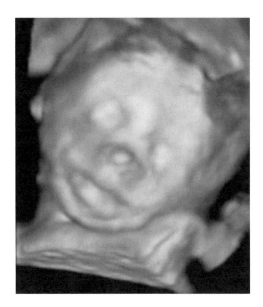

FIGURE 15.38
(See colour insert.) A 3D ultrasound rendering of a foetal face.

rendering algorithms are implemented, to give 3D impressions of the anatomy within the volume with varying degrees of transparency, or to display surfaces of structures such as the foetal face (Figure 15.38).

15.16 Ultrasound Elastography

This is a tissue-characterisation technique that has been under development for 20 years or so, but is now becoming commercially available. It provides an alternative to more conventional ultrasound techniques for the detection and assessment of tissue lesions. It aims to measure or image the elastic properties of tissue, of which hardness is known in many cases to correlate with abnormality. Although the bulk modulus (stiffness) of tissue plays a part in B-mode imaging (Section 15.5), it is difficult to extract this property from the echo data. The various emerging elastography techniques (Garra 2007) aim to provide greater image-contrast of tissues with different elasticity; or to measure relevant properties such as Young's elastic modulus E (different from the bulk modulus K), or simply the strain of the tissue under the influence of an applied stress (which is related to the elastic modulus). They are often combined with B-mode ultrasound because the same probe can be used for both. These systems might provide a 2D strain image ('elastogram') alongside a B-mode scan of the same region, allowing the techniques to complement each other.

The tissue is deformed (compressed) by an applied pressure. This is often applied externally to the body, in some cases by pressing down on the ultrasound probe. Ultrasound echo data stored before the compression are compared with those after the compression. A correlation process reveals how much the tissue at each depth has moved, from which images and measurements of elasticity may be derived, subject to considerable assumptions.

15.17 The Doppler Effect

In a pulse-echo system, waves reflected or backscattered from moving features exhibit a frequency shift. This is the Doppler effect. Its major application in diagnostic ultrasound is in the investigation of vascular conditions using a probe placed on or within the patient and measuring the frequency changes of the echoes backscattered by the blood cells moving within blood vessels. It is chiefly the red blood cells (RBCs) that produce scattering, and although the backscattered echoes are very weak, they exhibit a frequency shift which is proportional to the velocity v of the cells, allowing information about blood flow to be obtained.

The Doppler effect arises because the RBC changes its position *during* the scattering process, as illustrated in Figure 15.39. All similar parts of the ultrasound pulse, such as compressions, are emitted from the probe with period $t_e = 1/f_e$, and this is the period with which they would meet any stationary scatterer in the medium. However, in the illustration, the period with which they are incident on the RBC is shorter than t_e, because the RBC has a component of motion towards the probe, and so the following compression has to travel a shorter distance to meet it. Therefore the period t_r with which the compressions are backscattered by the RBC is shorter than t_e, and the frequency $f_r = 1/t_r$ with which they return to the probe is increased ($f_r > f_e$). Figure 15.39 shows how easily this explanation can be re-cast in terms of the change in wavelength between emitted and received pulses, based on the fact that the propagation speed c of the compressions is constant throughout.

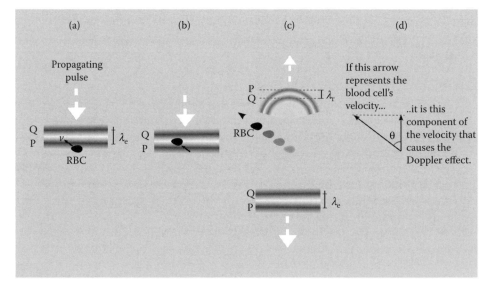

FIGURE 15.39

The Doppler effect. An emitted ultrasound pulse, with wavelength λ_e separating compressions P and Q, encounters a red blood cell RBC moving with velocity v. In (a), compression P is incident on the cell and will be scattered as well as transmitted. (b) Q is incident on the RBC after a time that is shorter than the period of the wave, because the RBC has moved forwards to meet it. (The movement is greatly exaggerated here.) (c) The backscattered wave therefore has a shorter period and shorter wavelength λ_r than the emitted wave. (d) Illustrates that it is the component of the cell's velocity in the direction of the ultrasound emission, that is, $v \cos \theta$, that influences the change in period and wavelength.

Further consideration shows that the change in period, wavelength and frequency of the wave is caused by the component of the RBC's motion in the direction of the ultrasound propagation, not by its true velocity v. This component is expressed mathematically as $v\cos\theta$, where θ is the angle between the RBC and ultrasound direction vectors. If the RBC had a net motion away from the probe (i.e. $v\cos\theta$ negative), then the period and wavelength of the backscattered pulse would be increased, and its frequency would be less than f_e. If the RBC moves at right angles to the pulse's propagation, there is no change in period, wavelength or frequency. As far as the pulse-echo system is aware, it is stationary.

The quantity of interest is the difference $f_r - f_e$ which is known as the Doppler shift and given the symbol f_D:

$$f_D \; (=f_r - f_e) \; = \; \frac{2v\cos\theta \times f_e}{c}$$

Note that the magnitude of the Doppler shift, if expressed as a fraction f_D/f_e of the emitted frequency, is proportional to the RBC's component of motion $v\cos\theta$, expressed as a fraction of the speed of sound c. Note also that it is usual for Doppler applications to consider v positive if the scatterer is moving towards the sound source so that positive Doppler shifts are associated with positive velocities.

Substitution of typical values for v, c and f_e shows that f_D is typically a few kHz or less. For example, a blood cell moving at 0.5 m s^{-1} directly towards a probe emitting a frequency of 3 MHz would produce a Doppler shift f_D of 2 kHz. Since the magnitude of $\cos\theta \le 1$ (i.e. ignoring its sign), this factor will reduce the magnitude of the Doppler shift unless $\theta = 0°$ or $180°$.

In practice, the Doppler equation is rearranged to have v or $v\cos\theta$ on the left-hand side, because the purpose of most Doppler ultrasound systems is to present information about blood movement from their estimate of f_D in the incoming echoes.

The $\cos\theta$ factor in the general Doppler equation represents a significant limitation of Doppler ultrasound. If the angle θ is not known, the true velocity of the blood flow cannot be estimated, only the component $v\cos\theta$. Furthermore, if $60°< \theta < 120°$, that is the magnitude of $\cos\theta < 0.5$, the system's estimate of f_D will not be sufficiently accurate for a reliable calculation of v or $v\cos\theta$. The small component of motion producing Doppler shift will be too error-prone. Indeed, if θ is close to $90°$, then there will be no Doppler shift and no indication that any blood is flowing.

15.17.1 The Doppler Spectrum of Blood

Since the millions of blood cells in even a small volume within a vessel will have a range of speeds, the Doppler system will, in effect, detect millions of Doppler signals with a corresponding range of Doppler frequencies. Ideally, the distribution of Doppler power versus Doppler frequency (the Doppler power spectrum) in the overall Doppler signal would match the distribution of the number of blood cells versus their velocity component $v\cos\theta$. In practice the match is only approximate, due to a number of factors discussed in Section 15.17.7.

Since in an artery the velocity of blood flow varies throughout the cardiac cycle, the Doppler spectrum in the echoes from arterial blood varies with time.

It is the function of Doppler systems to impart information about the Doppler spectrum, its variation with time, and its point(s) of origin within the body. There is a variety of systems that convey different aspects of this information.

15.17.2 Continuous-Wave Doppler Systems

In these systems, the probe contains two transducers, one which continuously emits a particular ultrasound frequency f_e, and the other which continuously receives any returning echoes, with frequencies $f_r = f_e + f_D$. (Remember that f_D is a spectrum of frequencies.) The initial echo-processing of a CW Doppler system is shown diagrammatically in Figure 15.40. The demodulator converts the received signal's frequency components into components with the same amplitude but with frequencies f_D. It does this by eliciting a 'beat signal' between the received RF signal and a reference signal with frequency f_e (by multiplying the two signals together). The beat signal has components which have the difference (beat) frequencies $f_r - f_e$ $(=f_D)$. A high-pass filter rejects frequency components in the Doppler signal below typically 100 Hz. This eliminates the large-amplitude, low-frequency echoes with little or no Doppler shift which are produced by stationary or slow-moving tissue, and prevents them from swamping the much weaker Doppler-shifted echoes from blood.

A CW Doppler system is sensitive only to moving targets in the region common to both the transmit and receive beams. By adjusting the sizes, shapes and relative positions and orientations of the two transducers, the manufacturer can make this region long or short, superficial or deep, narrow or wide, as appropriate to the intended application.

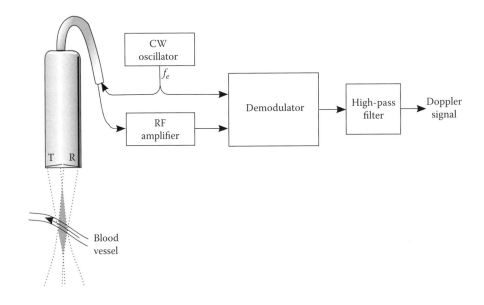

FIGURE 15.40
Main components of a continuous-wave (CW) Doppler system. The angle between the transmission (T) and reception (R) transducers is chosen to produce a short superficial cross-over region or one that extends deeper, as required. Moving blood within the cross-over region will result in a Doppler signal, as described in the text.

15.17.3 Audio Doppler Blood-Flow Indicators

These are the simplest CW Doppler devices, usually small hand-held units with a probe connected. Since the Doppler signals derived by the demodulator described above have frequencies of at most a few kHz, they can be presented to the operator via a loudspeaker. The human ear-brain combination is practised at analysing complex sound signals in this frequency range, so the operator can appreciate the relative amount of blood flowing at low and high speeds from the relative loudness of the low and high frequencies. Directional Doppler systems use two loudspeakers, filtering Doppler signals with negative Doppler shifts (flow away) into one channel and those with positive shifts (flow towards) into the other.

15.17.4 Sampling, Digitisation and Spectral Analysis of Doppler Signals

A more sophisticated type of Doppler system called a *spectrum analyser* provides a visual representation of the distribution of power and frequency in the Doppler signal. It presents the Doppler spectrum as a 2D frequency-time plot (a spectrogram), showing how its frequency content varies with time (Figure 15.41). The grey scale effectively adds a third dimension to the plot, showing the instantaneous strength of each frequency component.

As with B-mode processing, modern Doppler systems tend to digitise the signal at an early stage within the receiver. The signal would certainly be digitised before the spectrum analyser, and it is quite likely that demodulation would be done digitally as well. CW Doppler signals are readily digitisable because of their low spectral bandwidth. Since Doppler shifts are unlikely to exceed 10 kHz, a sampling rate f_s of 20 kHz is adequate (Nyquist's theorem).

Analysing the frequency content of the demodulated Doppler signal is usually achieved via the Fourier transform, and digital systems use an efficient computational algorithm known as the fast Fourier transform (FFT). The spectrum of a digitised signal is the same as that of its continuous analogue counterpart within the *principal interval* $(-f_s/2, +f_s/2)$ of

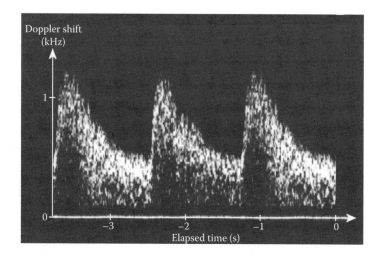

FIGURE 15.41

A typical Doppler spectrogram due to pulsatile blood velocity in an artery. Distance above or below the base line represents positive or negative Doppler frequency, respectively. A grey scale or colour scale is used to indicate the relative power of the Doppler signal at each Doppler frequency.

frequencies, but that spectrum is repeated along the frequency axis at intervals equal to f_s. In effect, however, the spectrum analyser ignores these extra frequency components and assumes that the frequency content of the signal is confined to the principal interval.

The frequency resolution Δf achieved by spectrum analysis is equal to the inverse of the duration T of the signal taken for analysis, thus $\Delta f = 1/T$. Doppler systems divide the Doppler signal into segments of typical duration 10 ms. A longer segment cannot usually be used because the blood flow in an artery will be changing over this sort of timescale, and it is not meaningful to analyse the spectrum of a segment while that spectrum is changing. The resulting 100 Hz resolution is not good enough for a detailed spectral histogram, but Fourier theory allows the segment to be extended artificially by adding periods of 'silence' to either end of it, and thus a more appropriate frequency resolution is achieved. Spectral analysis must be continuously repeated on new 10 ms segments of the Doppler signal to provide the real-time spectrogram.

15.17.5 Pulsed-Wave Spectral Doppler, Duplex Scanners and the Aliasing Artefact

In many applications it is desirable to be able to observe Doppler signals from a specific location in the body, and pulsed-wave Doppler allows this. It is usually combined with B-mode scanning in *Duplex mode*, so that the location of interest can first be identified on the B-mode image. The operator chooses a suitable direction on the image for a Doppler beam to travel from the array probe through the region of interest, and then, along that direction at the appropriate depth, marks a sample volume covering an axial range of typically 1–5 mm (Figure 15.42a). The chosen Doppler beam axis does not have to be normal to the face of the array probe. Many systems allow it to be steered away from the normal using phased array principles, if that provides a more suitable angle of intersection with the blood flow that is being investigated.

The operator then selects pulsed-wave Doppler mode and the B-mode scanning emissions are interleaved with Doppler emissions in the chosen direction. Usually, the same group of transducer elements that emits the Doppler pulses is used to receive their echoes, in conjunction with the digital beamformer (Section 15.9.2), but only those echoes from the chosen range interval are admitted to the receiver. A demodulator (Section 15.17.2) then converts the received RF signal to a baseband signal containing the Doppler frequencies. As in a CW system, this can be fed to a loudspeaker and, after spectrum analysis, to a spectral display (Figure 15.42b).

Although demodulation and spectral analysis of pulsed-wave Doppler appears similar in overview and implementation to that of CW Doppler, the theory is more complicated. First, the emitted pulse is of short duration, to allow the echoes from the chosen range interval to be discriminated. It may be longer than a B-mode pulse, but it would not be useful if longer than typically 10 μs. However, demodulated echoes even of this duration are much too short to allow useful spectral resolution (Section 15.17.4). It is therefore necessary to emit a sequence of pulses in the chosen direction, obtain a sequence of echoes from the sample volume, and process that sequence as a single long-duration entity. Although the sequence of echoes gives an interrupted rather than continuous demodulated Doppler signal, it can be made long enough to allow adequate spectral resolution. Typically, a sequence of 100 pulses might be used, with a total duration of the order of 10 ms which is the same as that used in CW spectral analysis.

The fact that the demodulated Doppler signal is interrupted means that it is effectively *sampled*, and therefore its spectrum repeats at frequency intervals equal to the sampling frequency f_s. In this case, $f_s = \text{PRF}$, the pulse repetition frequency of the emitted

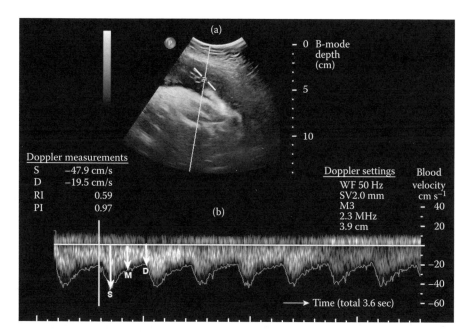

FIGURE 15.42

A duplex ultrasound display consists of (a) a B-mode image, allowing the operator to select the sample volume, and (b) the spectrogram obtained from the sample volume. In this case, the B-mode image is of a foetus *in utero*, and the sample volume is on the umbilical artery. The operator has aligned a cursor with the blood flow, allowing the machine to measure the angle θ in the Doppler equation, and hence indicate blood speed on the spectrogram rather than just f_D. The scanner has automatically drawn the envelope of the spectrogram, and identified the systolic (S), diastolic (D) and mean (M) Doppler shifts, which are involved in calculating pulsatility index (PI) and resistance index (RI—see text). The results are shown at the left, and some of the spectral Doppler settings at the right.

pulse-sequence. The PRF is limited by the need to wait for echoes to return from the deepest part of the B-mode image, in case the Doppler sample volume was placed there. It might be much less than the typical 20 kHz required to satisfy Nyquist's sampling theorem for Doppler signals. The consequence is that the principal frequency interval of Doppler shifts that can be evaluated by pulsed-wave Doppler is much more restricted than that by CW Doppler. Thus aliasing (see Sections 8.4.2 and 17.3.1) is more likely, and will result in an artefact. Any Doppler frequencies f_D that are outside the principal frequency interval defined by the Nyquist limits ±½ PRF are inadequately sampled, and will be misinterpreted (aliased) by the spectrum analyser, that is, they will be reconstructed with erroneous frequencies f_D. For example, as soon as an increasing Doppler frequency exceeds +½ PRF, the displayed frequency abruptly jumps to just above –½ PRF (see Figure 15.43), that is, the displayed frequency is reduced by an amount equal to the PRF.

Aliasing may not be a problem for moderate blood speeds in superficial blood vessels since a high PRF can be used, but it may be harder to avoid at larger depths (e.g. deep vessels, the heart and the foetus). Operators attempt to eliminate aliasing in various ways, such as increasing the PRF of the Doppler emissions. Specialised machines for vascular and cardiac applications have CW Doppler or high-PRF modes available to overcome aliasing.

The final complication of pulsed-wave Doppler processing is that the emitted pulse is not a single frequency as it was for CW. It contains a broad band of frequencies f_e, and all

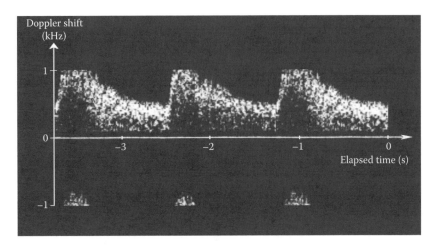

FIGURE 15.43
The aliasing artefact. In this example the PRF of Doppler pulses is 2 kHz, allowing only the range of Doppler shifts between ±1 kHz to be accurately displayed. The positive Doppler shifts at peak systole are outside this range, and are aliased into the wrong vertical position on the spectrogram.

of these are subject to Doppler shift according to the Doppler equation. Interestingly, both demodulation and Doppler spectral analysis are possible with these broadband Doppler-shifted received echoes. The sequence of broadband pulses at the output of the demodulator, forming the sampled signal containing the spectrum of Doppler shifts is similar, apart from its lower sampling rate, to the sequence of samples of the digitised demodulated signal in a CW system. In either case the spectrum of this sampled signal consists of the spectrum of Doppler shifts, repeated at frequency intervals of f_s, as far as the frequency-limit (bandwidth) of the individual samples; and the spectrum analyser displays only that part within the principal interval.

A simpler view of the functions of the demodulator and spectrum analyser is that they are correlating the received broadband echoes with the broadband emissions. Due to the Doppler effect, the former are slightly shifted in frequency by all the frequency shifts of the Doppler spectrum. Using the unshifted emission pulses as a reference, the correlation process is able to identify all these Doppler shifts, and also their relative strengths, that is, it can evaluate the Doppler spectrum.

An advantage of Duplex scanning is that it allows the machine to display blood speed rather than Doppler frequency. To do this the operator aligns an 'angle cursor' with the vessel or lumen axis, thereby allowing the machine to measure the angle θ needed to find v from f_D using the Doppler equation.

15.17.6 Interpretation of Doppler Signals

In some vessels the absolute blood speed is used to assess disease. For example, a peak systolic flow velocity of more than that expected in an artery or through a valve suggests a partial blockage (stenosis) at or near that point. A diastolic speed which is slower than expected generally indicates increased distal vascular resistance. Sometimes the shape of the outline of the Doppler spectrum (the 'Doppler waveform') can give diagnostic information. Even the audio output of the Doppler system gives some useful clinical information.

Often the ratio of the blood speeds at one site but at two different times in the cardiac cycle can be of diagnostic value. There is then no need to know the angle of the ultrasound beam to the blood flow, since it is constant with time, and the required ratio of blood speeds is the same as that of the corresponding Doppler shifts on the spectrogram. Thus the *pulsatility index (PI)* is defined as the difference between the maximum ('systolic' S) and the minimum ('diastolic' D) Doppler shifts in a cardiac cycle divided by the mean Doppler shift (M) over a cardiac cycle (see Figure 15.42b). It is used in the lower limb, for example, to assess proximal disease on the basis of the waveform damping it produces. The *resistance index (RI)* is a similar ratio, except that the denominator is the maximum rather than the mean Doppler shift. It is used to assess the resistance of the distal vasculature to blood flow.

15.17.7 Doppler Artefacts

As mentioned earlier, the Doppler power spectrum does not provide an accurate histogram of the number of blood cells versus speed. One reason is that, since the sensitivity within the sample volume is not uniform, those targets that happen to be in the most sensitive part will have a dominating effect on the power spectrum. Another artefact is *intrinsic spectral broadening*, whereby targets moving with a single well-defined velocity produce a range of Doppler frequencies rather than the single frequency one might expect. One cause of this is that ultrasound arrives at a target from all points on the transducer aperture and therefore from a range of different directions. This leads to a range of values for $\cos\theta$ in the Doppler equation, and hence for f_D.

Many of the artefacts described for B-mode imaging in Section 15.10 also occur in Doppler modes.

15.17.8 Doppler Imaging

By automatically and repeatedly measuring the Doppler signal at a large number of closely spaced sites within a B-mode scan plane, either the mean frequency or the power of the Doppler signal at each site may be plotted as a real-time colour overlay on the B-mode image (Figure 15.44). The mean frequency from each sample volume is of course related to the mean value of $v\cos\theta$, the speed of blood flow towards the probe in each volume.

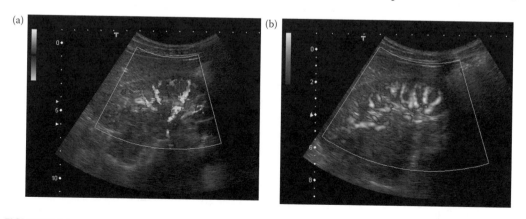

FIGURE 15.44
(See colour insert.) Doppler ultrasound images showing vasculature in the kidney. There are two main ways of mapping Doppler shifts detected within the 'colour box' on the B-mode image: (a) A colour flow map (CFM) shows mean Doppler shifts at each point; (b) A power Doppler map shows the strength of Doppler-shifted echoes at each point. Notice the reference colour bars to the left of each image.

The power is related to the amount of blood within each sample volume, that is, the perfusion at that point. The time needed to acquire the Doppler information means that only a restricted area of the B-mode field of view can be interrogated for Doppler information. The position and size of this 'colour box' is controlled by the operator.

A sequence of adjacent pulsed Doppler beams is emitted into the tissue defined by the colour box, each beam having contiguous sample volumes along it. Thus each sample volume corresponds to a 'Doppler pixel' in the colour box. The mean frequency or power estimates are stored in a Doppler image memory, where each stored number determines the colour of the corresponding pixel.

If the mean Doppler frequency for each Doppler pixel is represented, the method is called *colour flow maping* (CFM) or *colour flow imaging* (CFI) (see Figure 15.44a). Although this is a pulsed Doppler technique, the processing after the demodulator and wall filter is different from that in spectral Doppler. Instead of a spectrum analyser, a simpler autocorrelator assesses the phase shift between pairs of consecutive echoes from a sample volume to estimate Doppler shift. By averaging its estimate from say nine such pairs in a sequence of 10 echoes from each sample volume, the mean Doppler shift is obtained. This process is an order of magnitude faster than spectral analysis which requires around 100 pulse echoes from each sample volume. Furthermore, a single sequence of 10 pulse emissions along a particular Doppler line within the colour box provides a sequence of echoes from all sample volumes along that line. As a result of this efficiency, real-time CFM can be achieved. Scanning of the colour box alternates with B-mode sweeps of the whole image sector.

A colour scale at the side of the image shows how colour represents Doppler shift. There is no universal convention, but a typical colour 'map' might use red through to yellow to indicate increasing positive Doppler frequencies (flow towards the probe) and dark blue through to light blue to indicate increasingly negative Doppler shifts (flow away from the probe). Zero Doppler shift is usually represented by black.

Although useful clinically, CFM has limitations. Using only 10 echoes from each sample volume rather than 100 means that the accuracy and resolution (Section 15.17.4) of its Doppler-frequency estimate is not as good as that achieved by spectral Doppler. Remember also that the colours represent $v\cos\theta$, and there is no practicable way of correcting for the many different directions in which blood might be flowing at different points in the colour box. There may also be blood flow that produces no colour because it is at right angles to the scan line. Because of its poor frequency resolution, CFM may not display slow flow, being unable to distinguish it from zero Doppler shift.

Since CFM is a pulsed technique, there is also the possibility of aliasing (Section 15.17.5). If the Doppler frequency exceeds the Nyquist limit of half the PRF, the displayed colour will abruptly jump from that at one end of the colour scale to that at the opposite end.

For all these reasons, CFM does not provide a complete image of blood flow within the colour box, and should be regarded as a qualitative rather than a quantitative technique. It is usually used to image flow in discrete blood vessels or in the heart, where blood regurgitating through leaking valves or spurting through septal defects shows clearly as flow (colour) in an unusual direction.

Some CFM systems offer the option of showing the variability of the Doppler shift from each sample volume, by mixing an appropriate strength of another colour such as green. A large variability indicates turbulence in the blood flow, perhaps the sign of stenosis in the vessel.

Tissue Doppler imaging is the use of CFM to visualize soft tissue movement rather than blood flow, for example, to study heart muscle movements. In this case the filters are changed to preserve the soft tissue Doppler signals and reject those from blood.

If the power of the Doppler signal is mapped throughout the colour box, this is called *power Doppler* mode (see Figure 15.44b). Again, a colour scale is provided, but it is not related to Doppler shift—it simply indicates the power of the Doppler-shifted echoes at each point within the colour box. Power Doppler therefore shows where there is flowing blood, and how much, but gives no indication of direction or speed. It gives a clearer indication of perfusion than does CFM, since it does not produce a confusing mixture of colours. It also gives a more complete image of perfusion, being more sensitive to weak and slow flow. Even flow at right angles to the ultrasound beam can usually be seen, since the convergence or divergence of the beam means at least part of the beam will be non-perpendicular to the flow and will produce a Doppler signal with measurable power. Power Doppler is not subject to aliasing.

15.18 Ultrasound Safety

If used prudently, diagnostic ultrasound is rightly considered to be a safe technique. However, modern machines are capable of greater acoustic output than before (Hoskins et al 2003, Chapter 13) and, if used without due care, can produce hazardous thermal and non-thermal biological effects *in vivo*. The potentially hazardous physical effects of diagnostic ultrasound fall into the main categories of heating, cavitation, and mechanical forces. For a comprehensive review, see ter Haar and Duck (2000).

15.18.1 Physical Effects and Their Biological Consequences

Tissue heating, as a result of the absorption of ultrasound energy, could cause adverse biological effects and tissue damage. Its effect on dividing cells would be of particular concern, notably in the foetus during the first 8 weeks of gestation and in the foetal spine and brain up to the neonatal period. The World Federation of Ultrasound in Medicine and Biology has concluded that a temperature increase to the foetus in excess of 1.5°C is hazardous if prolonged, while 4°C is hazardous if maintained for 5 minutes. Diagnostic equipment can produce temperature rises of over 1°C in B-mode and theoretically up to 8°C in spectral Doppler mode. Spectral Doppler has the greatest heating potential because of the high pulse repetition frequencies and pulse lengths used, and the fact that the beam is held stationary. The highest temperature rises are associated with bone, since this has a very high absorption coefficient. Thus tissue in proximity to bone might be affected. The adult eye is also susceptible to heating, because it is not perfused with a blood supply which in other organs can carry away excess heat. Any pre-existing temperature elevation, such as in a febrile patient, and possible self-heating of the probe, must be considered as additive to the heating due to ultrasound absorption.

To help operators to be aware of any thermal hazard, scanners display a *thermal index* *(TI)* on the screen. This is a number that changes in response to the settings of the controls, and is intended to provide an approximate guide (within a factor of two) to the maximum temperature rise that is being produced in the tissue, according to one of three simple theoretical models. Where only soft tissues are involved, or where scanned modes are employed, the soft tissue thermal index (TI_S) is appropriate. The bone thermal index (TI_B) applies if the exposed tissues include bone. Thus, for example, TI_S is relevant for scanning a foetus in the first trimester, while TI_B would be appropriate for a pulsed Doppler scan in the second or third trimester, when bone is present. For adult trans-cranial imaging, with bone close to the probe, the cranial thermal index TI_C is used.

Cavitation refers to the energetic behaviour of small gas bubbles in a liquid medium due to the pressure variations of a sound wave. *Stable cavitation* (sometimes called *non-inertial cavitation*) refers to the vigorous radial oscillations of bubbles when the frequency of the sound is close to the resonant frequency of the bubble. Stable cavitation is considered to be a useful mode of action of ultrasound in physiotherapy, but is not considered relevant to diagnostic ultrasound. Diagnostic pulses are too short to establish it, and CW Doppler has too small a pressure amplitude.

Inertial or *transient cavitation* is a more violent and damaging form of cavitation that can be produced even by short pulses, provided the acoustic pressure amplitude is large enough. Here the bubble increases to many times its original size within one or two high negative-pressure excursions (rarefactions) and then violently implodes a short time later (usually around 1 ms). During the rapid collapse the bubble temperature may rise to several thousand degrees. Visible and ultraviolet light is emitted, shock waves are generated, and water is dissociated into free radicals. If the bubble is near a surface, for example, a blood vessel wall, the rapid inrush of liquid occurring during the bubble collapse can cause impact damage to the surface. Collapse cavitation can cause tissue damage and cell death, a fact that is used to advantage in ultrasonic sterilising tanks, and some forms of therapeutic ultrasound.

Cavitation is more likely to occur at low frequencies than high. Theory suggests that the probabilistic threshold for inertial cavitation is described by the *mechanical index (MI)*:

$$MI = \frac{p_-}{\sqrt{f}}.$$

For this index alone, the peak negative pressure p_- is expressed in MPa and the frequency f in MHz. (p_- is also *derated*, that is, calculated allowing for a nominal tissue attenuation of 0.3 dB cm^{-1} MHz^{-1}.) *MI* is displayed on machines as a continuously updated on-screen indicator of cavitation hazard, along with the TI mentioned above.

Because the likelihood of cavitation depends on many factors, it is not easy to determine exact values of *MI* at which damage may occur. Diagnostic ultrasound is not believed to be capable of producing cavitation *in vivo* unless gas-bubble nuclei are already present. For example, cavitation-like damage from diagnostic levels of ultrasound has been demonstrated in animal tissues containing gas cavities, such as the lung and bowel. It is not felt that bubble growth will occur for $MI < 0.7$, and this is often regarded as an unconditionally safe value. However, the collapse of contrast-agent bubbles (see Section 15.14) can be provoked at $MI = 0.3$, and so this value is used as a precautionary level if a contrast agent is in use.

Radiation force is a direct mechanical force produced by ultrasound on any object that has a smaller intensity on one side than on the other. It usually acts in the direction in which the ultrasound is propagating. It can produce bulk motion (*streaming*) of an absorbing liquid, due to the radiation force acting on each small region of the liquid. Streaming can sometimes be observed in real-time ultrasound images of cysts or other liquid masses containing particulate matter, particularly if spectral or colour flow Doppler is being used. Streaming speed depends on the local intensity of ultrasound and the energy-absorption of the fluid which is itself a function of the ultrasound frequency. The speed is typically only a few centimetres per second, but cells caught up in it and those at the walls of the containing structure will be subject to mechanical stresses.

At the moment, radiation force is not thought to cause harm, but it does provide a way of measuring the total power of an ultrasound emission. A target of sufficient size and composition that it totally reflects or absorbs the emission experiences a radiation force proportional to the total power of the ultrasound.

Other biological effects that are of increasing interest seem to be due to the vibration of the medium by ultrasound. The effects are not properly understood, but are believed to involve the cells of tissue. They are believed to occur at low intensities such as might be associated with diagnostic ultrasound, and may be beneficial (Martin 2009). However, the research has mainly been carried out with physiotherapy ultrasound equipment which emits longer-duration pulses than diagnostic systems. Whilst any biological effect of a diagnostic modality would be undesirable, at least this one, if it is produced by the short pulses of diagnostic ultrasound, is not considered harmful.

15.18.2 Minimising Hazard

The basic rules for operators are

1. Avoid scans for which there is no medical reason.
2. Ensure good training in scan technique and interpretation.
3. Keep up to date with safety guidelines.
4. Ensure the scanner is performing at its optimum via a quality-assurance programme.
5. Keep output power as low as possible, using increased overall gain to maintain sensitivity as far as noise allows.
6. Constantly monitor the safety indices, *TI* and *MI*.
7. Keep exposure times as low as possible.
8. Avoid holding the probe stationary for more than a few seconds.
9. Freeze the image rather than holding an emitting probe stationary.

Spectral Doppler should not be used in early pregnancy. If used in the foetal or neonatal skull or spine, control TIB to an appropriate level; and control TIS in the adult eye. In scanned modes, selecting a deep write zoom box or a deep transmit focus can often mean an increase in output. The instruction manual for the scanner should contain advice on how to set the controls to reduce output.

Conclusion

Diagnostic ultrasound is now a widely used tool in medicine, with the total number of examinations carried out per year probably exceeding that of general radiography. It provides various types of imaging and measurement of the body, for example of anatomy in two and three dimensions, of abnormally stiff tissue ("elastography"), and of internal blood flow using the Doppler effect. All are based on pulse-echo techniques and are provided in real time. Contrast agents are available to enhance the imaging of vascular or other fluid-filled structures.

The techniques are free from the harmful effects associated with even low doses of ionising radiation (although the production of free radicals is theoretically possible via a process called inertial cavitation). The equipment is also relatively cheap, and mobile or even portable. Thus it can be readily available both within and outside radiology departments.

From the start of its serious clinical application around 1970, the techniques and technology have developed significantly, and so therefore has the quality of the information provided, and the range of application. This development shows no sign of slowing.

Potential disadvantages are that skill is required in positioning and manipulating the probes and in interpreting the images, which are prone to artefacts. The operator is also responsible for minimising the risks due to warming or mechanical disruption of tissue. Nevertheless, in the hands of properly-trained operators, diagnostic ultrasound is considered to be safe and of great benefit to society.

References

Behar V. and Adam D., Parameter optimization of pulse compression in ultrasound imaging systems with coded excitation, *Ultrasonics*, 42(10), 1101–1109, 2004.

Claudon P., EFSUMB Study Group, Guidelines and good clinical practice recommendations for contrast enhanced ultrasound (CEUS) – Update 2008. See http://www.efsumb.org/mediafiles01/ceus-guidelines2008.pdf or *Eur J Ultrasound (Ultraschall in Med)*, 29(1), 28–44, 2008.

Duck F.A., *Physical Properties of Tissue*, Academic Press Ltd., London, 1990.

Duck F.A., Nonlinear acoustics in diagnostic ultrasound, *Ultrasound in Med & Biol*, 28(1), 1–18, 2002.

Garra B.S., Imaging and estimation of tissue elasticity by ultrasound, *Ultrasound Quart*, 23(4), 255–268, 2007.

Hoskins P.R., Thrush A., Martin K., and Whittingham T.A. (eds.), *Diagnostic Ultrasound: Physics and Equipment*, Greenwich Medical Media, London, 2003.

Lanza G.M. and Wickline S.A., Targeted ultrasonic contrast agents for molecular imaging and therapy, *Prog Cardiovasc Dis*, 44(1), 13–31, 2001.

Martin E., The cellular bioeffects of low intensity ultrasound, *Ultrasound*, 17(4), 214–219, 2009.

Nowicki A., Litniewski J., Secomski W., Lewin P. A., and Trots I., Estimation of ultrasonic attenuation in a bone using coded excitation, *Ultrasonics*, 41(8), 615–621, 2003.

Quaia E. (ed.), *Contrast Media in Ultrasonography: Basic Principles and Clinical Applications*, Springer Verlag, 2005.

ter Haar G. and Duck F.A. (eds.), *The Safe Use of Ultrasound in Medical Diagnosis*, British Institute of Radiology, London, 2000.

Tranquart F., Grenier N., Eder V., and Pourcelot L., Clinical use of ultrasound tissue harmonic imaging, *Ultrasound in Med. & Biol.* 25(6), 889–894, 1999.

Acknowledgements

Andrew Fairhead, who revised this chapter, wishes to thank Tony Whittingham, the original author, for laying such good foundations, not only in his writing, but also in the education of countless students of ultrasound. Grateful thanks also to Eleanor Hutcheon for help with the illustrations, and Manjula Sreedharan and Kathleen Anderson for providing images.

Exercises

1. Explain what actually travels when a sound wave propagates. Why is a short burst of vibration the essential waveform for diagnostic applications?

2. Why is the low megahertz range used for ultrasonic imaging? Give, with reasons, the frequencies and transmission focal lengths you would choose to make ultrasound examinations of (a) an eye, and (b) a liver.

3. Describe two ways in which values of the characteristic acoustic impedance of body components are significant to diagnostic ultrasound. Why is it necessary to use gel between the probe and the patient?

4. A source of pulsed ultrasound and a target are separated by normal soft tissue. Discuss the effect of each of the following on the amplitude of the ultrasonic pulse reflected back to the source:

   (a) The size and shape of the target

   (b) The characteristic acoustic impedances of the target substance and intervening tissue

   (c) The distance between source and target

   (d) The range to which the transmission focus has been set

   (e) The frequency of the ultrasound

   (f) The acoustic power of the source

5. What is the purpose of the backing layer, the impedance-matching layer and the lens in a typical ultrasound transducer? What would be the consequences for image quality of leaving out each of these in turn?

6. Explain briefly how a focussed beam of ultrasound can be obtained from the excitation of a number of transducer elements in a linear array probe. Why is the active group of a linear array progressively enlarged during dynamic focussing in reception?

7. Show that, in the scan plane, the beam of a single 1 mm wide element of a 3 MHz linear array has a near field length of only about 0.5 mm and then diverges between the angles +30° and −30° to the axis. How is the beam modified if it is produced by an active group of 20 adjacent elements?

8. Describe the different types of B-mode real-time scanning probe, and list their advantages and disadvantages. What might be the advantage of having probes that are applied internally rather than to the outside of the body?

9. Describe the function of the TGC facility in an ultrasonic scanner. In what way should this be regarded as artefact-correction? Which two artefacts can be caused by inappropriate operation of the TGC?

10. Out of the following B-mode controls—transmission focus, depth of field of view, overall gain, output power, TGC, write zoom, dynamic range, frame averaging—which affect the lateral resolution, which affect contrast resolution, which affect sensitivity and which affect temporal resolution?

11. Describe and explain a speckle pattern, as seen in a B-mode image. Which features of such a pattern are determined by the tissue and which by the machine?

12. Describe and explain two artefacts that cause the image echoes of genuine targets to be shown at incorrect positions.

13. Explain the cause and consequences of beam aberration in superficial tissue. Why is it difficult to correct for this aberration?

14. List the artefacts that tissue-harmonic imaging reduces, and in each case explain how the reduction works.

15. Explain the Doppler effect in medical ultrasound. What is the importance of the following parts of a Doppler receiver—(a) the demodulator; (b) the high-pass filter?

16. Distinguish between the terms: Doppler shift, Doppler signal, Doppler frequency, Doppler power, Doppler spectrum, Doppler waveform.

17. Describe how a pulsed Doppler device isolates the Doppler signals from a particular depth interval. What limitation is associated with Doppler frequency shift measurement by pulsed Doppler?

18. Why does CFM mode involve a lower frame rate than B-mode? Why is it possible for the same blood speed to be represented by different colours on the same CFM image? What effect does aliasing have on a CFM image? Can aliasing occur in power Doppler mode?

19. Explain what is meant by the TI and the MI, as displayed on the screens of scanners. How are these numbers meant to help the operator assess the possible hazard risk associated with a scan?

20. Which ultrasound mode has the greatest potential for thermal hazard, and why? Describe the measures that should be taken to ensure the prudent use of ultrasound.

16

Magnetic Resonance Imaging

Elizabeth A Moore

SUMMARY

Magnetic resonance imaging provides the best soft tissue contrast available in radiology and is an essential part of many diagnostic workups. This chapter will explain the following:

- Basic principles (magnetic moment of protons, precession, Larmor equation)
- Excitation and relaxation (RF pulses, spin-lattice and spin-spin relaxation)
- Spatial localisation using magnetic field gradients (frequency and phase encoding, slice selection)
- Image reconstruction using k-space
- Artefacts (effects of movement, flow, metal, encoding artefacts)
- Magnetic resonance (MR) angiography and diffusion-weighted imaging

CONTENTS

16.1 Introduction

The phenomenon of nuclear magnetic resonance (NMR) was discovered independently by two groups of workers in 1946 headed by Bloch at Stanford and Purcell at Harvard. The techniques developed were primarily used to study the structure and diffusion properties of molecules and subsequently Bloch and Purcell shared the Nobel Prize for physics in 1952. In the early 1970s the idea of spatial encoding using gradients was proposed, again by two independent researchers, Lauterbur and Mansfield, who shared the Nobel Prize for medicine and physiology in 2003. The 'nuclear' part of the name was dropped, leaving 'magnetic resonance' as the name used most often in medical imaging. Several variants of the name are used: MRI (imaging), MRS (spectroscopy), MRA (angiography), and so on.

This chapter concerns itself purely with the concepts of using MR for diagnostic imaging. It will be necessary to use some difficult ideas from quantum mechanics to explain the processes. However, we will keep this to a minimum, and concentrate on providing helpful analogies from classical physics.

16.2 Basic Principles of Nuclear Magnetism

Magnetism and electricity are closely associated with each other. Recall that when an electric current flows along a conductor, a magnetic field is generated (Figure 16.1). Similarly if a magnetic field changes its strength or direction, an electric current is generated in any conductor in the field. For instance, if a bar magnet is moved through a loop of wire a current will be induced. The movement is relative, meaning that either the magnet or the loop of wire can move; either scenario will generate an electric current.

Both the movement of the conductor, and the magnetic field, are described using vectors. A vector has both magnitude and direction; not only that, but either magnitude or direction or both may be changing (varying) in time. Vectors are denoted in bold typeface, for example, $\mathbf{B}_0$; their magnitude is shown between two upright lines, for example, $|\mathbf{B}_0|$. In MRI, we are always observing the effects of a dynamic process between nuclei and the magnetic field.

Atoms are composed of electrons associated with a nucleus which contains protons and neutrons. Modern MRI is only concerned with the nucleus (remember the original 'NMR'), and in fact uses only one particular element, hydrogen ^{1}H. The nucleus of the hydrogen atom contains a single proton and no neutrons, making it the simplest possible nucleus. It is the most abundant nucleus in the human body, present in water and long-chain lipids, which is part of the reason for the success of MRI. We often talk about 'protons' in MR, but the reader should keep in mind that this does not mean protons in other nuclei. The six protons in carbon nuclei, for example, although abundant in the body, play no part in conventional MRI.

Under the laws of quantum physics, all sub-atomic particles have a 'spin'; it is convenient to think of them as tiny spheres spinning on their own axes. Since protons each have a single positive charge, it is easy to imagine that this moving charge creates a tiny magnetic field. This is known as the magnetic moment of the proton, $\boldsymbol{\mu}$. (The bad news is that this simple view is not quite accurate: neutrons, which have no charge, also have a magnetic moment. The good news is that since ^{1}H doesn't have a neutron, it can be ignored.)

Wherever possible similar atomic particles 'pair up' so that their spins and therefore their magnetic moments will be cancelled out. Any nucleus with an unpaired spin (either

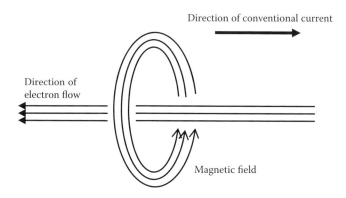

FIGURE 16.1
A magnetic field is generated when electric charges move along a conductor.

proton or neutron or both) will have a residual magnetic moment. In atoms with large atomic number there are a lot of electrons around the nucleus, which tends to shield the effect of the unpaired spins. However, the hydrogen nucleus is a single proton, and its single orbiting electron does not provide much shielding. ^{1}H therefore has the largest magnetic moment of all the elements, which is the other important factor making MRI such a sensitive technique. From now on, we will only consider the ^{1}H nucleus; this may also be called the proton or a spin.

16.3 Effect of an External Magnetic Field

The concept of precession is fundamental in MRI. To assist in understanding this it is helpful to use an analogy of a spinning top (or gyroscope) in the Earth's gravitational field. The gyroscope spins on its axis and has an angular momentum, shown as a vector. Consequently a torque is experienced by the gyroscope causing it to precess around the gravitational field, which is also a vector (Figure 16.2). The rate of precession depends on the angular momentum (i.e. rotational speed and mechanical characteristics of the gyroscope) and the magnitude of the gravitational field. A similar effect happens with protons, which have both angular momentum and a magnetic moment, when they are placed in a static magnetic field ($\mathbf{B_0}$). Each nucleus precesses about $\mathbf{B_0}$, or more precisely (Figure 16.2) its magnetic moment ($\mathbf{\mu}$) precesses around $\mathbf{B_0}$.

However, the analogy is not quite complete, since quantum physics allows for two stable spin orientations. These two orientations correspond to a low energy state (where $\mathbf{\mu}$ is aligned almost parallel with $\mathbf{B_0}$) and a high energy state (i.e. aligned anti-parallel to the magnetic field $\mathbf{B_0}$). These states are often called *spin-up* and *spin-down*, respectively; by convention, the direction of $\mathbf{B_0}$ is defined as the z-axis in all diagrams (Figure 16.3).

The two spin orientations do not have equal populations, due to the small difference in energy between them. Boltzmann statistics show that at equilibrium there is a small excess of protons in the low energy (i.e. parallel) state. At 37°C and in a $\mathbf{B_0}$ of 1 Tesla there is a population difference of about 6 ppm (parts per million). This excess depends on the temperature, which is difficult to modify *in vivo*, and more importantly on $\mathbf{B_0}$. An increase

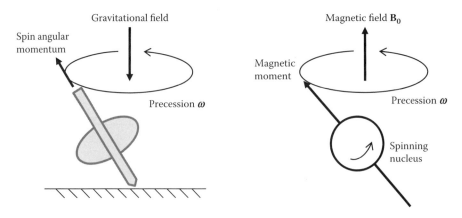

FIGURE 16.2

(Left) A spinning top precesses in the Earth's gravitational field. Similarly the ^{1}H nucleus precesses in an external magnetic field (right).

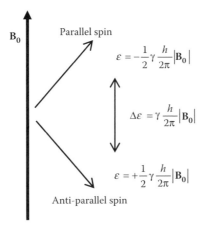

FIGURE 16.3
Spin orientations in a magnetic field. The energy difference is proportional to the magnetic field strength.

of $\mathbf{B}_0$ gives an increase in the population excess, which proportionately increases the signal to noise ratio (SNR) of the MR image.

16.3.1 The Larmor Equation

Since the proton obeys quantum physics, the energy difference $\Delta\varepsilon$ between the two population states is quantised, meaning that it has a precise and unchangeable value. $\Delta\varepsilon$ depends on the gyromagnetic ratio and the magnetic field strength

$$\Delta\varepsilon = \gamma \frac{h}{2\pi} |\mathbf{B}_0| \tag{16.1}$$

where γ is the gyromagnetic ratio. The gyromagnetic ratio is a constant for each nucleus, and for protons it has a value of 2.7×10^8 rad s^{-1} T^{-1}.

It is possible to cause a transition between the two states by the emission or absorption of a packet of electromagnetic energy (a photon). The photon's energy ε depends on its frequency ω, given by the equation

$$\varepsilon = \frac{h}{2\pi} |\omega| \tag{16.2}$$

Combining equations 16.1 and 16.2 yields

$$\frac{h}{2\pi} |\omega| = \gamma \frac{h}{2\pi} |\mathbf{B}_0|$$

$$\therefore \ |\omega_0| = \gamma |\mathbf{B}_0| \tag{16.3}$$

Via classical physics, the same result can be reached by considering the relationship between the rate of angular precession of the protons and the centripetal force between $\mathbf{B}_0$ and $\mathbf{\mu}$. This important relationship is known as the Larmor equation (after Sir Joseph

Larmor, Irish physicist 1857–1942), and ω_0 is called the *Larmor frequency*. It is more convenient to know that γ is 42.57 MHz T^{-1}, so that the Larmor frequency of protons in a 1 T scanner is approximately 42 MHz, and at 3 T it is approximately 128 MHz. These frequencies lie in the radiofrequency (RF) portion of the electromagnetic spectrum.

16.3.2 Net Magnetisation M_0

Once a collection of protons is placed in $\mathbf{B_0}$, it quickly reaches an equilibrium population of the two energy states. There is no phase coherence between protons, that is the spins are equally distributed in space around $\mathbf{B_0}$. Each magnetic moment (μ) can be resolved into two vector components: one along the z-axis, parallel to $\mathbf{B_0}$, and another located somewhere on the xy plane which is orthogonal to the z-axis. (We do not need to create separate x and y vector components; the protons only 'know' about being parallel or perpendicular to $\mathbf{B_0}$, so the difference between x and y is not important at this stage.) Due to the lack of phase coherence, the sum of the vector components in the xy plane is zero. However, in the z direction it is obvious that there is a net excess in the spin-up direction creating a net magnetisation ($|\mathbf{M_0}|$) in the +z direction (Figure 16.4). The magnitude of the net magnetisation vector is directly related to the population difference between the two energy states and consequently to the size of the magnetic field ($|\mathbf{B_0}|$).

16.3.3 From Quantum to Classical

There are many billions of protons in even 1 ml of human tissue. This means that MRI will observe the averaged behaviour of the protons, instead of quantised behaviour. Conveniently, this means we can leave behind the difficult ideas associated with quantum

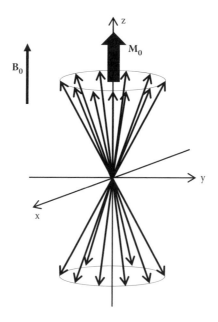

FIGURE 16.4
The two spin populations are randomly distributed around the z-axis. Each dipole moment has components along z and in the xy plane. The net magnetisation ($\mathbf{M_0}$) is along z.

physics and concentrate instead on classical models to continue the explanation of the basics of MRI. Each spin vector in the diagrams will now represent a large collection of protons which all happen to share the exact same Larmor frequency and the same phase.

These explanations will also make frequent use of a 'rotating frame of reference'. Since the protons are continuously precessing about $\mathbf{B_0}$, it is easier to show the changes that happen during excitation and relaxation if we can ignore the precession. As an analogy, the motion of a bouncing ball is much easier to describe on the earth (because we are on a rotating frame of reference) than from a distant planet, which would have to include the rotation of the earth. Consider a frame x'-y'-z' rotating at ω_0, the Larmor frequency, about its z'-axis, and with the z'-axis aligned with the z-axis in the laboratory frame (i.e. defined by the direction of $\mathbf{B_0}$). The magnetisation at equilibrium thus appears as a static vector of magnitude $|\mathbf{M_0}|$ along the z'-axis.

During excitation, and especially during relaxation, it is necessary to consider components of the magnetisation along the longitudinal direction, that is, z (z'), and in the transverse plane, that is, x' y'. These components are labelled $\mathbf{M_z}$ and $\mathbf{M_{xy}}$, respectively.

16.4 Excitation and Signal Reception

Although the individual magnetic moments are precessing, the equilibrium net magnetisation $\mathbf{M_0}$ is fixed along the z-axis and is too small to measure directly (it is of the order of tens of micro-Tesla for the average body, compared with 1.5 T or 3.0 T for $\mathbf{B_0}$). This equilibrium has to be disturbed by the process of excitation before a signal can be obtained. Excitation is achieved by applying an electromagnetic wave at the Larmor (RF) frequency. Electromagnetic waves consist of oscillating electric and magnetic fields; for MRI we are only interested in the magnetic part, the electric field is not used.

16.4.1 RF Excitation

The RF magnetic field is circularly polarised, so that it creates an additional magnetic field $\mathbf{B_1}$ aligned perpendicular to $\mathbf{B_0}$ and rotating at ω_0. In the rotating frame, this appears as a static magnetic field aligned, for example along the x'-axis. Just as an individual proton precesses about $\mathbf{B_0}$ in the laboratory frame, so $\mathbf{M_0}$ begins to precess about $\mathbf{B_1}$ in the rotating frame (Figure 16.5). The magnitude of $\mathbf{B_1}$ is much smaller than $|\mathbf{B_0}|$, typically a few μT, so the precession frequency is slower. After a certain time t_p, $\mathbf{M_0}$ will be aligned along the y'-axis, that is it has turned through exactly 90°. This amount of RF energy, measured by $|\mathbf{B_1}|$ and its duration t_p, is called a 90° pulse, or an excitation pulse. Leaving the RF on for double the time, or doubling the strength of the $|\mathbf{B_1}|$ field, would turn the magnetisation through 180°; the pulse would then be called a 180° pulse, also called a refocusing or inversion pulse. Intermediate sized pulses are easily produced, and are described by the angle through which the magnetisation is turned—the flip angle α.

For those interested in quantum physics, the RF field has exactly the right energy to stimulate transitions between energy states. Protons in the parallel state can absorb energy to jump to the anti-parallel state; those in the (higher energy) anti-parallel state can be stimulated to give up some energy and drop to the parallel state. Both transitions happen with equal probability, but since there are more spin-up protons at equilibrium, there will be a net absorption of energy.

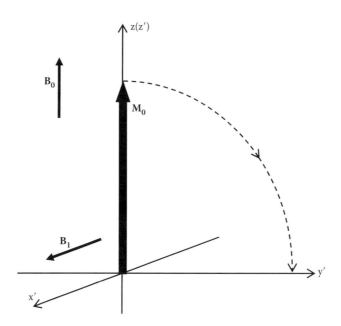

FIGURE 16.5
In the rotating frame **M** moves around the magnetic portion of the applied RF pulse.

When the energy states have equal populations, there will be no net magnetisation in the z' direction. However, using quantum physics it is difficult to explain how M_0 rotates, and in particular how the phase of the protons is affected by the RF energy. In addition to 'flipping' M_0 through the desired flip angle, the RF wave has the effect of bringing all the protons into phase with each other. Once again we leave the quantum physics explanation and return to the more useful classical concepts.

16.4.2 Signal Reception

Immediately after a 90° pulse, there is a transverse magnetisation vector (M_{xy}) aligned along the y'-axis, that is rotating at ω_0 in the laboratory frame and with magnitude $|M_0|$. With no external influence, the system returns to equilibrium, with M_0 aligned along the z'-axis, by losing the energy absorbed from the wave and phase coherence gradually being lost. To pick up the MR signal, a receiver coil is used with its axis perpendicular to B_0, which is fixed in the laboratory frame. The coil experiences a changing magnetic field as M_{xy} rotates, and a current is generated in the coil. The loss of extra energy and phase coherence are both random processes, leading to an exponential decay of $|M_{xy}|$. The MR signal in the coil, known as the free induction decay (FID), therefore oscillates at $|\omega_0|$ and shows an exponential decay (Figure 16.6). Once again it is convenient to 'ignore' the Larmor frequency, and the FID is often shown as a simple exponential decay.

16.5 Relaxation Processes

Two separate processes work on the longitudinal and transverse components of the magnetisation to return it to equilibrium. These are known as spin-lattice and spin-spin

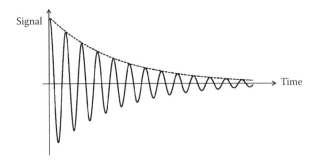

FIGURE 16.6
The free induction decay (FID). The signal oscillates at ω_0 and decays exponentially to zero (dotted line).

relaxation, respectively. Relaxation times are characteristic of particular tissues, and have a wide range *in vivo*, spanning 2 or 3 orders of magnitude. These properties can be exploited to control the contrast in the image, by manipulating a series of RF pulses.

16.5.1 Spin-Lattice Relaxation

Spin-lattice relaxation controls the loss of excess energy absorbed by the protons during excitation. It only affects the recovery of $|\mathbf{M}_z|$ from zero (after a 90° pulse) to the equilibrium $|\mathbf{M}_0|$. Energy is transferred from protons to the surrounding environment, or lattice, in which they are embedded. Due to its much larger size, the lattice can absorb the energy without raising its temperature. The recovery is described by the equation

$$\mathbf{M}_z = |\mathbf{M}_0| \left[1 - \exp\left(\frac{-t}{T_1} \right) \right] \tag{16.4}$$

where t is time and T_1 is known as the spin-lattice relaxation time (Figure 16.7a). This is a variant of the exponential decay equation discussed in Section 1.5 where the variable $\mathbf{M}_z$ increases with time. In tissues T_1 is typically of the order of a few hundred milliseconds (Table 16.1). It becomes very long in solids (such as cortical bone) or in free water, gets longer with increasing temperature, and also increases with $|\mathbf{B}_0|$.

16.5.2 Spin-Spin Relaxation

As the protons move around their environment, they experience varying magnetic fields due to their neighbours. The distribution of protons, even in a small volume of tissue, is not uniform, and a clump of protons can slightly alter $|\mathbf{B}_0|$ by means of their locally averaged magnetic moments. Since the protons are in constant motion (almost all body tissues are effectively liquid, not solid), each proton experiences a continuously varying magnetic field. The resonant frequencies also change dynamically, causing them to spread out in the rotating x'-y' plane. The net component of magnetisation $\mathbf{M}_{xy}$ is gradually reduced as the protons dephase. $|\mathbf{M}_{xy}|$ has an exponential decay governed by T_2, the spin-spin relaxation time (Figure 16.7b).

$$|\mathbf{M}_{xy}| = |\mathbf{M}_0| \exp\left(\frac{-t}{T_2} \right) \tag{16.5}$$

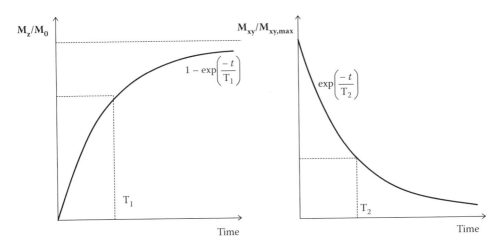

FIGURE 16.7
(Left) spin lattice relaxation time (T_1) is the time taken for 63% of $|M_0|$ to recover along the z-axis. (Right) spin-spin relaxation time (T_2) describes the exponential decay of the transverse magnetisation.

TABLE 16.1

T_1 and T_2 Values at Commonly Used Field Strengths for *In Vivo* Human Tissues

Tissue	T_1 Relaxation Time (ms)		T_2 Relaxation Time (ms)	
	1.5 T	3.0 T	1.5 T	3.0 T
White matter	560	832	82	110
Grey matter	1100	1331	92	80
CSF	2060	3700	–	–
Muscle	1075	898	33	29
Fat	200	382	–	68
Liver	570	809	–	34
Spleen	1025	1328	–	61

The value of T_2 like T_1 depends on the structure of the environment, and particularly on the mobility of protons. *In vivo* T_2 has a similar range to T_1, ranging from tens to hundreds of milliseconds. It is very short in solids (e.g. cortical bone), and much longer in free fluids. It also increases with temperature, and increases slightly with $|B_0|$ up to 1.5 T, then tends to decrease at high field strengths (Table 16.1).

16.5.3 Inhomogeneity Effects

The varying magnetic fields experienced by protons due to their neighbours may be thought of as an 'intrinsic' or microscopic magnetic field inhomogeneity. In addition, there is a macroscopic inhomogeneity in B_0 due to the technical impossibility of creating a perfectly uniform field. Macroscopic inhomogeneities cause the protons to dephase in the transverse plane more quickly, as they move physically around the volume of the field. The FID seen after a 90° pulse is subject to these inhomogeneity effects and is described by the characteristic relaxation time T_2^*. T_2^* may be thought of as the 'effective transverse' relaxation time, and is calculated by combining the magnetic field inhomogeneity ΔB_0 with T_2:

$$\frac{1}{T_2^*} = \frac{1}{T_2} + \frac{1}{2}\gamma\Delta B_0 \qquad (16.6)$$

In modern MR systems, ΔB_0 is often very small; however, the human body introduces its own inhomogeneity when the patient is placed in the system. This is caused by the varying diamagnetic and paramagnetic properties of different tissues, and ΔB_0 will be particularly bad wherever there is a pocket of air (bony sinuses, lungs, bowels) or dense bone (posterior fossa, vertebral column, etc.).

Insight

Understanding T_1 and T_2 Together

In many textbooks, T_1 and T_2 graphs are only shown separately. This can lead to the impression that the relaxation processes occur on similar time scales, whereas in fact T_1 is typically 10 times longer than T_2. If T_1 and T_2 are shown together (Figure 16.8), it can be seen that M_{xy} (the signal which we can measure) rapidly decays. However, it takes much longer for M_z to recover back to full equilibrium. The net magnetisation therefore has a magnitude less than $|M_0|$ for a long time; if another excitation pulse is applied during this time, only the current M_z is flipped into the xy plane to create the next signal.

16.6 Production of Spin Echoes

Spin echoes are commonly used in MRI, partly because they allow us to separate ΔB_0 effects from the intrinsic T_2, but also because they are least prone to artefacts. A spin echo is generated using a combination of a 90° excitation pulse, shortly followed by a 180° refocusing pulse. The echo is formed entirely in the transverse plane, so we can ignore M_z for

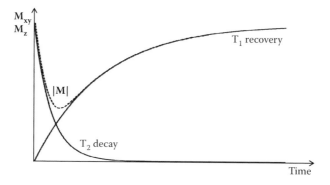

FIGURE 16.8
T_1 and T_2 relaxation shown on the same graph. Notice that the total magnetisation is not equal to $|M_0|$ at all times.

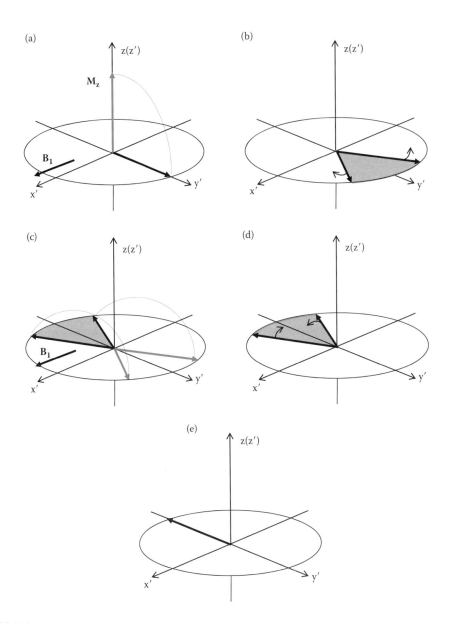

FIGURE 16.9
(a) The initial 90° pulse flips the magnetisation into the transverse plane, where (b) it begins to dephase. (c) At time TE/2 the 180° pulse ($\mathbf{B}_1$) is applied which flips all the magnetisation over. (d) The protons now begin to rephase, producing a spin echo (e) at time TE. Magnetic field gradients are superimposed on the main magnetic field, for example, in the X direction. The strength of B varies with distance (X) but not its direction.

now. Immediately after the 90° pulse $\mathbf{M}_{xy} = |\mathbf{M}_0|$, assuming the system was in equilibrium before excitation (Figure 16.9a). For a short time TE/2 $\mathbf{M}_{xy}$ decays as usual with T_2^* relaxation time producing a 'fan' of vectors in the transverse plane (Figure 16.9b). At this point the 180° pulse is applied along the y'-axis, flipping the fan over (Figure 16.9c). Protons which were ahead of $|\boldsymbol{\omega}_0|$ are suddenly behind $|\boldsymbol{\omega}_0|$, but they are in the same physical location and are experiencing the same $\Delta\mathbf{B}_0$. They continue to precess faster than $\boldsymbol{\omega}_0$, but now they appear to be catching up instead of pulling ahead. Similarly, protons which were falling behind find themselves ahead of the y-axis. Still with the same $|\omega|$ thanks to their local $|\mathbf{B_0}|$, they begin to lose their advantage and fall back towards the y'-axis. The protons are rephasing instead of dephasing (Figure 16.9d). After a further time TE/2, they will come back exactly into phase along the y'-axis, forming the spin echo (Figure 16.9e). The amplitude of the echo is governed by T_2, since the intrinsic dephasing cannot be reversed, only the macroscopic extrinsic dephasing. The echo lasts only for a fraction of a second before the protons continue to dephase as before. The shape of the signal in the receiver coil gives rise to the name 'echo'.

Insight

Runners-on-a-Track Analogy

To aid understanding of spin echo formation, there is a common analogy with runners on a track. Suppose the runners race in a clockwise direction until they reach the finishing line: they will arrive at different times according to each individual's 'local strength' and other factors. Those other factors might include something about the local environment (i.e. the lanes) in which they run. If there are differences in the quality of the running track in the various lanes we would expect to see this reflected in the arrival times. However, we cannot unmask these factors from the local strengths of the athletes.

Imagine now a different race is started in which the athletes first run clockwise but after a certain time they have to turn round and run back to the starting line. Ignoring local factors (such as the individual lanes) the runners would all arrive back at the starting line at exactly the same time. If, however, there were variations in the lanes around the track that affected each runner differently, they would arrive close together but not at exactly the same time. The lane variations are analogous to T_2 dephasing, causing the echo to have a lower intensity than the original FID height.

16.7 Magnetic Field Gradients

The MR signal described thus far is obtained from the system (or body) as a whole. To produce an image, some kind of spatial encoding is required. This is achieved using magnetic field gradients, which are superimposed on the uniform main field $\mathbf{B_0}$. Gradients are produced by passing electric currents through additional coils built in to the bore of the magnet. It is important to note that gradients modify $|\mathbf{B_0}|$ in a linear fashion, but that they do not change the direction of $\mathbf{B_0}$.

When a gradient field $\mathbf{G_r}$ is applied along the direction $\mathbf{r}$, it creates a varying field dB/dr which is zero in the centre of the magnet (Figure 16.10). When superimposed on the main field $\mathbf{B_0}$ the total field strength at point $\mathbf{r}$ is given by

$$\mathbf{B_r} = |\mathbf{B_0}| + \mathbf{r} \cdot \mathbf{G_r} \tag{16.7}$$

Protons at this position will resonate with frequency $|\omega_r| = \gamma \cdot |\mathbf{B_r}|$ instead of $|\omega_0|$. The effect of the gradient is to encode the spatial position of the protons, as the MR signal frequency.

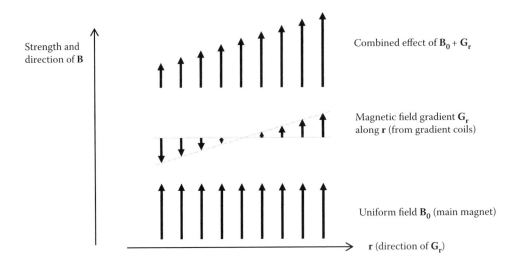

Strength and direction of **B**

Combined effect of **B**$_0$ + **G**$_r$

Magnetic field gradient **G**$_r$ along **r** (from gradient coils)

Uniform field **B**$_0$ (main magnet)

r (direction of **G**$_r$)

FIGURE 16.10
Magnetic field gradients are superimposed on the main magnetic field. The strength of **B** varies with distance (**r**) but not its direction.

Gradients along the three principal axes are switched on and off in a repeated pattern, called a pulse sequence. The protons respond to changes in the magnetic field strength immediately, changing their resonant frequencies within a few picoseconds (10^{-12} s).

Throughout this section we shall use upper case (X, Y and Z) when referring to the axes of the image volume. Z is conventionally (in cylindrical magnets) the direction of the magnetic field (superior-inferior in the body), Y is the vertical axis (anterior-posterior) and X is the remaining horizontal axis (right-left). Since all three directions cannot be encoded simultaneously, there are three slightly different ways of using gradients for spatial encoding.

16.7.1 Frequency Encoding Gradient and Fourier Transforms

Frequency encoding is the simplest spatial encoding to understand. The MR signal is acquired in the presence of a gradient in one of the three directions. The gradient is switched on just before data collection begins, and off again when the signal has been acquired. During this time, the frequency of signals is proportional to the protons' position along the frequency encoding direction. The MR signal is no longer a simple oscillation at ω_0, but a more complicated sum of peaks. In the simple setup of five partly-filled containers shown in Figure 16.11a, there are five signals each with different frequencies and amplitudes. These five signals occur simultaneously, not one after the other, and their sum is a complicated waveform. However, there is an elegant mathematical tool to separate the signals, Fourier transformation.

The basic idea of Fourier analysis is that any complex time-varying waveform can be decomposed into a series of signals with fundamental frequencies and amplitudes. Conversely, when a series of signals with different frequencies and amplitudes is combined, the result is a complex signal as a function of time. When the MR signal is measured with a gradient switched on, the Fourier transform untangles the frequencies, which represent spatial locations, and the amplitudes, which represent the number of protons at each location (Figure 16.11b).

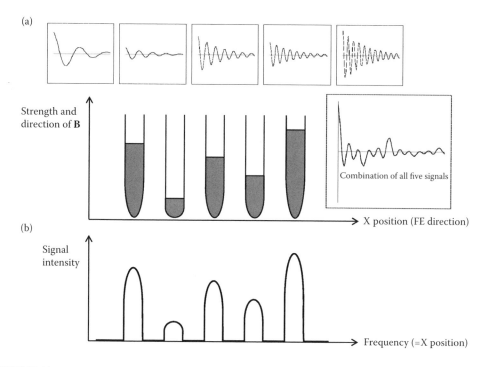

FIGURE 16.11

Signals acquired while a gradient is applied. (a) Each container has a different frequency due to its location along X; the signal amplitude depends on the total number of protons in each container. The combined signal has a complicated shape. (b) After a Fourier transform, the signals are shown as amplitude versus frequency, which is equivalent to X position.

16.7.2 Phase Encoding Gradient

Phase encoding is the most difficult technique to visualise and requires careful explanation. It is important to realise that phase and frequency are linked by time:

$$\Delta\phi = |\omega| \cdot \Delta t \qquad (16.8)$$

Begin by considering a row of protons perpendicular to the frequency encoding direction, all in phase due to the effect of the excitation pulse (Figure 16.12a). The phase encoding gradient is applied along the row of precessing protons; each proton will begin to precess at a slightly different frequency in direct proportion to the local magnetic field experienced (Figure 16.12b). The differences in precessional frequency between adjacent protons will cause relative phase shifts which are directly proportional to the differences in frequency, which in turn are directly related to the proton's position in the magnetic field gradient. By applying a gradient for a fixed time t_{PE}, the degree of phase shift may be predicted. When the gradient is switched off, all protons will return to the original precessional frequency ω_0 but will retain their relative phase shift (Figure 16.12c). This phase shift can be measured via the Fourier transform.

Unfortunately a single phase encoding (PE) gradient is not enough to uniquely localise protons along the PE direction. This is because phase is a scalar quantity, and it is impossible to distinguish between a phase angle of +30° and one of +390°, or indeed one of –120°. To overcome this, many PE gradients are applied, each one after a new excitation pulse

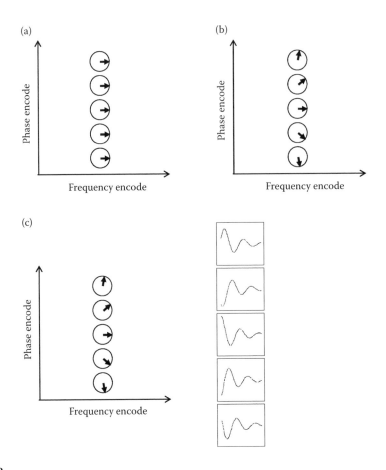

FIGURE 16.12
(a) The RF pulse brings all the protons into phase with each other. (b) The phase encode gradient along Y causes variations in precessional frequency and when it is switched off the columns have phase angles proportional to their position. (c) During the readout gradient the rows now have different frequencies as shown. The combination of phase and frequency information describes the position of each proton.

which resets the phase to zero. To get a range of phase angles, it is possible to vary t_{PE}, but that would cause differences in echo times (TEs), so in practice only G_{PE}, the strength of the gradient, is varied.

Now consider the set of phase angles measured for just one of the protons shown in Figure 16.12. Since G_{PE} has been varied slowly, the phase angles will correspond to a slowly varying signal in the PE direction (Figure 16.13). This looks just like the measurements for the frequency-encoded signal, except that the time scale is very different. Whereas the FE signal was measured in real time, the PE signal was measured in 'pseudo-time'. Mathematically the FE and PE directions are identical, except for this time scale. The Fourier transform can be used in the PE direction to separate the 'pseudo-frequencies' and their amplitudes.

16.7.3 Selective Excitation

The third dimension of the imaging volume is encoded using selective excitation. This is achieved by using an RF pulse containing a narrow band of frequencies, instead of only ω_0, and applying a magnetic field gradient in the direction perpendicular to the plane of the slice.

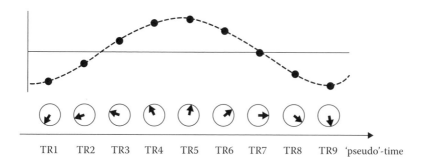

TR1 TR2 TR3 TR4 TR5 TR6 TR7 TR8 TR9 'pseudo'-time

FIGURE 16.13

The series of phase encode angles, acquired in 'pseudo-time', can be considered as a slowly varying (low-frequency) signal. One phase angle is measured in each TR period.

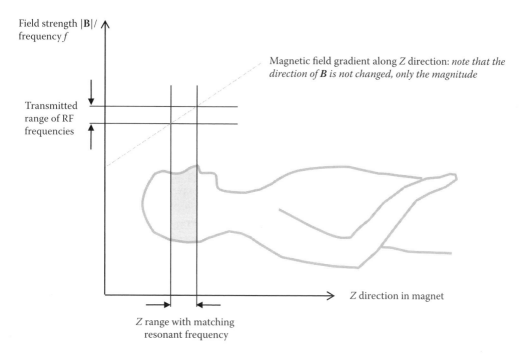

FIGURE 16.14

Selective excitation of a slice of tissue, achieved by applying a narrow bandwidth RF pulse whilst a gradient is on (in the Z direction in this example).

Only protons whose resonant frequencies match those in the RF pulse will absorb energy, that is, only those within the slice of tissue shown (Figure 16.14). The gradient is switched off after the RF pulse finishes: protons within the slice will then emit an MR signal at $|\omega_0|$, which can then be encoded for the two dimensions within the slice (frequency and phase encoding).

By changing the centre frequency of the RF pulse, a different slice location can be excited independently of the first. In fact signals from many slices can be excited and acquired in an inter-leaved fashion (multi-slice imaging).

Ideally, all protons within the slice should receive a 90° pulse, while all those outside should remain unexcited. In practice, however, this is impossible, and protons at the edges of the excited slice receive a smaller flip angle. Thus there is a slice profile which depends

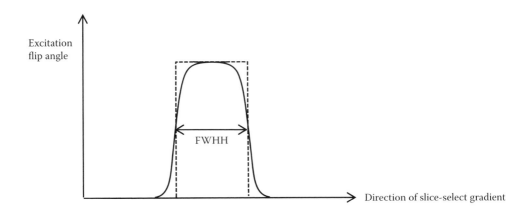

FIGURE 16.15
The ideal slice profile is rectangular (dotted lines). In practice the slice is shaped according to the characteristics of the RF pulse. Slice width is defined by the full width at half height (FWHH).

on the characteristics of the RF pulse and the slice select gradient. The slice width is defined as the full width of the slice at half the height (FWHH) of the profile (Figure 16.15).

16.7.4   Review of Image Formation

Consider a volume of tissue elements in the XYZ imaging volume. The system starts at equilibrium with $\mathbf{M_0}$ in the $+z'$ direction and $|\mathbf{M_{xy}}|$ equal to zero. When an RF pulse is applied with an appropriate slice select gradient G_{SS} in the Z direction the protons within the corresponding slice are flipped through 90° and are brought into phase (Figure 16.16a). When the RF and G_{SS} are switched off the protons will remain coherent (ignoring relaxation processes for now), precessing at the Larmor frequency ω_0.

Now a phase encoding gradient is applied in the Y direction. This causes 'columns' of vectors to precess at different frequencies according to the local magnetic field. When the phase gradient is turned off all vectors return to precessing at ω_0 but now with a phase shift across the columns which depends on their position along Y (Figure 16.16b). So far the signals have not been measured.

Finally a frequency encode gradient is used in the X direction which causes all the 'rows' to precess at distinct frequencies, and the signal is measured (Figure 16.16c). Following Fourier transformation the signal is decomposed into a series of peaks with different phases as well as different frequencies.

After some time, a new RF pulse is applied together with G_{SS}. This time a different G_{PE} is used, followed by frequency encoding. A new series of peaks with different phases is measured. The process is repeated many times, using a different G_{PE} each time. Finally a second Fourier transform is applied along the columns, resolving the phase patterns into pseudo-frequencies. The amplitudes of frequencies and pseudo-frequencies are plotted as the brightness of individual pixels, forming the MR image.

16.8   *k*-space or Fourier Space

k-space is a conceptual tool which is extremely useful in understanding the contributions of signal-to-noise ratio (SNR) and resolution to the MR image. It is based on a mathematical

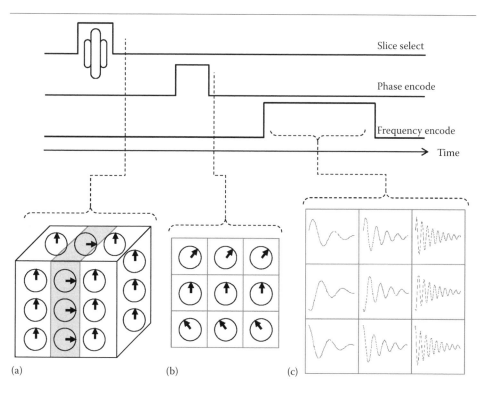

FIGURE 16.16
Only protons within the selected slice are at the right resonance frequency for the RF pulse, and see the 90° pulse. (a) Arrows indicate the direction of each packet of magnetisation. (b) The phase encode gradient produces a range of phase angles. Arrows now indicate the phase angle within each voxel. (c) With the frequency encode gradient on, the signals are acquired. The simplified pulse sequence is shown at the top of the diagram.

formulation; however, it can be understood without any maths, simply by considering the effect of gradients on proton phase. What makes k-space useful is that it separates signal from resolution. Information about SNR and contrast is encoded in the middle of k-space, while resolution (information about edges) is located in the edges. To acquire an image, it is essential to sample k-space uniformly, extending the sampling as far as necessary to achieve the desired spatial resolution. *How* we cover k-space to sample data—by rectilinear scanning, or by radial sampling, or by a continuous sweep—defines the main groups of pulse sequences in MRI.

There are three basic 'rules' for using k-space.

- An excitation pulse (90° or a smaller flip angle) 'resets' all the protons' phases to zero, and brings us to the centre of k-space.
- A refocusing pulse (180°) rotates the current k-space location through 180° about the centre of k-space.
- A gradient moves the current k-space location in the direction of the gradient; the speed of movement depends on the strength of the gradient.

Insight

Using *k*-Space

Use the *k*-space rules to see how the spin echo sequence collects an image. Refer to Figure 16.17 for the pulse sequence diagram. For this exercise we will ignore the slice select (Z) gradients, and just start with the 90° pulse, which locates the *k*-vector in the middle of *k*-space (Figure 16.17b). The first gradient is the phase encode on Y, which is applied simultaneously with a dephase gradient on the X-axis. The combined effect is to move diagonally to the upper right corner of *k*-space (Figure 16.17c). Next there is the 180° refocusing pulse, which rotates the position to the lower left corner of *k*-space (Figure 16.17d). Then the frequency encode (readout) gradient is applied on X, simultaneously with data sampling. While travelling horizontally across *k*-space, the signals are sampled and stored (Figure 16.17e). After a repetition time (TR) waiting time (to allow some relaxation, or multi-slice acquisition), another sequence is applied. This time the phase encode gradient is slightly smaller, so the diagonal travel ends up at a lower k_y position. After the 180° pulse, the second readout line is a little above the first one, so a new row of *k*-space is sampled (Figure 16.17f). This process is repeated, each time changing the phase encode gradient slightly, until we have sampled the whole of *k*-space with a set of horizontal lines. These data points can now be Fourier transformed in both directions, to produce the image.

It is possible to play some games with the *k*-space data, and see how the final image is affected. First, most of the edge information is discarded (Figure 16.18a,b). The reconstructed image has all the main features of signal and contrast, but is very blurred—low resolution. Next data at the edges of *k*-space are retained but the central region is discarded (Figure 16.18c,d). Some edges are visible, but the SNR is very low. Note that these manipulations are very similar to using low pass and high pass filters in the 'insight' on image processing (Section 6.11).

16.9 Production of Gradient Echoes

16.9.1 Dephasing Effects of Gradients

Magnetic field gradients may be considered as large magnetic field inhomogeneities, which cause more rapid dephasing of the magnetisation in the transverse plane. If this were not compensated for, the MR signal would quickly decay to zero. However, by reversing the current in the gradient coils, an inhomogeneity is created in the opposite sense, which allows the signal to be rephased. The signal will be exactly rephased when

$$\varphi_{rephase} = -\varphi_{dephase} \qquad (16.9)$$

that is, when the time integrals of the gradient pulses are equal and opposite. On the pulse sequence diagram, this is when the areas of the gradient pulses are equal. Because it is the area which is important, it is even possible to split a gradient pulse and move part of it to a different time in the sequence.

For the frequency encoding gradient, the maximum signal (the spin echo) should occur in the middle of the data capture window; it is important that the net phase effect is zero at this time. Thus, a 'frequency dephase' gradient pulse is applied before the frequency

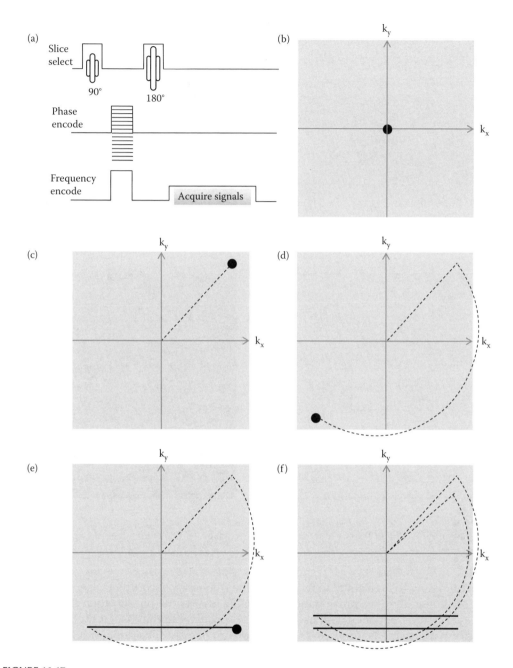

FIGURE 16.17
(a) The pulse sequence diagram. (b) The 90° pulse puts us in the middle of *k*-space. (c) The phase encode on Y, simultaneously with dephase on X, takes us diagonally to the upper right corner. (d) The 180° pulse rotates us to the lower left corner. (e) The frequency encode gradient on X, simultaneously with data sampling, takes us horizontally picking up information along the row. (f) The next phase encode gradient is slightly smaller, so we end up at a lower k_y position and sample a new row of *k*-space.

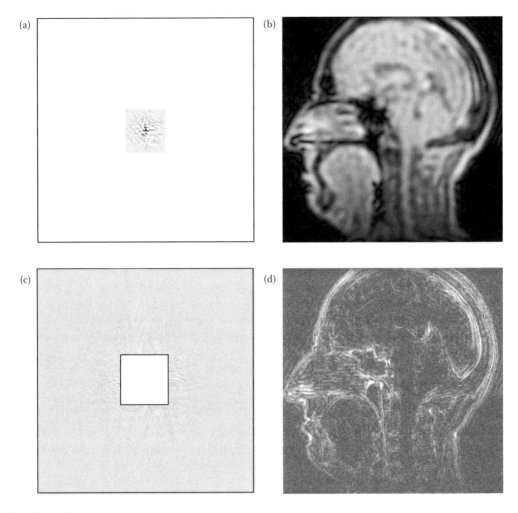

FIGURE 16.18
(a) Only the central section of k-space and (b) the resulting image. (c) The outer edges of *k*-space and (d) the resulting image.

encoding pulse, with area equal to half that of the frequency encoding pulse. The dephase lobe comes before the 180° pulse which effectively inverts the phases: therefore in a spin echo sequence the frequency dephase gradient is positive. In the case of the phase encoding gradient, rephasing is unnecessary (the phase angles give positional information).

The situation for the slice selection gradient is slightly more complex, because the RF pulse has a changing shape. By convention, however, the shaped RF pulse is considered to act as a sharp spike at the middle of the pulse; that is, during the first half of the pulse there is no excitation, then all protons in the slice receive a 90° pulse instantaneously. Whilst there are no excited protons, the gradient can have no dephasing effect, so the period up to the middle of the RF pulse is ignored. The remaining area of the slice selection gradient will cause dephasing of the newly excited protons, which may be remedied by applying a negative gradient with approximately equal area (the 'slice rephase' gradient). It is necessary slightly to adjust the size of the rephase gradient since the RF pulse does not act instantaneously.

16.9.2 Production of Gradient Echoes

From the previous section, it should be clear that it is possible to create a signal echo without using a 180° pulse, but simply using the gradients. Such an echo is known as a 'gradient echo' or sometimes a 'field echo'.

A gradient echo (GRE) sequence uses slice selection and phase encoding gradients as usual. However, since there is no 180° RF pulse, a negative frequency dephase gradient is used, instead of the positive dephase gradient in the spin echo sequence (Figure 16.19). When the frequency encoding gradient is switched on to start data capture, the effect of the original dephasing is gradually reversed, until at the centre of the data capture window it is zero, and the signal is a maximum. The continuing gradient then dephases the signal in the opposite direction decreasing the signal again; thus an echo has been formed. In k-space, the trajectories are simply a diagonal traverse to the left side, followed by the horizontal readout during data acquisition.

Due to the absence of the 180° pulse, GRE signals are described by T_2^*, not T_2. In practice, T_2^*-weighted images have similar contrast to T_2-weighted images, and are often described as 'gradient-echo T_2-weighted'. Magnet design has greatly improved since the early days of MRI so GRE images have very high image quality, comparable to spin echo in many cases, except where there are large inhomogeneities introduced by the human body.

Gradient echo sequences can be very fast, and they have an additional flexibility with the excitation RF pulse. It is often less than 90° and has an impact on both SNR and the image contrast, as we will see in the next section.

16.10 Image Contrast

In addition to the fundamental contrast differences between gradient and spin echo images, we can control the image appearance via sequence parameters such as TR, TE and flip angle $\alpha°$. In addition we can use extra RF and gradient pulses before the imaging sequence to 'prepare' tissues.

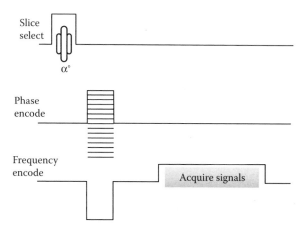

FIGURE 16.19
A gradient echo pulse sequence. The initial excitation pulse is usually less than 90°; there is no 180° refocusing pulse. The initial frequency dephase gradient is negative instead of positive.

Repetition time is defined as the time to repeat the sequence. Repetition is necessary for a number of reasons: to acquire signals with different phase encoding gradients, or to increase SNR by averaging signals together. In a GRE sequence TR is the time between two $\alpha°$ pulses; in a spin echo sequence it is the time between two $90°$ pulses.

Echo time is defined as the time from the excitation pulse to the centre of the echo (whether gradient or spin echo). For spin echo sequences, the $180°$ pulse must be exactly halfway between the $90°$ excitation and the desired TE.

16.10.1 Spin Echo Image Contrast

The spin echo pulse sequence only gives us two parameters with which to control contrast: TR and TE. Provided $TR > 3 \times T_2$, the influence of TR is purely related to $\mathbf{M_z}$ and therefore to T_1 relaxation. At short TRs, there is not enough time for full T_1 relaxation, so that $\mathbf{M_z}$ is reduced when the next excitation pulse is applied. At these short TRs, there is more contrast—signal difference—between tissues, than at long TRs. However, the signal is also reduced. We call this effect 'saturation'. At long TRs there is time for complete relaxation of T_1.

If TE is short, there is little time for T_2 relaxation and $\mathbf{M_{xy}}$ will be close to its starting value. At longer TEs, T_2 decay reduces the height of the echo, which means reduced SNR. However, there is also most contrast between different tissues, with fluids (long T_2) staying bright.

There are four possible combinations of TR and TE, but only three are useful (Figure 16.20).

- Long TR, short TE: contrast is dominated by the proton density (water content) of the tissues
- Long TR, long TE: contrast depends on T_2 (spin-spin relaxation time)
- Short TR, short TE: contrast is dictated by T_1 (spin-lattice relaxation time)

The fourth combination, short TR with long TE, has very low SNR and contrast which depends both on T_1 and T_2, which is of little value.

Since both proton density (PD) and T_2-weighted images have long TRs, it is possible to use repeated $180°$ pulses to produce more than one spin echo. With two spin echoes, one at a short TE and the other with long TE, both PD and T_2 images are produced within the same TR (and therefore the same scan time). Such 'dual echo' sequences are less common today than a decade ago; with modern speed-up techniques such as 'parallel imaging' and 'turbo spin echo', scan times are much reduced, and it is possible to create separate T_2 and PD scans in much less time than the old dual echo SE scans. (Explanation of these faster sequences is beyond the scope of this chapter.)

16.10.2 Gradient Echo Image Contrast

Gradient echo sequences offer an additional parameter, the flip angle $\alpha°$, to control image contrast. TE in GRE sequences is usually shorter than spin echo sequences, because T_2^* is usually shorter than T_2. Compared with spin echo images, TR is always 'short'. However, it's still possible to avoid saturation by reducing the flip angle.

When the flip angle $\alpha°$ is used, $\mathbf{M_{xy}}$ becomes $\mathbf{M_z^-}\sin\alpha$, where $\mathbf{M_z^-}$ is the longitudinal magnetisation immediately before the excitation pulse is applied. $\mathbf{M_{xy}}$ decays as usual with T_2 relaxation, but since it starts at a lower value, it effectively reaches zero quicker. The longitudinal component $\mathbf{M_z}$ is left with $\mathbf{M_z^-}\cos\alpha$ (Figure 16.21). $\mathbf{M_z}$ recovers back towards $|\mathbf{M_0}|$, controlled by T_1 as usual, and as successive RF pulses are applied, equilibrium is reached.

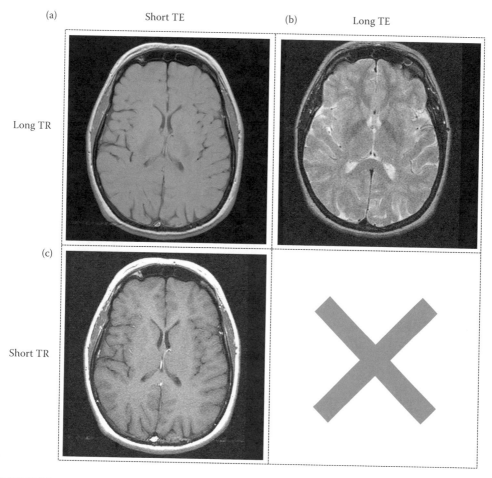

FIGURE 16.20
Different combinations of TR and TE give proton density weighting (a), T_2 weighting (b), or T_1 weighting (c). The combination of short TR and long TE is not useful.

So a small flip angle can be used to avoid saturation, even when TR is very short. We therefore can ignore TR as a contrast control mechanism. Flip angle becomes the parameter which controls $\mathbf{M}_z$, while TE still controls $\mathbf{M}_{xy}$.

Once again there are four possible combinations of flip angle and TE, but only three are useful (Figure 16.22).

- Small α°, short TE: contrast is dominated by the proton density (water content) of the tissues
- Small α°, long TE: contrast depends on T_2 (spin-spin relaxation time, or to be more correct T_2^* the effective spin-spin relaxation time)
- Large α°, short TE: contrast is dictated by T_1 (spin-lattice relaxation time)

The fourth combination, large α° with long TE, has very low SNR and contrast which depends both on T_1 and T_2, which is of no value, just like SE images with short TR and long TE. Since GRE images can be acquired in very short times, dual echo GRE (to produce PD and T_2^* images within the same scan) is never used. However, a multi-echo GRE is often

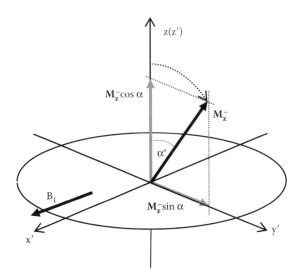

FIGURE 16.21
Gradient echo sequence use flip angles less than 90°. This leaves a significant $\mathbf{M_z}$, which makes full T_1 relaxation much quicker. The penalty is a smaller $\mathbf{M_{xy}}$, giving less SNR.

used with three or five echoes, with all the images summed together at the end of the scan, to provide an image with high SNR and good T_2^* contrast.

16.10.3 T_1w, T_2w and PDw Images

PDw images have the highest SNR, since they make best use of the available magnetisation. There is relatively little variation in PD between different tissues (typically ranging from 75% to 90%). There are only a few clinical applications where PDw images are useful; the most notable are fat-suppressed PDw images for musculoskeletal imaging.

T_1w images have relatively low SNR; fluids (such as cerebrospinal fluid [CSF] and synovial fluid) appear dark. Due to the wide range of T_1s in tissues, there is excellent contrast and T_1w images are usually known as 'anatomy' scans. They are clinically valuable in most applications, and particularly when using gadolinium (Gd) contrast agent.

T_2w or T_2^*w images may also have lower SNR. Due to their long T_2, fluids, appear bright, which makes the contrast particularly good for meniscal tears, oedematous collections, or regions with high vascularity. They are often called the 'pathology' scans and are used in all clinical applications.

It is important to note that image contrast is nowadays defined by these appearances, and not necessarily by the values of TR/TE and so on. What matters for clinical diagnosis are the relative signals from pathological and normal tissues, not the precise sequence parameters used to create them.

16.10.4 Inversion Recovery Sequences (STIR and FLAIR)

Short TI inversion recovery (STIR) and fluid attenuated inversion recovery (FLAIR) are widely used. Both are based on a spin echo sequence, but with an additional 180° pre-pulse. Acting on equilibrium magnetisation $\mathbf{M_0}$ this pulse inverts the magnetisation ($\mathbf{M_z} = -|\mathbf{M_0}|$). After a delay time called TI (inversion time), the 90° pulse of the spin echo sequence is applied. During TI, the tissues partially recover their longitudinal magnetisation with

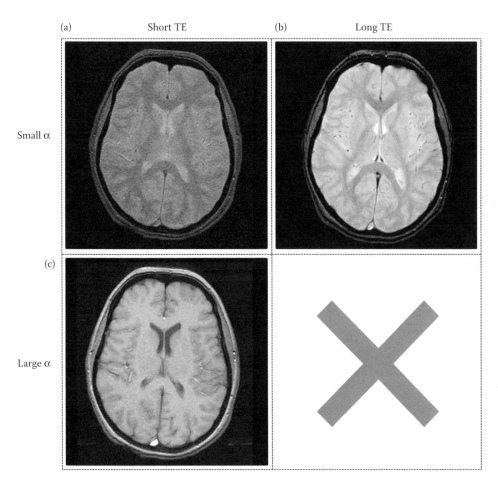

FIGURE 16.22
Different combinations of $\alpha°$ and TE give proton density weighting (a), T_2 (T_2^*) weighting (b), or T_1 weighting (c). The combination of long TE and large $\alpha°$ is not useful. TR is always short.

T_1 relaxation. If the inversion delay is such that one of the tissues is at the null point (i.e. $\mathbf{M_z} = 0$), its signal after the 90° pulse will be zero—the tissue will be suppressed in the final image. The appropriate TI is approximately $0.7 \cdot T_1$ of the tissue to be nulled, bearing in mind that T_1 depends on field strength (Figure 16.23).

With STIR, the aim is to create a fat-suppressed T_2w image. Fat has a short T_1 and so a short TI is needed (120 ms at 1.5 T, 150 ms at 3.0 T). At these short TIs, most other tissues still have a negative $\mathbf{M_z}$ which means they create negative spin echoes. When the image is formed however, we ignore the sign of the echo and only use its magnitude. Fluids, with the longest T_1s, have the most negative signal and therefore have the highest signal on the final images, giving the required T_2w contrast. A short TE is needed, to maximise the signal from fluids.

FLAIR is used exclusively in brain and spine imaging, to null CSF signals in a T_2w image. This is helpful to distinguish periventricular lesions from the high signals in the ventricles. Since CSF has a very long T_1, a long TI is needed (1700–2500 ms). There is a wide range of TIs because the signal is changing only slowly; a few tens of ms difference can still give adequate CSF suppression. At such long TIs, all other tissues already have positive $\mathbf{M_z}$ and in fact may be close to $|\mathbf{M_0}|$. A long TE is used to create T_2w contrast.

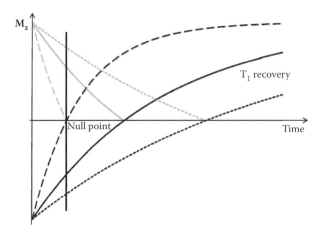

FIGURE 16.23

T_1 recovery after an initial $180°$ inversion pulse. If the excitation pulse is applied at the null point for a certain tissue, there will be no signal from that tissue in the resulting image. If other tissues have negative $\mathbf{M_z}$ at that null point, the imaging process takes the magnitude of all signals, so the longest T_1 tissues will appear brightest in the image.

16.11 Contrast Agents

Although MRI has a large range of intrinsic contrast, certain pathologies are not easily visible (perhaps it would be more correct to admit that we do not have time to find the optimum imaging sequence in these cases). There are two groups of contrast agents, by far the most common being Gd administered intravenously to enhance a range of tumours. Since Gd is extremely toxic it is chemically 'isolated' within a large molecule such as diethylene triamine penta-acetic acid (DTPA). Gd contrast agents do not pass through the healthy blood-brain barrier, but leak out of the blood supply through a disrupted barrier. Gd also accumulates in other pathological tissues such as breast and liver tumours and myocardial scars. Gd is strongly paramagnetic and changes the $|\mathbf{B_0}|$ on a microscopic scale. It reduces the T_1 relaxation times of tissues where it collects, so on imaging a T_1w scan after injection of Gd, pathological tissues have an enhanced signal. Manganese agents may also be used in this way.

Iron-oxide contrast agents, known as super-paramagnetic iron oxides (SPIOs) are also available. These work by causing local field inhomogeneities in tissues where they accumulate. The T_2 and T_2^* relaxation times are reduced in these regions, so that tissues which take up the contrast agent lose signal on T_2 or T_2^* weighted images. Administration may be as an oral suspension for abdominal imaging, or intravenous for liver and spleen imaging.

16.12 Artefacts and Avoiding Them

Refer to Figure 16.24 for samples of the most common artefacts in MRI. The most important artefact group is the effect of motion. This may be gross patient motion (which may be involuntary, e.g. in Parkinson's or similar disorders), or physiological motion—respiration, the cardiac cycle or peristalsis. All motion artefacts appear as 'ghost' images of the moving

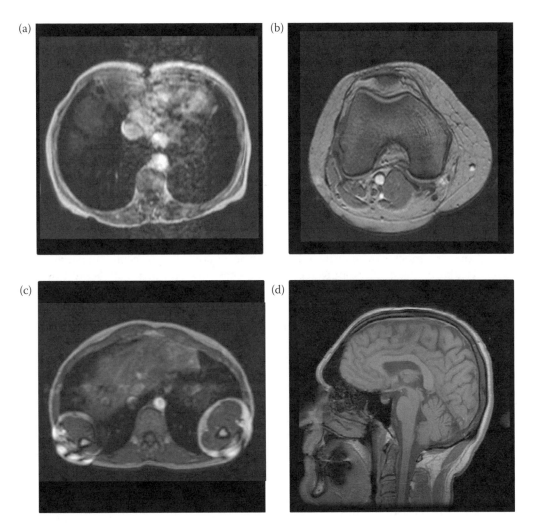

FIGURE 16.24
Examples of common MRI artefacts. (a) Motion during the scan produces 'ghosts' (cardiac in this case). (b) Inflowing arterial blood generates high signal in gradient echo images. (c) Phase wrap occurs when the field of view is too small in the phase encode direction, here the arms have wrapped around. (d) Metal implants cause signal dropout on most images, shown here in the common example of dental fillings.

structure, displaced in the phase encode direction (even if the actual motion is through the slice, e.g. blood flowing in a vessel).

Since they have (relatively) regular cycles, it is possible to compensate for respiratory and cardiac motion. In all cases, it is necessary to have some method of measuring the motion, and then to either adjust the phase encode gradient sequence, or temporarily suspend image acquisition.

16.12.1 Cardiac Gating

To 'freeze' the motion of the heart, the pulse sequence is triggered using the echocardiogram (ECG) waveform instead of a regular TR. Electrodes are attached to the patient in the usual way, and the waveform is electronically filtered and processed to detect the R wave. As each R peak is detected the imaging sequence is started, so that the effective TR is the RR interval which of course depends on the heart rate of the patient. Each line of data for a particular image is acquired at the same time relative to the cardiac cycle and thus motion artefacts do not appear.

Echocardiogram gating is essential when imaging the thorax for cardiac or pulmonary disease. Gated imaging can also be useful in situations where a major artery is in the field of view, or to remove artefacts due to CSF pulsation in spinal imaging. In the latter cases it is more convenient to use a photoplethysmographic (PPU) detector on the finger, which detects the arrival of arterial blood in the digit and thus provides an R-wave trigger. Since the 'effective TR' is determined by the RR interval the contrast and SNR in an image will to some extent be dependent on pulse rate, and images are usually T_1 weighted.

16.12.2 Respiratory Gating

Respiration has a much slower cycle but tends to be less regular, especially in patients with cardiovascular problems. The movement of the chest or abdomen is measured by strapping an expandable bellows around the patient. The scanner is able to detect the expansion and collapse of the bellows by measuring the pressure in the bellows. A 'peak' point in the cycle can be used to trigger the sequence, in much the same way as cardiac gating. However, the effective TR will be several seconds because of the slow respiratory cycle, and this produces an extremely long scan time.

An alternative method of respiratory compensation is phase re-ordering. This relies on the fact that the largest phase encode gradients (the edge of *k*-space) are least sensitive to motion. Respiration is measured for two or three cycles to determine the breathing pattern. Instead of varying the phase encode gradient from positive to negative in a regular way, the scanner now re-orders the sequence so that the largest gradients occur during the periods of most respiratory motion. In *k*-space, the rows are acquired in a respiratory-ordered fashion, instead of the usual linear order. This is much more time efficient, but its success does rely on the patient's breathing remaining regular.

Navigator gating is also available as part of specialist cardiac software packages. A navigator is simply a narrow column of tissue located over the right hemi-diaphragm, encoded only along the superior-inferior direction. Acquisition of this signal is very fast, and because of the high contrast between the lung and liver, it is possible to detect the signal edge. The navigator is measured during every TR and the lung-liver interface is measured at each point. An acceptance window is defined by the user; data which is acquired within the acceptance window goes into *k*-space, whereas data outside of the window is rejected and those phase encode steps are re-acquired.

16.12.3 MR Angiography

Moving blood within vessels causes distinctive artefacts, always in the phase encoding direction. On spin echo images, the lumen of through-plane blood vessels appears black, because the blood receives either the 90° or the 180° pulse but not both, and therefore cannot produce an echo (Figure 16.25a). However, in-plane or slow-flowing blood will have an intermediate signal, and may then cause ghost artefacts. GRE sequences produce the opposite appearance ('bright blood') because the echo is rephased by the gradients not an RF pulse (Figure 16.25b). These bright signals often produce many ghosts, especially where the flow is very pulsatile. Any turbulence or reduction in flow rate changes the appearance of moving fluid in the image, usually by reducing intensity.

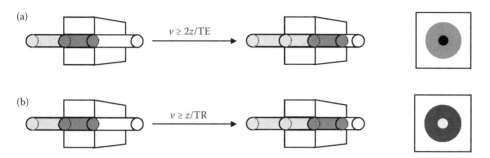

FIGURE 16.25
Appearance of flowing blood. v = blood velocity, z = slice thickness. (a) In spin echo images, when the velocity is higher than $2z/\text{TE}$, the flowing blood does not experience both RF pulses and does not generate an echo. Vessels are black. (b) In gradient echo images, when the velocity is higher than z/TR, inflowing vessels are bright because they contain 'fresh' magnetisation.

To remove the flow artefacts, it is necessary to use balanced gradients. Known as 'gradient moment nulling' or simply 'flow comp', additional gradient lobes are added to the pulse sequence, to ensure that the phase of moving protons as well as static tissue is properly rephased at the TE. These extra gradients tend to increase the minimum TE and TR of the sequence.

Insight

Turning Flow Artefacts into Useful Images: MRA Angiography

These flowing artefacts can be exploited to produce MR angiograms, which show only the blood vessels. Time-of-flight (TOF) angiography uses a flow-compensated GRE sequence with a very short TR. The signal from static tissues within the slice is almost zero, due to T_1 saturation. Blood flowing perpendicular to the slice, however, has 'fresh' magnetisation for each excitation and produces a high signal. The flow compensation prevents ghosting from the blood vessels. The high contrast between blood vessels and static tissue is usually viewed using a 'maximum intensity projection' (MIP). This post-processing technique generates a 2-D projection view from a volume of data. The volume may be acquired as a true 3-D acquisition, or as a stack of thin 2-D slices. Since it is basically a T_1w image, short-T_1 tissues can leave residual signals in the MIPs which obscure the vessels, so it is common practice to 'cut away' the unwanted regions.

Phase-contrast MRA is a rather different sequence, which uses the flow-compensation gradients to encode the velocity of the moving protons in the phase angle of the signal. This is separate from phase encoding for spatial position; repeated acquisitions are used with the flow-comp gradients reversed to separate these phase signals. Usually in PCA there is a magnitude and up to 3 phase images, the latter containing velocity information in each direction. These images can be combined into a single image for display purposes, and do not require MIPs. The advantages of PCA are that static protons have zero phase, and thus produce no signal on the final image, and that flow may be quantified. The gradient strengths may be adjusted to make the sequence sensitive to slow or fast flows, allowing arterial blood to be easily distinguished from venous return. The disadvantages are that to visualise flow in all directions, gradient moment nulling must be applied to all three gradients, which extends the imaging time.

16.12.4 Digital Imaging Artefacts

This group of artefacts arises from the imperfection of data sampling and Fourier transforms. Phase wrap, or fold-over artefact, occurs when the prescribed field of view (FOV) in the phase encode (PE) direction does not include all the tissues. A simple explanation uses the fact that phase encoding expects one side of the FOV to be at $+180°$ and the other side at $-180°$. However, tissues which are just outside the FOV will still be phase-encoded, for example, at $+200°$. Since it is impossible to distinguish between $+200°$ and $-160°$, those tissues will be wrongly shown in the final image on the opposite side. Phase wrap can be avoided by careful planning of the FOV, for example, adding saturation slabs to null tissue, or by increasing the FOV in the PE direction.

Ringing artefacts arise when the resolution in the PE direction is insufficient to resolve a high contrast boundary. A typical example is in T_2w spine with PE in the anterior-posterior direction. If the resolution is not high enough, the bright-CSF/dark-cord boundary gives an artefact in the middle of the cord which can mimic a syringomyelia. Since the number of PE steps has a direct impact on the scan time, it is often frequently reduced to speed-up examinations, but this should be done with care.

Due to small differences in the chemical structure of fat and water, they have slightly different resonant frequencies, separated by 220 Hz at 1.5 T and by 440 Hz at 3.0 T. This difference can lead to fat signals being encoded into the 'wrong' voxel by the frequency encoding process. At boundaries where the fat signal overlaps the water signal, a bright edge is seen, while the opposite edge shows a dark edge. This chemical shift artefact is only found in the frequency encode direction.

16.12.5 Dephasing Artefacts

This group of artefacts arises from the inhomogeneities in one of the magnetic fields—$\mathbf{B}_0$, $\mathbf{B}_1$ or $d\mathbf{B}/dt$. They are often due to natural properties of tissues, or to metal implants within the patient. Ferrous metals obviously create large $\Delta\mathbf{B}_0$ areas, which has an impact on almost every stage of image production. Typically black holes are seen around such implants, disproportionately large compared to the size and mass of the object itself. Non-ferrous metals also create problems for $\mathbf{B}_1$, and also lead to black holes. Often bright white edges are seen around these artefacts, along with geometric distortions. Natural inhomogeneities associated with blood breakdown products cause the same appearance.

Gradient echo images are most prone to these artefacts, so spin echo is preferred for good image quality in these cases. A new class of images known as susceptibility weighted imaging uses T_2^*w images combined with information from the signal phase to enhance the appearance of susceptibility artefacts. These images are very sensitive in detecting cerebral micro-bleeds, for example.

16.13 Technical Considerations

Magnetic resonance systems comprise a large number of complex components, the specifications of which change very rapidly. However, there are three main hardware components. First of these is a large-bore, high field magnet with good $\mathbf{B}_0$ homogeneity. There are three types of magnet, with advantages and disadvantages as shown in Table 16.2, but by far the most common are superconducting magnets. These rely on a bath of liquid helium

TABLE 16.2

Comparison of Various Magnet Types

Magnet Type	Properties	Advantages	Disadvantages
Permanent	Made of iron alloy blocks (permanently magnetised)	Cheap; zero running costs Open design	Low field (<0.6 T) Weight > 40 tonnes
Electromagnetic	Water-cooled copper coils	Field can be turned off in emergency Open design Mid cost	High power requirements Poor stability
Superconducting	Liquid helium cooled wires	High field Good homogeneity	High capital cost Cost of helium Relatively enclosed bore

at 4 K (–269°C) to cool the current-carrying wires so that they have almost zero resistance. Once the current is set to provide the correct magnetic field, no electrical power is necessary and the major running cost is that of topping up the liquid helium. Very good homogeneity is achievable, often to less than 0.2 ppm over the volume of the head coil.

The second major component is a gradient set containing three sets of coils, one for each orthogonal direction, which produce linear magnetic field gradients when current is passed through them. Gradient sets are characterised by a maximum gradient strength, expressed in mT m^{-1}, and a slew rate, T m^{-1} s^{-1}. The maximum gradient has implications for resolution, while both strength and speed are important for fast imaging. Modern gradient systems are often cooled with water or air, and the amplifiers which provide the high voltage and currents needed may also be water-cooled.

Finally there is the RF sub-system. All modern MR systems have a built-in 'body coil' which is used to transmit the RF pulses with homogeneous $\mathbf{B_1}$ over a large volume. It may also be used to receive the signal (it is a 'transmit/receive coil'), but it is more common to use a variety of RF coils, each designed to optimise SNR from a particular part of the body. These are known as 'receive-only' or 'surface' coils. Since the surface coil is close to the area of interest, it is much more sensitive to the signals of interest.

Radiofrequency coils are also described as quadrature, linear or phased array. The simplest 'linear' coils are single loops of copper, rarely circular but shaped appropriately for a particular part of the body. Quadrature coils have two loops, usually overlapping, and are able to improve the SNR by reducing the amount of noise detected. Phased array coils are the most common surface coils, using several coil elements to pick up signal. Each element forms an image, and during reconstruction they are combined together to form a single image. Such coils are able to create images over a large FOV with high SNR. Phased array coils are particularly important due to a relatively new technique called parallel imaging, which uses information about the spatial sensitivity of coil elements to recreate missing phase encoding information. This allows scan times to be reduced, by skipping PE lines.

The whole system is controlled by a central computer, with dedicated processors to do complex tasks such as reconstruction or pulse sequence generation. The software through which the user controls the system is nowadays very user friendly, using graphical interfaces and automated processing wherever possible to reduce examination times and avoid artefacts. The final item to mention in an MRI installation is the RF-screened room (or Faraday cage). Sheets of copper or aluminium built into the walls, floor and ceiling of the main magnet room form a totally closed metal box around the system. This prevents the tiny MRI signal from being swamped by the ambient RF noise.

16.14 MRI Safety

MRI is often described as a safe imaging technique, and it is certainly much less hazard-ous than methods involving ionising radiation. There are three areas to be considered for MRI safety: the large main magnetic field, the gradient fields which are switched at low frequencies, and the RF radiation. In addition, guidelines differentiate between patients (or volunteers), health-care workers, and the general public.

Most countries have some kind of guidelines for the safe use of diagnostic MRI, notably the Food and Drug Administration (FDA, 2003) in the United States and the Medicines and Healthcare Regulatory Authority (MHRA, 2007) in the United Kingdom. Although there are subtle differences, they all follow guidelines compiled by the International Commission for Non-Ionising Radiation Protection (ICNIRP, 2004) and bodies such as National Radiological Protection Board (NRPB, 1991) and the International Electrotechnical Commission (IEC, 1995). Every MR Unit should have local safety rules which are reviewed regularly and all workers in the MRI environment should be familiar with these.

16.14.1 Main Magnetic Field

The field strength of the main magnetic field ($|\mathbf{B}_0|$) ranges from 1.0 T to 3 T in most clini-cal systems. It is generally accepted that fields of up to 2 T produce no harmful bioeffects, including no chromosomal effects. In higher fields, effects are related to the induction of currents in the body (which is a conductor) when it moves through the field. These may cause visual sensations (magneto-phosphenes), nausea, vertigo and a metallic taste. However, these effects are short term, usually disappearing when the body is no longer in the magnetic field.

16.14.2 Projectile Effect of B_0

Although the main field is not intrinsically hazardous, its powerful effect on ferromag-netic objects constitutes a major problem. There is necessarily a 'fringe field', a spatial area of gradually changing magnetic field strength from the centre of the magnet. This is characterised by a set of iso-contours at various field strengths, but the most impor-tant one is the 0.5 mT iso-contour (also known as the 5 gauss line, the gauss being an alternative unit of $|\mathbf{B}_0|$). Magnetic bodies within the fringe field are attracted towards the centre of the magnet, and if they are free to move, they can acquire high velocities and cause significant damage to equipment and persons in their path. It is important to be aware that the fringe field strength may change very rapidly over small distances; the force on an object may become uncontrollable if it is brought just 50 cm too close to the magnet.

With modern superconducting magnets, the projectile hazard is present all the time, even when it is not scanning patients. The MRI unit typically has a series of physical bar-riers to prevent access especially when it is unoccupied, and the instructions of the safety officer should be heeded at all times.

It is essential that all potential projectiles are kept outside the magnetic field. The list of such objects is extensive, and includes keys, scissors, paper clips, stethoscopes, gas cylin-ders, drip stands, wheelchairs and so on. Apart from being a potential hazard and causing personal injury or damage to the equipment, small metal objects within the bore of the magnet will also cause image artefacts. Medically or accidentally implanted objects in the patient's body may also be ferromagnetic. The degree of hazard depends both on the type of implant and its location. Of particular concern are intra-cranial aneurysm clips which can cause fatal haemorrhage if they move.

Other items may not become missiles but will be damaged by the magnetic field and should be removed. Examples include analogue watches and cards with magnetic strips. Various magnetically activated devices also fall into this category such as cochlear implants and cardiac pacemakers. Persons with these implanted devices may not enter the 0.5 mT fringe field and therefore cannot be scanned.

16.14.3 Gradient Fields

When gradient fields are switched on and off, they may induce currents in the body. The size of the current will depend, among other factors, on the maximum gradient field strength and on the switching time; it is important to remember that the gradient fields increase away from the centre of the magnet, so the area for concern is not necessarily the part being scanned. Induced currents may be large enough to stimulate nerves, muscle fibres or cardiac tissue; effects may include magneto-phosphenes, muscle twitches, tingling or pain, or in the worst case ventricular fibrillation. Theoretical calculations and experimental evidence indicate that such effects will be avoided if the gradient switching is kept below 20 T s^{-1}. Most existing clinical scanners are well below this limit, but as scans become faster and technology improves the capabilities of the gradient sets, it is important that this hazard is carefully monitored. Note that only patients (or volunteers) undergoing a scan are exposed to gradient fields.

16.14.4 RF Fields

Radiofrequency waves contain both electric and magnetic fields oscillating at MHz frequencies. At these rates, the induction of circulating current in the body is minimal, as there are high resistive losses. Most of the RF power is therefore converted to heat in the body, and bioeffects and safety limits are considered accordingly. In healthy tissues, a local temperature rise caused by RF power deposition will trigger the thermoregulatory mechanism of increased perfusion to dissipate the heat around the rest of the body. If the rate of power deposition is very high, or the thermoregulation system is impaired in some way, heat will accumulate locally, eventually causing tissue damage. Some areas of the body are particularly heat sensitive, for example the eyes, the testes and the foetus, and extra care should be taken when scanning such patients. The safety limits are designed to limit the temperature rise of the body to 1°C. For the whole body in a 30-minute examination, the specific energy absorption rate (SAR) limit is 1 W kg^{-1}, while for a head scan of similar duration it is 2 W kg^{-1}.

Particular care must be taken when metal objects are in the imaging field, for example, ECG leads or non-removable metallic implants. Such objects absorb RF energy very efficiently and may become hot enough to burn the skin if in direct contact. It is worth noting that RF burns to patients form the majority of MRI-related accidents reported to either the FDA or MDA.

16.14.5 MRI in Pregnancy

There is some evidence that exposure to switched magnetic fields during early organogenesis may cause abnormal development. The effects of excess heat on the foetus are well established and evidence that exposure to RF radiation may cause heat-induced damage is available. Currently the evidence concerning the safety of MRI during pregnancy is not conclusive; indeed there are many centres which provide foetal scanning for abnormalities, usually following a suspect ultrasound scan.

Since the safety of MRI cannot be proved, the guidelines recommend that scans are not performed in the first trimester. However, if other non-ionising imaging methods are inadequate or the alternative diagnostic method involves ionising radiation, and if the information is regarded as clinically important, a scan may take place at any stage of pregnancy.

16.15 Conclusions and Future Developments

It is impossible for a single chapter in a book such as this to explain all the difficult concepts involved in MRI. However, there are many other textbooks (see *Further Reading* for three affordable choices), and each offers a slightly different view of the topic. Readers are encouraged to explore as many books as possible to improve their understanding.

MRI is a valuable addition to the range of radiological methods, and has proved superior to other techniques in many applications. The complexity of the underlying science is balanced by the flexibility of the information which may be deduced, from high resolution anatomy to flow, diffusion and other physiological processes. The pace of development in MRI has not slowed over the last 20 years, with major research areas expanding all the time. The future of MRI in clinical radiology can only expand as technology improves and new techniques gain acceptance in the medical community.

References

FDA: Food and Drug Administration (2003) Criteria for Significant Risk Investigations. http://www.fda.gov/MedicalDevices/DeviceRegulationandGuidance/GuidanceDocuments/ucm072686.htm [accessed 5 July 2009].

MHRA: Medicines and Healthcare Devices Regulatory Agency (2007) Safety Guidelines for Magnetic Resonance Imaging Equipment in Clinical Use. http://www.mhra.gov.uk/home/idcplg?IdcService=GET_FILE&dDocName=CON2033065&RevisionSelectionMethod=LatestReleased [accessed 5 July 2009].

ICNIRP: International Commission on Non-Ionizing Radiation Protection (2004) Medical Magnetic Resonance (MR) Procedures: Protection of Patients. Health Physics 87:197–216. http://www.icnirp.de/documents/MR2004.pdf [accessed 5 July 2009].

NRPB: National Radiological Protection Board (1991) Principles for the Protection of Patients and Volunteers during Clinical Magnetic Resonance Procedures. Documents of the NRPB Vol 2 No 1 (ISBN 0859513394). http://www.hpa.org.uk/webw/HPAweb&Page&HPAwebAutoListNameDesc/Page/1219908766891?p=1219908766891 [accessed 5 July 2009].

IEC: International Electrotechnical Commission (1995) Medical Electrical Equipment – Part 2: Particular Requirements for the Safety of Magnetic Resonance Equipment for Medical Diagnosis Edition 2.2. IEC 60601-2-33. http://www.iec.ch/webstore/ [accessed 5 July 2009].

Further Reading

Elster AD & Burdette JH, *Questions and Answers in Magnetic Resonance Imaging*, 2nd ed, Mosby, 2001.
Hashemi RH, Bradley WG & Lisanti CJ, *MRI the Basics*, 2nd ed, Lippincott Williams & Wilkins, 2003.
McRobbie DM, Moore EA, Graves MJ & Prince MR, *MRI from Picture to Proton*, 2nd ed, Cambridge University Press, 2007.

Exercises

1. What is meant by the term 'resonance' in relation to MRI?
2. Describe what is meant by the Larmor frequency and how it is related to the applied magnetic field?
3. What are the Larmor frequencies for water nuclei in magnetic fields of 1.0 T, 1.5 T and 3.0 T?
4. The Tesla (T) is a unit used to measure the magnetic field strength in MRI. What is the strength of the earth's magnetic field?
5. What is meant by 'net magnetisation'?
6. Classically, the spin population states are divided into 'parallel' and 'anti-parallel': what is the approximate population difference between these two states at 1.5 T and body temperature?
7. What determines the frequency of the rotating frame of reference?
8. What is meant by flip angle (α)?
9. Explain what is meant by FID of the MR signal?
10. Explain what is meant by spin-lattice relaxation and how is this characterised mathematically?
11. Explain what is meant by spin-spin relaxation and how is this characterised mathematically?
12. Explain the spin echo sequence ignoring the imaging process. How might the timing parameters be adjusted to reflect T_1, T_2 and proton density in the image?
13. What is the difference between a spin echo and a gradient echo sequence and how are these differences useful?
14. Explain how magnetic field gradients are used in the imaging process.
15. What is meant by frequency and phase encoding?
16. Describe how you would recognise motion and chemical shift artefacts and how are they related to the magnetic field gradients?
17. Explain the inversion recovery sequence and how might a modification of this sequence be used to remove the fat signal from the image?
18. What is the purpose of RF screening and where might it be found?
19. What is the fringe field?

20. What is the maximum value in the fringe field that is generally regarded as safe for a person with a cardiac pacemaker to stand?

21. List the main contraindications for MRI.

22. What precautions should be taken for staff or patients who may be pregnant in relation to magnetic fields?

23. What is the biological effect of applying radio frequencies to tissues?

17

Digital Image Storage and Handling

G Cusick

SUMMARY

The aim of this chapter is to explain the technology underlying Picture Archiving and Communications Systems (PACS) at a level that will help clinical users of such systems to understand the key factors that determine the performance of the systems.

- The stages in the imaging chain from the patient to reporting are identified.
- The requirements for acquiring faithful digital representations of images are discussed, with some reference to the mathematics that underpins these processes.
- Computer network architectures and some of the technicalities of networks are discussed.
- The different architectures of PACS are discussed.
- The essential features of modern display systems are described and the need for a customised display device to be matched to its purpose (diagnostic reporting, high-quality images in a non-radiological environment, generic displays) is explained.
- The importance of standards that ensure both compatibility of hardware for communication of images and patient safety from the potential impact of unsuitable software is emphasised.
- Methods of data compression to facilitate the ever-increasing amount of data collected in some imaging protocols are discussed.

CONTENTS

17.1 Introduction

The storage and transmission of images in digital form has had a profound effect on most aspects of medical imaging. The emergence of Picture Archiving and Communications Systems (PACS) over the last 20 years or so has transformed the way that images are captured, interpreted and reported upon, and used clinically. PACS is widely accepted to be one of, if not the, most important and influential applications of information technology (IT) in medicine.

The aim of this chapter is to explain the technology underlying PACS at a level that will help clinical users of such systems to understand the key factors that determine the performance of the systems. This will help users of the systems to communicate effectively with the range of technical and clinical specialists who develop, deploy and maintain them.

Conventionally, this sort of chapter would open with an explanation of bits, bytes, binary coding and computer fundamentals. IT is now so well established, and so ubiquitous that this is no longer necessary, or helpful. References to some basic computer science texts are included in the 'Further Reading' section for those readers who wish to pursue this further. There are important IT concepts that still bear explanation, though, and these are covered in the next sections. The purpose of this, at the minimum, is to establish a common vocabulary.

17.2 The Imaging Chain

Figure 17.1 shows a completely general imaging chain, from the patient to a report. The two key technological steps are image acquisition and image presentation. In a film-based imaging system, the image acquisition phase would encompass the exposure and processing of the film, the image presentation, generally by displaying the film on a lightbox or

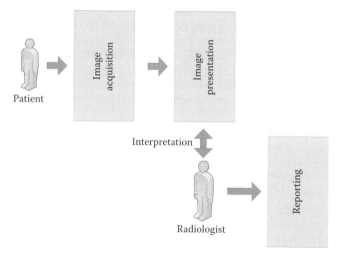

FIGURE 17.1
The imaging chain, from patient to report.

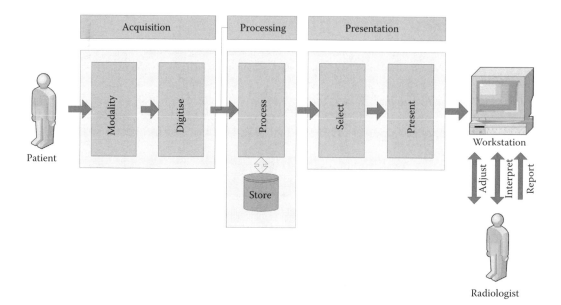

FIGURE 17.2
The digital imaging chain. This separates the acquisition phase, in which the image is converted to digital form, the processing phase, in which the image is stored and possibly re-formatted and compressed, and the presentation phase where selected images are properly formatted for sending to display or printing devices.

similar device. The options at that point for interacting with the image are limited—the presentation of the image is essentially fixed.

Figure 17.2 shows a simplified version of the equivalent digital system. Whilst the overall flow is similar, from patient to report, there are intermediate steps where the technology can affect the nature and quality of the image seen by the radiologist at the workstation. There is also the facility for the radiologist to interact with the image during the interpretation process.

The remainder of this chapter is devoted to examining the features of the digital imaging chain that have an impact on the quality of the image available for review, and with the system components that influence the performance and usability of the overall imaging system.

17.3 Image Acquisition—Digital Representation of Images

This section is concerned with the factors that limit the accuracy with which a digital image can represent the 'true' image. In designing digital systems, there are always compromises to be made in this area, driven by a mix of economic, functional and technical factors. It is normal to set out to minimise sample rates, the size of image matrices and the digitisation accuracy consistent with producing images that are of sufficient quality to be diagnostic.

Images are represented digitally as two-dimensional arrays of exposure or intensity values. The size of the array and the precision of the array elements have to be chosen to meet the requirements of the particular imaging system.

17.3.1 Sampling

To represent an image digitally, it is necessary to assign discrete values both to the coordinates in the image (x, y, and possibly in the third dimension, z), and to the value of the parameter being imaged. The process of assigning these values is referred to as digitisation. The digital values represent *samples* of the continuous, analogue image. The sampled digital version of the image can then be transmitted and stored just as any other set of digital data.

To view the image, it must be reconstructed from the sampled version. The reconstruction process entails reading back the image samples, placing them in the correct sequence and position, and removing artefacts introduced by the sampling process.

Accurate reconstruction requires that there is sufficient information in the sampled signal to specify the original signal completely. This requirement is codified in the Nyquist-Shannon sampling theorem (Nyquist 1928; Shannon 1948):

> If a function $x(t)$ contains no frequencies higher than v hertz, it is completely determined by giving its ordinates at a series of points spaced $1/(2v)$ seconds apart.

Although the Nyquist–Shannon theorem is written in terms of signals where a parameter is a function of time, the same principle holds where the variables sampled are functions of position: in these cases, frequency is specified in cycles per mm, and the sampling interval can be specified in mm. It is common to restate the theorem as the *Nyquist criterion* which sets the minimum sampling rate as twice the highest frequency present in the signal.

The choice of sampling frequency is essential to the acquisition of digital signals; it is easier to understand the consequences of under-sampling, in particular, by looking at the sampling and reconstruction processes in the frequency domain (see Insight).

Insight

Fourier Analysis

Any signal which is a function of a variable against time or space can be represented as a sum of sinusoidal signals. A plot of the amplitudes of the component sinusoids against frequency is known as the *amplitude spectrum* of the signal: this is a frequency-domain representation of the signal. The spectrum is derived from the time- (or space-) domain signal, $f(x)$, using the Fourier Transform, in which the frequency-domain representation $f(v)$ is calculated thus:

$$f(v) = \int_{-\infty}^{\infty} f(x)e^{-2\pi ixv}\, dx$$

This is the *Continuous Fourier Transform*; more commonly, the *Discrete Fourier Transform* is computed. The form of this is similar, substituting summation of discrete frequency components for the integration of the continuous transform:

$$X_k = \sum_{n=0}^{N-1} x_n e^{(-2\pi i/N)kn}$$

The values X_k are complex numbers, representing the amplitude and phase of the sinusoidal components of the original signal x_n. The amplitude spectrum, illustrated in Figure 17.3, is calculated by taking the amplitudes of the complex values. This application of the Fourier transform is one-dimensional; the signal is a function only of time in this example. It is straightforward, though to generalise the transform to more than one dimension, for example, in mapping the spatial frequencies present in an image. Further related transforms, notably the discrete cosine transform (DCT) and the discrete wavelet transform, are the basis for common compression methods, discussed later in the chapter.

Figure 17.3 shows the steps in sampling and reconstruction for an arbitrary signal. The negative frequencies shown in the figure arise from the mathematical analysis; there are several conventional ways to try to rationalise negative frequency as a physical concept, none of them particularly satisfactory. Figure 17.3a shows the spectrum of the original signal. Figure 17.3b shows the spectrum after sampling at a sample rate f_s; in Figure 17.3b, the sample rate is more than twice the highest frequency present in the original signal. The effect of the sampling process is to produce images of the original spectrum centred on the sample frequency. The original signal can be reconstructed accurately by passing the sampled signal through a low-pass filter which removes all signal components with frequency greater that $f_s/2$. This is shown in Figure 17.3c.

Figure 17.4 shows a similar sampling process, but in this case, the sample rate is below the Nyquist rate. In the sampled signal, then, the image spectra overlap the original signal spectrum (the arrowed areas in Figure 17.4a). Reconstructing the signal using a low-pass filter produces the result shown in Figure 17.4b. Note that the parts of the signal at frequencies above $f_s/2$ are folded back into the signal spectrum. This is known as *aliasing*; its importance is that the aliased signals are not distinguishable from genuine signals in the reconstruction.

The discussion above has assumed idealised conditions: the signals being sampled are entirely contained within a limited range of frequencies (i.e. the signal is *band limited*), and the reconstruction filter ideal in that it passes all signals below the cut-off frequency and none above. Real signals will generally contain some components (e.g. noise) which cover

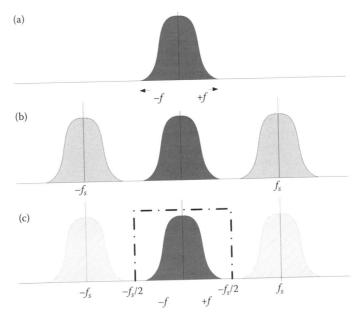

FIGURE 17.3
Sampling of signals, where the sampling frequency exceeds the Nyquist frequency. (a) The spectrum of the original signal. (b) Sampling creates 'image' spectra, shifted by the sampling frequency f_s. (c) Reconstruction of the signal uses a filter, indicated by the broken line, to remove the image spectra.

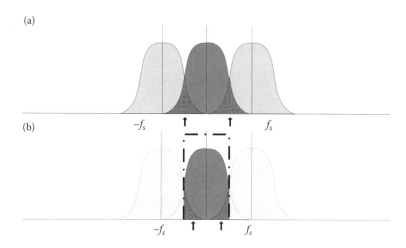

FIGURE 17.4
Sampling of signals, where the sampling frequency is less than the Nyquist frequency. (a) The image spectra now overlap the original signal in the arrowed regions. (b) When reconstructed using the appropriate filter, parts of the image spectra remain (arrowed) as aliases.

a range of frequencies that is, in principle, infinite, and ideal filters are hard to implement. Consequently, the Nyquist criterion must be taken as a theoretical limit: in practice, sampling rates 2.5 or 3 times the maximum expected signal frequency are required for reliable reconstruction.

It is conventional to express the quality of an image in terms of the spatial resolution, expressed in line-pairs per mm. This translates more or less directly to spatial frequency; thus, to achieve a resolution of 5 line-pairs per mm, a sampling rate in excess of 10 samples per mm would be required in theory.

17.3.2 Encoding and Storage

Each sample in an image is assigned a numerical value, with a limited range and precision. The limits on both are linked to the way in which the computing system represents numbers; these are known generally as data types; some representative types are listed below.

- Unsigned integers, representing positive whole numbers in the range 0 to 65,535. Unsigned integers are 16-bit binary values, requiring 2 bytes of storage.
- Unsigned long integers, representing positive whole numbers in the range 0 to 4,294,967,295. Unsigned long integers are 32-bit binary values, requiring 4 bytes of storage.
- Floating-point numbers, representing real numbers in the range $\pm 3 \times 10^{38}$. Floating Point numbers are 32-bit binary values, encoding the mantissa and exponent of the number, requiring 4 bytes of storage.

Grey-scale images can be represented by a single numerical value for each sampled point in the image, representing the intensity at that point. Within the limits of the detector, digital systems exhibit a linear relationship between exposure and grey scale. This is different from the s-shaped characteristic of film.

The number of bits available for each sampled point sets a limit on the fineness of the grey-scale representation possible. Figure 17.5 shows four versions of a grey-scale image

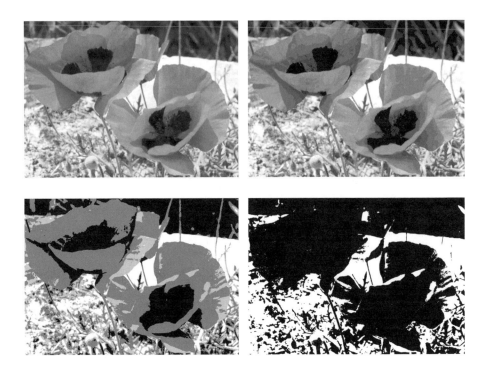

FIGURE 17.5
The effect of bit depth on image appearance. The four images are rendered at accuracies of 8, 4, 2 and 1 bits, from top left to bottom right. Note that 8 bits corresponds to 256 grey shades (2^8= 256), 4 bits to 16 shades, 2 bits to 4 shades, and 1 bit to 2 shades only.

of flowers, at steadily reducing grey-scale resolution at 256, 16, 4 and 2 levels from top left to bottom right.

A consequence of reducing the grey-scale resolution is the production of artefactual contours in the image. For example, there are visible artefacts in the 16-level image in Figure 17.5. Compare the appearance of the two image sections in Figure 17.6. The lower (4 grey shade) image shows some clear contours where there are none in the higher-resolution image. In medical images, where interpretation often depends on detecting subtle changes of density, this is particularly important.

Whilst floating-point number representations offer a greater range of values and accuracy, this is obtained at a cost in the processing requirement. It is not generally practicable to manipulate large image arrays of floating-point values sufficiently fast to be usable.

17.3.2.1 Files

Computer systems use files to store data. This concept is familiar to anyone who has used a desktop computer for word processing. In general, a file is no more than a structured block of digital data stored, for example, on a computer disk. The file structure is determined by the application.

The image representation discussed above, in which the image is represented as a rectangular array of intensity values, is often known as a *bitmap*. The term derives from the display of strictly black and white images, in which each pixel was either 'on' or 'off', and

FIGURE 17.6

Creation of artefactual boundaries due to insufficient accuracy. The top image is digitised at 8 bits. Note the boundaries that appear in the lower image, digitised at 2 bits accuracy (only 4 grey shades).

could thus be represented by a single bit. Bitmap files have some of the simplest structures, but even then, have data additional to the raw image data.

For example, the outline of the Microsoft bitmap (bmp) file format is shown in Table 17.1.

The information in the first three elements in the file is used by the system that displays the file to determine how to interpret the bitmap itself. The reason for doing this is that it increases flexibility by allowing the same file type, and the same basic readback mechanism, to handle a range of different detailed files. By reading the information in this header, the system can determine how to render the information in the bitmap that follows it.

The DICOM standard (see Section 17.6.1 below) includes detailed definitions for the storage of Pixel Overlay and Waveform data. The definitions are complex, and much more extensive than the simple example above. But the purpose is exactly the same: to provide detailed rules by which a device or system creating an image file must operate, in order that another device or system can operate on the file, to render it for display or printing, for example.

TABLE 17.1

Fields in the Header of a Bitmap File

BMP File Header	Stores General Information about the BMP File
Bitmap information	Stores detailed information about the bitmap image, such as: • The sizes of the elements in the file • The height and width of the image in pixels • Colour depth which determines the number of bits per pixel • The compression method used
Colour palette	Stores the definition of the colours being used for indexed colour bitmaps
Bitmap data	Stores the actual image, pixel by pixel

17.4 PACS System Architectures

By far the most important context for the digital storage and manipulation of medical images is in PACS. The clinical and operational benefits of PACS are huge, and, where PACS have been implemented, practice throughout the healthcare organisation has been affected. Many of the benefits are rather prosaic, and stem from the elimination of the need to store, transport and locate X-ray films.

PACS rely heavily on the IT infrastructure on which they depend. The rate of development and evolution in this field continues to be dizzying. Equally, the rate of technological extinction means that any text that describes particular devices is likely to be obsolete before the ink is dry! Technologies, such as magneto-optical disks, that were the white hope for the future in 2003 are now technological dinosaurs.

With that in mind, this section deliberately avoids describing specific devices or technologies. It tries to explain some of the established technologies in terms of their performance, and in terms of the kinds of choices that affect the implementation of a PACS.

The key components in the digital imaging chain were shown in Figure 17.2. They fall, functionally into three groups, concerned with image acquisition, processing and storage, and presentation or rendering. Whilst Figure 17.2 shows these elements connected together as a daisy chain, they would normally all connect to a common data network.

17.4.1 Networks

The data network is fundamental to the operation of a digital image storage system, since it provides the channel by which the elements of the system communicate. The performance of the network is thus a key factor determining the performance of the system. This section explains the structure of modern data networks, referring particularly to the features which either enhance or limit performance in one way or another. Network design is a large and complex field. This section necessarily simplifies things: the goal is to give an appreciation of some of the issues, and to remove some of the mysteries.

The design of an 'enterprise' network intended to meet all the IT requirements of, for example, the whole of a modern hospital is complex. The single physical network installed in modern buildings is expected to serve a wide range of applications. Different applications place different demands on the network: digital telephony, often known as voice over IP, for example, requires relatively low network bandwidth, but requires that the delay in transmitting the information (latency) is minimised. PACS can consume large bandwidth, but because the large data volumes are transferred between intelligent devices, latency is less of an issue. Good practice mandates the appointment of a *network design authority*, with responsibility for designing and implementing the network, and all changes to it, to deliver the required performance.

17.4.1.1 Functions of the Network

The network serves a single function: it provides communications channels between sources and recipients of information. At the simplest level, a network comprises just that: a source of data, a receiver and a link—essentially a piece of wire—between them (Figure 17.7). Indeed, the early incarnations of the DICOM standards were written to help to standardise the protocols used at either end of such a link.

In principle, most communications on data networks follow this logical model: a source transmits a package of information to a receiver. The essential difference is that, whilst the logical communication is point to point, the *physical* communication medium is shared amongst a group of transmitters and receivers. The sharing of the medium and the establishment of links is governed by network protocols.

17.4.1.2 Open Systems Interconnection

The early development of computer networks was carried out in the early 1970s, either in an experimental setting, or by computer manufacturers. Whilst it was accepted that the networks would have to be capable of interconnecting a heterogeneous set of devices, each manufacturer nonetheless developed its own network architecture, conventions for interconnecting its own equipment. It very soon became apparent that there was a need to be able to interconnect systems from different manufacturers. This led the International Organisation for Standardisation (ISO) to establish subcommittee SC16, with the aim of producing standards for 'Open Systems Interconnection'. The result is a series of international standards, ISO/IEC 7498-n, first published in 1984 which define a Reference Model for the interconnection of systems. The Abstract to the current (1994) edition of ISO/IEC7498-1, the basic standard (ISO 1994a), sets out its purpose thus:

> The model provides a common basis for the coordination of standards development for the purpose of systems interconnection, while allowing existing standards to be placed into perspective within the overall Reference Model. The model identifies areas for developing or improving standards. It does not intend to serve as an implementation specification.

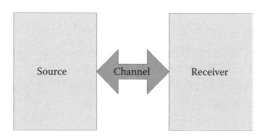

FIGURE 17.7
A point-to-point network connection.

TABLE 17.2

The Layers of the ISO Open Systems Interconnect Model

	Data Unit	Layer	Function
		7. Application	Network process to application
Host layers	Data	6. Presentation	Data representation and encryption
		5. Session	Interhost communication
	Segment	4. Transport	End-to-end connections and reliability
	Packet	3. Network	Path determination and logical addressing
Media layers	Frame	2. Data Link	Physical addressing
	Bit	1. Physical	Media, signal and binary transmission

The model divides the networking system into layers. Functionality within a layer is implemented by one or more entities, each of which can interact directly only with the layer directly below it and can provide facilities for use by entities in the layer immediately above it. The benefits of this are, first, that it divides overall network functionality into smaller and simpler components, facilitating their design, development and troubleshooting; and second, that the effects of changes in functionality in one layer are restricted to that layer alone.

Table 17.2 shows the layers in the ISO Open System Interconnect (OSI) model. The seven layers cover all aspects of communication between systems, from the physical connection to the network (layer 1) to the communications aspects of the application seen by the user. The two layers with the most importance in the design and implementation of networks are layer 2, the Data Link Layer, and layer 3, the Network Layer.

17.4.2 Ethernet

The predominant networking technology is now based upon developments of the Ethernet technology, developed by the Xerox Corporation in the early 1970s. The original Ethernet used a common coaxial cable to which all attached machines were connected. Any connected machine could transmit on the cable, and access was therefore governed by a scheme known as carrier sense multiple access with collision detection (CSMA/CD). When an attached machine wanted to send a package of data, it first waited for the cable to be free for a minimum time. During transmission of the data, it checked continuously for corruption of the data due to collision with data transmitted by another machine. If a collision was detected, the transmission was terminated, and restarted after a short, randomly chosen delay.

This scheme was much simpler than the competing technologies at the time, and was rapidly adopted as the basis of the Institute of Electrical and Electronics Engineers (IEEE) standard 802, for the standardisation of local area networks (LAN). The standard became IEEE802.3, and was later adopted by the ISO as ISO/IEEE 802/3.

17.4.2.1 Network Topology—How Devices Are Connected Together

In its original incarnation, Ethernet relied on connecting all devices to a single, shared coaxial cable which served as a common communication medium. Contention for the medium was managed by the CSMA/CD process outlined above. The network bandwidth on the cable was limited (to 10 Mb/s), and the maximum length of a cable segment limited by the propagation time to 500 m. A segment could accept a maximum of 100 connections.

Add to these limitations the high cost of installing the thick coaxial cable and the difficulties of expanding the network incrementally, and it is easy to see why this particular technology has not survived.

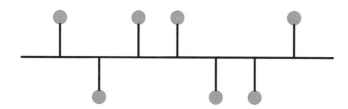

FIGURE 17.8
A network arranged in a bus topology. All nodes on the network connect identically to the network cable.

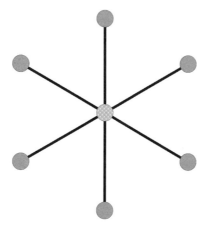

FIGURE 17.9
A star topology. All the outer nodes connect to the central node of the star.

The original Ethernet topology was a bus, as shown in Figure 17.8. The term 'bus' derives from electrical engineering, where a single conductor connecting together a number of circuits is known as a *busbar.* All nodes are connected onto a common backbone connection. The simplicity of this arrangement is more than offset by the disadvantages. Damage to the cable anywhere on the segment can result in signal reflections from the resulting impedance discontinuity, which mimic collisions and can therefore prevent valid transmissions. Failure of any node on the segment can cause the whole segment to cease to function.

The star topology, shown in Figure 17.9, avoids some of these problems. The hub, at the centre of the star, essentially isolates the connections to the nodes from one another. Only a fault in the hub itself can disrupt the network partitioning. The initial applications of star topologies in Ethernet networks still used coaxial cables, though often 'thin' (RG/58) cable.

A key development, though, was the move to using unshielded twisted-pair cables. In these simple networks, the hub is a simple device that just re-transmits every packet to every connected node. The total throughput of the hub is limited to that of a single link, and all must operate at the same speed. Obviously, as network sizes rise, the capacity of the hub becomes a significant restriction.

17.4.2.2 Bridging and Switching

In simple networks arranged as in Figure 17.9, the entire network is a single collision domain: every node must be able to detect collisions throughout the network. This problem can be alleviated by *bridging*, where the network is divided into multiple segments, and only well-formed Ethernet packets are passed between segments. Bridges learn where devices are by building a map of the unique hardware identifiers (MAC addresses) of the devices, and only forward packets to a segment when the destination is known to be on that segment, or is unknown.

Developments of the bridge devices which inspected the entire packet before forwarding it, have become known as *switches*, though this name does not appear in the standards. Modern Ethernet networks are generally constructed around a hierarchy of switches, allowing segmentation of the network to increase performance, and meshing, where there are multiple potential paths between parts of the network to increase reliability.

Switches also allow different parts of the network to operate at different speeds, allowing high-bandwidth (and hence high cost) links to be used where they are required. Typically, the network 'backbone', connecting the switches at the highest level of the hierarchy, will be connected using optical fibres which offer the highest bandwidth, but are relatively costly to install. In a network supporting PACS, high-bandwidth (1 Gb/s) copper links will be concentrated in the imaging department, connecting the acquisition systems and diagnostic workstations.

17.4.2.3 Data Packaging on the Network

Layer 1, the Physical Layer, of the OSI model (see Table 17.2) is concerned in part with the conversion between the representation of digital data in user equipment and the corresponding signals transmitted over the communications channel. Data is transmitted on an Ethernet network in *packets*. The packet of data can be likened to a letter. The start and end of the letter are marked by conventional phrases ('Dear Sir,' 'Yours faithfully'), and the whole is contained in an envelope. Each Ethernet packet has the same basic structure, illustrated in Figure 17.10, comprising a frame of data surrounded by data used by the network to control the handling of the packet.

An *octet* is eight bits of data, corresponding to a single byte on modern computers.

The *Preamble* and *Start-of-Frame Delimiter* are fixed elements, used by the hardware to detect the start of a new frame. The two addressing fields identify, respectively the destination and source of the frame. The Ethertype field specifies either the length of the frame, or its interpretation type.

Frame setup		Addressing		Frame type	Data	Error check
Preamble	Start-of-frame delimeter	MAC destination	MAC source	Ethertype/ Length	Payload (Data & Padding)	CRC32
7 Octets of 10101010	1 Octet of 10101011	6 Octet	6 Octets	2 Octets	46–1500 Octets	4 Octets
			72–1526 Octets			

FIGURE 17.10
The Ethernet frame.

The *Payload* is the data actually carried in the packet. The maximum length, about 1500 bytes, was chosen as a compromise between maximising throughput (the larger the better), and re-transmission overhead (the smaller the better). Current developments will probably lead to the adoption of longer frames in due course.

The *CRC32* field contains a 32-bit Cyclic Redundancy Code (CRC), calculated from all the data fields in the frame. The CRC is calculated by the transmitting machine, and appended to the frame. The receiving machine carries out the same calculation and compares the result with the CRC in the packet. Virtually all transmission errors will result in a difference between the received and calculated values, and the receiver can request re-transmission of the packet.

A gap of at least 12 octets must be left idle after each packet.

Each packet can carry a maximum of approximately 1500 bytes of 'useful' data; the remaining 26 bytes (plus a minimum 12-byte interpacket gap) are an overhead of the basic transmission mechanism.

17.4.2.4 Layers 2 and 3

Layer 2, the Data Link Layer provides the functional and procedural means to transfer data between network entities, and to detect and possibly correct errors that may occur in the Physical Layer. The Data Link Layer is concerned with local delivery of frames between devices on the same LAN. Data Link frames, as these protocol data units are called, do not cross the boundaries of a local network. Inter-network routing and global addressing are higher layer functions, allowing Data Link protocols to focus on local delivery, addressing, and media arbitration between parties contending for access to a medium. The Data Link Layer protocols respond to service requests from the Network Layer (Layer 3) and they perform their function by issuing service requests to the Physical Layer (Layer 1).

The interpretation of the Ethertype field in the packet is carried out in Layer 2.

The Network Layer (Layer 3) provides the functional and procedural means of transferring variable length data sequences from a source to a destination host via one or more networks while maintaining the quality of service (QoS) and error control functions. The Network Layer responds to service requests from the Transport Layer (Layer 4) and issues service requests to the Data Link Layer (Layer 2). It is concerned with the source-to-destination routing of packets, including routing through intermediate hosts. By contrast, Layer 2 is concerned with shorter range, node-to-node delivery on the same link.

Layers 2 and 3 are important in network design, as it is in these layers that virtual LANS (VLAN), switching and routing operate.

17.4.3 Internet Protocol

The transmission of data over networks is specified in a number of protocols. The protocols are sets of rules that describe the content and formatting of the data transmitted and associated control data, and specify how the network hardware and software must interpret, particularly, the control data.

The internet protocol (IP) is particularly important, since it is so widely used, with over 1.5 billion users worldwide. The IP is part of a suite of protocols, TCP/IP which is itself described in terms of a layer model. The abbreviation TCP/IP stands for Transmission Control Protocol over IP. Whilst there is a fairly natural correspondence between TCP/IP and OSI layers, there is not a one-to-one mapping. Since the OSI model is defined in international standards, it is helpful to relate the TCP/IP model to it.

17.4.3.1 IP Addressing

In an IP network, every attached device is assigned a unique address (its IP address); this is used in the protocol to direct packets of data to that device. In the most widely used version of the protocol (IPv4), addresses consist of 32 bits, usually expressed for readability in a dot-decimal notation as shown in Figure 17.11.

A revision of the IP (IPv6) has been agreed which increases the size of addresses from 32 to 128 bits, in response to the huge growth in the number of devices and items of equipment that now require an IP address. Whilst modern operating systems generally support IPv6 addressing, it is not widely implemented in other devices. The remainder of this section will therefore concentrate on IPv4.

The allocation of IP addresses is managed globally by an organisation called The Internet Assigned Numbers Authority (IANA). The allocation of addresses is a key factor in keeping the internet working properly. Of the 2^{32} (4,294,967,296) available addresses, about 18 million are reserved for private networks, a further 270 million for multicast addresses, such as that used by the network time protocol, NTP. Networks implemented within an organisation will normally use addresses from the private network pool. This should not cause conflict because these addresses will not propagate outside the organisation. Where such a corporate network connects to the internet, a device, generally referred to as a *proxy*, is required that translates the internal, private addresses into public addresses.

It is simplest to consider the IP address for a particular device to be fixed, assigned when the device was configured. In networks with few devices, or where there is minimal change, this is practicable, and has the advantage that the IP address can be used as a prime identifier for the device. In larger networks, however, or where devices may join or leave the network, managing a fixed addressing scheme becomes difficult. An alternative is that the 'network' assigns an IP address from an available pool at the time the device connects to the network. This is governed by a specific protocol, the Dynamic Host Configuration Protocol, DHCP, and is widely used. To minimise 'churn' of addresses when devices join and leave the network repeatedly, DHCP addresses are assigned a 'lease'; having been assigned to a particular device, the assignment is retained for the duration of the lease. If the device reconnects during that time, the same address will be assigned.

The IP address cannot then be used as the primary identifier, as it may change from time to time. Instead, a unique *Host Name* is assigned to each device. But because IP relies on using the IP address for routing data packets, the host name must be converted to an IP address before the device can communicate on the network. This process, host name

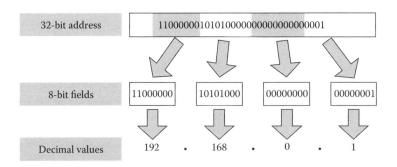

FIGURE 17.11
IP addressing. The 32-bit address is broken into four 8-bit fields, each of which can be represented as a decimal number in the range 0–255.

resolution, is required at the start of any communication with the device, and adds an overhead to that process.

It is common practice for some devices on a network to have fixed IP addresses.

17.4.3.2 Bandwidth and Latency

Network bandwidth is the rate at which data crosses the network. Ethernet networks are described in terms of their raw speed—100BaseT networks, for example, operate at a raw data rate of 100 megabits per second. The maximum actual rate at which useful data can be transferred is always lower than this, to take account of the transmission overheads incurred through the communications protocols. In applications, such as PACS, where large volumes of data are transferred regularly, bandwidth is an important consideration.

Latency refers to the delay incurred by a packet of data in traversing the network. There is a fundamental lower limit on latency, determined by the speed of light. Data cannot arrive at its destination in less time than the physical distance divided by the speed of light in the medium. There are other sources of delay, including delays caused by collisions or other transmission errors on the network that require re-transmission of packets, and by delays in store-and-forward routing devices. Latency is particularly important in systems handling real-time signals: a common example is IP telephony. In this application, voice telephone signals are converted to digital form and transmitted over an IP data network. For a telephone conversation to remain intelligible there must not be long transmission delays which means that excessive network latency can have a significant detrimental effect on call quality.

17.4.3.3 Shared Networks

Modern network infrastructure generally uses a single, common physical network which is shared between a number of applications. This sensible and economic approach does bring some problems, in particular:

- Allocating network resources so that a single, demanding application does not unduly compromise the performance of the network as a whole
- Isolating applications from one another, to limit the consequences of failure in a particular device or application
- Providing security, so that access to information can be controlled at the application level

17.4.3.4 Subnets and Virtual Networks

In a TCP/IP network, it is usual to divide the network into a number of *subnets*. The subnets are smaller and more efficient than the entire network, through preventing excessive rates of packet collisions. Subnets are defined by a routing prefix comprising the n most significant bits of the IP address. In IPv4 networks, this is often expressed as a *subnet mask* which is a bit mask where the n most significant bits are set to 1. For example, 255.255.255.0 would be the subnet mask for a subnet of 256 addresses based on network IP address 192.168.1.0, as shown in Figure 17.12.

A virtual LAN (VLAN) is a group of network hosts with a common set of requirements that communicate as if they were connected to their own private network, regardless of their position on a physical network. A VLAN has the same attributes as a physical

Subnet mask	255	·	255	·	255	·	0	·
	11111111		11111111		11111111		00000000	
	11000000		10101000		00000001		00000000	
IP address	192	·	168	·	1	·	0	·

FIGURE 17.12

Subnet masking. The subnet mask 255.255.255.0 selects a group of 256 network addresses.

TABLE 17.3

Quality of Service-Related Terms

Bit rate	The total number of physically transferred bits per second over a communication link, including useful data as well as protocol overhead
Delay or latency	The time from the *start* of packet *transmission* to the *start* of packet *reception*
Packet dropping probability	Probability that a packet will be dropped (i.e. fail to be received), expressed as a percentage. A value of 0 means that a packet will never be dropped, and a value of 100 means that all packets will be dropped
Bit error rate	The proportion of bits that have errors relative to the total number of bits received in a transmission, usually expressed as ten to a negative power

network, but it allows for a logical grouping of network devices that is independent of their physical location. VLANs are OSI Layer 2 constructs, whereas subnets are implemented at Layer 3, as part of the IP. In an environment using VLANs, there is often a one-to-one relationship between VLANs and subnets, though there can be multiple VLANs on one subnet and vice versa.

VLANs offer the possibility of separating logical position on the network from physical location, and also mean that network reconfiguration can often be done without physically moving connections. In complex networks, these are important benefits.

17.4.3.5 Quality of Service

In this context, the term Quality of Service (QoS) refers to the mechanisms used to allocate network resources, rather than the measured (or perceived) QoS. If network resources were boundless, QoS mechanisms would not be required, but in real networks, the need to provide, for example, a required bit rate, delay, jitter, packet dropping probability and/or bit error rate means that they are necessary (see Table 17.3 for explanations of these terms). QoS is the ability to provide different priority to different applications, users, or data flows, or to guarantee a certain level of performance to a data flow. QoS guarantees are especially important for real-time streaming applications such as voice over IP, patient monitoring and IP-TV, since these often require fixed bit rate and are delay sensitive.

QoS mechanisms make use of either subnets, at the IP layer, or VLANs at Layer 2 of the OSI model, to define coherent service groups.

17.4.4 Servers

Figure 17.13 shows the principal server components of a typical PACS concerned with the acquisition, storage and processing of images. In a normal implementation, there will be

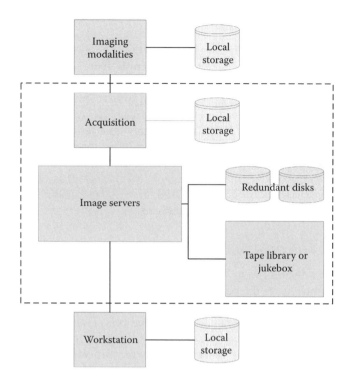

FIGURE 17.13
A typical PACS server architecture. The servers shown are 'logical' devices; the distribution amongst physical machines may be different.

additional servers, particularly those required to provide interfaces to other hospital systems such as the patient master index and radiology information system. There may also be a separate server supporting web presentation of images, though some systems incorporate this function into the main image servers.

Server functions are discussed in the next sections.

17.4.4.1 Image Acquisition

The image acquisition server manages the process of receiving images from the modalities. In this process, the acquisition server:

- Acquires image data from the radiological device, managing communication with the devices and carrying out any error recovery tasks required
- Where required, it converts the image data to a standardised, PACS-compliant format, meeting the requirements of the relevant DICOM standards
- Forwards the image to the image servers and display workstations

Images will generally be buffered by the acquisition server, until the image servers confirm that the image is safely and correctly stored. This provides insurance against the failure of either the image server or the communications channel.

17.4.4.2 Image Database

The Image Servers create and maintain a database of all images stored on the system, and are thus at the core of PACS. Logically, this is a single database containing every image ever acquired; physically, the database is normally split into a 'current' portion, comprising images acquired or accessed recently, and an archive containing other images. The current images are stored on high-speed disks, the archive on slower storage, typically an automated tape library. This split is made for largely economic reasons: fast disk storage has historically been substantially more costly than tape storage, though the difference is steadily diminishing.

The primary function of the image servers is to manage this storage, ensuring that images are properly stored as they are acquired, and that requested images are retrieved from storage and delivered to display workstations.

17.4.4.3 Integration with Other Systems

PACS relies heavily on integration with other hospital information systems to be useful. The most obvious of these are the patient master index which is, as a minimum, the source of patient demographic information, and the radiology information system which provides scheduling services for patients and examinations. The number and complexity of the systems with which PACS must integrate is steadily growing, driven by a combination of technical feasibility and clinical demand. The communication with these systems will normally be managed by one or more integration servers which translate and format the messages between systems.

17.4.5 Virtualisation

In computing, *virtualisation* refers to a range of techniques intended to facilitate the sharing of computing resources—processing, network or storage, for example—between separate and independent applications. The techniques all work by abstraction, breaking the connection between a physical resource, for example, a specific hard disk, and its application, a unit of storage used by an application.

Virtualisation is becoming widespread throughout commercial computing. In the case of medical systems, a caveat is required. Since many such systems, PACS included, may be classified as medical devices, the decision to virtualise components of a PACS must be supported by the system manufacturer.

17.4.5.1 Storage

In a hospital IT context, PACS is likely to place by far the largest demand on storage. Annual data volumes for a filmless hospital can easily exceed 1 TB. Experience in centres which have established PACS is that the annual volume continues to grow. New modalities, for instance spiral CT, generate large numbers of high resolution, high dynamic range images. Storage and its management constitute a large part of PACS costs, and are central to good performance.

Virtualisation technologies affect the implementation of storage through systems such as the storage array network (SAN). A SAN is an architecture to attach remote computer storage devices (such as disk arrays, tape libraries and optical jukeboxes) to servers in such a way that the devices appear as locally attached to the operating system. Sharing storage in this way usually simplifies storage administration and adds flexibility since cables and storage devices do not have to be physically moved to shift storage from one server to another. SANs also tend to enable more effective disaster recovery processes. A SAN could span a primary data centre and a distant location containing a secondary storage array. This enables storage replication either implemented by disk array controllers, by server software, or by specialized SAN devices.

17.4.5.2 Sizing Storage

The calculation of the amount of storage needed by PACS is based on an analysis of the workload to be supported, in terms of the number and types of studies to be stored. For example, a standard plain X-ray study will generate about 30 MB of data, whilst a multi-slice CT examination could create over 1 GB of image data.

17.4.5.3 Strategies

The quantity and type of storage used will have marked effects on system performance, and informed choices have to be made. The factors to consider in this include the following:

- The number of studies that are required for immediate, rapid access. This will be affected by the patient population served, the types of study in use, and the patient throughput
- Availability and cost of storage

There can be no hard-and-fast rules determining how much of what sort of storage or how much is ideal. Technical developments in this area continue at a surprising rate; storage decisions need to be considered carefully both at the time of PACS installation, and throughout the life of the system.

17.4.5.4 Backup and Security

The requirements for backup and security can be simply stated:

- No image may be lost as a result of system failure. It should be possible to recover additionally from human error.
- Access to images must be restricted to those explicitly authorised.
- Protection from malicious software.

The first requirement translates into a need to implement a robust and comprehensive backup strategy, so that up-to-date copies of image data are regularly and frequently created. It is equally essential that these backup copies are capable of being restored correctly.

The second requirement is addressed through the authentication and sign-on processes for PACS, and through encryption of data outside the secure cordon. Such security measures have to be implemented within an appropriate management framework to ensure that all users comply with the requirements.

Protecting against malicious software requires a range of protective measures such as firewalls that prevent unauthorised access to the network system, and anti-virus software that detects and disables malicious software. Again, management measures are necessary to minimise the risk of introducing malicious software from portable storage, such as the ubiquitous USB stick.

17.5 Display Devices

17.5.1 Classes of Workstation

It is convenient, particularly for planning purposes, to classify the workstations on which images are viewed. Just as X-ray images on film are viewed in different ways at different times, there are several ways in which clinicians will use images stored and presented by PACS.

The workstation is a part of the imaging chain, and should be covered by a relevant DICOM compliance statement.

17.5.1.1 Diagnostic Reporting

Diagnostic reporting workstations are used within the imaging department for reviewing and reporting on images. These offer the highest image quality and the widest range of image manipulation facilities. In general, diagnostic and reporting workstations will use native (i.e. not derived) images, avoiding artefacts created, for example, by lossy compression.

They require at least two large, high resolution displays. Early PACS implementations experimented with larger numbers of displays, mimicking the widely used 4 + 4 film display, but, with familiarity, two has been found to be sufficient.

Diagnostic and reporting workstations will generally require rapid access to many images, and the network connections used will reflect this. The ergonomics of these workstations require careful attention.

17.5.1.2 High-Quality Clinical Workstations

These are similar to the diagnostic and reporting workstations, and are used where detailed examination of the images is required. Examples include the workstations used in an orthopaedic operating theatre, where the surgeon requires full access to the images.

Since these workstations may be located in sensitive clinical settings, such as operating theatres, there may be additional factors. In an operating theatre, for instance, the infection control requirements are particularly stringent, and specific designs of workstation are required. In other clinical settings, the workstation may be within the patient environment which creates special requirements for electrical safety.

17.5.1.3 Generic

Images are widely used outside the imaging department, generally in association with the radiologist's report. In these circumstances, the requirements for image quality are less stringent, but the cost of the high-grade workstations and displays is prohibitive. It is common, therefore, to provide a view of images via a web browser on more or less standard desktop computers.

Web transmission requires that images are compressed, with the attendant artefacts and loss of quality. Web displays will permit some limited manipulation of the image.

17.5.2 Properties of Electronic Displays

By far the most common type of display now in use is the liquid crystal display (LCD). These have largely supplanted the cathode ray tube (CRT) displays that were the standard. LCDs have numerous advantages over CRTs, from which they are fundamentally different.

Liquid crystals belong to a class of materials that have properties between those of a conventional liquid and those of a solid crystal. For instance, liquid crystal may flow like a liquid, but its molecules may be oriented in a crystal-like way. LCDs rely on the fact that the molecules of some liquid crystals will align with an applied electrical field.

Figure 17.14 shows the structure of a simple LCD. The six layers are, from the front of the display:

(a) A linear polariser

(b) Glass window with a pattern of electrodes deposited on it

(c) A liquid crystal layer, typically 10 μm thick

(d) A second glass window with electrodes

(e) A rear linear polariser, with its axis of polarisation perpendicular to the other polariser

(f) Backlight or reflector

The display generates the image as a physical pattern in the liquid crystal layer which is viewed by the light from the backlight transmitted through the layers in front of it. The

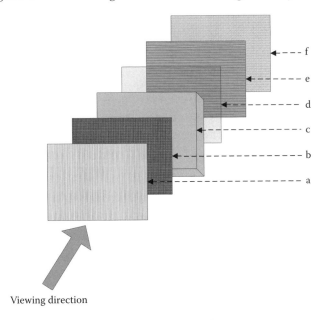

Viewing direction

FIGURE 17.14
Liquid crystal displays operate by modulating the light transmitted through the liquid crystal layer and a pair of crossed linear polarisers.

display's operation relies on the fact that, in one state, the liquid crystal layer (c) rotates the polarisation of light by 90 degrees. The liquid crystal is chosen so that, in its relaxed state (i.e. with no electric field applied), the molecules are twisted and rotate the polarisation of light passing through it. Thus, polarised light passing through the rear polariser is rotated so that it will also pass through the front polariser. Applying an electric field, by placing a voltage between electrodes on the front and rear windows, aligns the molecules and stops the rotation of polarisation. As a result, the pattern of the electrodes shows as dark.

The matrix displays used in computer applications have 'row' and 'column' electrodes. In early displays, these were driven directly by external electronics (so-called passive-matrix displays). All modern displays are *active-matrix* displays, where the electrode drive circuitry is integrated on the display itself. Active-matrix addressed displays look 'brighter' and 'sharper' than passive-matrix addressed displays of the same size, and generally have quicker response times, producing much better images.

A colour display can be created by overlaying a set of colour filters over groups of pixels.

The key characteristic of LCDs is that they are transmissive (or transflective) displays, rather than emissive displays like CRTs.

17.5.2.1 Dynamic Range

The dynamic range of a display is the ratio of the brightness of the brightest and darkest tones that can be displayed. Currently, good-quality LCDs can produce a dynamic range of about 1000:1. This is comparable with the density range for film, but rather less than that obtainable from a well-adjusted CRT.

A limitation on LCDs is the visibility of the image off-axis. Modern LCDs have a usable viewing angle of ±85 degrees in both horizontal and vertical planes, though contrast will fall off perceptibly at the extremes.

17.5.2.2 Stability and Reliability

Controlling image geometry on a CRT display required complex circuitry to compensate for the intrinsic errors in the positioning of the electron beam on the screen. These errors do not occur with LCDs which are essentially addressable pixel displays. Thus the requirement to monitor and recalibrate display geometry does not arise.

LCDs rely on a backlight for their operation. The backlight illuminates the back of the display, and is transmitted selectively by display pixels. The commonest backlight technology uses a number of cold-cathode fluorescent lamps. These lamps have finite life, and display brightness will thus change over time. The lamps are subject to some production spread which can affect display uniformity.

17.6 Standards

17.6.1 DICOM

17.6.1.1 Overview

The DICOM (Digital Imaging and Communications in Medicine) standard originated in the 1980s from a joint initiative between the American College of Radiology (ACR) and the US National Electrical Manufacturers Association (NEMA), stimulated by the difficulties arising from the proprietary communications methods used in the growing number of computed tomography and magnetic resonance imaging devices. The first standard resulting from the initiative, ACR/NEMA 300, was published in 1985. The 1993 revision of the standard V3.0 was the first to be given the DICOM title, and is the basis of the current standard. DICOM remains a standard published and maintained by NEMA through its Medical Imaging & Technology Alliance.

The DICOM standard (DICOM 2009) is freely downloadable from the NEMA website.

17.6.1.2 What Is DICOM?

DICOM is a standard for handling, storing, printing and transmitting information in medical imaging. It includes a file format definition and a network communications protocol. The communication protocol is an application protocol that uses TCP/IP to communicate between systems. DICOM files can be exchanged between two entities that are capable of receiving image and patient data in DICOM format. The standard defines:

- *A large number of information objects.* The information object definition is an abstract data model used to specify information about real-world objects. For example, the standard includes an information object definition for a computed tomography image. This ensures that two systems, or components within a system, that comply with the DICOM standard will have a common view of the information to be exchanged.

- *A number of services, grouped as service classes.* Each service class definition provides an abstract definition of real-world activities applicable to communication of digital medical information. Services, most of which entail transmission of data over the network, include, e.g. (a) Storage, used for the transmission of images; (b) Printing and (c) Query/Retrieve which allows a workstation to find lists of images or other such objects and then retrieve them from a PACS.

- *A range of underlying data definitions.* These include, for example, conventions for the description of patient orientation.

- *A detailed description of the contents of a conformance statement.* The manufacturer of a device or system is responsible for declaring conformance with the standard. This contrasts with the process for most ISO standards, where conformance is assessed by an accredited test house. DICOM conformance is more complex, and the Conformance Statement reflects this, setting out which elements of the standard a device meets, and containing detailed information on the way it operates. Consequently, a simple statement that a device is DICOM compliant is not really meaningful.

17.6.2 The Medical Devices Directive

17.6.2.1 Overview

Since 1998, medical devices placed on the market in Europe have been required to meet the essential requirements of the Medical Devices Directive. The Directive (European Parliament 2007) was last amended in 2007, and defines a Medical Device as follows:

> … any instrument, apparatus, appliance, software, material or other article, whether used alone or in combination, together with any accessories, including the software

intended by its manufacturer to be used specifically for diagnostic and/or therapeutic purposes and necessary for its proper application, intended by the manufacturer to be used for human beings for the purpose of:

- diagnosis, prevention, monitoring, treatment or alleviation of disease,
- diagnosis, monitoring, treatment, alleviation of or compensation for an injury or handicap,
- investigation, replacement or modification of the anatomy or of a physiological process,
- control of conception,

and which does not achieve its principal intended action in or on the human body by pharmacological, immunological or metabolic means, but which may be assisted in its function by such means.

The 2007 revision of the Directive contains an important clarification on the classification of software systems:

It is necessary to clarify that software in its own right, when specifically intended by the manufacturer to be used for one or more of the medical purposes set out in the definition of a medical device, is a medical device. Software for general purposes when used in a healthcare setting is not a medical device.

The importance of this is that it makes a clear separation between those systems actively used in diagnosis or treatment, and those that would normally be thought of as 'administrative'. There is then little doubt that PACS, used as they are for acquiring, viewing and processing medical images, are medical devices under the Directive.

17.6.2.2 Consequences of Classification

By classifying PACS as a medical device, the manufacturer certifies that, when used for its intended purpose, the system will perform correctly and safely. In general, the manufacturer will place some conditions on this undertaking, typically in the form of a requirement for a 'warranted environment'. This will set out the requirements for servers, networks and workstation components for the system, and might, in some cases, mean that all components for the PACS must be supplied by the PACS manufacturer.

17.7 Availability and Reliability

17.7.1 Availability

17.7.1.1 Importance of PACS

The imaging department is a critical resource in most hospitals. Departments of Emergency Medicine will rely heavily on images in the assessment of critically ill patients; image availability is thus critical to the ability to deal with emergencies.

One consequence of moving to filmless imaging centred on PACS for the digital storage and transmission of images is that the failure of PACS can have far-reaching consequences.

PACS has to be considered, then, in the context of the whole hospital's operation, and measures taken to minimise the risk of PACS failure.

17.7.1.2 Management Consequences

One important consequence of this is that PACS cannot be seen as purely the responsibility of the imaging department. This has to be reflected in the management arrangements for the system which should include representatives from key areas which will rely on imaging.

17.7.2 Reliability

PACS must be viewed as a high-reliability system. This implies some key design requirements:

- *Elimination of single points of failure.* A careful analysis of the system and the infrastructure that it uses, may reveal particular resources—items of equipment pieces of software, or even people—failure of which would result in failure of PACS.
- *Resilience.* An analysis of external factors will reveal those non-PACS factors with the potential to bring PACS down.

Single points of failure can generally be eliminated by redundant design, where critical items are duplicated. Resilience factors include the use of backup and uninterruptible power supplies. It is important, though, to ensure that both redundant and resilient design are based on rigorous analysis.

17.8 Data Compression

17.8.1 Background—Reasons for Needing Compression

A single plain X-ray image can create 15 MB of data, and a patient study will contain a number of images. This imposes loads on both the data network, for transferring the images, and the image store. Even though storage costs continue to fall, the transfer time for images across the data network remains a problem. For these reasons, it is beneficial to compress the image file.

There are two fundamentally different approaches to compression. Lossless compression creates a compressed version that contains all the information present in the original image. The decompressed image is then identical to the original. The file compression programs widely used on desktop computers (e.g. to produce Zipped archives on Windows systems) are lossless.

Lossy compression produces a compressed version that, when decompressed, gives a result similar to the original, but with errors. The JPEG compression schemes used in digital cameras fall into this category.

The DICOM standards include attributes for the compression status of images, in particular to ensure that images that have been subject to lossy compression are properly identified.

17.8.2 Lossy Compression

The most widespread lossy compression method is the *JPEG technique* which takes its name from the Joint Picture Experts Group who drafted the standard for the technique. The standard was ratified and published (ISO 1994b) by the International Organization for Standardisation as ISO/IEC10918–1:1994.

  The objective of the JPEG process is to produce compressed images which look acceptable to a human observer. The compression process exploits some limitations on the human perceptual system, to reduce the information content of the compressed image. The process can be summarised as follows:

1. The representation of the colours in the image is converted from RGB to YCbCr, consisting of one luma component (Y), representing brightness, and two chroma components, (Cb and Cr), representing colour. This step is sometimes skipped.

2. The resolution of the chroma data is reduced, usually by a factor of 2. This reflects the fact that the eye is less sensitive to fine colour details than to fine brightness details.

3. The image is split into blocks of 8×8 pixels, and for each block, each of the Y, Cb, and Cr data undergoes a discrete cosine transform (DCT). A DCT is similar to a Fourier transform in the sense that it produces a kind of spatial frequency spectrum.

4. The amplitudes of the frequency components are quantised. Human vision is much more sensitive to small variations in colour or brightness over large areas than to the strength of high-frequency brightness variations. Therefore, the magnitudes of the high-frequency components are stored with a lower accuracy than the low-frequency components. The quality setting of the encoder affects the extent to which the resolution of each frequency component is reduced. If an excessively low quality setting is used, the high-frequency components are discarded altogether.

5. The resulting data for all 8×8 blocks is further compressed with a lossless algorithm.

The JPEG compression algorithm works best on images of realistic scenes with smooth variations of tone and colour. For web usage, where the bandwidth used by an image is important, JPEG is very popular. JPEG is not as well suited for line drawings or other images with the sharp contrasts between adjacent pixels where noticeable artefacts are created. As the compression ratio is increased, JPEG typically achieves compression ratios of 10:1, without perceptible loss of quality, on suitable images. It is important to remember, though, that the compressed image is not identical to the uncompressed.

17.8.3 Lossless Compression

17.8.3.1 Attributes

The distinguishing feature of lossless compression is that it is reversible: the image produced by compression followed by decompression is identical to the original. In diagnostic applications, this is intuitively preferable to lossy compression. There are many lossless data compression techniques, often suited to particular types of data set. For example, line images comprising large areas of discrete tones are easily and efficiently compressed using run-length encoding.

  Figure 17.15 illustrates this process on a section of a continuous-tone image. The values of the grey shade in each pixel along the indicated line are shown below the image: there

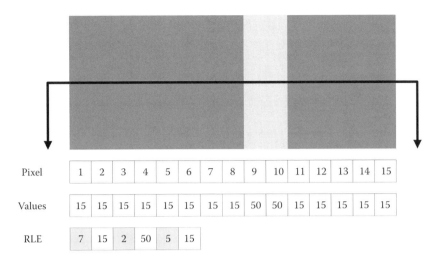

Pixel	1	2	3	4	5	6	7	8	9	10	11	12	13	14	15

Values	15	15	15	15	15	15	15	15	50	50	15	15	15	15	15

RLE	7	15	2	50	5	15

FIGURE 17.15
Run-length encoding is a simple lossless compression technique which works well with line images.

are 14 values in all, requiring 14 bytes to store them. But if, instead of storing the value of every pixel we store pixel count/pixel value pairs, we obtain the result in the RLE row. In this trivial example, the storage requirement has been reduced from 14 to 6 bytes. For line images, this process is very effective, but where there are continuous variations of grey shade, as in many medical images, the compression achieved is not great.

The requirement for a lossless compression technique should not require prior classification of the image to select the optimum method to use.

17.8.3.2 JPEG 2000

The DICOM standards recognise compression mechanisms compliant with the JPEG2000 standard for lossless compression, ISO/IEC 15444–1:2004 (ISO 2004). JPEG2000 compression can be either lossy, where it gives a modest improvement in compression performance, but increased flexibility in the processing of images, or lossless. As a lossless scheme, it has the advantage of not requiring significant prior knowledge of the image content. The remainder of this section is concerned only with lossless compression.

In the JPEG2000 process, the image is subjected to similar pre-processing to that used in JPEG compression: separating the signal into luminance and chrominance components. In the case of pure grey-scale images, of course, there is no chrominance component.

The image is then broken into tiles which are rectangular areas of the image that are then processed separately. In contrast to the JPEG process that uses a fixed 8×8 pixel tile, the tiles can be of any size, up to the whole image. All tiles are the same size, normally powers of 2 to simplify computation, with the possible exception of those on the right and lower boundaries of the image. Breaking the image up in this way has the advantage of reducing the storage required in processing the image, but introduces an additional source of noise that will degrade the reconstructed image.

The tiles are then *Wavelet Transformed*. Wavelet transformation involves passing the data recursively through a set of high- and low-pass filters, as shown in Figure 17.16. In the diagram, the pairs of blocks labelled HPF and LPF are pairs of complementary high- and low-pass filters, the circles labelled DS down-sample the filter outputs by a factor of two, essentially by taking the mean of pairs of samples.

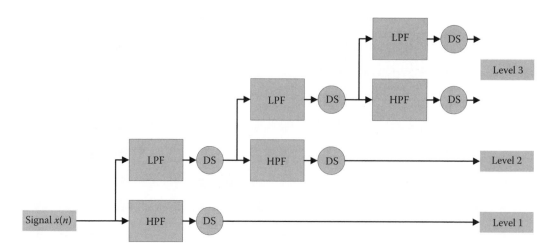

FIGURE 17.16
The wavelet transform process.

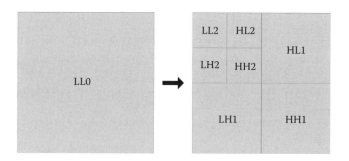

FIGURE 17.17
The core of the JPEG2000 compression process.

The filtering processes are one-dimensional; applying one-dimensional transforms in the horizontal and vertical directions produces two-dimensional transforms, yielding four smaller image blocks; one with low resolution, one with high vertical resolution and low horizontal resolution, one with low vertical resolution and high horizontal resolution, and one with all high resolution. This process of applying the one-dimensional filters in both directions is then repeated recursively on the low-resolution image block.

The total amount of information present is identical to that in the source image, so perfect reconstruction is possible; the reason for carrying out this step is that it makes it possible in subsequent processing to remove redundant information, and thereby reduce the size of the image data set.

The results of the wavelet transform can be presented as an array identical in size with the image tile they represent. Figure 17.17 shows the result of two stages of decomposition. The areas in the output, LL2, HL2, and so on are referred to as *sub-bands*, each of which is a set of real coefficients representing the content of the image in a particular geometric region and frequency band.

The sub-bands are further subdivided and then coded to remove the redundancy in the data, using a process known as Embedded Block Coding with Optimal Truncation.

JPEG 2000 lossless compression produces, on average, a factor of 2 reduction in image sizes, while permitting true reconstruction of the image. The performance of the technique is particularly good in images with low-contrast edges.

Conclusion

As in many other areas of life, digital technology has brought huge changes in the way that we acquire, use and exchange information. In medical imaging, PACS affects all aspects of the radiologist's working environment: the entire clinical team relies on the availability, facilities and performance of the system. Whilst it is possible to operate in this environment without an understanding of the underlying principles, it is surely better to work with and appreciation of the technical and scientific principles which allow PACS to work at all, but, more importantly, limit what can be expected of it. Laying the basis for this appreciation has been the goal of this chapter.

References

DICOM (2009) DICOM Standards. Online Available from http://medical.nema.org/ [accessed 24th January 2011].

European Parliament (2007) Directive 2007/47/EC of the European Parliament and of the Council of 5 September 2007. Online Available from http://ec.europa.eu/enterprise/medical_devices/revision_mdd_en.htm [accessed 24th January 2011].

ISO (1994a) ISO/IEC 7498–1:1994: Information technology—Open systems interconnection—Basic reference model: The basic model. Online Available from http://www.iso.org/iso/catalogue_detail.htm?csnumber=20269 [accessed 24th January 2011].

ISO (1994b) ISO/IEC 10918–1:1994: Information technology—Digital compression and coding of continuous-tone still images: Requirements and guidelines. Online Available from http://www.iso.org/iso/catalogue_detail.htm?csnumber=18902 [accessed 24th January 2011].

ISO (2004) ISO/IEC 15444–1:2004: Information technology—JPEG 2000 image coding system: Core coding system. Online Available from http://www.iso.org/iso/iso_catalogue/catalogue_tc/catalogue_detail.htm?csnumber=37674 [accessed 24th January 2011].

Nyquist H (1928) Certain topics in telegraph transmission theory. *Trans. AIEE*, vol. 47, pp. 617–644, April, 1928.

Shannon CE (1948) A mathematical theory of communication. *Bell System Technical Journal*, vol. 27, pp. 379–423, 623–656, July, October, 1948.

Further Reading

Brookshear J G (2008) *Computer Science: An Overview*. Pearson Education (A good introduction to computer science concepts, for those interested in exploring the basics of computing and computers).

Oakley J D (ed.) (2003) *Digital Imaging*. Cambridge University Press (This is a good, introductory level text covering many of the practicalities of selecting and using PACS).

Huang H K (2004) *PACS and Imaging Informatics—Basic Principles and Applications (2nd edition).* Wiley Blackwell (An in-depth, authoritative book covering the technical and operational aspects of PACS).

Exercises

1. What are the consequences of the Nyquist-Shannon theorem for the digital acquisition and storage of images? What are the differences in requirements between a CT scan image and a gamma camera image?

2. Why is it common practice in modern data networks to use VLANs? How does a VLAN differ from a subnet in an IP network?

3. Why could a PACS be classified as a 'Medical Device', and what are the implications of this?

4. How are the servers used in a typical PACS? How may the use of virtual servers affect this, and what are the advantages and disadvantages of server virtualisation?

5. What is the DICOM standard and what should the standard define? How does the form of the DICOM standard differ from that of the ISO standards?

6. Explain how a liquid crystal display works. What are the advantages and disadvantages for displaying diagnostic images when compared to a cathode ray tube?

7. Why is data compression important for radiological images and how can this be achieved? What are the essential differences between lossy and lossless compression?

18

Multiple Choice Questions

CONTENTS

The following multiple choice questions (MCQs) have been constructed so as to test your knowledge of the subject. Most answers are either given in the book or may be obtained by deductive reasoning. A few may require some additional reading, particularly for a fuller explanation.

The questions have been grouped in the same order as the chapter headings. There is no constraint on the number of parts of any question that may be right or wrong.

Unless otherwise stated, assume in all practical radiographic situations that any change in conditions is accompanied by an adjustment of mAs to give a similar signal at the receptor, for example, optical density on the film. Note that without this constraint many of the questions would be ambiguous.

18.1 MCQs

1.1 The nucleus of an atom may contain one or more of the following:
   a) Photons
   b) Protons
   c) Neutrons
   d) Positrons
   e) Electron traps

1.2 The atomic number of a nuclide:
   a) Is the number of neutrons in the nucleus of an atom
   b) Determines the chemical identity of the atom
   c) Will affect the attenuation properties of the material at diagnostic X-ray energies
   d) Is the same for 11-C as for 14-C
   e) Is increased when a radioactive atom decays by negative beta emission

1.3 Orbital electrons in an atom:
a) Contribute a negligible fraction of the mass of the atom
b) Are equal in number to the number of neutrons in the nucleus
c) Can be raised to higher energy levels when the atom is excited
d) Are sometimes ejected when atoms decay by electron capture
e) Play an important part in the absorption of X-rays

1.4 The following statements refer to K shell energies
a) An electron in the K shell has less energy than a free electron
b) The K shell energy is the same for all isotopes of an element
c) For elements with atomic numbers less than 20, the K shell energy has no role in diagnostic radiology
d) They are responsible for characteristic X-rays in an X-ray emission spectrum
e) They influence the sensitivity of some image receptors

1.5 All isotopes of an element:
a) Have the same mass number
b) Have the same physical properties
c) Emit radiation spontaneously
d) Have the same number of extra nuclear electrons in the neutral atom
e) When excited emit characteristic X-rays of the same energy

1.6 With respect to radioactive decay processes:
a) The activity of a radionuclide is 1 MBq if there are 1000 nuclear disintegrations per second
b) The half-life is 0.693 multiplied by the decay constant
c) After 10 half-lives the activity of a radionuclide will have decreased by a little more than 1000 times
d) Different radioisotopes of the same element always have different half-lives
e) The physical half-life is independent of the biological half-life in the body

1.7 The following are properties of *all* ionising radiations:
a) Decrease in intensity exponentially with distance
b) Have energy of at least 8 keV
c) Produce a heating effect in tissue
d) Produce free electrons in tissue
e) Act as reducing agents

1.8 The following statements are correct in respect of radionuclides:
a) They have energy levels in the nucleus
b) There is no change in mass number when gamma rays are emitted
c) There is a change in mass number when negative beta particles are emitted
d) Metastable states have half-lives of less than a minute
e) Orbital electron capture results in the emission of characteristic X-rays

1.9 The exponential process:
  a) Applies to both radioactive decay and absorption of monoenergetic gamma rays
  b) Is expressed mathematically by the inverse square law
  c) When applied to radioactive materials means constant fractional reduction in activity in equal intervals of time
  d) When applied to absorption of monoenergetic gamma rays means the half value thickness will be the same at any depth in the absorber
  e) Will never reduce the intensity of a beam of gamma rays to zero

2.1 The anode angle of a diagnostic X-ray tube:
  a) Is the angle the face of the target makes with the direction of the electron beam
  b) Is usually about 25 degrees
  c) Influences the size of the effective focal spot
  d) Determines the largest X-ray receptor area that can be adequately covered at a given distance from the focus
  e) Affects the amount of heat produced in the target for a given kV and mAs

2.2 Tungsten is used as the target material in an X-ray tube because it has the following desirable properties:
  a) A high atomic number
  b) A high density
  c) A high melting point
  d) A relatively low tendency to vaporise
  e) A high thermal conductivity

2.3 The effective focal spot size of an X-ray tube depends on:
  a) The size of the filament
  b) The anode material
  c) The anode angle
  d) The kV and mAs selected
  e) The amount of inherent filtration

2.4 Heat produced in the target of a rotating anode tube:
  a) Is a potential source of damage to the tube
  b) Helps the process of X-ray emission
  c) Accounts for a very large percentage of the incident electron energy
  d) Is dissipated mainly by radiation
  e) Reduces the speed of rotation of the anode

2.5 The photon energy of the characteristic radiation emitted from an X-ray tube:
  a) Is independent of the type of generator used
  b) Depends on the mA

    c) Decreases as the atomic number of the target material increases

    d) Must be less than the maximum energy of the electrons striking the target

    e) Depends on the material used for filtration

2.6 The quality of the X-ray spectrum emerging from an X-ray tube depends on:

    a) The peak voltage applied to the tube

    b) The waveform of the applied voltage

    c) The atomic number of the anode

    d) The angle of the anode

    e) The atomic number of the tube window

2.7 The following statements relate to the continuous part of X-ray spectrum:

    a) The maximum energy of X-rays generated at a fixed kVp is independent of the voltage profile applied to the tube

    b) The intensity of X-rays is influenced by the anode-cathode distance

    c) The intensity of X-rays is influenced by the atomic number of the anode

    d) There are two maxima

    e) It is the result of deceleration of electrons in the target

2.8 In carrying out quality assurance checks on X-ray equipment:

    a) An ionisation chamber may be used to detect leakage radiation

    b) Light beam alignment may be checked by imaging a rectangular metal frame.

    c) Total filtration may be estimated by measuring the half value thickness of aluminium for the beam

    d) An error in kVp setting will affect both image quality and patient dose

    e) With a digital receptor tube output is not checked because the software adjusts image quality

2.9 The following statements are true regarding the rating of an X-ray tube:

    a) It specifies maximum safe electrical and thermal operating conditions for a long tube lifetime

    b) Full-wave and half-wave rectified tubes have the same rating curves

    c) The maximum tube current is inversely proportional to the area of the focal spot target

    d) The limiting mAs increases with time of exposure because of cooling

    e) During fluoroscopy, a radiographic exposure causes a sharp increase in the temperature of the anode

2.10 The following are features of modern X-ray equipment:

    a) Pulse counting timers are more accurate than electronic timers

    b) A higher output is achieved with a fine focus spot than with a broad focus

    c) A rotating envelope tube increases heat loss from the anode by radiation

    d) A metal-walled tube envelope reduces extra-focal radiation

    e) A graphite block brazed onto the back of the anode increases the heat radiating efficiency

2.11 Advantages of a high-frequency generator compared to a single phase full-wave rectified generator are as follows:

a) A higher tube rating for short exposures

b) Near maximum loading may be applied to the tube throughout the exposures

c) Shorter exposure times

d) Higher repetition rate

e) A higher maximum energy of photons.

2.12 Two X-ray tubes, both with tungsten anodes, are set to operate at 80 kVp. One has 100% ripple, the other has 5% ripple. Which statements are true?

a) The tube with 100% ripple is driven by a high-frequency generator

b) With 100% ripple, the whole spectrum is shifted to the right relative to the 5% ripple spectrum

c) The tube with 100% ripple gives the higher output (area under the curve)

d) Both spectra show characteristic lines

e) The tube with 5% ripple will give more intense characteristic lines

2.13 The following statements are true concerning X-ray beam filtration:

a) The inherent filtration of a conventional X-ray tube is about 0.5–1.0 mm

b) High inherent filtration is essential for mammography

c) Added filtration becomes unnecessary above 125 kV

d) It reduces the skin dose

e) It reduces the integral absorbed dose

3.1 The photoelectric attenuation process:

a) Occurs when photons collide with atomic nuclei

b) Decreases continuously with increasing radiation energy

c) Is negligible in water at 120 keV

d) Gives rise to scattered radiation outside the patient's body

e) Is the main reason for lead being such a good protective material for diagnostic X-rays

3.2 If the first half value thickness in aluminium for a beam of X-rays from an X-ray tube operating at 80 kVp is 2 mm:

a) The intensity of the beam will be halved if 2 mm of aluminium is placed in the beam

b) The intensity of the beam will be reduced by 75% if 4 mm of aluminium is placed in the beam

c) The half value thickness would be greater if measured in copper

d) The half value thickness in aluminium would be less if the tube were operated at 100 kVp

e) The half value thickness in aluminium would be less if the tube current (mA) were reduced

3.3 When X-rays interact with matter, the following statements are true:

a) Absorption of X radiation by the photoelectric effect results in annihilation radiation

b) The photoelectric effect usually predominates over the Compton effect in X-ray computed tomography

c) In the Compton effect the energy of the incident photon is shared between an electron and a scattered photon

d) In the Compton effect the majority of photons are scattered backwards at the scattering centre at diagnostic energies

e) Pair production is always accompanied by the release of 0.51 MeV radiation

3.4 The first half value thickness for 30 keV gamma rays in water is 20 mm. The following statements are true:

a) The second half value thickness is also 20 mm

b) 60 mm of water will stop 95% of 30 keV X-rays

c) The half value thickness in healthy lung for 30 keV gamma rays is more than 20 mm

d) The half value thickness of water for X-rays from a tube operating at 30 kVp is 20 mm

e) The second half value thickness of water for X-rays from a tube operating at 30 kVp is greater than the first half value thickness

3.5 When an X-ray photon in the diagnostic energy range passes through matter it may:

a) Be deflected through more than 90° and lose energy

b) Be deflected without any change of energy

c) Be transmitted without any change in energy or direction

d) Cause the production of characteristic radiation

e) Undergo an interaction resulting in the creation of two particles with the same kinetic energy

3.6 The tenth value layer of a diagnostic X-ray beam is as follows:

a) The depth at which the beam intensity has fallen to one-tenth

b) The depth at which the maximum X-ray energy has fallen to one-tenth

c) Approximately five times the half value layer in soft tissue

d) Greater in materials of higher atomic number

e) Independent of the voltage supply

3.7 The following statements refer to the attenuation, scatter and absorption of X-rays:

a) Scatter may occur without loss of energy

b) Scatter to the image receptor has to be eliminated in diagnostic radiology

c) Attenuation represents the combined effects of scatter and absorption

d) The Compton effect causes both scatter and absorption

e) The mass attenuation coefficient determines the radiation dose to the patient

3.8 When 60 keV photons of gamma ray energy interact with a crystal of sodium iodide:

a) All the photons undergo a photoelectric interaction

b) Each gamma ray photon releases about 10 visible light photons

c) The number of visible light photons may be increased by adding certain impurities to the crystal

d) Visible light is released from many different depths in the crystal

e) The wavelength of light produced is directly proportional to the wavelength of the absorbed gamma photon

3.9 The following statements refer to X-ray beam filtration:

a) The quality of the X-ray beam is affected

b) Molybdenum K edge filters are always used in mammography

c) The principal purpose is to reduce patient dose

d) Copper together with aluminium may be used in high kV equipment

e) High atomic number materials are unsuitable as filters in diagnostic X-ray sets

4.1 Absorbed dose can be specified in the following:

a) Joules/kg

b) Millisieverts

c) Gray

d) Coulombs/kg

e) Electron volts

4.2 Absorbed dose:

a) Is a measure of the energy absorbed per unit mass of material

b) Is the energy released in tissue per ion pair formed

c) May be determined from an ionisation chamber measurement

d) Is different in different materials exposed to the same X-ray beam

e) Decreases exponentially as the X-ray beam from an X-ray tube passes through soft tissue

4.3 Ionisation in air is chosen as the primary radiation standard because:

a) The mean atomic number is close to that of soft tissue

b) Free air ionisation chambers can be compact

c) Air is readily available and its composition is close to being universally constant

d) A large response is produced for a small amount of radiation energy

e) The voltage applied to the capacitor plates to collect the charge can be kept very constant

4.4 A free air ionisation chamber is totally irradiated with 100 kV X-rays. It is more sensitive, that is, produces a larger current, for a given exposure rate, if:

a) Its air volume is increased

b) The ambient temperature is greater

c) The ambient pressure is greater

d) The effective atomic number of the wall material is less than that of air

e) The voltage across the electrodes is less than that necessary to saturate the chamber

4.5 The following are features of practical ionisation chamber monitors:

a) A secondary ionisation chamber uses an 'equivalent wall' to provide a more compact instrument

b) Filling the chamber with high atomic number inert gas at high pressure provides a more sensitive instrument

c) Dose area product meters measure the total energy deposited in the patient

d) Pocket exposure meters cannot be used for personnel dosimetry below about 50 keV

e) A dose calibrator in nuclear medicine has 4π geometry for radiation detection

4.6 If the dose rate in air at 100 cm from an X-ray machine when operating at 100 kV and 100 mA is 600 mGy/min then:

a) The dose rate at 50 cm will be 2.4 Gy/min

b) The dose rate at 25 cm will be 4.2 Gy/min

c) The dose rate at 100 cm will be halved if the machine is operated at 100 kV and 25 mA

d) The dose rate will remain unchanged if the kV is altered

e) Placing an extra filter in the beam will decrease the dose rate

4.7 The following are features of a Geiger-Müller counter used for measuring radioactivity:

a) It has a rather poor efficiency for detecting gamma rays

b) A thin window is necessary to detect beta radiation

c) The collection volume is open to the atmosphere

d) The operating voltage may be similar to that of an ionisation chamber

e) It is better than a scintillation crystal monitor for checking surface contamination in the diagnostic nuclear medicine department

4.8 Semiconductor detectors have the following properties:

a) Amorphous silicon is a commonly used material

b) The forbidden band of energy levels must be small

c) Electrons are promoted from the valence band into the forbidden band

d) The thickness of the depletion layer in an n-p diode detector must be large to achieve high sensitivity

e) The thickness of the depletion layer in an n-p diode detector is affected by the polarity of the voltage supply across it

4.9 Uses of semiconductor diode detectors include the following:

a) Measuring and checking the time of an exposure

b) Non-invasive measurement of tube kV

c) Accurate measurement of mixed beta and gamma radiation fields

d) High precision measurement of X- and gamma-ray photon energies

e) Personal dosimetry over several orders of magnitude of dose

4.10 In comparison with a Geiger-Müller counter, a scintillation counter:
   a) Is more sensitive to gamma rays
   b) Is more portable
   c) Can be adapted more readily for counting radiation from one radionuclide in the presence of another
   d) Is better for dealing with high count rates
   e) Requires less complex associated apparatus

5.1 The following statements relate to the characteristic curve of an X-ray film:
   a) The curve is plotted on axes of optical density against exposure
   b) Film speed is the exposure to cause an optical density of 1.0
   c) The maximum slope of the curve is the film gamma
   d) The film latitude is the X-ray exposure range that will produce useful optical densities
   e) At very high exposures the optical density may start to decrease

5.2 When X-rays interact with photographic film the process depends on:
   a) Photoelectric interactions with the gelatine
   b) Impurities in the crystal structure
   c) Heat generation by the X-rays
   d) Formation of clusters of neutral silver atoms
   e) Removal of excess silver by the developer

5.3 Film fog is increased by the following:
   a) Storage of film at high temperature
   b) Prolonged storage of film before use
   c) Use of rare earth screens
   d) Excessively strong fixation
   e) Poor viewing conditions

5.4 Intensifying screens:
   a) Emit electrons when bombarded with X-rays
   b) Contain high atomic number nuclides to improve X-ray stopping efficiency
   c) Reduce the patient dose by at least a factor of 10
   d) Increase the sharpness of a radiograph
   e) Reduce the amount of quantum mottle in the image

5.5 In an image intensifier:
   a) Caesium iodide is chosen for the input phosphor because its light emission is well matched to the spectral response of the eye
   b) Sensitivity is improved by increasing the input phosphor thickness
   c) Electrons emitted by the input phosphor are focused onto the output phosphor
   d) Electrons are accelerated by potential differences of about 25 kV
   e) Image brightness is increased by minification by a factor of about 10

5.6 The following statements refer to the phosphor layer in an image receptor:

a) Matching the K shell edge to the X-ray spectrum increases absorption efficiency

b) Addition of dye to the phosphor increases absorption efficiency

c) Increasing the kVp increases the absorption efficiency

d) A thicker phosphor increases efficiency of conversion of X-ray energy into light

e) Rare earth phosphors increase the efficiency of conversion of X-ray energy into light

5.7 Receptors for digital radiography (DR) have the following properties:

a) Detector arrays can cover a large area (30 cm × 40 cm)

b) In an indirect detector the initial X-ray detector is a phosphor

c) In a direct detector the initial X-ray detector is amorphous silicon

d) Image blurring by visible light spreading is only a problem with indirect detectors

e) Electronic noise and quantum noise may be a problem with all receptors

5.8 The following statements refer to receptors for computed radiography (CR)

a) The phosphor screen is a re-usable plate

b) The wavelengths of the laser stimulating light and the output light are similar

c) For general radiology the spatial resolution of CR is better than that of film screen

d) The CR receptor latitude is better than film-screen latitude

e) CR receptors and DR receptors have similar matrix sizes and resolutions

5.9 Charge coupled devices (CCDs) have the following advantages over television cameras:

a) A single device can cover a larger field of view

b) The resolution does not vary with field size

c) A scanning beam of electrons is not necessary

d) The output is a digital image

e) CCDs can record images directly from X-rays

6.1 The following statements are true regarding the radiographic image:

a) Increasing the field size increases markedly the geometric unsharpness

b) The heel effect is most pronounced along the anode side of the X-ray beam

c) The visible contrast on a fluorescent screen between two substances of equal thickness depends on the differences between their mass attenuation coefficients and densities

d) For an X-ray generator operating above 150 kV bone/soft tissue contrast is less than soft tissue/air contrast

e) The limiting spatial resolution between different parts of the image depends on the contrast

6.2 The exposure time for taking a radiograph:
   a) Is decreased if the kV is increased
   b) Is increased if the mA is increased
   c) Is decreased by increasing the added filtration
   d) Depends on the thickness of the patient
   e) Is decreased if a grid is used

6.3 In a normal radiographic set up:
   a) A shorter focus-receptor distance reduces the magnification
   b) The smaller the object the greater the magnification
   c) The image of an object increases in size as its distance from the image receptor increases
   d) The image of an object increases in sharpness as its distance from the image receptor decreases
   e) The magnification depends on the focal spot size

6.4 The following statements relate to the correct use of linear grids:
   a) An increased grid ratio will reduce the scattered radiation reaching the image receptor
   b) An increased grid ratio will reduce the field of view
   c) The width of the lead strips in the grid must be less than the required resolution
   d) The optical density on a film will be reduced towards the edges
   e) The dose to the patient will increase by the ratio lead strip thickness/interspace distance

6.5 In abdominal radiography the amount of scattered radiation reaching the image receptor may be reduced, relative to the primary beam by:
   a) Increasing the kV
   b) Increasing beam filtration
   c) Increasing the tube-patient distance
   d) Increasing the patient-receptor distance
   e) Abdominal compression

6.6 Contrast on an analogue radiographic image is increased by the following:
   a) Increasing the distance from the patient to the film
   b) Decreasing the focal spot size
   c) Increasing the filtration
   d) Decreasing the field size
   e) Using a film of higher gamma

6.7 The following statements refer to the processing of digital images:
   a) Smoothing filters reduce the impact of noise due to counting statistics
   b) Edge enhancement is an example of a point operation

    c) Image segmentation might help to decide if a tumour nodule had increased in size between sequential images

    d) Histogram equalisation can assist automatic collimation

    e) Inverting an image results in negative values for some pixels

6.8 Regarding noise in a digital image:

    a) Adding two images will reduce the quantum noise

    b) Quantum mottle is a consequence of dividing the image into pixels

    c) High pass image filtering increases noise

    d) Increasing the matrix size increases the noise

    e) Increasing the dose of radiation reduces electronic noise

6.9 The following statements relate to digital images:

    a) Histogram equalisation adjusts the grey scale such that all pixel values are equal

    b) Windowing selects a limited range of pixel values for display

    c) Look-up tables condense the data levels to the number the eye can readily distinguish

    d) The images are unsuitable for ROC analysis

    e) Image edge enhancement can reduce the effect of noise

6.10 The following statements relate to digital radiography:

    a) Digital information is any information presented in discrete units

    b) A variable window width facility is a means of manipulating contrast

    c) If a uniform source is examined with an ideal imager, variations in grey scale across the image may be reduced by increasing the radiation dose

    d) If N frames from a digitised set of images are added, the fractional variation in statistical noise is reduced by a factor N

    e) Digital data must be converted to analogue form before storage

7.1 In relation to the radiographic image:

    a) Contrast between two parts of the film is the ratio of their optical densities

    b) Full width half maximum is a measure of resolution

    c) Modulation transfer function is a method of measuring contrast

    d) Noise increases with increasing patient dose

    e) Fine detail corresponds to high spatial frequencies

7.2 The following statements correctly relate to the radiographic image:

    a) The ability to see fine detail depends on ambient light intensity

    b) The minimum perceptible contrast between an object and its background decreases with increasing object size

    c) Definition can be related to the image of a sharp edge

    d) Systematic noise has no effect on the task of perception

    e) Resolution can be derived from the modulation transfer function measured on a low contrast object

7.3 The amount of quantum noise in the image produced by an image intensifier and television system for fluoroscopy can be reduced by the following:
a) Increasing the screening time
b) Increasing the exposure rate
c) Use of thicker intensifier input screens
d) Increasing the brightness gain on the image intensifier
e) Reducing electronic noise in the television chain

7.4 In the analysis of the spatial frequency response of an imaging system which uses an image intensifier as the receptor:
a) The modulation transfer function (MTF) is normalised to one at zero spatial frequency
b) The resultant MTF is the sum of the MTFs of the components
c) MTF and line spread function (LSF) contain the same information
d) The MTF always decreases with increasing spatial frequency
e) The resultant MTF is independent of the design of the X-ray tube

7.5 The following changes will improve the contrast to noise ratio of a digital image:
a) Increasing the focus to receptor distance
b) Decreasing the matrix size
c) High pass filtering
d) Frame averaging
e) ROC analysis

7.6 In the assessment of a diagnostic imaging procedure:
a) The line spread function is an effective measure of resolution
b) The modulation transfer function measures the performance of the complete imaging system
c) The sum of the true positive and false positive images will be constant
d) A strict criterion for a positive abnormal image is a better way to discriminate between two imaging techniques than a lax criterion
e) Bayes Theorem provides a way to allow for the prevalence of disease

7.7 Receiver operator characteristic (ROC) curves:
a) Provide a means to compare the effectiveness of different imaging procedures
b) Are obtained when contrast is varied in a controlled manner
c) Normally plot the false positives on the X axis against the false negatives on the Y axis
d) Require the observer to adopt at least five different visual thresholds
e) Reduce to a straight line at 45 degrees when the observer guesses

7.8 Computer aided diagnosis:
a) Can only be used with digitised images
b) Has been used both as a 'stand alone' system and to provide prompts to an expert observer

    c)   Incorporates image enhancement techniques

    d)   Is dependent on a library of normal images

    e)   Gives no information on the clinical nature of the abnormality

7.9 In optimisation studies, the relationship between dose and image quality for a film screen and a DR receptor may be different because:

    a)   One is a digital receptor, the other analogue

    b)   The materials used in the two receptors are different

    c)   The systems are operating at different kVps

    d)   The MTFs of the two systems are different

    e)   K shell energies vary with atomic number

8.1 In X-ray computed tomography:

    a)   Compton interactions predominate in the patient

    b)   A semiconductor detector is more sensitive than a xenon detector

    c)   Filtered back projection is a method of data reconstruction

    d)   The Hounsfield CT number for water is zero

    e)   Small differences in contrast are more easily seen by decreasing the window width

8.2 In X-ray computed tomography:

    a)   X-ray output must be more stable than in conventional radiology

    b)   The detectors are normally photomultiplier tubes

    c)   Pixel size places a limit on spatial resolution

    d)   A pixel may have a range of attenuation values if it contains several different tissues

    e)   Hounsfield numbers may be changed by altering the window level and window width

8.3 In longitudinal tomography:

    a)   The thickness of plane of cut increases with angle of swing

    b)   The contrast decreases with increasing angle of swing

    c)   The image of the plane of cut is magnified

    d)   Objects equidistant from the plane of cut, above and below, are equally blurred

    e)   Digital techniques cannot be used on the images

8.4 CT is an abbreviation for Computed Axial Transmission Tomography. Which of the following terms applies to digital tomosynthesis?

    a)   Computed

    b)   Axial

    c)   Transmission

    d)   Tomography

    e)   All of these

8.5 X-ray computed tomography:
   a) Is more sensitive than conventional radiography to differences of atomic number in tissues
   b) Can only resolve objects greater than 10 mm in diameter
   c) Depends on the different attenuation of X-rays in different tissues
   d) Is usually performed with X-rays generated below 70 kV
   e) Was originally used to obtain pictures of transaxial body sections

8.6 In X-ray computed tomography:
   a) Modern CT scanners have more than 1000 detectors
   b) The partial volume effect occurs when the X-ray beam does not pass through part of the patient cross section
   c) Filtered back projection provides additional attenuation between the patient and the detector
   d) Hounsfield CT numbers may be either positive or negative
   e) Decreasing the window width is a mechanism for visualising small differences in contrast

8.7 In X-ray computed tomography:
   a) The majority of X-ray interactions in the patient are inelastic collisions
   b) Xenon gas detectors are no longer used because they show too much variation in sensitivity
   c) Afterglow is a serious problem when quantum noise is high
   d) Resolution is improved by increasing the pixel size
   e) The amount of noise in the image can be estimated by imaging a water phantom

8.8 In spiral CT:
   a) Anodes with a high heat storage capacity are essential.
   b) If the table feed/rotation is 10 mm s^{-1} and the nominal beam thickness 5 mm, the pitch is 0.5
   c) The image is reconstructed from data collected in several different planes
   d) Tube current modulation cannot be applied
   e) Retrospective selection of slice starting point facilitates detection of small lesions

8.9 In multi-slice CT:
   a) Adaptive arrays use combinations of detectors of different widths.
   b) Data collection is slower than in single-slice CT
   c) The X-ray focal spot moves to ensure that all beam projections are parallel
   d) An important feature is improved resolution in the z-direction
   e) Iterative methods allow image reconstruction to begin before all profiles have been collected

8.10 In computed tomography the causes of certain artefacts in the image are as follows:
  a) Streak artefacts—patient movement
  b) Ring artefacts—mechanical misalignment
  c) Rib artefacts—beam hardening
  d) Low-frequency artefacts—under-sampling high spatial frequencies
  e) Partial volume effects—beam width not including the whole pixel

9.1 In asymptomatic mammographic screening:
  a) X-rays generated below 20 kV must be used
  b) The X-ray tube may have a rhodium target
  c) A focus film distance of at least 1 m is necessary
  d) Dose to the normal-sized breast must be less than 2 mGy per view
  e) A DR system (flat panel detector) can be fitted retrospectively to a film-screen mammography unit

9.2 The spectrum of X-rays emitted from an X-ray tube being used for mammography:
  a) Will contain characteristic X-rays if a molybdenum anode is being used
  b) Is likely to contain X-rays with a maximum energy of 50 keV
  c) At fixed kilovoltage, has a total intensity that depends on the atomic number of the anode
  d) Is independent of the tube window material
  e) Will pass through additional filtration before the X-rays fall on the patient

9.3 The following statements relate to high kV imaging:
  a) Scattered radiation is a major problem
  b) Exposures can be shorter than at low kV
  c) Problems associated with the limited dynamic range of flat panel digital detectors can be eliminated
  d) Photoelectric interactions contribute very little to subject contrast
  e) Increasing the kV has no effect on magnification or distortion

9.4 If the focus film distance for a particular examination is increased from 80 cm to 120 cm with the image receptor kept as close to the patient as possible:
  a) The image is magnified more
  b) Geometric unsharpness is reduced
  c) A larger film should be used
  d) The entrance skin dose to the patient is reduced
  e) The exit skin dose to the patient is unchanged

9.5 Requirements for good magnification radiography include the following:
  a) A tube-patient distance of at least 1 m
  b) A fine focus X-ray tube
  c) A higher mAs than conventional radiography of the same body part at the same kV

d) Stationary grids

e) Image receptors with higher intrinsic resolution

9.6 In digital subtraction angiography the logarithm of pixel values is taken before subtraction to remove the effect of variation in the following:

a) Overlying tissue

b) Underlying tissue

c) Scatter

d) X-ray tube output

e) Noise

9.7 In digital subtraction angiography the input dose rate to the patient:

a) Is reduced by using 0.2 mm copper

b) Is increased as the preferred method to improve image quality

c) Can be monitored using a dose area product meter

d) Must be increased with decreasing field size to maintain image quality

e) Is reduced when faster frame rates are used

9.8 The following would be features of a state-of-the-art system for interventional work:

a) 2 mm or 4 mm selectable focal spot size

b) Adjustable pulse frequencies

c) Variable source-detector distance

d) 0.2–3 mm variable aluminium filtration

e) Higher dose rates for digital cine mode than fluoroscopy

9.9 For a PA projection of the chest of a small child or neonate:

a) A grid should normally be used to reduce scatter

b) Exposure times may be as short as a few milliseconds

c) More added filtration is used than for the corresponding adult examination

d) Criteria of image quality for an adult must be modified

e) An acceptable entrance surface dose for a neonate would be 200 µGy

10.1 Technetium-99m is a suitable radionuclide for imaging because:

a) Its half-life is such that it can be kept in stock in the department for long periods

b) Its gamma ray energy is about 140 keV

c) It emits beta rays that can contribute to the image

d) It can be firmly bound to several different pharmaceuticals

e) A high proportion of disintegrations produce gamma rays

10.2 A molybdenum-technetium generator contains 3.7 GBq of molybdenum-99 at 0900 hours on a Monday. The activity of technetium-99m in equilibrium with the molybdenum-99:

a) At any time depends on the half-life of technetium-99m

b) Is 3.7 GBq at 0900 hours on the Monday

c) Depends on the temperature

d) Will be about 370 MBq at 09:00 hours on the following Friday

e) At any time will be reduced if the generator has been eluted

10.3 Desirable properties of a radionuclide to be used for skeletal imaging are as follows:

a) Adequate beta ray emission

b) Gamma ray emissions of two distinct energies

c) A non-radioactive daughter

d) A medium atomic number

e) A half-life of less than 10 min

10.4 The effective dose to an adult patient from a radionuclide bone scan:

a) Is affected by radiochemical purity

b) Increases with increasing body weight

c) Is reduced if the patient empties their bladder before the scan

d) Is typically about 15 mSv per examination

e) Is high enough for the patient to be designated as a classified person

10.5 The amount of activity administered to a child is scaled down from adult activity:

a) By body weight or body surface area

b) Maintaining image quality and noise levels

c) To reduce the effective dose

d) According to the child's age

e) Should not go below a recommended minimum

10.6 When using a collimator with a gamma camera:

a) The main purpose of the collimator is to remove scattered radiation

b) A high resolution collimator is required for dynamic studies

c) High energy collimators give poorer resolution than low energy collimators

d) With a parallel hole collimator resolution is independent of depth

e) A pin-hole collimator will magnify a small object

10.7 In a gamma camera:

a) Resolution increases with increasing thickness of the crystal

b) Non-linearity is caused more by the scintillation detector than by the photo-multiplier tubes

c) The pulse height analyser eliminates all scattered photons

d) The pulse height analyser enables two radionuclides to be detected at the same time

e) Monoenergetic gamma rays produce light flashes of fixed intensity in the detector

10.8 The resolution of a gamma camera image is affected by the following:

a) The septal size of the collimator

b) The energy of gamma rays

c) The count rate

d) The number of photomultiplier tubes

e) The window width on the pulse height analyser

10.9 The resolution of a SPECT image is as follows:

a) Better than the resolution of a planar radionuclide image of the same object

b) Dependent on the radius of rotation

c) Independent of the acquisition time and count rate

d) Worse than a CT image of the same object

e) Dependent on the reconstruction filter applied to the projection data

10.10 The design of a new tomographic gamma camera should include the following:

a) Small field of view detectors to optimise resolution

b) Multiple detectors to improve detection efficiency

c) Slip-ring technology to increase flexibility and speed of rotation

d) A 15 mm thick scintillation crystal to improve sensitivity

e) Dedicated software for resolution recovery

11.1 The following radionuclides are suitable for PET imaging:

a) Tritium (H-3)

b) Carbon-11

c) Fluorine-18

d) Rubidium-82

e) Molybdenum-99

11.2 The following are correct descriptions of detection events in PET:

a) Single events describe the response of the detectors to a small amount of gamma emitter injected for calibration

b) Scattered events are the result of elastic scattering in the patient

c) For true coincidences, a small time interval between the two detector responses is permissible

d) In a scattered event, both photons may arise from the same disintegration

e) Random coincidences are the result of interactions between cosmic radiation and an annihilation photon

11.3 Regarding the resolution limit of a clinical PET imager:

a) It is the sum of the contributions from the detectors, positron range and non-colinearity of the annihilation photons

b) The non-colinearity contribution is greater than for a small animal imager

c) For F-18 the positron range makes little contribution to the system resolution

d) The system spatial resolution is comparable with that of a gamma camera

e) Unlike a gamma camera, the resolution is not uniform over the field of view of the detector

11.4 In quantitative PET, corrections to the data may be required for the following:

   a) Detector dead time

   b) Scatter effects

   c) Radioactive decay

   d) Partial volume effects

   e) Quantum effects

11.5 In PET/CT:

   a) The patient is moved from one imager to the other as quickly as possible.

   b) The CT scan is used to provide attenuation data

   c) The CT scanner is adjusted to operate at the same effective keV as the radiation from the positron emitter

   d) Patient movement in the two scans will be similar since the imaging times are comparable

   e) For radiotherapy treatment planning the two imaging methods give good agreement on gross tumour volume

11.6 The following statements relate to the whole body dose to the operator from a PET scan using 375 MBq of F-18 FDG:

   a) Injecting and positioning the patient are major sources of dose

   b) The operator would probably have to be classified if engaged in more than 3 scans/week

   c) Ingestion of radioactivity will contribute negligibly to the dose

   d) Doses in the control room are at safe public levels

   e) Doses from samples taken from the patient can be disregarded because of the short half-life of the radionuclide

12.1 Equivalent dose is:

   a) Affected by radiation weighting factors

   b) The absorbed dose averaged over the whole body

   c) For X-rays numerically equal to the absorbed whole body dose in grays

   d) Used to specify some dose limits

   e) Used to specify doses to patients from X-ray examinations

12.2 The following statements refer to the biological effects of radiations on cells and tissues:

   a) The risk of fatal cancer is higher than the risk of severe hereditary disease

   b) A tissue reaction increases steadily in severity from zero dose

   c) Stochastic effects may be caused by background radiation

   d) Chromosomal aberrations are strong evidence that ionising radiation causes mutations in humans

   e) Equal equivalent doses of different radiations cause equal risk to different tissues

12.3 With respect to the responses of cells to ionising radiation:

   a) Stem cells are generally more sensitive than differentiated cells

b) Cellular repair is apparent by 6 h after irradiation

c) Dose rate effects are evidence of recovery

d) The RBE of neutrons (*in vitro*) will increase with increasing dose

e) Cells show more evidence of recovery after neutron irradiation than after X-ray irradiation

12.4 The following statements relate to the long-term effects of radiation:

a) Stochastic risks are additive when radiation doses are fractionated

b) Mutagenic effects have been demonstrated statistically following radiation exposure of humans

c) Early transient erythema is a tissue reaction

d) There is no evidence of radiation-induced malignancy from the diagnostic use of X-rays

e) Radiation-induced solid tumours may not appear until 40 years after exposure

12.5 The dose and dose rate effectiveness factor (DDREF):

a) Is derived from radiobiological models for which there is no direct evidence

b) Is applied when extrapolating risk from high doses to low doses

c) Is used to calculate risk in radiological examinations when only part of the body is irradiated

d) Is a correction applied by ICRP when estimating nominal risk coefficients at low doses

e) Is more important for diagnostic X-rays than for high energy X-rays

12.6 The following are early effects of ionising radiation:

a) Toxic chemicals produced

b) Double strand breaks in DNA

c) Homeostatic repair of stem cells

d) Reactive free radicals generated by photoluminescence

e) Apoptosis

12.7 Tissue reactions are characterised by the following:

a) The dose threshold varies from one tissue to another

b) The response is a consequence of interaction of radiation with germ cells

c) Repair and recovery

d) The bystander effect

e) The dose threshold varies from one patient to another

12.8 The survey of Japanese Atomic Bomb survivors is the most important source of quantitative data on low dose radiation risk to humans for the following reasons:

a) A large population of all ages and both genders was exposed

b) Exposure was to X-rays only

c) Cancer incidence and mortality data are available

    d)   Many survivors were exposed to doses in the diagnostic range

    e)   Cancers appeared earlier in children than in adults

13.1   The exit dose from a patient in the primary X-ray beam:

    a)   Is caused primarily by backscattered electrons

    b)   Depends on X-ray field size

    c)   Depends on X-ray focal spot size

    d)   Decreases if tube voltage is increased with mAs constant

    e)   Increases if tube filtration is increased but other factors remain constant

13.2   The following statements relate to attenuation of an 80 kVp X-ray beam in the patient:

    a)   It is independent of mAs

    b)   It results in variations in exit dose for different investigations

    c)   It may result in different doses to the critical organ for AP and PA projections

    d)   It is independent of field size

    e)   It gives a dose at 20 cm depth that is about half the dose at 10 cm depth

13.3   Absorbed dose:

    a)   Is directly proportional to the mAs

    b)   Is typically a few micro Gy at the skin surface in a diagnostic examination of the abdomen

    c)   Is less in fat than in muscle for a given exposure of 30 kVp X-rays

    d)   Is about the same in bone and muscle for a given exposure of 30 kVp X-rays

    e)   Is the only factor determining biological effect

13.4   If the gonads are outside the primary beam during the taking of a radiograph, the dose to them will be reduced if:

    a)   A grid is used

    b)   A more sensitive image receptor is used

    c)   The field size is reduced

    d)   A smaller focal spot size is selected

    e)   The patient-receptor distance is increased

13.5   A typical P-A chest radiograph will:

    a)   Include the female gonads in the direct beam

    b)   Give a dose to the male gonads of about 10 µGy

    c)   Give a skin dose to the patient in the direct beam of about 2 mGy

    d)   Use X-rays generated above 60 kV

    e)   Give a lower effective dose than an A-P chest radiograph

13.6   In computed radiography using photostimulable screens the entrance dose to the patient on the axis of the beam is reduced by the following:

    a)   Increasing the kV

    b)   Using a grid

    c) Patient compression

    d) Using a small focal spot

    e) An air gap technique

13.7 Tissue weighting factors:

    a) Allow for the fact that all tissues are not equally sensitive to radiation

    b) Must sum to 1.0 for all tissues

    c) Have SI units of sievert

    d) Allow effective doses to be calculated very accurately

    e) Are different for males and females

13.8 The effective dose to a patient from a single plane film radiograph is as follows:

    a) Decreased at lower mA

    b) Decreased by the use of grids

    c) Decreased by using a faster film

    d) Decreased by decreasing the field size

    e) Always less than 5 mSv

13.9 The following dosimetric quantities are approximately correct:

    a) The effective dose from a barium enema is 7 mSv

    b) The annual whole body equivalent dose limit for members of the public is 15 mSv

    c) A chest X-ray entrance skin dose is 150 µGy

    d) The effective dose to a patient from a lung perfusion scan with 100 MBq Tc-99m is 1 mSv

    e) The average annual effective dose to the UK population from medical exposures is 380 µSv

13.10 The risk from injected radioactivity:

    a) Is independent of the biological half-life in the body

    b) Depends on both the radionuclide and on the pharmaceutical form

    c) Decreases steadily with time after injection

    d) May be primarily due to the dose to a single organ

    e) Is a factor limiting the quality of radionuclide images

13.11 The following statements relate to risks from radiological examinations:

    a) The risk of fatal cancer from a lumbar spine examination is about 5 in 10^5

    b) An imposed risk of 1 in 10^4 is likely to be challenged by the general public

    c) A PA examination of the chest involves a smaller risk than a single anaesthetic

    d) The risk from a CT examination varies from about 1 in 10^4 to 1 in 10^5

    e) Diagnostic doses do not cause tissue reactions

13.12 The following statements relate to the use of diagnostic X-rays during pregnancy:

    a) The dose to the uterus is usually calculated to obtain an estimate of the dose to the foetus

    b) Severe mental retardation is the most serious risk during the first few weeks of pregnancy

    c)   Hereditary effects may be no higher than after birth

    d)   The effective dose to the foetus cannot be higher than the effective dose to the mother

    e)   The risk of cancer to age 15 will be about 1 in 13,000 for each 1mGy of absorbed dose to the foetus

13.13 Diagnostic doses of radiation to the uterus during pregnancy may reach levels where the following adverse effects to the foetus are a cause for concern:

    a)   Carcinogenesis

    b)   Lethality

    c)   Mental retardation

    d)   Malformations and developmental effects

    e)   Hereditary effects

14.1 Thermoluminescent dose meters:

    a)   May be made of lithium chloride

    b)   Have a longer useable period than film

    c)   Can only be used for measuring doses over 10 mGy

    d)   Can be used for *in vivo* measurements in body cavities

    e)   Have a response which is a good measure of tissue dose over a wide range of energies

14.2 Photographic film as a dose meter:

    a)   Measures dose rather than dose rate

    b)   Is commonly used for personnel monitoring

    c)   Can cover a wide range of dose

    d)   Has a response which is independent of the radiation quality

    e)   Requires carefully controlled development conditions

14.3 Thermoluminescent dosimetry:

    a)   Depends on the emission of light from an irradiated material after heating

    b)   Can only be used in a very restricted dose range

    c)   Gives an indication of the type of radiation causing the exposure

    d)   Is an absolute method of determining absorbed dose

    e)   Is used for routine personal monitoring

14.4 In a film badge:

    a)   The film is single coated

    b)   The open window enables beta ray doses to be assessed

    c)   The lead-tin filter is to protect the film from fogging

    d)   Doses in air of 100 mGy diagnostic X-rays can be measured

    e)   Alpha-particles will not be detected

14.5 The lead equivalent of a protective barrier:

    a)   Is the amount of lead the barrier contains

    b)   Depends on the density of the barrier material

    c)   Increases as radiation energy increases in the range up to 5 MeV if the atomic number of the barrier material is less than that of lead

    d)   Must be at least 2 mm of lead in diagnostic radiology

    e)   Varies with the amount filtration in a diagnostic X-ray beam

14.6 The following statements concerning the UK Ionising Radiations Regulations (1999) are true:

    a)   Local rules are required for any work notifiable to the Health and Safety Executive under the Regulations

    b)   Staff must be designated as classified radiation workers if the dose to their hands is likely to exceed 150 mSv per annum

    c)   Radiation Protection Supervisors are appointed to assist the Radiation Protection Adviser with their work

    d)   All radiation workers must have their personal dose monitored at regular intervals

    e)   Comforters and carers are subject to the same dose limits as members of the public

14.7 The following are consequences of the UK Ionising Radiations Regulations (1999):

    a)   Radiation Protection Supervisors require more training than other radiation workers

    b)   Written systems of work sometimes provide a suitable alternative to staff classification

    c)   Local rules must address foreseeable accident situations

    d)   In a supervised area a worker is likely to exceed three-tenths of a dose limit

    e)   Boundaries to controlled areas must correspond to physical barriers such as walls, doors and so on

14.8 The following statements refer to the UK Ionising Radiations (Medical Exposures) Regulations (2000):

    a)   All persons who can act as 'Referrers' for a particular procedure must be identified by the employer

    b)   Only a clinician can be assigned the responsibility for justifying a patient exposure

    c)   Diagnostic dose reference levels must be established by the employer for every imaging procedure involving ionising radiation

    d)   For every medical procedure one person must be identified as the 'Operator'

    e)   The regulations do not apply to medical research involving ionising radiation

14.9 The following statements relate to staff doses in a radiology department:

    a)   They are high enough for most staff to be classified

    b)   Doses to the hands can sometimes be a reason for classification

    c)   Staff involved in interventional radiology receive higher doses (on average) than those working in general radiology

d) Lead curtains reduce staff doses more for over-couch than for under-couch tubes

e) For some interventional procedures two or more personal monitors may be necessary

14.10 The following precautions are taken to minimise staff doses in nuclear medicine:

a) Keeping bottles containing radionuclides in lead pots

b) Using tongs to handle bottles containing radionuclides

c) Working over a tray when transferring material between containers

d) Wearing leaded gloves when giving injections

e) Staying as far away from a patient who has received radioactivity as good patient care permits

14.11 The following are desirable design features for a diagnostic X-ray room:

a) All walls should be lined with lead sheet of a suitable thickness

b) No special care is required over the design of the door into the room because the kV is low

c) Extra shielding may be required wherever the primary beam can strike the wall

d) The dose rate should not exceed 1 µSv/h in the radiographer's cubicle

e) Suitable hanging rails for lead-rubber aprons should be provided

14.12 Responsibilities of an employee working in the radiology department under the UK Ionising Radiations Regulations (1999) include the following:

a) Only to use X-ray equipment after adequate training

b) Wearing personal protective equipment provided

c) Wearing personal monitors as instructed

d) Notifying the employer immediately of any suspected over-exposure to radiation

e) Notifying the employer immediately of any malfunction of equipment

14.13 UK Ionising Radiations (Medical Exposure) Regulations (2000):

a) Do not apply to departments with only one X-ray room

b) Require training records to be kept for radiologists and radiographers

c) Require that the Dose Reference Level is never exceeded

d) Require a Local Ethics Committee to be set up

e) Do not allow research on children involving ionising radiations

14.14 Under the UK Ionising Radiations (Medical Exposure) Regulations (2000):

a) The same person cannot act as referrer, operator and practitioner

b) The responsibility for implementation lies with the head of the radiology department

c) The practitioner must not justify a request if insufficient information is given

d) The medical physics expert is responsible for carrying out the investigation if a dose 'greater than intended' is given to a patient

e) Only a radiologist is authorised to tell a patient if a dose 'greater than intended' has been given

15.1 With reference to an ultrasound wave:

a) It is a combination of potential and kinetic energy

b) Although it causes the particles of the medium to move, no energy is propagated

c) A high-frequency wave travels faster than a low-frequency one

d) Particle velocity reaches a peak of 1540 ms^{-1} in tissue

e) Excess pressure is in phase with particle velocity

15.2 Attenuation of ultrasound in soft tissue:

a) Is independent of frequency

b) Depends on the thermal index of tissue

c) Is principally due to reflection and scattering at boundaries

d) Causes the centre frequency of the spectrum of an ultrasound pulse to reduce as the pulse propagates through the tissue

e) Is compensated for by swept gain in the receiver

15.3 Concerning tissue properties relevant to diagnostic ultrasound:

a) The scanner assumes the speed of sound is constant in all tissues

b) Blood is anechoic to ultrasound

c) Fluids in the body appear black on a B-mode image because of their high attenuation

d) An ultrasound pulse of nominal frequency 5 MHz retains only about one thousandth of its energy after travelling through 10 cm of liver

e) As ultrasound propagates through tissue, the rate of build-up of harmonic signals depends primarily on the tissue non-linearity coefficient

15.4 The principles and practice of B-mode imaging depend on the following:

a) Megahertz-frequency waves are used in order to achieve real-time imaging

b) Lateral resolution and axial resolution are governed by different mechanisms

c) A typical value for the dynamic range of echoes returning to the transducer is 100 dB

d) 'Write zoom' improves lateral resolution

e) Frame averaging reduces noise in the image

15.5 Decreasing the pulse length from a conventional B-mode scanner has the effect of increasing:

a) The dynamic range

b) The frequency band width

c) The axial resolution

d) The lateral resolution

e) The penetration

15.6   Concerning focussing of diagnostic ultrasound:

  a)   It improves lateral resolution

  b)   It improves contrast resolution

  c)   The beamformer moves the receive-beam focus through the tissue at half the speed of sound

  d)   Lateral focussing of the ultrasound beam is achieved by a lens

  e)   A two-dimensional array probe is necessary for slice-width focussing

15.7   Concerning the ultrasound probe:

  a)   The mechanically scanned probe is obsolescent because of the shape of its field of view

  b)   The beam from a phased-array is steered by altering the relative delay between its transducer elements

  c)   An ultrasound aperture that is n wavelengths across has a near-field length of n^2 wavelengths

  d)   The damping layer at the back of the transducer decreases its efficiency

  e)   Array probes emit a sequence of beams from each individual transducer in turn

15.8   The following statements relate to B-mode artefacts:

  a)   The fact that the beam scans across the field of view results in beam-width artefact

  b)   Acoustic shadows are caused by boundaries where there is a large difference in characteristic impedance

  c)   The speckle pattern on an image is caused by scattering features that are too close together for the scanner to resolve

  d)   Reverberation can be the result of using high amplitudes to overcome attenuation effects

  e)   Spatial compounding reduces shadowing

15.9   In Doppler ultrasound:

  a)   Continuous waves must be used

  b)   A Doppler-shifted received echo lies in the MHz frequency range

  c)   The Doppler shift depends on the acoustic properties of the medium between the probe and the moving target

  d)   Doppler systems measure the component of blood velocity that is normal to the ultrasound beam

  e)   Power Doppler is the most hazardous mode of diagnostic ultrasound

15.10 Concerning the Doppler effect in medical ultrasound:

  a)   It causes returning echoes to contain harmonics of the emitted wave created by the Doppler effect

  b)   It works best if the blood is flowing at an angle of at least 60° to the ultrasound beam

  c)   In a pulsed Doppler system, the shorter the emitted pulse, the better the Doppler resolution

d) In pulse wave Doppler, the pulse repetition frequency is in the kHz range

e) In a pulsed Doppler system, aliasing occurs if a negative Doppler shift exceeds half the pulse repetition frequency

15.11 Concerning new ultrasound techniques:

a) In tissue harmonic imaging, the second harmonic is the most useful

b) Pulse-inversion harmonic methods are superior to frequency-based harmonic filtering methods

c) Compound imaging is a combination of B-mode imaging and spectral Doppler

d) Infusing liquids with bubbles is an important technique for creating hyper-echoic contrast

e) Ultrasound elastography allows the non-invasive assessment of tissue hardness

15.12 Concerning ultrasound bio-effects and safety:

a) Manipulation of the receiver controls makes no difference to the safety of a scan

b) The thermal index indicates approximately the temperature rise in centigrade degrees being produced in the tissue

c) Bone in the field of view decreases the thermal effect of ultrasound by attenuating its energy

d) Ultrasound can produce ionising effects via inertial cavitation

e) An ultrasound beam exerts a radiation force which can cause mechanical stresses in blood cells

16.1 Imaging by magnetic resonance:

a) Requires at least a 1 T static magnetic field

b) Depends on excitation of nuclei by a time varying RF magnetic field

c) Can demonstrate blood flow without injection of contrast medium

d) Produces a digital image

e) Can use diamagnetic materials to enhance contrast

16.2 In magnetic resonance imaging:

a) The strength of the signal increases as the strength of the static magnetic field increases

b) Short T_1 values are associated with highly structured tissues

c) Field gradients must be applied to obtain spatial information

d) A gradient echo sequence with long repetition time (TR) and long echo time (TE) will produce T_1 weighted contrast

e) The major hazard to health limiting the static magnetic field is the associated temperature rise in the tissues

16.3 In magnetic resonance imaging of protons:

a) The SI unit of magnetic induction (magnetic flux density) is tesla per metre

b) A magnetic field varying with the Larmor frequency is used to define the slice to be imaged

    c)   Magnetic field gradients may be applied in more than one direction

    d)   The strength of the static magnetic field is a factor in determining the duration of the time varying field

    e)   Local variations in the static magnetic field affect T_2^* values

16.4 In magnetic resonance imaging:

    a)   Theoretically, the static magnetic field can act in any direction

    b)   The magnitude of the magnetic field gradient determines the thickness of the slice to be imaged

    c)   A superconducting magnet has almost no electrical resistance

    d)   The gyromagnetic ratio is one of the factors that determines the strength of the signal from a given volume

    e)   Switching on the gradient field induces currents in the body

16.5 In magnetic resonance imaging, the Larmor frequency:

    a)   Measures the gyroscopic rotation of the nuclei

    b)   Is the radiowave frequency at which the magnet system can absorb energy

    c)   Is one of the factors contributing to the strength of the net magnetic moment

    d)   Is directly proportional to the Boltzmann constant

    e)   Is the frequency of the detected radio signal

16.6 In magnetic resonance imaging, for a fixed pixel matrix, the signal intensity within a pixel is affected by the following:

    a)   The gyromagnetic ratio

    b)   The amplitude of the applied radiofrequency magnetic field

    c)   The repetition time

    d)   Flow

    e)   The field of view

16.7 Magnetic resonance imaging:

    a)   May utilise the carbon content of the body to generate the image

    b)   Uses signals which depend on the magnitude of the magnetic vector

    c)   Needs a radiofrequency wave to induce a signal

    d)   Is contra indicated for patients with aneurism clips

    e)   Produces stray fields proportional to the static field strength

16.8 In magnetic resonance imaging

    a)   The net spin angular momentum for protons should be zero

    b)   The ratio of nuclei in the low/high energy state depends on the static magnetic field

    c)   Precession occurs when the resultant magnetisation vector of the nuclei is parallel to the external field

    d) T_2 is always shorter than T_1 in biological tissues

    e)   The STIR sequence is used to suppress the high signal from fat

16.9 In magnetic resonance imaging the magnetic field gradient:
   a) Is generated by a superconducting magnet
   b) Is applied together with a 90° pulse to form the image slice
   c) Is applied perpendicular to the slice plane in selective excitation
   d) Inverts the nuclear magnetisation
   e) Produces a linear variation of resonant frequency with position

17.1 The minimum sampling frequency for an X-ray image is determined by the following:
   a) The X-ray energy used
   b) The maximum spatial frequency present in the image
   c) The resolution of the image
   d) The physical dimensions of the image
   e) The capabilities of the sampling device

17.2 In a network using IPv4, the *IP Address* of a device is as follows:
   a) Always fixed
   b) One of about 4 billion addresses
   c) The same as the host name
   d) Unique across the entire Internet
   e) Unique in a single (e.g. hospital) network

17.3 Modern Ethernet networks:
   a) Require all devices to be connected to a single cable
   b) Segment the network using switches and routers
   c) Use coaxial cable
   d) Require that all devices operate at the same speed
   e) Are the subject of an ISO standard

17.4 The DICOM standards:
   a) Are ratified by the International Organisation for Standardisation
   b) Are mandatory standards for equipment manufacturers
   c) Provide standardised definitions of the way that devices interact
   d) Define information objects and services operating on them
   e) Must be met by all devices placed on the market

17.5 Lossless data compression:
   a) Involves the loss of information from the image
   b) Means that the decompressed image is identical with the original
   c) Provides a greater degree of compression than lossy compression
   d) Is defined in the JPEG2000 standard
   e) Is permitted under the DICOM standards for original images

17.6 In a PACS System, images are stored:
   a) In the acquisition server
   b) In the integration server
   c) In an offline archive
   d) In an image database
   e) By an image server

18.2 MCQ Answers

Question	a)	b)	c)	d)	e)
1.1	F	T	T	F	F
1.2	F	T	T	T	T
1.3	T	F	T	T	T
1.4	T	T	T	T	T
1.5	F	F	F	T	T
1.6	F	F	T	T	T
1.7	F	F	T	T	F
1.8	T	T	F	F	T
1.9	T	F	T	T	T
2.1	F	F	T	T	F
2.2	T	T	T	T	F
2.3	T	F	T	T	F
2.4	T	F	T	T	F
2.5	T	F	F	T	F
2.6	T	T	T	T	T
2.7	T	F	T	F	T
2.8	T	T	T	T	F
2.9	T	F	F	T	T
2.10	T	F	F	T	T
2.11	T	T	T	T	F
2.12	F	F	F	T	T
2.13	T	F	F	T	T
3.1	F	F	T	F	T
3.2	T	F	F	F	F
3.3	F	F	T	F	T
3.4	T	F	T	F	T
3.5	T	T	T	T	F
3.6	T	F	F	F	F
3.7	T	F	T	T	F
3.8	F	F	T	T	F
3.9	T	F	T	T	T
4.1	T	F	T	F	F
4.2	T	F	T	T	F
4.3	T	F	T	T	F
4.4	T	F	T	F	F

Question	a)	b)	c)	d)	e)
4.5	T	T	T	T	T
4.6	T	F	F	F	T
4.7	T	T	F	T	F
4.8	T	T	F	F	T
4.9	T	T	F	T	T
4.10	T	F	T	T	F
5.1	F	F	T	T	T
5.2	F	T	F	T	F
5.3	T	T	F	F	F
5.4	F	T	T	F	F
5.5	F	T	F	T	F
5.6	T	F	F	F	T
5.7	T	T	F	T	T
5.8	T	F	F	T	T
5.9	F	T	T	F	F
6.1	F	T	T	T	T
6.2	T	F	F	T	F
6.3	F	F	T	T	T
6.4	T	T	F	T	F
6.5	F	F	F	T	T
6.6	T	F	F	T	T
6.7	T	F	T	T	F
6.8	T	F	T	T	F
6.9	F	T	T	F	F
6.10	T	T	T	F	F
7.1	F	T	F	T	T
7.2	T	T	T	F	F
7.3	F	T	T	F	F
7.4	T	F	T	T	F
7.5	F	F	F	T	F
7.6	T	T	F	F	T
7.7	T	F	F	F	T
7.8	T	T	T	T	F
7.9	F	T	T	F	T
8.1	T	T	T	T	T
8.2	T	F	T	F	F
8.3	F	T	T	F	F
8.4	T	F	T	T	F
8.5	F	F	T	F	T
8.6	T	F	F	T	T
8.7	T	F	F	F	T
8.8	T	F	T	F	T
8.9	T	F	F	T	F
8.10	T	F	T	T	F
9.1	F	T	F	T	F
9.2	T	F	T	F	T

(*continued*)

Question	a)	b)	c)	d)	e)
9.3	T	T	F	T	T
9.4	F	T	F	T	T
9.5	F	T	T	F	F
9.6	T	T	F	T	F
9.7	T	F	T	T	F
9.8	F	T	T	F	T
9.9	F	T	T	T	F
10.1	F	T	F	T	T
10.2	F	T	F	F	F
10.3	F	F	T	F	F
10.4	T	F	T	F	F
10.5	T	T	T	F	T
10.6	F	F	T	F	T
10.7	F	F	F	T	F
10.8	T	T	T	T	T
10.9	F	T	F	T	T
10.10	F	T	T	F	T
11.1	F	T	T	T	F
11.2	F	F	T	T	F
11.3	F	T	T	F	T
11.4	T	T	T	F	F
11.5	F	T	F	F	F
11.6	T	F	T	T	F
12.1	T	F	T	T	F
12.2	T	F	T	T	F
12.3	T	T	T	F	F
12.4	T	F	T	F	T
12.5	F	T	F	T	F
12.6	T	T	F	F	F
12.7	T	F	T	F	T
12.8	T	F	T	T	F
13.1	F	T	F	F	F
13.2	T	F	T	F	F
13.3	T	F	T	F	F
13.4	F	T	T	F	F
13.5	F	F	F	T	T
13.6	T	F	T	F	F
13.7	T	T	F	F	F
13.8	F	F	T	T	T
13.9	T	F	T	T	T
13.10	F	T	F	T	T
13.11	T	T	T	F	T
13.12	T	F	T	F	T
13.13	T	F	F	F	F
14.1	F	T	F	T	T
14.2	T	T	T	F	T
14.3	T	F	F	F	T
14.4	F	T	F	T	T

Question	a)	b)	c)	d)	e)
14.5	F	T	T	F	T
14.6	T	T	F	F	F
14.7	T	T	T	F	F
14.8	T	F	F	F	F
14.9	F	T	T	T	T
14.10	T	T	F	F	T
14.11	F	F	T	T	T
14.12	T	T	T	T	F
14.13	F	T	F	T	F
14.14	F	F	T	F	F
15.1	T	F	F	F	T
15.2	F	F	F	T	T
15.3	T	F	F	T	F
15.4	F	T	T	T	T
15.5	F	T	T	F	F
15.6	T	T	T	F	F
15.7	F	T	T	T	F
15.8	T	T	T	T	T
15.9	F	T	F	F	F
15.10	F	F	F	T	T
15.11	T	T	F	T	T
15.12	F	T	F	T	T
16.1	F	T	T	T	F
16.2	T	T	T	F	F
16.3	F	F	T	F	T
16.4	T	T	T	T	T
16.5	T	T	F	F	T
16.6	T	F	T	T	T
16.7	F	T	T	T	T
16.8	F	T	F	T	T
16.9	F	T	T	F	T
17.1	F	T	T	F	F
17.2	F	T	F	F	T
17.3	F	T	F	F	T
17.4	F	F	T	T	F
17.5	F	T	F	T	T
17.6	T	F	T	T	T

18.3 Notes

Some of these notes illustrate one of the weaknesses of MCQs in this subject area. It is often very difficult to set non-trivial questions that are not potentially ambiguous with deeper knowledge. Thus MCQs are a good teaching aide for testing an understanding of the subject and sometimes promote further discussion. They are less satisfactory as a method of examination.

1.3d) As Auger electrons (see Persson L, The Auger effect in radiation dosimetry, *Health Phys* 67, 471–6, 1994.

1.7e) Ionising radiations are oxidising agents, for example, ferrous ions to ferric ions.

2.2b) The high density prevents too much electron penetration into the anode.

2.3d) kV and mAs will affect the effective spot size because they will influence the performance of the cathode focussing cup in forming a small target on the anode surface.

2.6d) There is substantial self-absorption of X-rays in the anode and this will be affected by anode angle.

2.8e) Patient dose may be unacceptably high.

2.10c) The extra heat is lost mainly by conduction.

2.10e) Black bodies are good emitters of radiation.

2.12e) Actual kV, and hence keV, is above the K shell energy for longer.

3.1b) Because of absorption edges.

3.1d) Any such radiation will be of such low energy that it is absorbed within the body, not scattered.

3.2b) The second half value thickness will be greater because of beam hardening.

3.3d) This statement would normally be true for the whole body because of the effects of attenuation, but is not true for the Compton process itself.

3.4c) Healthy lung tissue is less dense than water.

3.5b) There will be a small amount of elastic scattering.

3.5d) Characteristic radiation will be produced if a K shell (or higher shell) vacancy is created. If this occurs in the body (with low atomic number elements) the radiation may be of too low energy to escape.

3.7b) Scatter may be reduced but is never eliminated.

3.8a) Some X-rays will be Compton scattered. Note also that many photons will not interact at all, but these are excluded by the stem of the question.

3.9b) See Section 9.2.3.

3.9e) There are several reasons but the most important is that the filter, in practice, would be too thin.

4.2b) Some energy will be deposited as excitation.

4.3e) The voltage does not need to be very constant on the ionisation plateau.

4.7d) Although the electric field will be much higher than for an ionisation chamber, the operating voltage may be similar.

5.4e) Although the intensifying screen stops a higher fraction of photons than film, the number of incident photons will be greatly reduced.

5.9d) Both outputs are digital images so this is not an advantage.

6.3e) If the penumbra are excluded the larger focal spot will give a smaller image, if the penumbra are included it will give a larger image.

6.5c) This assumes the field size is set at the cassette, if it were set on the patient surface there might be less scatter because of less beam divergence.

6.6a
and d) Both reduce the scatter reaching the film.

6.10c) A major contribution to the variations in grey scale will be quantum mottle.

7.1d) Quantum noise = $N^{\frac{1}{2}}$ so it increases with dose. Signal to noise ratio (usually the key consideration in imaging) also equals $N^{\frac{1}{2}}$ and improves with increasing dose.

7.9e) The dominant factor will be the way in which the quantum efficiency of the receptor varies with keV. This will depend on the K shell absorption edges of the constituent material.

8.7c) The two are unrelated. Quantum noise is caused by too few photons, after-glow is a property of the receptor.

9.7b) Increasing the concentration of iodine contrast is more effective.

9.7e) Each frame needs the same photon density so photon flux must be increased.

9.9e) 50 µGy should not normally be exceeded, 30 µGy should suffice.

10.2d) 4 days is less than two half-lives for Mo-99.

10.2e) Note the question says 'equilibrium', by the time the generator has returned to equilibrium, it will not matter whether the generator has been eluted or not.

10.10d) This thickness of crystal will cause loss of resolution with minimal increase in sensitivity for Tc-99m.

11.2b) Compton scattering, not elastic scattering.

11.3a) The contributions are added in quadrature.

11.3d) It is better for the PET imager.

11.5d) PET imaging time is much longer.

12.5e) X-rays of all energies are low LET radiations, so the quadratic term in the model for double strand breaks (see Section 12.7.1) is important for both.

12.8b) At Hiroshima there was a substantial neutron component.

13.1e) Filtration cannot increase any component of the spectrum, so the exit dose cannot increase with the specified conditions.

13.2b) Variation is in input dose, exit dose is governed by the receptor sensitivity.

13.2e) For this to be true the half value thickness would have to be 10 cm—it is much less.

13.6c) See general instructions at the beginning of the questions. Compression will actually reduce the amount of tissue in the beam.

13.7 Many values of w_T are only given to one significant figure.

13.8d) See general instructions.

13.9c) A reasonable mean from quite a wide range of quoted values.

13.11e) Interventional procedures may cause skin reactions but they are an adjunct to therapy, not diagnostic.

13.13e) Although hereditary effects cannot be positively excluded, the risk will be very low and no higher than for adults.

14.1e) The mean atomic number is similar to that of air and soft tissue.

14.7e) Although this is recommended good practice, it may not be feasible—for example, mobile radiography.

14.9e) See Section 9.6.4.

14.10d) it is preferable to retain dexterity and work quickly.

14.12e) Only if a patient received a dose 'much greater than intended'.

15.6e) Full 2D is not necessary. Either a lens or a few rows of detectors covering the slice-width direction (to allow focussing) will suffice.

15.9b) It is the difference in frequency between the outgoing and returning signals that is in the audible range.

15.9e) The word 'power' in the name of this mode refers to the measurement of echo power and is not a reference to the acoustic power used, which is less than in most other Doppler modes.

15.12a) Improving the quality of the received image reduces scan time.

16.7a) The amount of C-11 is too small to give an adequate signal, C-12 gives no signal.

Index